Gels Horizons: From Science to Smart Materials

Series editor

Vijay Kumar Thakur, School of Aerospace, Transport and Manufacturing, Cranfield University, Cranfield, Bedfordshire, UK

This series aims at providing a comprehensive collection of works on the recent advances and developments in the domain of *Gels*, particularly as applied to the various research fields of sciences and engineering disciplines. It covers a broad range of topics related to *Gels* ranging from *Polymer Gels, Protein Gels, Self-Healing Gels, Colloidal Gels, Composites/Nanocomposites Gels, Organogels, Aerogels, Metallogels* & *Hydrogels* to *Micro/Nano gels*. The series provides timely and detailed information on advanced synthesis methods, characterization and their application in a broad range of interrelated fields such as chemistry, physics, polymer science & engineering, biomedical & biochemical engineering, chemical engineering, molecular biology, mechanical engineering and materials science & engineering.

This Series accepts both edited and authored works, including textbooks, monographs, reference works, and professional books. The books in this series will provide a deep insight into the state-of-art of *Gels* and serve researchers and professionals, practitioners, and students alike.

More information about this series at http://www.springer.com/series/15205

Vijay Kumar Thakur · Manju Kumari Thakur
Editors

Hydrogels

Recent Advances

Springer

Editors
Vijay Kumar Thakur
Faculty in Manufacturing, Enhanced Composites and Structures Centre, School of Aerospace, Transport and Manufacturing
Cranfield University
Cranfield, UK

Manju Kumari Thakur
Division of Chemistry, Government Degree College Bhoranj
Himachal Pradesh University
Shimla, Himachal Pradesh, India

ISSN 2367-0061 ISSN 2367-007X (electronic)
Gels Horizons: From Science to Smart Materials
ISBN 978-981-10-6076-2 ISBN 978-981-10-6077-9 (eBook)
https://doi.org/10.1007/978-981-10-6077-9

Library of Congress Control Number: 2018940412

This Springer imprint is published by the registered company Springer Nature Singapore Pte Ltd.
The registered company address is: 152 Beach Road, #21-01/04 Gateway East, Singapore 189721, Singapore

Preface

A gel is described as a soft, solid, or liquid-like unique condensed material that has a three-dimensional network composed of several components such as long polymers, species of small molecules, and a large amount of solvent. These 3D network condensed materials usually form through chemical, physical, or supramolecular crosslinking. The weight and size of gels are more like a liquid, but they are treated like a solid. Two important characteristics of gels are phase state and their rheological properties. On the other hand, a polymer is defined as a large molecule (macromolecules) composed of repeating structural units that comprise of multiple assemblies of simple structural units. In gels, the polymer network can be physically or chemically crosslinked. In case of physical gels, the network formation occurs due to various weak interactions, like the entanglement of the polymer chains, hydrogen bonds, or van der Waals interactions. Such structures are usually not permanent, and they dissolve over the time when immersed in their solvents. However, the polymer chains can also be crosslinked through chemical reactions, leading to strong covalent bonds. The chemically crosslinked network is much more stable and cannot be dissolved without the degradation of the polymer. Therefore, chemical gels are usually preferable in the majority of the application fields. Polymer gels comprise a great variety of different polymeric components that present innumerable industrial applications. Polymers can be naturally produced (sometimes referred as biobased polymers), in which case the most representative group is polysaccharides. Natural polymers' demand is expected to grow 7.1% every year. Moreover, their low toxicity and excellent biodegradability have also attracted researchers to pay attention toward the widespread application of natural polymers. Polymer obtained from natural sources such as chitosan, alginate, dextran, starch, pectin, cellulose, lignin has shown excellent potential for biomedical and other applications in the form of microsphere, nanoparticles, crosslinked hydrogels, beads, membranes, and granules. On the other hand, a wide variety of synthetic polymers capable of forming gels presents different industrial applications, such as polyacrylamide and polyvinyl alcohol-based gels. Both synthetic and natural polymer-based gels find applications from health sciences such as agents for controlled drug delivery, sustained drug delivery, targeted drug delivery, and various other types of novel drug delivery systems to water purification.

Polymer gels due to their several unique characteristics have become an indispensable part of new advanced and smart materials in the twenty-first century for numerous applications including but not limited to biological, biomedical, electronic, and environmental. Keeping in mind the immense advantages of polymer gel-based materials, this volume of the series is solely focused on the *Hydrogels*. It provides a comprehensive collection of works on the recent advances and developments in science and fundamentals of both synthetic and natural polymer-based hydrogels particularly as applied to the various research fields of science and engineering disciplines. The development of new hydrogel-based novel materials that respond to external stimuli, such as temperature, pressure, pH, light, biochemicals, or magnetic, represents a new class of intelligent materials. With their multiple cooperative functions, hydrogel-based materials exhibit different potential applications which range from biomedical to water purification systems. *Hydrogels: Recent Advances and Applications* provides a cutting-edge resource for all polymer chemists, materials scientists, organic chemists, and biotechnologists working in different interdisciplinary fields. Some of the important topics include but not limited to superabsorbent polymer hydrogel; intelligent hydrogels for drug delivery; hydrogels from catechol-conjugated materials; nanomaterial-loaded hydrogel; electrospinning of hydrogels; biopolymer-based hydrogels; injectable hydrogels; interpenetrating polymer network hydrogels: radiation and sonochemical synthesis of micro-/nano-/macroscopic hydrogels; hydrogel-based strategies for stem cell therapy; protein- and nanoparticle-loaded hydrogels studied by small-angle scattering and rheology techniques; hydrogel-based stimuli-responsive functionalized graft copolymers for the controlled delivery; electrospinning of hydrogels for biomedical applications; self-assembling hydrogels from pH-responsive ionic block copolymers; cellulose hydrogels as sustainable materials regenerated from waste; DNA-based hydrogels; macroporous hydrogels; and multifunctional applications of hydrogels, to name a few.

In editing and organizing this volume *Hydrogels: Recent Advances and Applications* of the book series "Gels Horizons: From Science to Smart Materials," we have made our best efforts to cover the growing field of polymer gels and related technologies. It reflects the recent theoretical advances and experimental results and opens new avenues for researchers as well as readers working in the field of polymer and functional materials. In addition, several critical issues and suggestions for future work are comprehensively discussed in this book with the hope that the book will provide a deep insight into the state of the art of "Polymer Gels." We express our sincere thanks to all the authors, who have contributed their extensive experience through their work for the success of this book. We would also like to thank Swati Meherishi and the rest of the team at Springer for invaluable their help in the publication process.

Cranfield, UK Vijay Kumar Thakur, Ph.D.
Shimla, India Manju Kumari Thakur, M.Sc., M.Phil., Ph.D.

Contents

About the Editors

Dr. Vijay Kumar Thakur, Ph.D. prior to commencing in the School of Aerospace, Transport and Manufacturing at Cranfield University, he was working as Staff Scientist in the School of Mechanical and Materials Engineering at Washington State University, USA (2013–2016). Some of his other prior significant appointments include being Research Scientist in Temasek Laboratories at Nanyang Technological University, Singapore (2009–2012), and Visiting Research Fellow in the Department of Chemical and Materials Engineering at LHU, Taiwan. He did his postdoctoral study in materials science and engineering at Iowa State University and received his Ph.D. in polymer chemistry (2009).

In his academic career, he has published more than 100 SCI journal research articles in the field of chemical sciences/materials science and holds one US patent. He has also published 33 books and 35 chapters on the advanced state of the art of polymer science/materials science/nanotechnology with numerous publishers. His research interests include the synthesis and processing of biobased polymers, composites, nanostructured materials, hydrogels, polymer micro-/nanocomposites, nanoelectronic materials, novel high dielectric constant materials, engineering nanomaterials, electrochromic materials, green synthesis of nanomaterials, and surface functionalization of polymers/nanomaterials. Application aspects range from automotive to aerospace, energy storage, water purification, and biomedical fields.

He is an editorial board member of several international journals, as well as a member of scientific bodies around the globe. Some of his significant appointments include Associate Editor for Materials Express (SCI), Advisory Editor for SpringerPlus (SCI), Editor for Energies (SCI), Editor for Cogent Chemistry (SCI), Associate Editor for Current Smart Materials, Associate Editor for Current Applied Polymer Science, Regional Editor for Recent Patents on Materials Science (Scopus); and Regional Editor for Current Biochemical Engineering (CAS). He also serves on the Editorial Advisory Board of Polymers for Advanced Technologies (SCI) and is on the Editorial Board of Journal of Macromolecular Science, Part A: Pure and Applied Chemistry (SCI), International Journal of Industrial Chemistry (SCI), Biointerface Research in Applied Chemistry (SCI), and Advances in Natural Sciences: Nanoscience and Nanotechnology (SCI).
e-mail: Vijay.Kumar@cranfield.ac.uk

Dr. Manju Kumari Thakur, M.Sc., M.Phil., Ph.D. has been working as Assistant Professor of Chemistry at the Division of Chemistry, Government Degree College Sarkaghat, Himachal Pradesh University, Shimla, India, since June 2010. She received her B.Sc. in chemistry, botany, and zoology; M.Sc. and M.Phil. in organic chemistry; and Ph.D. in polymer chemistry from the Chemistry Department at Himachal Pradesh, University, Shimla, India. She has rich experience in the fields of organic chemistry, biopolymers, composites/nanocomposites, hydrogels, applications of hydrogels in the removal of toxic heavy metal ions, drug delivery, etc. She has published more than 30 research papers in several international journals, has co-authored five books, and has also published 25 chapters in the field of polymeric materials. e-mail: chauhanmanjuchem@gmail.com

Chapter 1
Intelligent Hydrogels as Drug Delivery Systems

Katarina Novakovic, Simon Matcham and Amy Scott

A drug delivery system (DDS) can be defined as a formulation or a device that facilitates the release of a therapeutic substance in the body. Key parameters of interest in DDS are safety, delivery rate, efficiency, as well as time and place of release of drugs. Lately, hydrogels have attracted significant attention for application in drug delivery. ***Hydrogels*** are three-dimensional polymer networks consisting largely of water. They are characterised by a porous structure with porosity, pore size and geometry that can be varied during the hydrogel synthesis. Importantly, due to porous structure they have the ability to incorporate biomolecules. Hydrogels can be loaded with large amounts of drug relative to their volume which along with their biocompatibility are all features that make them good candidates as materials for the controlled and sustained release of therapeutics (Oh et al. 2008; Jiang et al. 2014). Vital characteristics of hydrogels include hydrophilicity, ability to form three-dimensional crosslinked polymeric networks that can absorb large amounts of biological fluids or water whilst staying insoluble. Studies have shown that the amount of water can reach up to 99% by weight (van der Linden et al. 2003), resulting in hydrogels resembling living tissue. As hydrogels can be made in different forms, such as films, macrogels, microgels, nanogels and coatings, they are considered to be advantageous for a range of different applications from medical industry and agriculture to enhanced oil recovery. The polymer network of hydrogels is comprised of long molecular chains attached to one another during chemical synthesis. The hydrophilicity of the hydrogels occurs when polymers contain hydrophilic groups such as carboxylic (–COOH), hydroxyl (–OH) or sulphonic (–SO_3H) (Ganji et al. 2010). However, it is worthwhile noting the existence of hydrophobic polymers, which can be used via blending or copolymerisation

K. Novakovic (✉) · S. Matcham · A. Scott
School of Chemical Engineering and Advanced Materials, Newcastle University, Newcastle upon Tyne, UK
e-mail: katarina.novakovic@ncl.ac.uk; katarina.novakovic@newcastle.ac.uk

V. K. Thakur and M. K. Thakur (eds.), *Hydrogels*, Gels Horizons: From Science to Smart Materials, https://doi.org/10.1007/978-981-10-6077-9_1

process with hydrophilic polymers. One of the reasons for their introduction is to slow hydrogel degradation process.

Crosslinking process is used to interconnect polymer chains to enhance the mechanical stability of the hydrogel. By altering the density of the crosslinks in the gel matrix, the porous structure can be tuned and the degree of swelling is adjusted. The degree and type of crosslinks used can also affect elasticity, viscosity and the insolubility of the polymer. Whilst small amounts of crosslinking can result in viscous hydrogels, able to flow freely, it can also lead to the hydrogels becoming soluble in solvents and having low stability. Conversely, too high a degree of crosslinking can result in brittle hydrogels although it can improve creep behaviour and improve insolubility (Maitra and Shukla 2014). Crosslinking can be carried out using physical or chemical methods, dependent on the type of polymer used. ***Physical crosslinks*** involve entwined chains, hydrogen bonding, hydrophobic interaction and crystalline formation and have weaker connections than chemical crosslinks. However, they can ensure that the hydrogel is insoluble in an aqueous media (Maitra and Shukla 2014). Still, physical crosslinking can result in hydrogels which are unstable and therefore may decompose fast and without prior warning (Ottenbrite et al. 2010), limiting their applications. Nevertheless, physically crosslinked hydrogels are becoming increasingly popular due to the simplicity of the processes by which they are produced. Like radiation-chemical crosslinking, all methods of physical crosslinking do not require the addition of crosslinking agents. This means that the separation of any excess agents is not required. This has consequently led to physical crosslinking becoming more popular in the food and drinks industry due to sterility reasons (Gulrez et al. 2011).

The first examples of physical crosslinking are ionic interactions between polyanions and polycations which can form coacervates as the polymer's opposing electrical charges bind them together. Proteins below their isoelectric point are positively charged and have been found to bind with anionic hydrocolloids to form polyanion complex hydrogels (Magnin et al. 2004). Similarly, polymers containing carboxyl groups such as carboxymethylcellulose (CMC) can form weak hydrogen bonds by reducing the pH of the solution which they are present in. An example of hydrogen-bonded hydrogels is multilayer compounds created by the crosslinking of polyvinylpyrrolidone (PVP) and polymethacrylic acid (PMAA) which have been used in the controlled release of proteins (Kharlampieva et al. 2007). Another method of physical crosslinking is via heating or cooling of a polymer solution. In some cases, the application of heat forms crosslinks. This is the case with polyethylene glycol–polylactic acid solutions which undergo block copolymerisation when heated (Hennink and Nostrum 2002). Cooling solutions of gelatin and carrageenan cause the solution of polymers to morph from a randomly coiled structure to form helixes which become intertwined, causing the gelation of the solution (Funami et al. 2007). Another method of physical crosslinking is via freeze–thawing. This is a less common method of crosslinking that also founded use in the post-synthesis treatments of chemically crosslinked hydrogels to improve their mechanical stability (Nwosu et al. 2015). In physical crosslinking, the freezing of a solution containing polymers creates microcrystals in the structure. An example of this technique is the crosslinking of poly(vinyl alcohol) (PVA) hydrogels by repeated freeze–thaw cycles (Hassan and Peppas 2000). The number and length of these cycles are reported as a key factor in controlling the degree of crosslinking (Paranhos et al. 2007). Maturation is the final form of physical crosslinking. It occurs when compounds containing proteins are heated. The heated proteins then change structure and bind together to form crosslinks between polymer chains. An example of heat-induced maturation is the crosslinking of gum arabic by exposing it to heat where molecules associate to form larger molecular mass aggregates (Phillips and Williams 2009). This method had been used to reduce the distribution of molecular weights in emulsification products.

Chemical crosslinks are through covalent bonds and provide stronger, more permanent bonds, increasing the mechanical strength. This method of crosslinking allows for swelling to occur without the loss of mechanical structure. Methods of chemical crosslinking include polymerisation, radiation, small molecule crosslinking and polymer–polymer crosslinking such as a condensation reaction (Ebara et al. 2014). Figure 1 shows the difference in molecular structure of hydrogels that are physically crosslinked (a) and chemically crosslinked (b).

One of the chemical crosslinking agents that are becoming increasingly popular is ***genipin*** (Matcham and Novakovic 2016). Genipin is an iridoid which is enzymatically hydrolysed from geniposide, an extract from the fruit of the *Gardenia jasminoides Ellis*, an evergreen bush which is found to grow in India, China, Japan, Vietnam and South America (Djerassi et al. 1960; Endo and Taguchi 1973; Park et al. 2002). Traditionally, it was used to treat pyrogenic infections, haematemesis, febrile diseases and also as an external treatment for sprains and swellings (Tsai et al. 2002). Genipin's chemical properties were first explored in the 1960s when simplistic plug flow hydrolysis methods of extracting genipin from the fruit were employed (Fujikawa et al. 1987). Genipin derivatives quickly found its use as blue food pigment named 'gardenia blue' via the reaction of genipin with primary amine groups in the presence of oxygen due to their high stability when exposed to heat, pH and light (Paik et al. 2001). It was theorised that the fruit's extracts could also have applications in choleretic, diuretic, antiphlogistic and tumour inhibitory drugs. It was also predicted that it could be used in the synthesis of biologically active cyclopentenone prostaglandins and triquinane sesquiterpenes (Naruto et al. 1978; Berkowitz et al. 1982; Bonini et al. 1984).

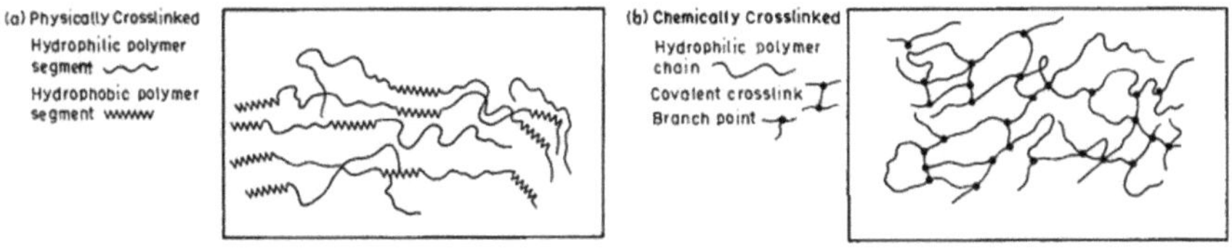

Fig. 1 Molecular structure of hydrogels **a** physically crosslinked and **b** chemically crosslinked (Begam et al. 2003)

The structure of genipin is displayed in Fig. 2 and shows that it is a bicyclic aglycone which contains a dihydropyran ring and an ester group. Its pure physical form is a clear, monoclinic, crystalline needle-like solid (Trevor et al. 2008). The key driving force behind investigating genipin's crosslinking behaviour is that it is 5000–10,000 times less cytotoxic than the most common crosslinking reagent, glutaraldehyde (Sung et al. 1999; Matcham and Novakovic 2016). Glutaraldehyde (Fig. 3) is colourless oily liquid and finds most of its uses in sterilisation and disinfectant applications. Due to its cytotoxicity, glutaraldehyde limits the uses of any hydrogel products which it is present in. Another advantage genipin has when compared to glutaraldehyde and numerous other synthetic crosslinkers is that it degrades much slower than glutaraldehyde (Delgadillo-Amendariz et al. 2014). This means that any products containing genipin in place of glutaraldehyde should theoretically have a longer shelf life.

The reaction mechanism of genipin crosslinking molecules of chitosan is well understood and has been explored extensively over the past two decades. There are several reaction mechanisms by which this can occur which are dependent on a number of factors. However, it has been reported that pH is the governing environmental influence that dictates which reaction pathways are inhibited (Mi et al. 2005; Delmar and Bianco-Peled 2015; Matcham and Novakovic 2016). Temperature has been found to affect mainly the speed and extent of reaction (Chen et al. 2005). The reaction mechanism that takes place under acidic conditions is described in several steps in Figs. 4, 5, 6 and 7.

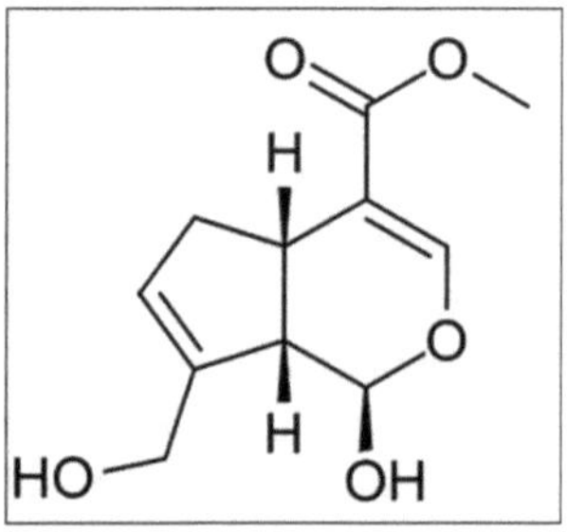

Fig. 2 Skeletal structure of genipin

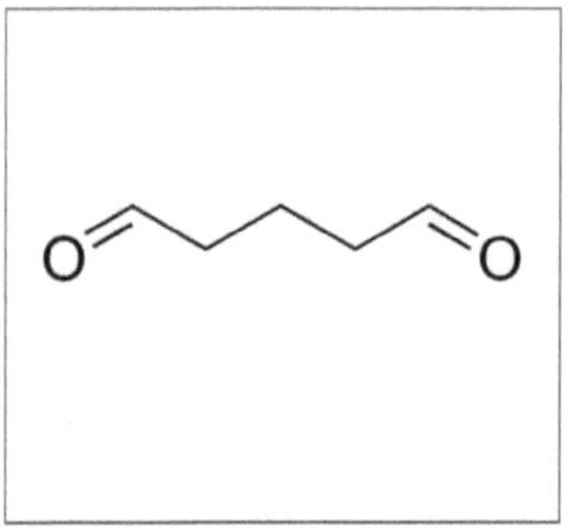

Fig. 3 Skeletal structure of glutaraldehyde

Figure 4 shows the S_N2 nucleophilic substitution reaction mechanism by which the ester group on the genipin molecule is replaced by a secondary amide linkage to a molecule of chitosan.

Figure 5 displays the second crosslinking reaction mechanism. Here, a nucleophilic attack of the C3 carbon atom by a primary amine group from a second molecule of chitosan forms an intermediate aldehyde group. The dihydropyran ring is then opened, resulting in another aldehyde group which is then attacked by the secondary amine formed in the first step of the reaction to form a heterocyclic amide linkage.

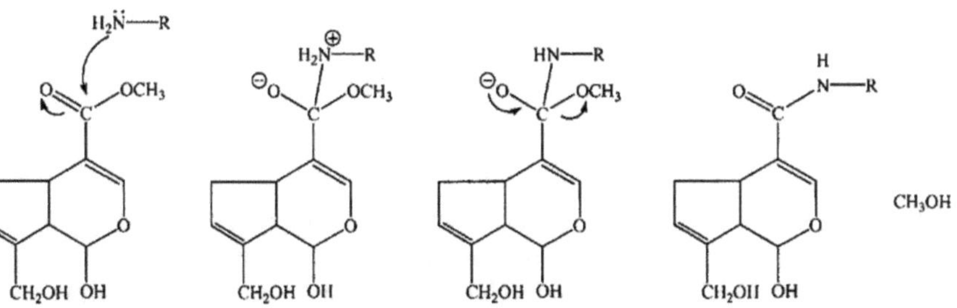

Fig. 4 Monosubstitution of an amide group (Butler et al. 2003)

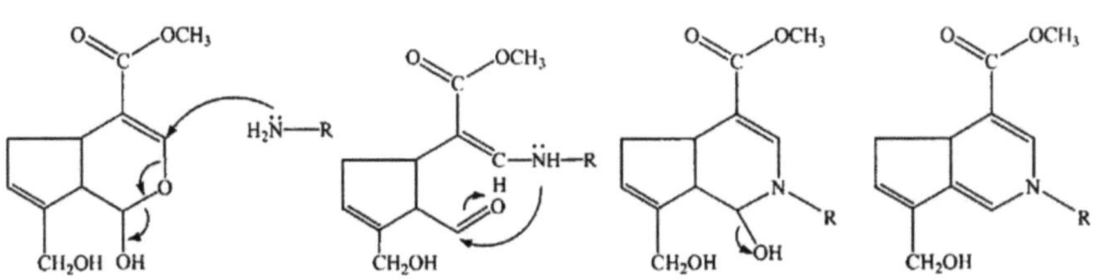

Fig. 5 Tertiary amine substitution (Butler et al. 2003)

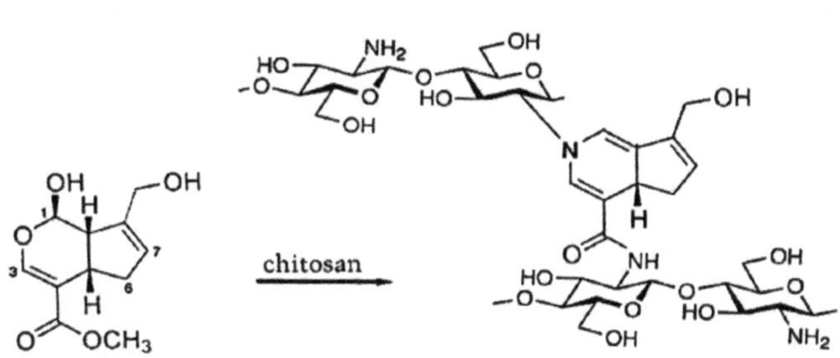

Fig. 6 Chitosan–genipin crosslinking mechanism (Muzzarelli 2009)

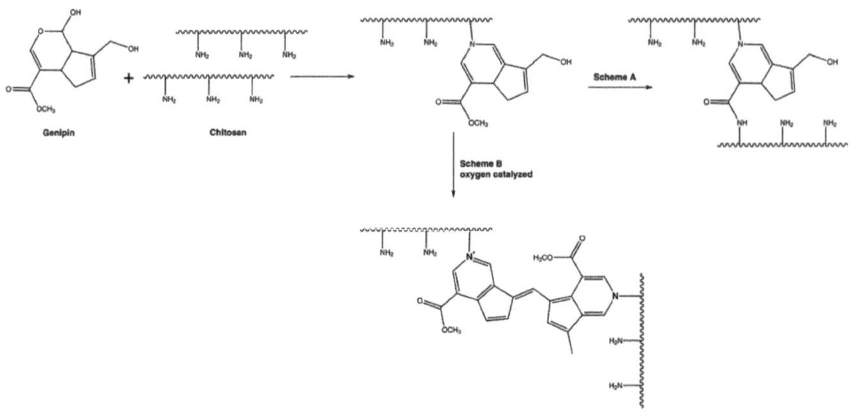

Fig. 7 Alternate crosslinking mechanisms of chitosan using genipin (Dimida et al. 2015)

Figure 6 shows the product of the reactions described by the mechanisms in Figs. 4 and 5. Here, one molecule of genipin forms a single bifunctional crosslink between two molecules of chitosan. Under more basic conditions, genipin molecules have been found to self-polymerise due to the nucleophilic attack by OH^- ions allowing aldol condensation reactions to occur. This typically takes place prior to crosslinking reactions with chitosan to form long bridges of genipin between chitosan chains (Mi et al. 2005; Matcham and Novakovic 2016). These genipin oligomer bridges are terminated by a Schiff base reaction with the amino groups on chitosan to form crosslinked networks with larger distances between chitosan chains, as shown by scheme B in Fig. 7.

These longer crosslinks created in the more basic environment should lead to the creation of genipin-crosslinked chitosan polymer chains with a different shape and behaviour to the structures of the crosslinked polymers which are produced in acidic environments due to the difference in length of the crosslinks. Conveniently, genipin fluorescence upon crosslinking offers a non-invasive in situ fluorescence imaging method that can be used for screening of hydrogel sol–gel transition (Matcham and Novakovic 2016) or screening of hydrogel swelling post-synthesis (Hurst and Novakovic 2013).

Hydrogels can be classified based on their mechanical and structural features, method of preparation, physical structure and responsiveness to environmental stimuli. One type of classification is based on whether natural or synthetic polymers have been used, both of which offer different advantages as discussed by Roldan (2003) and Vashist et al. (2014). Natural polymers such as chitosan, dextran and collagen have a high biocompatibility, are biodegradable and have low toxicity by-products. However, they can have a low mechanical strength and can vary between batches. Synthetic polymers such as polyethylene glycol–polylactic acid–polyethylene glycol (PEG–PLA–PEG) and polyvinyl alcohol offer more flexibility and can provide a broad range of properties. They also offer advantages of low immunogenicity and minimise the risk of biological pathogens or contaminants. However, they can have low biodegradability and contain toxic substances (Roldan 2003; Vashist et al. 2014). Hydrogels can also be defined by their degree of swelling, which is caused by several different forces, one of which is the polymer–water interactions. In general, the greater the hydrophilicity of the polymers, the stronger the interaction between the polymer and water and therefore the greater the extent of swelling. Osmosis also results in the swelling of hydrogels, as a result of the difference in ion concentration between the gel and the outside solution. The ions are a result of the ionisation of pendant ionic groups, and the greater the difference between these and the solution ions, the greater the osmotic pressure (Ottenbrite et al. 2010). The degree of swelling can be influenced depending on a variety of factors, including composition, structure and crosslinking density. For some applications, the objective of the hydrogel is to reach a specified volume swelling ratio to maximise the area taken up. Common applications of hydrogels with elevated swelling capability are in hygiene, for disposable short-term products with superior and prompt ability to swell. Highly absorbent hydrogels can also be used in agriculture, to enhance plant growth in soil (Ottenbrite et al. 2010) or to carry pesticides (Sakthivel et al. 2015). Other applications are as superdisintegrants added to disintegrate solid dosage forms (tablets, capsules) and eco-friendly water management materials (Ottenbrite et al. 2010).

Intelligent hydrogels, also known as smart hydrogels, are able to respond to change in environment, resulting in a significant volume change in response to a variety of physical and chemical stimuli. When there is an environmental change, intelligent hydrogels will swell or collapse depending on their properties and will then return to their original volume and shape after the environmental trigger has been removed. Environmental changes which trigger these conformational rearrangements include but are not limited to pH, temperature, electric fields, mechanical stress, chemical and biological agents, magnetic fields, radiation and ultrasound (Mahajan and Aggarwal 2011; Hurst and Novakovic 2013). The results of these microscopic changes in structure can lead to changes in solubility and surface characteristics of the respective smart polymer. The most attractive feature which makes smart polymers so promising is that these changes can nearly always be reversed by simply removing the environmental trigger which caused the response initially. These dramatic changes in chemical structure compared to relatively small changes in stimuli have seen smart polymers applied in the enhanced oil recovery and bioprocessing industries (Peppas et al. 2000; Wang et al. 2011). Novel applications for smart polymers include biomimetic actuators, chemical valves and thermoresponsive surfaces (Galaev and Mattiasson 1999). A significant amount of research has been directed towards the use of temperature-sensitive hydrogels (Qiu and Park 2001; Ebara et al. 2014). The presence of a hydrophobic group such as methyl, ethyl and propyl groups causes the hydrogel to swell or contract due to a variation in temperature. It is possible for thermosensitive hydrogels with physical crosslinks to also undergo sol–gel phase transitions at a critical solution temperature, rather than a change in volume (Ebara et al. 2014).

Other physical stimuli-responsive hydrogels include electric signal-sensitive hydrogels, which contain polyelectrolytes, and swell, shrink or bend when an electric field is applied, as well as light-sensitive hydrogels. In the presence of UV light or visible light, chromophores are absorbed resulting in an increased temperature of the hydrogel, which then induces swelling. Both hydrogels can be used as actuators or artificial muscles; however, several limitations including slow response time are present (Qiu and Park 2001). Throughout the body, pH varies significantly, from acidic to neutral, towards a more basic pH. The variations can be taken advantage of by the use of pH-responsive hydrogels, which swell or contract depending on the pH of their environment. They are formed from polymeric backbones with ionic pendant groups, which ionise in the relevant pH, developing fixed charges. The swelling or contracting is a result of the generation of electrostatic repulsive forces (Gupta et al. 2002). Depending on the pendant groups, pH-sensitive hydrogels can be classified as anionic or cationic. Anionic hydrogels contain acidic pendant groups such as carboxylic or sulphonic acids. They swell in high pH and therefore basic solutions, due to the pendant groups ionising above the pKa of the polymeric network resulting in large osmotic swelling force. Basic pendant groups of cationic hydrogels, such as ammonium salts, ionise below the pKa, resulting in swelling in acidic solutions where the pH is below the pKa of the network (Gupta et al. 2002; Qiu and Park 2001). This is shown in Fig. 8.

Hydrogels in medical applications The elevated presence of water in hydrogels makes them highly beneficial for medical applications, due to their similarity to natural living tissue. Sakthivel et al. (2015) argues that hydrogels resemble tissue more than any other type of synthetic biomaterials. They can be used for many medical applications including regenerative medicine, tissue engineering, diagnostics and drug delivery. As noted by Vashist and co-workers (Vashist et al. 2014), significant advancement has been made in the use of hydrogels for medical applications. Since 1960, the copolymers of 2-hydroxyethyl methacrylate (2-HEMA) and ethylene dimethacrylate have reached many applications including surgical gloves, contact

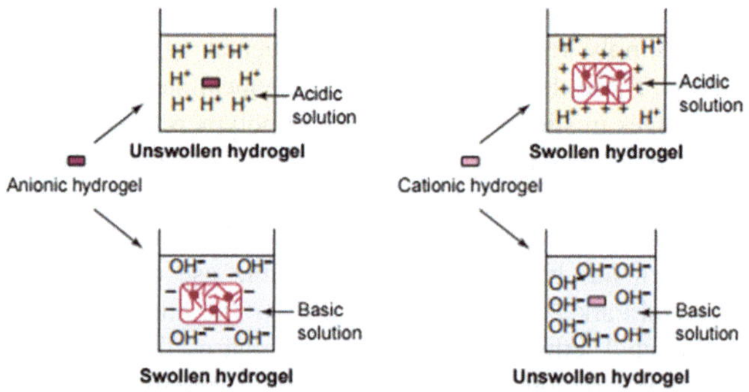

Fig. 8 pH-responsive swelling of anionic and cationic hydrogels (Gupta et al. 2002)

lenses and wound dressings, with current research being carried out into the use of AquaForm hydrogels for wound dressings, with reduced toxicity and increased viscosity.

Contact lenses were the initial application of hydrogels in the 1960s, with more complex polymers used in hydrogels for contact lenses today. Solid rods of hydrogel are polymerised in plastic tubes which are then cut into thin cylindrical slices called 'buttons'. These are then cut using lathes to give the lens the required dioptre of curvature. Other methods of production include spin casting followed by hydration of the resulting hydrogel. Hydrogels are uniquely suited for use as contact lenses because they are biocompatible, easily wetted, have excellent mechanical stability and are permeable to water (Robitaille et al. 2013). The most common type of hydrogels used in the production of contact lenses is silicon based as they have higher oxygen permeability and therefore a more comfortable fit (Nichols 2013). However, even the most prevalent formulations of hydrogels would benefit from further improvements when it comes to associated issues with the protein deposition which leads to lenses having a shortened period of time during which they can be worn comfortably.

Another application of hydrogels is in ***wound dressings***. The ideal wound dressing should absorb excess exudates and toxins and provide a physical barrier to protect from external sources of infection, all whilst keeping the wound moist. Wound dressings should also keep the wound cool, possess adequate permeability to gases, be fully sterile as well as easily changed (Caló and Khutoryanskiy 2015). Hydrogel wound dressings are currently available in two forms: sheets of gel and amorphous gel which is applied topically. Hydrogel dressings prevent the wound from drying out due to them being 'moisture donors'. This encourages granulation, epithelialisation and autolytic debridement which allows wounds to heal more rapidly (Stashak et al. 2004). One of the problems with conventional gauze is that it can dry onto wounds, damaging the new tissue when removed as the dressing is changed. The polymer network within the dressing can expand, containing any puss or exudate which is produced by the wound during the healing process which prevents further infection. Holowka and Bhatia (2014) discuss the use of hydrogels for wound dressings and wound closures. From the perspective of surgical wounds leaking, which can lead to infection, blood loss as well as death, hydrogels can be used to absorb any fluid leaving the wounds, preventing these complications. Adhesive hydrogels can also be used to protect wounds whilst delivering anti-inflammatory therapeutics or tissue-promoting growth factors. An example of this is polyethylene glycol-based adhesives which provide high biocompatibility and result in biodegradable, synthetic hydrogels. Hydrogel adhesives are also advantageous as sealants, with the FDA approving the use of FocalSeal®-L in 2000, for use after removal of lung tumours to seal air leaks. Hydrogels can be attached to an adhesive pad, woven into composite gauze materials or applied from a tube or syringe depending on the gels' physical properties. An example of a current commercially available hydrogel wound dressing is Burnshield™, a polyurethane foam spray containing 96% water and 1.06% *Melaleuca alternifolia* extract which is used as an emergency burn dressing. Further examples of commercially available hydrogel wound dressings are described in Table 1. Amorphous hydrogel dressings typically have to be changed daily, whilst hydrogel sheets have to be changed 2–3 times per week (Stashak et al. 2004).

Hydrogels can also be beneficial for ***diagnostic devices***, due to their ease of integration within a microdevice. PEG-based pH-sensitive hydrogels can be used for the actuation of valves within microfluidic channels, as well as to capture and localise cells from flowing solutions, which can be used for sensing, cell separation and cell-based microreactors (Peppas et al. 2006). Van der Linden et al. (2003) looks at the applications of hydrogel sensors in chemical (micro)analysis. Within hydrogel sensors, a transducer converts the swelling signal of the hydrogel to the electrical or optical domain. Van der Linden argues that stimulus-responsive hydrogels have good potential as microsensors and microactuators. This is due to advantages including high resilience which allows hydrogel actuators to have high dust tolerance and making them appropriate for cell handling applications.

Another application of hydrogels is in ***tissue engineering***. Owing to their structural similarity to the extracellular matrix (ECM) of natural tissues, biocompatibility and mild durability, hydrogels have been employed in the creation of synthetic tissue scaffolds to be used as tissue replacements. Hydrogels used in this capacity employ the three-dimensional form to coordinate cells and offer stimuli to direct the formulation of desired tissue (Drury and Mooney 2003). Early attempts at tissue engineering consisted of microfabricated gels which were coated with cells to try and stimulate new tissue growth. However, as further research has been carried out on both hydrogels and tissue engineering, it has become possible to integrate cells within the gels prior to gelation. The hydrogels composed of natural materials including collagen, alginate, gelatin, hyaluronic acid, chitosan and fibrin are all useful for the investigations of cell–cell and cell–ECM interactions and form the basis of most tissue

Table 1 Hydrogel wound dressing products (Caló and Khutoryanskiy 2015)

Product	Main constituents	Characteristics
Granugel	Pectin, CMC and propylene glycol	A clear, viscous hydrogel for the management of partial and full thickness wounds, may even be used as a filler for dry cavity wounds to provide a moist healing environment
Intrasite gel	Modified CMC (2.3%) and propylene glycol (20%)	Amorphous sterile hydrogel dressing for use in shallow and deep open wounds
Purilon gel	Sodium methylcellulose, >90% water	Indicated in conjunction with a secondary dressing for necrotic wounds as well as first- and second-degree burns
Aquaflo	Polyethylene glycol and propylene glycol	Disc-shaped wound dressing which helps to fill shallow cavities. Translucent gel that allows wound visualisation
Woundtab	Sulphonated copolymer, CMC, glycerol and water	Superabsorbent polymer gel able to absorb bacteria and retain them within its structure

engineering solutions which have been produced to date (Annabi et al. 2010; Li et al. 2012). Development of hydrogels in this direction is important due to current limitations of organ donors and useable organs. In the UK alone, there were 6943 patients waiting for a transplant at the end of March 2015, 429 patients had died whilst on the active waiting list, and 807 had been removed from the list mostly due to deteriorating health (NHS: Blood and Transplant 2015). Of these patients still waiting, 30% are on the list for more than 2 years, and approximately 500 are on the list for more than 5 years (Organ Donation 2015). By use of hydrogels for tissue engineering, more people could receive transplants without being limited by the number of donors as well as the uncertainty of whether the organs from these donors can be used. This is also a suitable path for applications such as skin grafts, for patients who have suffered severe burns or skin cancer. Here, hydrogels are beneficial for tissue engineering as they promote the formation of new blood vessels and skin (Vashist et al. 2014) and can support cell growth and function through sending signals to the cells. They are readily formed in situ during implantation and can perform encapsulation of the cells as they are crosslinked with tissue surfaces (Jiao et al. 2006). Lee and Mooney (2001) discuss the employment of hydrogels for tissue engineering as patient cells loaded scaffolds which can then be injected into the patient. These scaffolds can then bring the cells to the required location in the body, enabling the formation of new tissue, in a minimally invasive manner.

Within tissue engineering, hydrogels must fulfil several criteria to ensure appropriate functionality, including biocompatibility, biological performance parameters and physical parameters such as degradation and mechanics. Natural polymers also often provide good biocompatibility, whereas synthetic polymers can result in a negative response of the body. For that reason, there are limitations of application as to what type of hydrogel can be used. It is also essential that non-toxic crosslinkers are used and that the hydrogel used has sufficient mechanical properties which can create and maintain space for tissue development. Advantages and disadvantages of hydrogels as part of tissue engineering matrix have been summarised by Hoffman and co-workers (Table 2, Hoffman 2012).

Table 2 Advantages and disadvantages of hydrogels as tissue engineering matrices (Hoffman 2012)

Advantages	Disadvantages
Aqueous environment can protect cells and fragile biopolymers such as peptides, proteins and nucleic acids	Can be hard to handle
Good transport of nutrient to cells and products from other cells	Usually mechanically weak
May be easily modified with cell adhesion ligands	May be difficult to load drugs and then crosslink in vitro as a prefabricated hydrogel
Can be injected in vivo as a liquid that gels at body temperature	May be difficult to sterilise if contains radiation sensitive molecules
Usually biocompatible	Can have short shelf life

The degree of crosslinking within tissue engineering solutions is of importance as it governs the structural properties of the scaffold. The required properties are dependent on the environment which the scaffold will be used in. For example, a structure which will be placed within articular cartilage will be subjected to far greater loads than a structure being placed in soft abdominal tissue. Post-synthesis treatments can be used if required and have been shown to increase the lubricity of anterior knee cartilage which must be able to withstand loads of three to five times a person's body weight (Braithwaite 2013). Table 3 describes how the size of synthetic scaffold determines its role within an engineered tissue.

Either top-down or bottom-up processes can be used in the production of tissue engineering scaffolds. Top-down methods use microengineering approaches to control the features of relatively large hydrogels, whereas bottom-up methods build tissues from smaller hydrogel pieces (Khademhosseini and Langer 2007). Methods such as photolithography, emulsification, microfluidics and micromoulding have been used to try and control the crosslinking process with a high degree of spatial resolution. Photolithography is a particularly attractive method of crosslinking as it can be performed in situ in a minimally invasive manner after delivering the hydrogel via injection to the target area (Nguyen and West 2002). An example of a hydrogel tissue engineering scaffold is bioactive peptides which were attached to PEG backbones in order to create synthetic ECMs. This hydrogel was seeded with muscle cells and successfully encouraged the growth of smooth muscle as it offered many of the signals and interactions provided to cells by the naturally occurring ECM in tissues (Mann et al. 2001). Roughley et al. synthesised ECMs which mimicked the nuclei of intervertebral disc cells (IVD) from chitosan and anionic proteoglycan. It was found that in vitro experiments resulted in the successful production and retention of nucleus pulpous within the hydrogel (Roughley et al. 2006). This method could be extended to in situ repair of fibrocartilaginous tissue, but the authors noted that IVD damage was often due to a lack of nutrient supply to the cartilage. This method might only fix part of the problem, but shows the potential of hydrogels in synthetic ECM applications. Another example of a synthetic tissue scaffold is a hyaluronic acid–poloxamer hydrogel which is reversibly thermosensitive and undergoes the sol–gel transition at body temperature, making injection an ideal application method. Inspired by mussel adhesion in nature, this hydrogel was reported to be ideal for cell delivery due to its mechanical stability and tissue adhesion properties (Lee et al. 2010).

Mussels have been the inspiration for ***hydrogel actuators*** (Lee and Konst 2014). Novel hydrogel actuators have been developed which can act at the microscale as micromanipulators and as actuators in microfluidic systems. Analogous to

Table 3 Applications of various sizes of hydrogel tissue engineering structures

Structure size (μm)	Application
0.1–10	Microenvironment of individual cells
10–400	Structure of cell clusters
>400	Interactions between cell clusters

bimetallic strips, sheets of stimuli-responsive hydrogels can be attached to a substrate which is moved as hydrogels swell and contract in response to stimuli. Applications include smart optical lenses with tuneable focal lengths, smart valves for the control of liquid flow and materials with switchable colouration (Yu et al. 2001; Dong et al. 2006; Shin et al. 2012).

Hydrogels for drug delivery applications Important applications of hydrogels in medical industry are in drug delivery. Current methods of drug delivery can have poor bioavailability, as well as poor activity due to the amount and speed at which a drug gets to the targeted site (Bhattarai et al. 2010). Therefore, a variety of drug delivery systems (DDSs) is researched including nanoparticles, liposomes, implants and hydrogels which can be more suitable due to high biocompatibility and stimuli sensitivity. Drug delivery is a hydrogel application which has been explored extensively over the past two decades. Typically, active pharmaceutical ingredients are contained within a hydrogel structure which is then delivered orally or injected into the body. The benefits of using hydrogels for drug delivery are that they offer the sustained release of compounds which results in a high local concentration over long periods (Hoare and Kohane 2008).

There are a variety of DDS currently available to patients in a range of forms including injections, tablets and sprays. These systems must deliver the correct dosage of drug, at an optimal concentration to ensure that the drug is performing effectively. However, it is often the case that these systems reach a peak of drug concentration, which then falls requiring the patient to take another dosage. If the peak is above the required dosage, it can result in an overdose or increased toxicity, or if it falls below the required amount, the drug may become ineffective requiring re-administration (Bierbrauer 2005). This again can lead to overdose or increased toxicity if dosage guidelines are not followed. The use of a hydrogel for a DDS allows for a controlled release, resulting in site-specific and time-controlled optimum concentration. This reduces the toxicity by ensuring only the target site is exposed to the medication and improves efficiency. Furthermore, the use of hydrogels could limit any potential side effects, such as from peaks of high concentration, and allows the medication to be taken in a single dosage. These benefits have led to Vashist et al. (2014) claiming that hydrogel-based DDS has revolutionised the pharmaceutical world.

Drugs are usually loaded into the hydrogels by the use of physical entrapment. This can be carried out by positioning the hydrogel in a solution containing the specified drug at the required concentration or by allowing the hydrogel monomers and crosslinker to mix with the drug and then polymerise (Ninawe and Parulekar 2011). The pore size of hydrogels can be altered by varying the degree of crosslinking and the gel's water content. There are four main methods of drug release from hydrogels; swelling controlled, diffusion controlled, chemical controlled and stimuli controlled. In the swelling controlled release, hydrogels discharge the compounds encapsulated by the gel when they swell due to contact with biological material. This allows the active pharmaceutical ingredients to pass through the now widened gaps in the hydrogels polymeric structure as shown in Fig. 9.

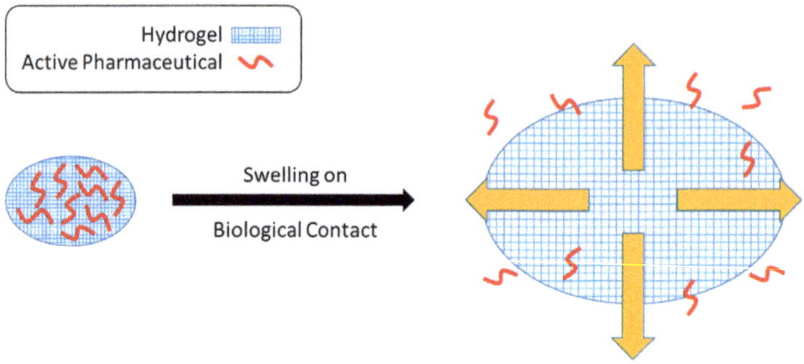

Fig. 9 Swelling-controlled drug release from a hydrogel

In the diffusion-controlled release, hydrogels are further categorised into two types, matrices and reservoir membranes. In reservoir membranes, the active pharmaceutical ingredient is enclosed by a hydrogel membrane. The drug in the centre of the hydrogel diffuses through the membrane over time. The centre of the active core has a higher concentration than its exterior to compensate in the reduction of surface area over time to ensure a constant concentration of released pharmaceutical (Bierbrauer 2005). This is shown in Fig. 10.

Hydrogen matrices which undergo diffusion-controlled drug release already contain the active ingredient dispersed throughout the hydrogel when they are produced. Drug release is achieved as the active pharmaceutical ingredient is released from the pores in the hydrogel. This process is shown in Fig. 11. Diffusion-based systems serve as reservoirs with a drug encapsulated within a hydrogel. These matrix delivery systems can benefit from drug being dispersed uniformly throughout the hydrogel, within a larger polymer structure (Wei et al. 2012).

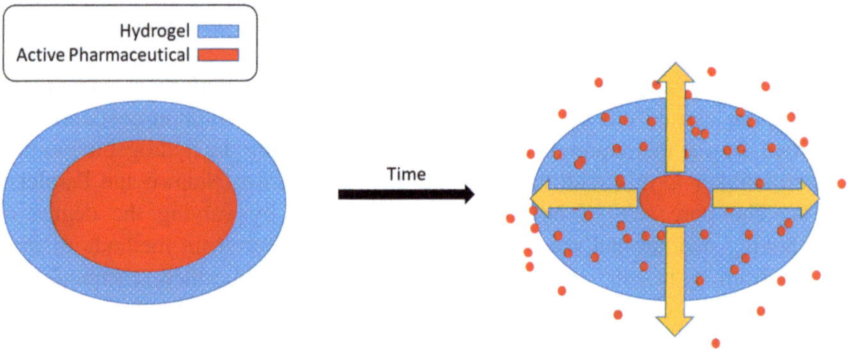

Fig. 10 Diffusion-controlled drug release through a hydrogel membrane from a reservoir

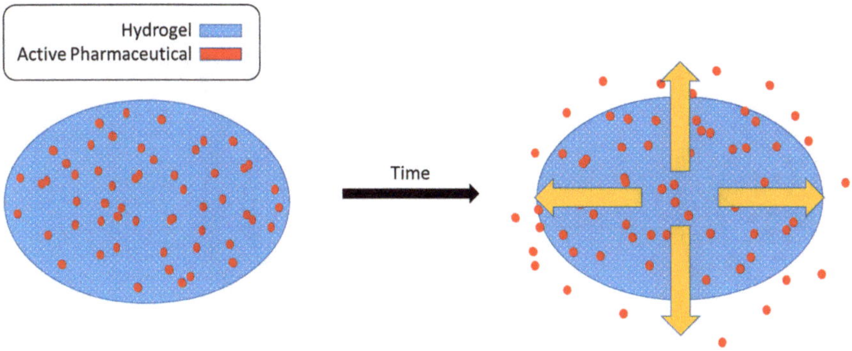

Fig. 11 Diffusion-controlled drug release from a hydrogel membrane matrix

Chemical-controlled diffusion is the bioerosion of hydrogels which contain active pharmaceutical ingredients by naturally occurring chemicals such as gastric fluid. As the hydrogel is broken apart, the drugs contained within it are released from the polymeric structure. Early examples of hydrogels being used in drug delivery were orally consumed hydrogels consisting of methacrylic acid (MAA) and PVP–CMC membranes. These were observed to release molecules through super porous membranes as swelling occurred when coming into contact with gastric fluid (Park 1988; Chen et al. 2000).

One example of a promising hydrogel drug delivery system is a dexamethasone punctum plug which is inserted into the tear duct of patients who have undergone eye surgery. The plug is designed to release corticosteroid to counteract any inflammation. This product has entered phase three clinical trials (AAO 2014). The hydrogel naturally decomposes into harmless products after it has finished the release cycle, eliminating the need for the hydrogel to be removed. Stimuli-controlled drug release has become increasingly popular due to the increased use of smart hydrogels throughout recent years. Drugs have been developed which are delivered as a result of changes in temperature or pH, causing breaks in stimuli-sensitive bonds or changes in hydrogel structure which allow for the previously trapped drug to be released. An example of temperature-stimulated delivery is shown in Fig. 12 where an increase in temperature causes a peptide to 'unzip', releasing the pharmaceutical ingredient contained within a liposome (Al-Ahmady et al. 2012).

An example of a drug delivery mechanism which is actuated by a change in pH is liposome vesicles which have both PEG chains and proteins attached to their exterior for the treatment of tumour cells (Koren et al. 2012). As the PEG chains are longer than the proteins which attack the tumour cells, proteins are inhibited from reacting until the PEG chains are removed. The bonds which attach the PEG chains to the liposome are pH sensitive and break once exposed to a pH 5–pH 6 environment (Fig. 13).

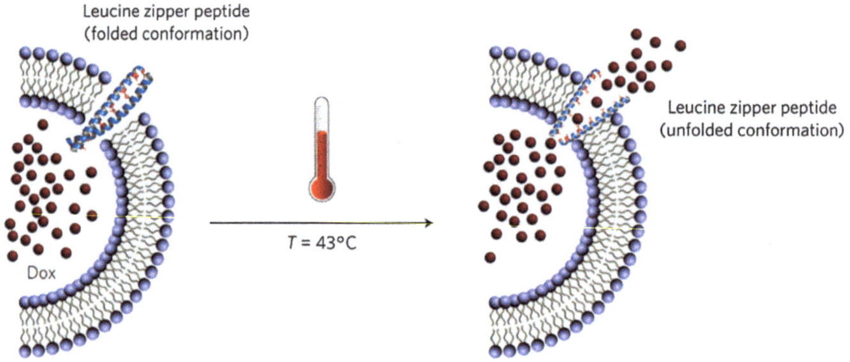

Fig. 12 Temperature-stimulated drug delivery (Al-Ahmady et al. 2012)

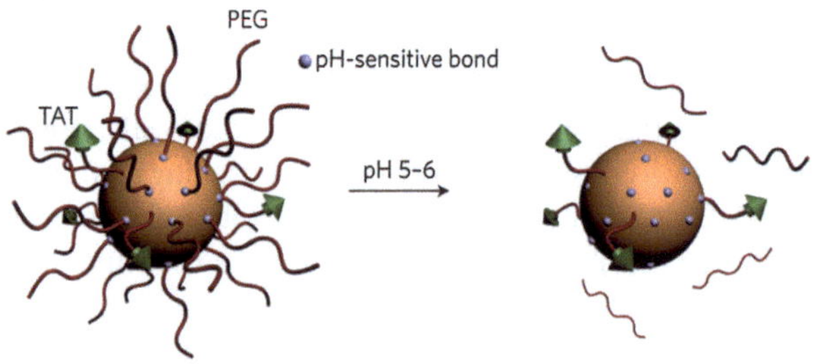

Fig. 13 pH-stimulated drug delivery (Koren et al. 2012)

The range of stimuli available to trigger-controlled drug release is dictated by the stimuli-responsive polymers which they contain. Novel stimuli-controlled methods have been reported using voltage, ultrasound, infrared light and radiation as drug release initiators (Rapoport et al. 2009; Yan et al. 2010; Schroeder et al. 2012; Xiao et al. 2012). Due to their benefits, several hydrogel DDSs reached the market and large number is in the stages of clinical and pre-clinical trials. Some of the products are detailed in Table 4.

A variety of research studies show promising results in the area of hydrogel-based DDS. Holowka and Bhatia (2014) discuss the use of hydrogels as implantable sensing and responsive devices. For diseases such as diabetes, patients have to consistently monitor and correct their insulin levels and often have to inject and take medication throughout the day. This can result in patient non-compliance, skin problems as well as high toxicity levels at the site of injection. Glucose-responsive hydrogels can be used to help treat diabetes by controlling the administration of insulin as a function of blood glucose levels. This can be achieved by combining the reaction of glucose oxidase with pH-responsive swelling which is

Table 4 Drug delivery applications of hydrogels

Product	Purpose	Target site	Developmental phase	Reference
TRETIN X Gel	Acne vulgaris	Topical	On the market	Vashist et al. (2014)
Aquatrix™ II	Implants, skin adhesive and drug delivery matrices	Topical	On the market	Gupta et al. (2002)
Moraxen	End-stage cancer pain: release product of morphine sulphate	Rectum	On the market	Gupta et al. (2002)
DuraSeal™ dural sealant system	Synthetic absorbable sealant and applicator for delivery of sealant	Brain	Clinical trials	Vashist et al. (2014)
Gelesis100 (previously Attiva)	Obesity—stretches the stomach walls to simulate being full	Non-invasive capsulated device for stomach	Clinical trials	Vashist et al. (2014)
PEG hydrogel	PEG loaded with insulin for diabetes	Diabetes	Pre-clinical trials	Vashist et al. (2014)

the most common approach, by use of phenylboronic acid or by the use of lectin (Ottenbrite et al. 2010). Use of a pH-sensitive hydrogel switch is studied for the treatment of diabetes, due to hydrogel ability to replicate the body's natural insulin release pattern (Cao et al. 2001). The device consists of a hydrogel switch cavity, which is sealed by a diaphragm, preventing interaction between the hydrogel and the drug. In high pH, the hydrogel swells, forcing the diaphragm to expand and therefore prevent any drug from passing through the release chamber. In low pH, the hydrogel collapses and therefore opens the release channel, resulting in a high drug release rate.

Tuberculosis treatment, usually conducted with anti-tubercular drugs which have short biological half-life, can also benefit from the use of hydrogels. Current methods of treatment by use of implants, microspheres and liposomes have high toxicity, are not cost effective, have low loading efficiencies and often require surgery. The lengthy treatment period of 6 months also results in patient non-compliance. The use of hydrogels can provide a non-invasive delivery and reduce the dosage frequency which is usually for a period of at least 6 months, and therefore increase patient compliance (Vashist et al. 2014). Research carried out on guinea pigs by Sharma et al. (2004) showed that three oral applications of loaded hydrogel beans provide the same therapeutic protection against tuberculosis as 45 daily doses of free antibiotics. This resulted in a more appropriate sustained release kinetic and drug bioavailability, as well as a reduced dosing frequency of every 10 days rather than daily.

Chitosan-based hydrogels can be particularly advantageous for drug delivery owing to their good biocompatibility, low toxicity and degradability by human enzymes. They also allow for targeted release due to pH sensitivity and can bind to the mucus lining of the gastrointestinal tract due to mucoadhesive properties, which is common for most polysaccharides (Bhattarai et al. 2010). Chitosan is also one of the most promising candidates for cancer therapy due to its potential antibacterial and antioxygenic properties (Wang et al. 2016). Bhattarai et al. (2010) discusses the use of hydrogels for the implantation of radio therapeutics adjacent to the target tissue. Without the use of hydrogels, this process called brachytherapy provides elevated doses of radiotherapy to the aimed site. Chitosan-based hydrogels can gel in vivo, therefore providing a less invasive method. Hydrogels based on chitosan, crosslinked with glutaraldehyde and loaded with I-norcholestrol have been studied for the prevention of tumour recurrence in a breast cancer mouse model (Azab et al. 2007). This research has currently been carried out on mice, but shows a reduction in progression rate and prevented 69% of tumour recurrence. Research also illustrated that there was small or no distribution of the drug following implantation. This reduces toxicity and increases the effectiveness of treatment. Ocular drug delivery can also be enhanced by the use of hydrogels. Vashist et al. (2014) indicates that hydrogels offer more efficiency and improved vision compared to conventional ocular delivery systems. The use of chitosan-based hydrogels is considered the best choice for ocular delivery due to mucoadhesive properties and excellent penetration nature. This is further supported by the work of Felt et al. (1999). One of the most widely used antibiotics for ocular diseases is aminoglycosides, which is used for a variety of treatments such as for conjunctivitis and bacterial keratitis which can result in reduced vision and light sensitivity. The aminoglycoside tobramycin has been proven to help against bacterial infections, but it often has short residence time and poor transcorneal penetration, resulting in multiple and frequent doses being required. Research carried out by Felt et al. (1999) shows that the use of chitosan-based hydrogels improves the residence time and allows for easier administration when low concentrations and therefore low viscosity chitosan solutions are used.

Significant prospective applications of genipin combined with chitosan are in the fabrication of elastic cartilage substitutes, the assembly of carriers for the controlled drug delivery, the encapsulation of biological products and living cells, the biofabrication of synthetic tissues such as muscle and arterial walls, and the dressing of wounds in animals and humans (Muzzarelli et al. 2015). Genipin is slowly replacing more cytotoxic crosslinking agents such as glutaraldehyde as all of the aforementioned applications are detrimentally affected by materials which provoke an undesirable immunogenic response. In recent years, genipin's ability as a crosslinking agent has also been investigated with compounds such as carboxymethyl chitosan (CMCht), collagen and bovine serum albumin (BSA). When CMCht molecules were exposed to genipin to create porous membranes, less crosslinking than predicted was experienced, which has been attributed to genipin self-polymerisation (Fiamingo and Campana-Filho 2016). However, the same research noted that the morphology, swelling characteristics and ability to absorb phosphate-buffered solution (PBS) were still affected by the amount of genipin present in the CMCht hydrogels. Butler et al. (2006) postulated that the formation of genipin dimers, trimers and oligomers was the cause for genipin-crosslinked chitosan hydrogels not swelling as predicted. However, the addition of genipin reportedly made the hydrogels much less brittle. Genipin-crosslinked BSA was found to swell significantly under acidic conditions. Genipin-crosslinked mixtures of chitosan and BSA were found to be able to withstand the 7 kPa pressures created by the grinding actions in the stomach, as well as swelling by up to 400%, making it a promising candidate for drug delivery (Butler et al. 2006). Viscoelastic properties of genipin-crosslinked chitosan hydrogels have also been investigated. Moura et al. (2007) noted that the addition of genipin as a crosslinking agent reduces the gelation time compared to a chitosan-only solution and improves the elasticity and three-dimensional stability of the gel. This study also proved that the viscoelastic features of chitosan hydrogels can be 'tuned' by changing the genipin loading and that the in situ formulation of the hydrogel made promising matrices for the encapsulation of cells and bioactive materials (Moura et al. 2007). Crosslinking density within genipin–chitosan hydrogels has also been found to have a marked impact on the macroporous structure and mechanical stiffness. An increase in the crosslinking density was found to improve the hydrogel's cell bioadhesion properties and viability (Gao et al. 2014). Again, the presence of genipin was found to have no undesired cytotoxic effects. Electrospun nanofibrous chitosan scaffolds crosslinked by genipin have seen use in both bone and cartilage tissue engineering, with early investigations into its application alongside natural alginate and collagen compounds proving promising (Muzzarelli et al. 2015). Genipin-crosslinked tissue scaffolds containing osteoblasts for the regeneration of bone tissue have also showed progressive adhesion, proliferation and colonisation in vitro (Vozzi et al. 2013). More recent endeavours have included seeding genipin-crosslinked hydrogels with stem cells in order to stimulate tissue regeneration (Muzzarelli et al. 2016). Genipin could also find further applications in hydrogels implanted into the pancreas to treat obesity and its related conditions. In obesity studies, genipin has been shown to inhibit the leaking of protons from uncoupling protein 2 (UCP2). In the same research, genipin was found to stimulate insulin production from pancreatic islets and type 2 diabetes in mice (Zhang et al. 2006). Overall, the majority of the potential applications for genipin-crosslinked hydrogels are for applications which currently exist but are already fulfilled by other, more cytotoxic crosslinking agents. Early research which has been conducted on genipin was to confirm that it could carry out the same role as other chemicals with the same level of flexibility. This has been achieved with a high level of success and unexpectedly surpassed expectations in some applications. Now that genipin has been seen to be able to meet and outperform other more cytotoxic crosslinking agents, it is being included in a myriad of new medical products which are being researched within drug delivery, wound dressings and tissue engineering.

Although drugs can be delivered in a number of different ways including injection and implants, **oral delivery** is the preferred method, accounting for approximately 90% of therapeutics used (Bhattarai et al. 2010), due to it being non-invasive and easy to administer. It also improves patient compliance and comfort, allowing the medication to be taken on a flexible schedule and can potentially reduce the cost compared to injection-based administration. However, oral administration can be limited to small molecule drugs and limited by those with low bioavailability and poor penetration into the bloodstream (Sharpe et al. 2014). To overcome this, hydrogels can be used especially for the drug delivery of proteins and peptides. One example is the protein insulin, which is injected for diabetes treatment, often at least twice a day. This can result in discomfort and non-compliance from patients, therefore making oral administration beneficial. Other difficulties with oral DDS include the harsh environments of the gastrointestinal tract such as the presence of enzymes, low residence time and a highly acidic environment that can be detrimental to the efficiency of medication orally taken. Despite these difficulties, drug delivery aimed to the stomach and colon is advantageous for many treatments including Crohn's disease, inflammations and infections (Bhattarai et al. 2010). For oral mucosal drug delivery, it has been suggested that mucoadhesive polymers such as chitosan are beneficial, to prevent the fast elimination of the drug due to flushing action of the saliva which is considered as a major challenge for this route of drug delivery (Tang et al. 2014). Furthermore, studies showed that chitosan-based films exhibited elevated water uptake and antibacterial efficacy, making them ideal candidates for oral DDS.

Challenging area of drug delivery that could benefit from hydrogel use is ***hydrophobic drug delivery***. This area is of particular importance due to the wide variety of applications. For example, a significant amount of drugs used for the treatment of cancer are poorly soluble in water, and current methods of injection can result in varied concentration of doses. By incorporating hydrophobic drugs into hydrogels, the solubility of the drug in water can be improved, and a prolonged release can be attained (McKenzie et al. 2015). Yet, due to the hydrophilicity of hydrogels, DDSs are mainly related to hydrophilic drugs. When a hydrophobic drug is encapsulated, rapid and uncontrolled drug release in the initial phase is often noted, resulting in hydrophobic drug delivery usually being carried out by other polymer mixtures (McKenzie et al. 2015; Peng et al. 2011; Martin et al. 2003; Petta et al. 2016; Gou et al. 2008). Despite hydrophobic drugs being beneficial, e.g. in increasing transport for buccal (Martin et al. 2003) and mucosal administration (Liu and Lin 2010), and hydrogels being proven to be good DDS, limited research has been carried out in the area of hydrophobic drug delivery. The use of a hydrogel for a versatile DDS, in which a relevant drug could be loaded depending on the clinical treatment, was assessed by Petta et al. (2016). The work specifically looked at the use of a hydrogel to deliver drugs after bone graft treatment, which often requires hydrophobic drugs. Usual methods have a high risk and are associated with high donor site pain, as well as morbidity and irregular shape formation. The results of using a thermoresponsive hyaluronan hydrogel showed that hydrogels provided a DDS that could be easily handled and that was readily recognised as a natural extracellular matrix component by the host tissue, reducing the risk of side effects caused by alternative methods. As mentioned previously, the use of hydrogels for cancer treatment has the potential to provide a more efficient and less toxic method than alternative DDS. A variety of work has been carried out looking into the possibility of using hydrogels including work carried out by Ye and Hu (2016) assessing the use of acetalated-β-cyclodextran injectable hydrogels, by use of physical crosslinkers, for the release of the hydrophobic anticancer drug camptothecin. Alternative methods were unable to travel through all of the blood capillaries before reaching the target site, making them inefficient. The research showed that the target site was reached and that sustained release was achieved in neutral and alkaline environments. In addition, studies of the thermosensitive hydrogel Pluronic F127 as injectable DDS of honokiol used for cancer therapy have been reported (Gou et al. 2008). The study showed that site-specific delivery could be achieved with a sustained release and that reduced dosages were required compared to conventional methods. Despite the reported success of current research in the use of hydrogels for hydrophobic drug delivery, limitations are still present and further studies are needed. Example of reported limitations is given in the study of crosslinked palmitoyl glycol chitosan hydrogels for the delivery of hydrophobic drug denbufylline (Martin et al. 2003). Sustained but slow drug release was achieved. It was found that the gel hydration was reduced, leading to a slow release time. Additional difficulty was measurement of the amount of drug released due to low solubility in the aqueous media.

Potential hydrogels having ***anti-inflammatory drug delivery*** has been suggested, and the need for further research has been identified (Barron et al. 2016). Variables to consider include the impact of using anti-inflammatory drugs of different sizes and delivering drugs via varied release profile. An example of anti-inflammatory drug is ibuprofen. Ibuprofen is a non-steroidal anti-inflammatory drug (NSAID), which is both commercially available to patients and available via prescription. It is poorly soluble in water. Once swallowed, ibuprofen is quickly, and nearly fully, absorbed from the upper gastrointestinal tract which comprises of the oesophagus, stomach and duodenum (Rainsford 2012). The over the counter form of ibuprofen has a racemic chemical structure, composing of a 50:50 mixture of the R(−) and S (+) enantiomers (Rainsford 2012) as shown in Fig. 14.

It has been proven that ibuprofen has comparable or better medicinal effects than other NSAIDs in a variety of conditions including dental pain, colds and influenza, various minor injuries and inflammatory arthritis (Rainsford 2012). Ibuprofen is therefore an essential drug, used for a variety of treatment which could benefit from the use of hydrogels. Research carried out by Djekic et al. (2016) looked into the

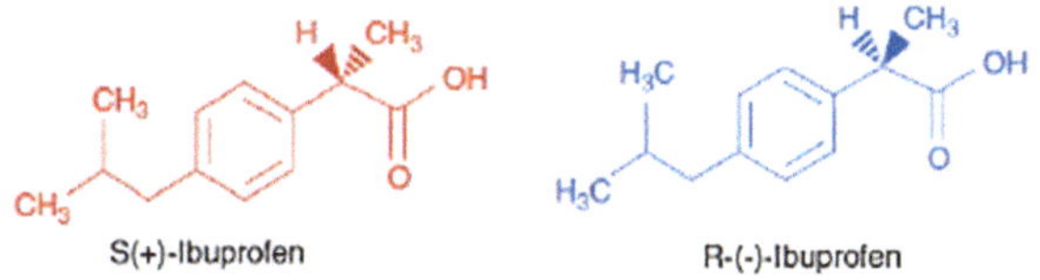

Fig. 14 Chemical structures of S(+) and R(−) isomers of ibuprofen

use of hydrogel-thickened emulsions for the release of ibuprofen. The method combines the advantages of both microemulsions and hydrogels to achieve increased drug solubility and loading capacity. The studies showed the successful in vitro release which followed zero-order kinetics for 12 h. Maskare et al. (2013) reported the use of hydrogel-thickened DDS forms for the release of ibuprofen. Nanoemulsions thickened with hydrogel were found to have an increased solubility, as well as an enhanced effect, making them more desirable for the treatment of musculoskeletal pain. Work performed by Jabeen et al. (2016) showed the consequence of varying pH on the loading capability and release profile of sodium alginate hydrogels loaded with ibuprofen. The hydrogels were prepared using sodium alginate, polyethylene oxide and acrylic acid, and three pH values of 1, 7 and 12 were studied. It was found that hydrogels prepared at a neutral pH showed faster release rate and a higher loading capacity than the hydrogels prepared at pH 1 and 12.

Slow response hydrogels tend to have slow drug release that can be in some cases use as an advantage. The application of hydrogels for DDS can therefore be considered for ***long-term sustained release***. For many treatments such as those against cancer and for diabetes, patients often have to take daily doses which can lead to patient non-compliance as well as high costs and high levels of toxicity. The use of a hydrogel would allow for multiple dosages to be released over a period of time, at the target site, with effective levels of therapeutic concentration. They could also allow for the release of multiple medications all from the same hydrogel. Long-term release could also be beneficial for the treatment of HIV in areas where medical facilities are limited. Often, treatment requires a daily dosage of medication, not accessible to some patients resulting in difficulties of carrying out therapy.

Torres et al. (2011) carried out a computation study, simulating the release of paclitaxel to brain tumours. This study showed that an effective concentration of paclitaxel was maintained for over 30 days. Ozeki et al. (2010) reported a study in which microspheres of PLGA containing camptothecin were injected into the brains of rats which had been implanted with C6 glioma cells. This resulted in long-term, sustained release of camptothecin and reduced the size of the tumour. Studies led by Oren Scherman at the University of Cambridge have also shown the potential of hydrogels for long-term release (Scherman 2012). An injectable hydrogel was prepared which could be loaded with proteins and therapeutics. Results showed a continual discharge of bovine serum albumin over a period of 160 days and were capable of delivering sustained release of proteins for up to 6 months.

Despite the reported success and the potential hydrogels have in the area of drug delivery, there are also recognised ***limitations*** and therefore further work is required. Two elements in need of further improvements are response time and mechanical strength (Gupta et al. 2002). Low mechanical strength of natural polymers such as chitosan causes difficulties in loading drugs into the hydrogel. For that reason, further studies are needed to achieve relevant payload of such hydrogels (Vashist et al. 2014). Studies for payloads greater than 70 kDa are primarily with non-pH-responsive hydrogels, in non-oral applications (Watkins and Chen 2015).

Studies into the use of pH-responsive, lysine crosslinked PLP hydrogels show that payloads of up to 2000 kDa can be delivered (Watkins and Chen 2015); however, further work in increasing payloads is required. At the same time, recognised challenge is to extend the duration of drug release to allow for hydrogel long-term applications in particular when it comes to delivery of hydrophobic molecules (Hoare and Kohane 2008). Whilst beneficial for some, slow hydrogel response can be limiting factor for other drug delivery applications. In such cases, although benefit of site-specific drug delivery still remains, slow response can be a significant drawback. As Hamidi et al. (2008) pointed out, this can be overcome by the use of thinner and smaller hydrogels; however, this would incur fragility and loss of mechanical strength. This could result in the hydrogel degrading before the desired delivery site, due to the harsh conditions inside the human body. Limitations also exist in current methods of drug delivery to the brain. A particular challenge is the blood–brain barrier, of which 98% of newly synthesised drugs fail to cross (Vashist et al. 2014) for which the use of crosslinked PEG and polyethyleneimine gels has shown potential for transport of oligonucleotides across this barrier.

References

AAO (2014) Tear duct implant effective at reducing pain and inflammation in cataract surgery patients. American Academy of Opthalmology

Al-Ahmady ZS, Al-Jamal WT, Bossche JV, Bui TT, Drake AF, Mason AJ, Kostarelos K (2012) Lipid-peptide vesicle nanoscale hybrids for triggered drug release by mild hyperthermia *in vitro* and *in vivo*. ACS Nano 6:9335–9346

Annabi N, Nichol JW, Zhong X, Li C, Koshy S, Khademhosseini A, Dehghani F (2010) Controlling the porosity and microarchitecture of hydrogels for tissue engineering. Tissue Eng Part B Rev 16(4):371–383

Azab AK, Kleinstern J, Doviner V, Orkin B, Srebnik M, Nissan A, Rubinstein A (2007) Prevention of tumor recurrance and distant metastasis formation in a breast cancer mouse model by biodegradable implant of I-Norcholestrol. J Control Release 123:116–122

Barron V, Killion JA, Pilkington L, Burke G, Geever LM, Lyons JG, McCullagh E, Higginbotham CL (2016) Development of chemically cross-linked hydrophillic-hydrophobic hydrogels for drug delivery applications. Eur Polym J 75:25–35

Begam T, Nagpal AK, Singhal R (2003) A comparative study of swelling properties of hydrogels based on poly(acrylamide-co-metyl methacrylate) containing physical and chemical crosslinks. J Appl Polym Sci 89:779–786

Berkowitz WF, Choudhry SC, Hrabie JA (1982) Conversion of asperuloside to optically active prostaglandin intermediates. J Org Chem 47(5):824–829

Bhattarai N, Gunn J, Zhang M (2010) Chitosan-based hydrogels for controlled, localized drug delivery. J Adv Drug Deliv Rev 62(1):83–99

Bierbrauer F (2005) Hydrogel drug delivery: diffusion models, internal report. School of Mathematics and Applied Statistics, University of Wollongong, Australia. www.bierbrauerf.weebly.com

Bonini C, Iavarone C, Trogolo C, Fabio RD (1984) One-pot conversion of 6-hydroxy-Δ^7-iridoid glucosides into cis-2-oxabicyclo[3.3.0]oct-7-enes and transformation into corey's lactone analogue. J Org Chem 50(7):958–981

Braithwaite G (2013) Hydrogels, polymers and plastics in medical devices. Boston, MA, USA. www.campoly.com/educational-resources/presentations/

Butler MF, Ng Y-F, Pudney PDA (2003) Mechanism and kinetics of the crosslinking reaction between biopolymers containing primary amine groups and genipin. J Polym Sci 41(24):3941–3953

Butler MF, Clark AH, Adams S (2006) Swelling and mechanical properties of biopolymer hydrogels containing chitosan and bovine serum albumin. Biomacromol 7(11):2961–2970

Caló E, Khutoryanskiy VV (2015) Biomedical applications of hydrogels: a review of patents and commercial products. Eur Polymer J 65:252–267

Cao X, Lai S, Lee LJ (2001) Design of a self-regulated drug delivery device. Biomed Microdevice 3(2):109–118

Chen J, Blevins WE, Park H, Park K (2000) Gastric retention properties of superporous hydrogel composites. J Control Release 64(1–3):39–61

Chen H, Ouyang W, Lawuyi B, Martoni C, Prakesh S (2005) Reaction of chitosan with genipin and its fluorogenic attributes for potential microcapsule membrane characterization. J Biomed Mater Res 75A(4):917–927

Delgadillo-Amendariz NL, Rangel-Vazquez NA, Marquez-Brazon EA, Rojas-DeGascue B (2014) Interactions of chitosan/genipin hydrogels during drug delivery: a QSPR approach. Quim Nova 37(9):1503–1509

Delmar K, Bianco-Peled H (2015) The dramatic effect of small pH changes on the properties of chitosan hydrogels crosslinked with genipin. Carbohyd Polym 127

Dimida S, Demitri C, Benedictis VMD, Scalera F, Gervaso F, Sannino A (2015) Genipin-cross-linked chitosan-based hydrogels: reaction kinetics and structure-related characterstics. J Appl Polym Sci 132(28):1–8

Djekic L, Martinovic M, Stepanovic-Petrovic R, Micov A, Tomic M, Primorac M (2016) Formulation of hydrogel-thickened nonionic microemulsions with enhanced percutaneous delivery of Ibuprofen assessed *in vivo* in rats. Eur J Pharm Sci

Djerassi C, Gray JD, Kincl FA (1960) Naturally occurring oxygen heterocyclics. IX. Isolation and characterisation of genipin. J Org Chem 25(12):2174–2177

Dong L, Agarwal AK, Beebe DJ, Jiang H (2006) Adaptive liquid microlenses activated by stimuli-responsive hydrogels. Nature 442:551–554

Drury JL, Mooney DJ (2003) Hydrogels for tissue engineering: scaffold design variables and applications. Biomaterials 24:4337–4351

Ebara M, Kotsuchibash Y, Narain R, Idota N, Kim Y-J, Hoffman JM, Aoyagi T (2014) Smart biomaterials. Springer, Tsukuba, Japan

Endo T, Taguchi H (1973) The constituents of gardenia jasminoides geniposide and genipin-gentibioside. Chem Pharm Bull 21:2684–2688

Felt O, Furrer P, Mayer JM, Plazonnet B, Buri P, Gurny R (1999) Topical use of chitosan in ophthalmology: tolerance assessment and evaluation of precorneal retention. Int J Pharm 180:185–193

Fiamingo A, Campana-Filho SP (2016) Structure, morphology and properties of genipin-crosslinked carbomethylchitosan porous membranes. Carbohyd Polym 143(1):155–163

Fujikawa S, Yokota T, Koga J (1987) The continuous hydrolysis of geniposide to genipin using immobilized β-glucosidae on calcium alginate gel. Biotech Lett 9(10):687–702

Funami T, Hiroe M, Noda S, Asai I, Ikeda S, Nishinari K (2007) Influence of molecular structure imaged with atomic force microscopy on the rheological behaviour of carrageenan aqueous systems in the presence or absence of cations. Food Hydrocolloids 21(1):617–629

Galaev IY, Mattiasson B (1999) Smart polymers and what they could do in biotechnology and medicine. Trends Biotechnol 17(1):335–340

Ganji F, Vasheghani-Farahani S, Vasheghani-Farahani E (2010) Theoretical description of hydrogel swelling: a review. Iran Polym J 19(5):275–298

Gao L, Gan H, Meng Z, Gu R, Wu Z, Zhang L, Zhu X, Sun W, Li J, Zheng Y, Dou G (2014) Effects of genipin cross-linking of chitosan hydrogels on cellular adhesion and viability. Colloids Surf B 117(1):398–405

Gou M, Li X, Dai M, Gong C, Wang X, Xie Y, Deng H, Chen L, Zhao X, Qian Z, Wei Y (2008) A novel injectable local hydrophobic drug delivery system: biodegradable nanoparticles in thermo-sensitive hydrogel. Int J Pharm 359:228–233

Gulrez SKH, Al-Assaf S, Phillips GO (2011) Hydrogels: methods of preparation, characterisation and applications. In: Carpi A (ed) Progress in molecular and environmental bioengineering—from analysis and modeling to technology applications. Hydrocolloids Research Centre, Wrexham, UK

Gupta P, Vermani K, Garg S (2002) Hydrogels: from controlled release to pH-responsive drug delivery. Drug Dis Today 7(10):569–579

Hamidi M, Azadi A, Rafiei P (2008) Hydrogel nanoparticles in drug delivery. Adv Drug Deliv Rev 60:1638–1649

Hassan CM, Peppas NA (2000) Structure and morphology of freeze/thawed PVA hydrogels. Macromolecules 33(1):2472–2479

Hennink WE, Nostrum CF (2002) Novel crosslinking methods to design hydrogels. Adv Drug Deliv Rev 54(1):13–36

Hoare TR, Kohane DS (2008) Hydrogels in drug delivery: progress and challenges. Polymer 49 (8):1993–2007

Hoffman AS (2012) Hydrogels for biomedical applications. Adv Drug Deliv Rev 64:18–23

Holowka EP, Bhatia SK (2014) Drug delivery: materials design and clinical perspective. Springer, New York

Hurst G, Novakovic K (2013) A facile *in situ* morphological characterization of smart genipin-crosslinked chitosan-poly(vinyl pyrrolidone) hydrogels. J Mater Res 28(17):2401–2408

Jabeen S, Maswal M, Chat OA, Rather GM, Dar AA (2016) Rheological behaviour and Ibuprofen delivery applications of pH responsive composite alginate hydrogels. Colloids Surf B Biointerfaces 139:211–218

Jiang Y, Chen J, Deng C, Suuronen EJ, Zhong Z (2014) Click hydrogels, microgels and nanogels: emerging platforms for drug delivery and tissue engineering. Biomaterials 35:4969–4985

Jiao Y, Liu Z, Ding S, Li L, Zhou C (2006) Preparation of biodegradable crosslinking agents and application in PVP hydrogel. J Appl Polym Sci 101(3):1515–1521

Khademhosseini A, Langer R (2007) Microengineered hydrogels for tissue engineering. Biomaterials 28:5087–5092

Kharlampieva E, Erel-Unal I, Sukhishvili SA (2007) Amphoteric surface hydrogels derived from hydrogen-bonded multilayers: reversible loading of dyes and macrmolecules. Langmuir 23 (1):175–181

Koren E, Apte A, Jani A, Torchilin VP (2012) Mulitfunctional PEGylated 2C5-immunoliposomes containing pH-sensitive bonds and TAT peptide for enhanced tumor cell internalization and cytotoxicity. J Control Release 160:264–273

Lee BP, Konst S (2014) Novel hydrogel actuator inspired by reversible mussel adhesive protein chemistry. Adv Mater 26(21):3415–3419

Lee KY, Mooney DJ (2001) Hydrogels for tissue engineering. Chem Rev 101(7):1869–1879

Lee Y, Chung H, Yeo S, Ahn C-H, Lee H, Messersmith PB, Park TG (2010) Thermo-sensitive, injectable, and tissue adhesive sol-gel transition hyaluronic acid/pluronic composite hydrogels prepared from bio-inspired catechol-thiol reaction. Soft Matter 6:977–983

Li Y, Rodrigues J, Tomás H (2012) Injectable and biodegradable hydrogels: gelation, biodegredation and biomedical applications. Chem Soc Rev 41:2193–2221

Liu T-Y, Lin Y-L (2010) Novel pH-sensitive chitosan-based hydrogel for encapsulating poorly water-soluble drugs. Acta Biomater 6(4):1423–1429

Magnin D, Lefebvre J, Chornet E, Dumitriu S (2004) Physiological and structural characterisation of a polyanionic matrix of interest in biotechnology, in the pharmaceutical and biomedical fields. Carbohyd Polym 55(4):437–453

Mahajan A, Aggarwal G (2011) Smart polymers: innovations in novel drug delivery. Int J Drug Dev Res 3(3):16–30

Maitra J, Shukla VK (2014) Cross-linking in hydrogels—a review. Am J Polym Sci 4(2):25–31

Mann BK, Gobin AS, Tsai AT, Scmedlen RH, West JL (2001) Smooth muscle cell growth in photopolymerized hydrogels with cell adhesive and proteolytically degradable domains: synthetic ECM analogs for tissue engineering. Biomaterials 22:3045–3051

Martin L, Wilson CG, Koosha F, Uchegbu IF (2003) Sustained buccal delivery of the hydrophobic drug denbufylline using physically cross-lined palmitoyl glycol chitosan hydrogels. Eur J Pharm Biopharm 55:35–43

Maskare R, Bajaj A, Jain D, Braroo P, Babul N, Kao H (2013) Hydrogel-thickened nanoemulsions for topical administration of Ibuprofen. J Pain 14(4):S86

Matcham S, Novakovic K (2016) Fluorescence imaging in genipin crosslinked chitosan–poly (vinyl pyrrolidone) hydrogels. Polymers 8:385

McKenzie M, Betts D, Suh A, Bui K, Kim LD, Cho H (2015) Hydrogel-based drug delivery systems for poorly water-soluble drugs. Molecules 20:20397–20408

Mi F-L, Syu S-S, Peng C-K (2005) Characterization of ring-opening polymerization of genipin and pH-dependent cross-linking reactions between chitosan and genipin. J Polym Sci 43 (10):1985–2000

Moura J, Figueiredo M, Gil H (2007) Rheological study of genipin cross-linked chitosan hydrogels. Biomacromol 8:3823–3829

Muzzarelli RAA (2009) Genipin-crosslinked chitosan hydrogels as biomedical and pharmaceutical aids. Carbohyd Polym 77(1):1–9

Muzzarelli RAA, Mehtedi ME, Bottegoni C, Aquili A, Gigante A (2015) Genipin-crosslinked chitosan gels and scaffolds for tissue engineering and regeneration of cartilage and bone. Mar Drugs 13(12):7314–7338

Muzzarelli RAA, Mehtedi ME, Bottegoni C, Gigante A (2016) Physical properties imparted by genipin to chitosan for tissue regeneration with human stem cells. Int J Biol Macromol. https://doi.org/10.1016/j.ijbiomac.2016.03.075

Naruto M, Ohno K, Naruse N (1978) The synthesis of useful chiral prostanoid intermediates and naturally occurring prostaglandins from aucubin. Chem Lett, 1419–1422

Nichols JJ (2013) Contact Lenses 2012. Contact Lenses Spectrum 28:24–29. https://www.clspectrum.com/issues/2013/january-2013/contact-lenses-2012

Ninawe PR, Parulekar SJ (2011) Drug loading into and drug release from pH- and temperature-responsive cylindrical hydrogels. Biotechnol Prog 27(5):1442–1454

Nguyen KT, West JL (2002) Photopolymerizable hydrogels for tissue engineering applications. Biomaterials 23:4307–4314

NHS: Blood and Transplant (2015) Organ donation and transplantation: activity report 2014/15. NHS

Nwosu CJ, Hurst GA, Novakovic K (2015) Genipin cross-linked chitosan-polyvinylpyrrolidone hydrogels: influence of composition and postsynthesis treatment on pH responsive behaviour. Adv Mat Sci Eng 1–10

Oh JK, Drumright R, Siegwart DJ, Matyjaszewski K (2008) The development of microgels/nanogels for drug delivery applications. Prog Polym Sci 33:448–477

Organ Donation (2015) NHS blood and transplant reveals nearly 49,000 people in the UK have had to wait for a transplant in the last decade. www.organdonation.nhs.uk/news-and-campaigns/news/nhs-blood-and-transplant-reveals-nearly-49-000-people-in-the-uk-have-had-to-wait-for-a-transplant-in-the-last-decade/

Ottenbrite RM, Park K, Okano T (2010) Biomedical applications of hydrogels handbook. Springer

Ozeki T, Hashizawa K, Kaneko D, Imai Y, Okada H (2010) Treatment of rat brain tumors using sustained-release of camptothecin from poly(lactic-co-glycolic-acid) microspheres in a thermoreversible hydrogel. Chem Pharm Bull 58(9):1142–1147

Paik Y-S, Lee C-M, Cho M-H, Hahn T-R (2001) Physical stability of the blue pigments formed from geniposide of gardenia fruits: effects of pH, temperature, and light. J Agric Food Chem 49 (1):430–432

Paranhos CM, Oliveira RN, Soares BG, Pessan LA (2007) Poly(vinyl alcohol)/sulfonated polyester hydrogels produced by freezing and thawing technique: preparation and characterisation. Mat Res 10(1):43–46
Park K (1988) Enzyme-digestible swelling hydrogels as platforms for long-term oral drug delivery: synthesis and characterization. Biomaterials 9(5):435–441
Park J-E, Lee J-Y, Kim H-G, Hahn T-R, Paik Y-S (2002) Isolation and characterization of water-soluble intermediates of blue pigments transformed from geniposide of gardenia jaminoides. J Agric Food Chem 50(22):6511–6514
Peng K, Tomatsu I, Kros A (2011) Hydrogel-based drug carries for controlled release of hyrdrophobic drugs and proteins. J Controll Release 152(1):e72–e74
Peppas NA, Bures P, Leobandung W, Ichikawa H (2000) Hydrogels in pharmaceutical formulations. Eur J Pharm Biopharm 50(1):27–46
Peppas NA, Hilt JZ, Khademhosseini A, Langer R (2006) Hydrogels in biology and medicine: from molecular principles to bionanotechnology. Adv Mater 18:1345–1360
Petta D, Fussell G, Hughes L, Buechter DD, Sprecher CM, Alini M, Eglin D, D'Este M (2016) Calcium phosphate/thermoresponsive hyaluronan hydrogel composite delivering hydrophilic and hydrophobic drugs. J Orthopaedic Transl 6:57–68
Phillips GO, Williams PA (2009) Handbook of hydrocolloids, 2nd edn. Woodhead Publishing, UK
Qiu Y, Park K (2001) Environment-sensitive hydrogels for drug delivery. Adv Drug Deliv Rev 53:321–339
Rainsford KD (2012) Ibuprofen: pharmacology, therapeutics and side effects. Springer, Sheffield
Rapoport NY, Kennedy AM, Shea JE, Scaife CL, Nam KH (2009) Controlled and targeted tumour chemotherapy by ultrasound-activated nanoemulsions/microbubbles. J Controll Release 138:268–276
Robitaille M, Shi J, McBride S, Wan K-T (2013) Mechanical performance of hydrogel contact lenses with a range of power under parallel plate compression and central load. J Mech Behav Biomed Mat 22:59–64
Roldan JE (2003) Hydrogels: introduction and applications in biology and engineering. Department of Biological Sciences, Louisiana Tech University, Louisiana
Roughley P, Hoemann C, DesRosiers E, Mwale F, Antoniou J, Alini M (2006) The potential of chitosan-based gels containing intervertebral disc cells for nucleus pulposus supplementation. Biomaterials 27:388–396
Sakthivel M, Franklin DS, Guhanathan S (2015) Intelligent hydrogels for controlled drug delivery system: a review. Int J Front Sci Technol 3(2):37–47
Scherman OA (2012) Biomaterials 33(18):4646–4652. http://www.pharmtech.com/formulation-development-forum-hydrogels-long-sustained-release
Schroeder A, Goldberg MS, Kastrup C, Wang Y, Jiang S, Joseph BJ, Levins CG, Kannan ST, Langer R, Anderson DG (2012) Remotely activated protein-producing nanoparticles. Nano Lett 12:2685–2689
Sharma A, Pandey R, Sharma S, Khuller GK (2004) Chemotherapeutic efficacy of poly(DL-lactide-co-glycolide) nanoparticle encapsulated antitubercular drugs at sub-therapeutic dose against experimental tuberculosis. Int J Antimicrob Agents 24(6):599–604
Sharpe LA, Daily AM, Horava SD, Peppas NA (2014) Therapeutic applications of hydrogels in oral drug delivery. Expert Opin Drug Deliv 11(6):901–915
Shin J, Han SG, Lee W (2012) Dually tunable inverse optical hydrogel colorimetric sensor with fast and reversible color changes. Sens Actuactors B Chem 168:20–26
Stashak TS, Farstvedt E, Othic A (2004) Update on wound dressing: indications and best use. Clin Tech Equine Pract 3(2):148–163
Sung HW, Huang RN, Huang LL, Tsai CC (1999) *In vitro* cytotoxicity of a naturally occurring cross-linking reagent for biological tissue formation. J Biomat Sci Polym Ed 10(1):63–78
Tang C, Guan Y-X, Yao S-J, Zhu Z-Q (2014) Preparation of Ibuprofen-loaded chitosan films for oral mucosal drug delivery using supercritical solution impregnation. Int J Pharm 473(1–2):434–441

Torres AJ, Zhu C, Shuler ML (2011) Paclitaxel delivery to brain tumors from hydrogels: a computational study. Biotechnol Prog 27(5):1478–1487

Trevor SL, Butler MF, Adams S, Laity PR, Burley JC, Cameron RE (2008) Structure and phase transitions of genipin, an herbal medicine and naturally occurring cross-linker. Cryst Growth Des 8(5):1748–1753

Tsai T-R, Tseng T-Y, Chen C-F, Tsai T-H (2002) Identification and determination of geniposide contained in gardenia jaminoides and in two preparations of mixed traditional chinese medicines. J Chromatogr A 961(1):83–88

van der Linden HJ, Herber S, Olthuis W, Bergveld P (2003) Stimulus-sensitive hydrogels and their applications in chemical (micro)analysis. Analyst 128(4)

Vashist A, Vashist A, Gupta YK, Ahmad S (2014) Recent advances in hydrogel based drug delivery systems for the human body. J Mat Chem B 2:147–166

Vozzi G, Corallo C, Carta S, Fortina M, Gattazzo F, Galletti M, Giordano N (2013) Collagen-gelatin-genipin-hydroxyapatite composite scaffolds colonized by human primary osteoblasts are suitable for bone tissue engineering applications: *in vitro* evidences. J Biomed Mat Res A 102(5):1415–1421

Wang Y, Lu Z, Han Y, Feng Y, Tang C (2011) A novel thermoviscosifying water-soluble polymer for enhancing oil recovery from high-temperature and high-salinity oil resevoirs. Adv Mat Res 306:654–657

Wang J, Wang L, Yu H, Zain-Ul-Abdin, Chen Y, Chen Q, Zhou W, Zhang H, Chen X (2016) Recent progress on synthesis, property and application of modified chitosan: an overview. Int J Biol Macromol 88:333–334

Watkins KA, Chen R (2015) pH-responsive, lysine-based hydrogels for the oral delivery of a wide szie range of molecules. Int J Pharm 478(2):496–503

Wei C-S, Kim C, Kim H-J, Limsakul P (2012) Hydrogel drug delivery: diffusion models

Xiao Z, Ji C, Shi J, Pridgen EM, Frieder J, Wu J, Farokhzad OC (2012) DNA self-assembly of targeted near-infrared-responsive gold nanoparticles for cancer thermo-chemotherapy. Angew Chem Int Ed 54:11853–11857

Yan Q, Yuan J, Cai Z, Xin Y, Kang Y, Yin Y (2010) Voltage-responsive vesicles based on orthgonal assembly of two homopolymers. J Am Chem Soc 132:9268–9270

Ye Y, Hu X (2016) A pH-sensitive injectable nanoparticle composite hydrogel for anticancer drug delivery. J Nanomat 1–8

Yu Q, Bauer JM, Moore JS, Beebe DJ (2001) Responsive biomimetic hydrogel valve for microfluidics. Appl Phys Lett 78:2589–2591

Zhang C-Y, Parton LE, Ye CP, Krauss S, Shen R, Lin C-T, Porco Jr JA, Lowell BB (2006) Genipin inhibits UCP2-mediated proton leak and acutely reverses obesity—and high glucose-induced β cell dysfunction in isolated pancreatic islets. Cell Metab 3(6):417–427

Chapter 2
History, Classification, Properties and Application of Hydrogels: An Overview

Sourbh Thakur, Vijay Kumar Thakur and Omotayo Ademola Arotiba

Abstract The term hydrogel was coined in 1894 as it was employed to explain a colloidal gel. The first report on the application of hydrogels was given by Wichterle and Lim in 1960, which was in the biomedical field. Hydrogels generally absorb a large amount of water, and this swelling is responsible for the rubbery and soft properties of hydrogel. Hydrogels have found applications in environmental, biomedical, food, etc., fields. This chapter presents a brief review of hydrogels—basic definition, classifications, preparations and applications. This chapter highlights among others, the application of polysaccharide-based hydrogels in adsorption and dye removal in water treatment.

Keywords Hydrogels · Hydrogel classification · Hydrogel application Polysaccharide-based hydrogels · Hydrogel as adsorbent · Dye removal

1 Historical Background

Polymer gel is composed of solid and liquid phases and thus possesses both liquid- and solid-like properties. Hydrophilic groups play an important role in gel chemistry. Hydrophilic groups such as hydroxyl, carboxylic and sulphonic functional groups are useful in the cross-linking of the polymer chains through physical or chemical bonds, which lead to the formation of a three-dimensional network

S. Thakur · O. A. Arotiba (✉)
Department of Applied Chemistry, University of Johannesburg,
Doornfontein, Johannesburg 2028, South Africa
e-mail: oarotiba@uj.ac.za

V. K. Thakur
School of Aerospace, Transport and Manufacturing, Enhanced Composites and Structures Center, Cranfield University, Bedfordshire MK43 0AL, UK

O. A. Arotiba
Centre for Nanomaterials Science Research, University of Johannesburg,
Doornfontein, 2028 Johannesburg, South Africa

V. K. Thakur and M. K. Thakur (eds.), *Hydrogels*, Gels Horizons: From Science to Smart Materials, https://doi.org/10.1007/978-981-10-6077-9_2

structure of the polymer. The elastic behaviour of polymer gels is due to the cross-linking among the polymer strands. Polymer chains can be cross-linked to each other via H-bonding, covalent bonds and van der Waals interactions. Polymer gels can be used in many areas like separation technology, biomedical industry, food packaging, agricultural industry, pharmaceutical applications and the cosmetic industry.

In the nineteenth century, the first scientific study of gels was undertaken by Thomas Graham who used sol–gel chemistry to make a silica gel (Campbell 2007). Generally, in the sol–gel process, the fine particles are formed by using a colloidal suspension and the final step involves the evaporation of solvent.

Considerable research on the development of gels was performed by various researchers/workers such as Flory (1941a, b), Huggins (1942, 1943) and Treloar (1951, 1952) between 1940 and 1950. Treloar (1952) is famous for the study of elasticity in polymer networks, whereas Flory carried out experimental as well as theoretical studies of macromolecules to produce modern polymer science and for which he received the Nobel Prize in 1974 (Kumar 2013). In recent years, attention has been focused on the synthesis, characterization and applications of synthetic polymer hydrogels in separation technology.

2 Hydrogels

Hydrogels are regarded as novel polymers utilized for new material growth (Klouda and Mikos 2008; Killion et al. 2011; Prettyman and Eddington 2011). Hydrogels are cross-linked, three-dimensional, hydrophilic polymer structures, which can absorb, swell and retain huge amounts of water or aqueous fluids (Brannon-Peppas 1990). The properties of hydrogels depend on various important parameters such as the hydrophilicity and the degree of cross-linking of the polymer chains (Jones et al. 2005; Hoffman 2002; Sastry et al. 1999). The water retaining ability of hydrogels is due to the functional groups such as hydroxylic (–OH), carboxylic (–COOH), amidic (–CONH–), primary amidic (–$CONH_2$) and sulphonic (–SO_3H) groups found within the polymer network (Ganji et al. 2010). The copolymerization and cross-linking of one or more functional monomers can be used to synthesize hydrogels. Generally, initiator, monomer and cross-linker are the main synthetic components of hydrogels. The characteristics of hydrogels can be regulated by changing the synthetic parameters such as initiator concentration, the concentration of monomer, temperature of reaction, reaction vessel, time of reaction and the ratio of cross-linker and monomer.

Great interest has been in hydrogels in the past 30 years because of their unique roles in separation technology. The high capacity of the hydrogel to absorb water by filling the porous spaces of polymer permits the selective diffusion of solutes via the matrix. Due to these features, hydrogels can be used in a membrane separation process (Bell and Peppas 1995). Hydrogels may be prepared from a variety of sources comprising both natural and synthetic systems; they may also be prepared for various applications in separation technology.

The hydrogel term came out first time in 1894 as it was employed to explain a colloidal gel (Bemmelen 1894). The first report on the application of hydrogels was given by Wichterle and Lim (1960), which was in the biomedical field. Hydrogels absorb a large amount of water which results in swelling and responsible for the rubbery and soft character. The rubbery and soft nature of hydrogels makes it useful in biomedical application. Later in 1980 further progress in hydrogels make it useful as diaper products in France and Germany. Many articles have been coming out to study the hydrogel materials thoroughly, each explaining own viewpoint (Mathur et al. 1996). Kulicke et al. (1989) briefly reviewed biopolymeric, synthetic and semi-synthetic hydrogels. Proteins and polysaccharide-based super-adsorptive materials were investigated by Ichikawa and Nakajima (1996).

In 1936, the first synthetic hydrogel named poly(2-hydroxyethyl methacrylate) (PHEMA) was synthesized by DuPont scientists (Nemours 1936). Wichterle and Lim (1960) established the significance of PHEMA hydrogels in contact lens applications in 1960. The use of hydrogel in contact lens introduced a new class of hydrogels called biomedical hydrogels. Some hydrogels are also called "smart hydrogels" because changes in hydrogels can be observed with respect to changes in environmental conditions (solvent composition, temperature, pH, and electrical field). Research in the past years indicates that hydrogels could be used in different areas such as agriculture (Rehab et al. 1991), biomaterials (Pulapura and Kohn 1992; Zhang et al. 1993), pharmaceutics (Yoshida et al. 1993; Kamath and Park 1993; Heller 1993; Peppas et al. 2000) and biotechnology (Daubresse et al. 1994). Recently, hydrogels have gained attention in the field of wastewater treatment (Mittal et al. 2016).

3 Classification of Hydrogels

Hydrogels can be classified into several categories on the basis of origin, composition, ionic charge, physical structure and cross-linking.

3.1 Classification Based on Origin

3.1.1 Natural Hydrogels

Natural hydrogels are of natural origin and are prepared using natural polymers such as proteins and polysaccharides like alginate, chitosan and dextran (Ma et al. 2003; Silva et al. 2010).

3.1.2 Synthetic Hydrogels

Synthetic hydrogels are of synthetic origin and are synthesized via chemical polymerization using man-made monomers. These hydrogels can be homopolymeric, copolymeric and multipolymeric. Poly(ethylene glycol)–polylactide-poly (ethylene glycol)-based hydrogel is an example of a synthetic hydrogel (Li et al. 2007).

3.1.3 Hybrid Hydrogels

Hybrid hydrogels are prepared by combining natural polymers with synthetic polymers. Wang et al. synthesized a protein cross-linked 2-hydroxypropyl methacrylamide hybrid hydrogel (Wang et al. 1999).

3.2 Classification Based on Composition

3.2.1 Homopolymer Hydrogels

Homopolymer hydrogels are cross-linked polymer networks derived from one type of monomer. The structural framework of these hydrogels is dependent on the nature of the monomer, polymerization technique and cross-linker. Polyethylene glycol-based hydrogel is an example of a homopolymer hydrogel, which is suitable for controlled release of drugs, biomolecules and protein and is widely used in drug delivery systems (Lim et al. 2009).

3.2.2 Copolymer Hydrogels

Copolymer hydrogels are produced from two types of monomer where one of the monomers is hydrophilic and responsible for the swellability of the hydrogel (Singhal and Gupta 2016). These are arranged in a different configuration like graft, random, block and alternate type. These hydrogels are chemically synthesized by polymerization or cross-linking of both monomers using an initiator and a cross-linker. These can also be physically cross-linked by different forces, namely chain aggregations, hydrogen bonding, ionic interaction and ion-polymer complexations. Carboxymethyl cellulose and carboxymethyl chitosan-based hydrogels, which are used for the adsorption of metal ion are examples of copolymer hydrogels (Mitsumata et al. 2003; Hirsch and Spontak 2002).

3.2.3 Multipolymer Hydrogels

Multipolymer hydrogels are produced from three or more monomers using polymerization and cross-linking reactions. An example of such hydrogel is poly(acrylic acid-2-hydroxy ethyl methacrylate)/gelatin hydrogel (Jaiswal and Koul 2013).

3.2.4 Interpenetrating Network (IPN)

Interpenetrating network is an important class of hydrogels, made up of two intertwined polymer networks without any chemical bond between the polymers. The network of the first polymer is linear, whereas the second polymer has a cross-linked network. The linear network of the first polymer diffuses into the second polymer (Miyata 2002). Acrylamide/acrylic acid copolymer hydrogels containing polyallylammonium chloride, which is used for the release of theophylline is an example of a semi-interpenetrating network. This interpenetrating hydrogel is pH sensitive, and *N*,*N'*-methylenebisacrylamide was used as a cross-linker to synthesize this hydrogel (Zhang et al. 2005).

3.3 Classification Based on Ionic Charge

Hydrogels are classified into three groups on the basis of nature of electric charge on the cross-linked chains (Singhal and Gupta 2016; Kabiri et al. 2011).

3.3.1 Neutral (Non-Ionic) Hydrogels

Neutral hydrogels carry no charge on their backbone or side groups.

3.3.2 Ionic Hydrogels

Ionic hydrogels can be cationic or anionic. Cationic hydrogels containing positively charged groups (e.g. amines and sulphonic acid) and exhibit an increase in the swelling at low pH, whereas anionic hydrogels containing negatively charged groups (carboxylic acid, sulphonic acid) and show an increase in swelling at high pH.

3.3.3 Ampholytic Hydrogels

Ampholytic hydrogels carry negative as well as positive charge on the same polymer chain, which balances at the isoelectric point.

3.4 Classification Based on Pore Size

Hydrogels are classified into three types on the basis of porosity, namely non-porous, microporous and superporous (Mastropietro et al. 2012).

3.5 Classification Based on Physical Appearance

On the basis of physical appearance, hydrogels can be a matrix, film or microsphere, depending on the polymerization method (Ahmed 2015).

3.6 Classification Based on Configuration

Depending on the physical structure and chemical composition, hydrogels can be classified as follows (Ahmed 2015).

3.6.1 Amorphous (Non-Crystalline)

In amorphous hydrogels, polymeric network contains randomly arranged macromolecular chains.

3.6.2 Semi-Crystalline

Semi-crystalline is a complex mixture of amorphous and crystalline phases and characterized by dense regions of ordered macromolecular chains.

3.7 Classification Based on Cross-Linking

On the basis of nature of cross-linking, hydrogels are of two types: physical hydrogels and chemical hydrogels. Physical hydrogels are cross-linked by various physical processes such as crystallization, hydrogen bonding and hydrophobic interactions, whereas covalent cross-linking is used to prepare chemical hydrogels (Zhu and Marchant 2011).

4 Synthesis of Hydrogels

Hydrogels are polymer networks prepared from natural or synthetic polymers using various polymerization techniques such as bulk, solution and suspension by physical and chemical cross-linking routes (Ahmed 2015; Maitra and Shukla 2014). Physical cross-linking involves hydrogen bonds, stereo-complexation and soft assembly, whereas chemical cross-linking involves cross-linking in the presence of different cross-linkers. Chemically cross-linked hydrogels are great in demand due to their good mechanical strength. These hydrogels contain covalent junction between polymeric chains, which can be introduced by the following cross-linking methods (Singhal and Gupta 2016):

- Physical method
- Chemical method
- Photochemical Grafting
- High energy irradiation
- Using enzymes

Of all the above-mentioned cross-linking methods, chemical methods are the most widely used. The hydrogels in this study were synthesized in the presence of an initiator and a cross-linker using the chemical method.

4.1 Chemical Method

In the chemical method, graft copolymerization of different vinyl monomers onto backbones was carried out using various radical initiators, viz. ammonium persulphate (APS), ceric ammonium nitrate (CAN) (Chauhan et al. 1999, 2000), potassium permanganate ($KMnO_4$) (Mondal 2003) and potassium persulphate (KPS) (Mittal et al. 2013a, b, c). These initiators assist in determining the track of grafting.

Various cross-linked hydrogels have been prepared using grafting (Hennink and Nostrum 2002). The mechanism is a three-step process: initiation, propagation and termination. Graft copolymerization has been used for the modification of chemical and physical properties of polymers (Battaerd and Tregear 1967). High molecular weight graft copolymers can be produced from free radical grafting, whereas low molecular weight copolymers are prepared from ionic grafting (Battaerd and Tregear 1967; Chauhan et al. 2012). In the chemical method, active sites are generated on the polymeric backbone for the incorporation of the polymeric chains (Battaerd and Tregear 1967). Various methods such as chemical methods, physical methods, enzymatic grafting, photochemical grafting and plasma radiation are used for the generation of active sites. Gum ghatti with acrylamide (Mittal et al. 2013a), poly (vinyl alcohol) (Singhal and Gupta 2016), gum ghatti with acrylamide and acrylic acid (Mittal et al. 2013b), poly(2-hydroxyethyl methacrylate) (Wichterle and

Lim 1960), gum ghatti with acrylic acid (Mittal et al. 2013c) are the examples of hydrogels prepared by the chemical method.

5 Characteristics of Hydrogels

An ideal hydrogel should have features such as high absorption capacity in aqueous fluids, good porosity, photo-stability, non-toxicity, pH-neutrality, re-wetting capability, lowest soluble content and residual monomer, highest absorbency under load (AUL) and high durability (Ahmed 2015). Stimuli-responsive hydrogels can respond to environmental changes like temperature (Ricka and Tanaka 1984), pH (Hoffman 1987), electric field (Eisenberg and Grodzinski 1984), solvent composition (Katayama et al. 1984) and salt concentration (Park and Hoffman 1993). The high stability and durability of hydrogels in swelling environment make it useful in separation processes, drug delivery, tissue engineering and cell encapsulation (Tasdelen et al. 2015).

5.1 Temperature-Sensitive Hydrogels

Temperature-sensitive hydrogels, also known as thermogels, have the ability to swell and shrink according to the surrounding temperature (Richter 2010). The common feature of these hydrogels is the occurrence of hydrophobic groups like methyl, ethyl and propyl groups. These hydrogels can be positive and negative temperature-sensitive systems (Laftah et al. 2011). Positive temperature hydrogels have an upper critical solution temperature (UCST) (Peppas et al. 2000). These hydrogels contract below the UCST and show swelling at the temperature higher than the UCST. Negative temperature-sensitive hydrogels have lower critical solution temperature (LCST); they contract when the temperature increases above the LCST and exhibit swelling at lower LCST (Ullah et al. 2015). Hydrogel chitosan–poly(methacrylic acid) synthesized by Khan et al. is sensitive to temperature and 55 °C was reported as the phase transition temperature for this hydrogel (Khan et al. 2013). The polymer poly(*N*-isopropylacrylamide) exhibits a phase transition temperature of 34.3 °C in water; these polymers are a well-known temperature-sensitive polymer (Ullah et al. 2015). Katono et al. reported poly (acrylic acid) and polyacrylamide as positive temperature hydrogel (Katono et al. 1991). Negative temperature-sensitive poly(*N*-isopropylacrylamide)–poly(vinylpyrrolidinone) hydrogels were developed by Geever et al. (2008).

5.2 pH-Sensitive Hydrogels

Some hydrogels exhibit different characteristics such as swelling with changes in pH. These hydrogels are of two types anionic and cationic (He 2006) on the basis of the pendant group. Anionic hydrogels contain carboxylic and sulphonic acid groups (Qiu and Park 2001). In these hydrogels, ionization takes place, when environmental pH is above the pK_a, which subsequently increases the swelling (He 2006). On the other hand, cationic hydrogels contain amine groups where ionization occurs below pK_b, which enhances swelling due to the increased electrostatic repulsion (Peppas et al. 2000). Polydiethylaminoethyl methacrylate and alginate-*N*, *O*-carboxymethyl chitosan are the examples of pH-responsive hydrogels (Ullah et al. 2015).

6 Applications of Hydrogels

As mentioned elsewhere in this thesis, hydrogels are used in various fields like separation technology, drug delivery, agriculture, horticulture, biomedical and pharmaceutical applications. Hydrogels are very useful in separation technology as adsorbent because of properties like porosity and high swelling capacity.

6.1 Separation Technology

Hydrogels show good swelling capacity under many external conditions, and this makes them a useful adsorbent for the adsorption of water-soluble contaminants such as dyes, metallic impurities, biomolecules. The first study on water adsorbent behaviour of polymer was carried out in 1938 by Buchholz and Graham (1998). Initially, the adsorptive and chelating features of maleic or itaconic acid-based hydrogels were studied. The different ratio of maleic acid and itaconic acid was used to adsorb bovine serum albumin (Saraydin et al. 1994; Karadag et al. 1994). It was found that an increase in the content of these acids raises the bovine serum albumin adsorption. In 1997, Karadag et al. produced acrylamide/itaconic acid hydrogels adsorbent for the adsorption of nicotine and nicotine derivatives (Karadag et al. 1997).

The complexation abilities of some hydrogels lend them to heavy metal adsorption. Chujo et al. successfully examined the complex formed between the polyoxazoline-based hydrogel and metal ions like iron (+2), cobalt (+3) in 1993 (Chujo et al. 1993a, b). The hydrogel of acrylamide was developed as an adsorbent for uranium, iron and copper adsorption by Saraydin et al. (1995). The presence of monomer like maleic or itaconic acid in hydrogels added chelating characteristic into hydrogels which increased the adsorption capacity of hydrogels for heavy metal ions. Then in the year of 1995, the hydrogel was used for uranyl ions adsorption from aqueous solution (Karadag et al. 1995) and later in 2000 some modification in hydrogel increased the intake of uranyl ion (Akkas and Guven 2000).

Adsorption capacities of acrylamide/itaconic and acrylamide/maleic acid hydrogels were investigated for the removal of cationic dyes such as basic violet, basic blue, basic red, cresyl violet and basic green (Saraydin et al. 1996; Karadag et al. 1996). It was found that diprotic acids like itaconic and maleic in acrylamide hydrogels increase the adsorption capacity towards dyes with respect to pure acrylamide hydrogels. The outcomes attained on metal ion intake, adsorption of dyes and bio-applications showed that the hydrogels synthesized from hydrophilic monomers have a unique characteristic.

Wastewater from the textile and colouring paper industries contains dyes in large amounts (Kyzas and Lazaridis 2009; Erdener et al. 2000). When released to the environment, these dyes pose health hazards to humans and other lives. Hydrogels are being examined in separation technology as an adsorbent for the extraction of dyes and toxic heavy metal ions from water because of the presence of hydrophilic group such as –COOH, –OH, $-CONH_2$ and $-NH_2$ in their structures (Kyzas and Lazaridis 2009; Erdener et al. 2000; Deepika et al. 2013; Kurniawan et al. 2006; Morris et al. 1997; Yetimoglu et al. 2007; Shukla et al. 2012; Mittal et al. 2015a, b).

6.2 *Biomedical Applications*

The rubbery and soft nature of hydrogels makes them useful in biomedical application (Ratner and Hoffman 1976). Hydrogels can mimic the biological, chemical and physical properties of biological organs and tissue in response to changes in temperature, electric field, pH and enzymes (Kloxin et al. 2010). Thus, hydrogels are used in artificial organ preparations like in kidney, robotic grippers, artificial muscle formation, biosensors, wound healing, bioadhesives, vocal cord replacement materials, sexual organ reconstruction material and artificial skin (Ratner and Hoffman 1976; Peppas and Langer 1994). Hydrogels are also ideal candidates for the preparation of contact lenses because of their good mechanical stability, high oxygen permeability and favourable refractive index (Tighe 1976). The antimicrobial activity of gelatin–carrageenan hydrogels was performed by Padhi et al. (2016). A hybrid hydrogel methoxylpoly(ethylene glycol)-poly(caprolactone)-acryloyl chloride/glycidylmethacrylated chitooligosaccharide/*N*-isopropylacrylamide/acrylamide was reported for use in postoperative of breast cancer patients (Qu et al. 2015). Wang et al. used micelle hydrogels of polycaprolactone–poly(ethylene glycol) for therapy of breast cancer (Wang et al. 2014).

6.3 Pharmaceutical Applications

Drug release profile, adsorption and dispersion of drug can be improved by drug delivery method. Transmucosal, non-invasive peroral and inhalation are the usual paths of drug delivery (Tan et al. 2012). Hydrogels have been used in drug delivery systems for a long time due to their swelling behaviour and highly porous structure. Hydrogels play an important role in the rate of drug release; they increase drug solubility and limits drug degradation and toxicity (Stojanovic et al. 2011). Glycidyl methacrylate-β-cyclodextrin-based hydrogel was developed for drug delivery in oculopathy therapy (Hua et al. 2016). The copolymer hydrogel of methyl acrylic anhydride/cyclodextrin–hydroxyethyl methacrylate was used for the delivery of ofloxacin and puerarin (Hu and Gong 2016). Das et al. synthesized dextrin/poly(acrylic acid) hydrogel for the release of ciprofloxacin and ornidazole (Das et al. 2015).

6.4 Agricultural and Horticultural Applications

Superabsorbent polymers are cross-linked network of hydrophilic polymers that absorb a large amount of water and are widely used in agriculture and horticulture (Zohuriaan-Mehr and Kabiri 2008). The superabsorbent polymer influences the permeability of the soil and improves the structure, density, texture and infiltration rate of soil and prevents soil erosion, enhances soil aeration, microbial activity and water holding capacity (Abd El-Rehirn et al. 2004). Various polysaccharides such as pectin, chitosan and carboxymethyl cellulose-based hydrogels are used as fertilizer to improve the soil (Jamnongkan and Kaewpirom 2010). Romero et al. (2016) formulated an autonomous smart scheme of copolymer hydrogel tris[(hydroxymethyl) methyl]acrylamide–methyl methacrylate for agriculture irrigation and successfully avoided the wastage of water. Half a litre of water used daily in irrigation was reduced to three litres of water for irrigation per month by using this hydrogel. Cellulose was used to stabilize the hydrogel throughout the swelling–deswelling period. Yan et al. (2016) proposed starch-grafted-poly(acrylic acid)/organo-zeolite hydrogel for effective application in agriculture. The chitosan hydrogel with nano-copper was applied in plant growth, and the use of nano-copper incorporated chitosan hydrogel increased the growth as well as the quality of tomato fruit (Antonio et al. 2016).

7 Biopolymers—Polysaccharides

7.1 Sodium Alginate

Alginate is an anionic natural polysaccharide obtained from cell walls of brown algae. The structure of alginate is consisted of linear copolymers blocks of 1–4 linked β-D-mannuronic acid (M) and α-L-glucuronic acid (G) in different ratio (Lee and Mooney 2012). The blocks can either have homopolymeric (GGGGGG, MMMMMM) or heteropolymeric (GMGMGM) sequences. The physical characteristics of alginates and its hydrogel depend on various factors such as the M/G ratio, molecular weight and block sequence (George and Abraham 2006). Alginate was discovered by British chemist E. C. C. Stanford in 1881. Alginate can be utilized in agriculture, drug delivery, food and cosmetic applications (Babu et al. 2007; Pongjanyakul and Puttipipatkhachorn 2007).

7.2 Xanthan Gum

Xanthan gum is a microbial polysaccharide produced from Xanthomonas campestris by fermentation. It consists of cellulosic backbone attached to alternate glucose residues at C-3 and side chains of the trisaccharide (D-glucose, D-mannose and D-gluconate) (Palaniraj and Jayaraman 2011). Scientists of the Northern Regional Research Laboratory of the United States Department of Agriculture discovered the xanthan gum in 1950. Xanthan gum is widely used in cosmetic, food and pharmaceutical industries (Pooja et al. 2014). The rheological properties of xanthan gum make it useful in industrial applications.

7.3 Gum Ghatti

Gum ghatti is an anionic polysaccharide, belongs to Combretaceae family and obtained from the Anogeissus latifolia tree. The structure of gum ghatti was thoroughly studied by Aspinall et al. (1955). The composition of gum ghatti consists of L-arabinose, D-galactose, D-mannose, D-xylose and D-glucuronic acid, which are in a 48:29:10:5:10 molar ratio (Aspinall 1980). The side chain of gum ghatti contains l-arabinofuranose residue and main chain comprise of alternating 4-*O*-substituted and 2-*O*-substituted α-D-mannopyranose units and chains of 1 → 6 linked β-D-galactopyranose units. Gum ghatti is being used in many industrial applications (Rani et al. 2012).

7.4 Gum Karaya

Gum Karaya branched anionic polysaccharide containing D-galacturonic acid, D-galactose, L-rhamnose and D-glucuronic acid. The main chain of gum karaya consists of rhamnogalacturonan with a α-(1 → 4)-linked D-galacturonic acid and α-(1 → 2)-linked-L-rhamnosyl residues. Gum Karaya has side chain comprising of (1 → 3)-linked β-D-glucuronic acid, or (1 → 2)-linked β-D-galactose on the galacturonic acid where rhamnose units linked by (1 → 4)-β-D-galactose (Anderson et al. 1982; Anderson and Wang 1994). The family name of gum karaya is Sterculiaceae family, and it is received from Sterculia urens. Gum karaya has been used in pharma industry and food. It is also used as a mucoadhesive polymer for drug delivery system (Prajapati et al. 2013).

7.5 Gum Arabic

Gum arabic is an exudate of the Acacia tree. This hydrocolloid is relatively inexpensive and complex hetero-polysaccharide. The structure of gum arabic consists of main chain with (1 → 3)-β-D-galactopyranosyl units and the side chain containing L-arabinofuranosyl, L-rhamnopyranosyl, D-galactopyranosyl and D-glucopyranosyl uronic acid units (Kulshrestha et al. 1962; Sarika et al. 2014). A status of acceptable daily intake (ADI-non-specified) is given to gum arabic by Joint WHO/FAO Expert Committee on Food Additives (JECFA) (Gils et al. 2010). According to US Food and Drug Administration (USFDA), gum Arabic has generally recognized as safe (GRAS) status. It is extensively used in various applications like confectionery, pharmaceuticals, beverage or liquid flavour emulsions and cosmetic products (Gerola et al. 2016).

7.6 Carrageenans

Carrageenans came into light in 1844 during the isolation of the mucilage from the red seaweed (Schmidt 1844). It consists of linear backbone composed of 3-linked β-d-galactopyranosyl and a 4-linked α-d-galactopyranosyl and 3,6-anhydro-α-d-galactopyranosyl moiety (Stortz and Cerezo 2000). This polysaccharide is rich in galactose. The hydroxyls of carrageenans are substituted by sulphate ester at respective positions (Usov 2011). Carrageenans found in red seaweed of Rhodophyceae class. There are three generic families of carrageenan (kappa, iota or lambda). The most significant commercial class of carrageenans is κ-carrageenan (Zhao et al. 2015). The texture of cottage cheese can be improved by the use of carrageenans. It can be used as stabilizers and binders for the preparation of sausages and low-fat hamburgers (Campos et al. 2009). Carrageenans are also useful in cosmetics, pharmaceutical, printing and textile formulations (Imeson 2000).

7.7 Chitosan

Chitosan is a linear polysaccharide containing β-(1–4)-linked d-glucosamine and *N*-acetyl-D-glucosamine units (Rinaudo 2006). It is obtained by the partial deacetylation of chitin (Zeng et al. 2004). Chitosan is a cationic polysaccharide and second most abundant biopolymer after cellulose on earth. The hydrophilic character of chitosan is due to the presence of amino ($-NH_2$) and hydroxyl (–OH) groups. The first thermosensitive hydrogel of chitosan was synthesized by Chenite in 2000 (Chenite et al. 2000). Chitosan has been used in environmental protection, biotechnology, pharmaceutical industry and food processing (Rinaudo 2006).

The materials based on polysaccharides are being continuously used for pollutants removal from aqueous solution (Mittal et al. 2016; Sharma et al. 2015a). These materials act as sorbents, and they have many advantages over conventional sorbents like synthetic ion-exchange resins and activated carbons. Some of these advantages are as follows:

- Most of the polymers are obtained from rock tar by some chemical treatment, which is not always risk-free and environmentally safe. Researchers are looking for natural inexpensive sorbents, and polysaccharides are the best choice for this.
- Adsorbents based on biopolymer are very cost-efficient. The amount of biosorbent required to remove pollutants is small in comparison with conventional sorbents. Also, less quantity of biopolymer is needed to form the composite material.
- Biosorbents possess versatile characteristics. Different forms of the sorbent can be used due to versatility.
- Different concentrations of adsorbents comprising polysaccharides are very effective for the extraction of pollutants. They are selective and have high adsorption rate and high adsorption capacity. They can be effectively applied in the removal of pollutants from both concentrated and dilute solutions. These adsorbents can decrease the concentrations of pollutant to parts per billion (ppb) scale. The applications are limited to parts per million (ppm) levels in case of synthetic resins and activated carbon.
- Polysaccharides show fine complexing and chelating abilities towards a number of pollutants such as dyes, aromatic compounds and heavy metals due to the presence of repetitive hydroxyl and carboxyl functional groups in their structure. The adsorption ability of polysaccharides can be increased in a selective manner by the inclusion of several functional groups in the backbone of the polysaccharides.

- Polysaccharides-based adsorbents swell in water due to their hydrophilic character, and they possess quick diffusion action for the pollutants in water.
- Adsorbents can be reused by simply washing with appropriate solvent because the adsorption process is controlled by electrostatic or hydrophobic interactions. The regeneration study is cheap and easy.

The polysaccharide like guar gum (Gupta et al. 2014; Sharma et al. 2015b; Hiremath and Vishalakshi 2015), gum karaya (Mittal et al. 2015a) and gum ghatti (Mittal et al. 2013c, 2015b; Mittal and Mishra 2014; Sharma et al. 2013, 2014) has been exploited in the treatment of water contaminated with dyes. These polysaccharides have been used for the adsorption of cationic dyes since most of them contain the anionic functional group. The introduction of an inorganic filler such as metal oxide (Fe_3O_4) and nanoparticles such as carbon nanotubes or graphene into the biopolymer matrix has led to an increase in the adsorption capacity of the biopolymers (Mittal et al. 2015a; Gupta et al. 2014; Mittal and Mishra 2014; Mittal et al. 2014; Banerjee and Chen 2007; Ghorai et al. 2012, 2013; Gils et al. 2010; Agorku et al. 2014; Mahdavinia et al. 2013; Ghorai et al. 2014; Hosseinzadeh 2015; Thakur et al. 2016; Kaur et al. 2013; Khullar et al. 2012).

8 Conclusions and Future Perspective

This chapter demonstrates the classification, properties and application of hydrogel. Some directions for the synthesis of hydrogel by chemical modification were given. Polysaccharide-based hydrogels as adsorbent particularly for dyes were discussed. The mechanism of dyes adsorption onto polysaccharide-based hydrogels generally proceeds via electrostatic attractions within oppositely charged pollutants and adsorbents. Hydrogel still a hot topic in material science because of technological applications. Hydrogels are innovative materials that swell to a large extent irrespective of their initial size and show excellent properties. Hydrogel has the ability to absorb large quantity of water or aqueous fluids in a very short period. For new applications, hydrogels need to improve mechanical properties, biocompatibility and durability. Although incorporation of nanoparticles into hydrogels may improve the swelling and mechanical properties, synthesis of hydrogels with low cost having super-absorption and good mechanical properties at the same time is still a challenge. The preparation of hydrogels in nanoscale having high water absorption capacity is still in concern. Also, nanoparticle can lower the sensitivity of hydrogel. Naturally, further extensive explanation will elaborate the understanding of structure–properties–applications in regard to hydrogels, which will open the new way for the class of soft material.

References

Abd El-Rehim HA, Hegazy ESA, Abd El-Mohdy HL (2004) Radiation synthesis of hydrogels to enhance sandy soils water retention and increase performance. J Appl Polym Sci 93:1360–1371

Agorku ES, Mittal H, Mamba BB, Pandey AC, Mishra AK (2014) Fabrication of photocatalyst based on Eu^{3+}-doped ZnS–SiO_2 and sodium alginate core shell nanocomposite. Int J Biol Macromol 70:143–149

Ahmed EM (2015) Hydrogel: preparation, characterization, and applications: a review. J Adv Res 6:105–121

Akkas P, Guven O (2000) Enhancement of uranyl ion uptake by pre-structuring of acrylamide/maleic acid hydrogels. J Appl Polym Sci 78:284–289

Anderson DMW, Mcnab CNB, Anderson CG, Braown PM, Pringuer MA (1982) Gum exudates from the genus sterculia (gum karaya). Int Tree Crops J 2:147–154

Anderson DMW, Wang W (1994) The tree exudate gums permitted in foodstuffs as emulsifiers, stabilisers and thickeners. Chem Ind Forest Prod 14:73–84

Antonio JM, Hortensia OO, Fabian PL, Gregorio CP, Adalberto BM (2016) Cu nanoparticles absorbed on chitosan hydrogels positively alter morphological, production and quality characteristics of tomato. J Appl Bot Food Qual 89:183–189

Aspinall GO (1980) Chemistry of cell wall polysaccharides. In: Preiss J (ed) The biochemistry of plants. Academic Press, New York, NY, pp 473–500

Aspinall GO, Hirst EL, Wickstrom A (1955) Gum ghatti (Indian gum). The composition of the gum and the structure of two aldobiouronic acids derived from it. J Chem Soc 1160–1165

Babu VR, Sairam M, Hosamani KM, Aminabhavi TM (2007) Preparation of sodium alginate–methylcellulose blend microspheres for controlled release of nifedipine. Carbohydr Polym 69:241–250

Banerjee SS, Chen DH (2007) Fast removal of copper ions by gum arabic modified magnetic nano-adsorbent. J Hazard Mater 147:792–799

Battaerd HAJ, Tregear GW (1967) Graft Copolymers. Interscience, New York

Bell CL, Peppas NA (1995) Measurement of the swelling force in ionic polymer networks. III. swelling force of interpolymer complexes. J Control Release 37:277–280

Bemmelen JMV (1894) Der Hydrogel und das kristallinische Hydrat des Kupferoxydes. Z Anorg Chem 5:466

Brannon-Peppas L (1990) Preparation and characterization of crosslinked hydrophilic networks. Absorbent Polymer Technology, Studies in Polymer Sci 8:46–66

Buchholz FL, Graham AT (1998) Modern superabsorbent polymer technology. Wiley, New York, pp 1–7

Campbell TD (2007) Synthesis and physical characterization of biocompatible hydrogels. PhD Thesis, Department of Chemistry and Biochemistry, The Florida State University

Campos VL, Kawano DF, Silva DB, Carvalho I (2009) Carrageenans: biological properties, chemical modifications and structural analysis—a review. Carbohydr Polym 77:167–180

Chauhan A, Chauhan P, Kaith BS (2012) Natural fiber reinforced composite: a concise review article. J Chem Eng Process Technol 3:1–5

Chauhan GS, Kaur I, Misra BN, Singha AS, Kaith BS (1999) Modification of natural polymers: graft co-polymerisation of methyl methacrylate onto rayon fibre initiated by ceric ions- a study in the swelling and thermal properties. J Polym Mater 16:245–252

Chauhan GS, Kaur I, Misra BN, Singha AS, Kaith BS (2000) Evaluation of optimum grafting parameters and the effect of ceric ion initiated grafting of methyl methacrylate onto jute fibre on the kinetics of thermal degradation and swelling behavior. Polym Degrad Stab 69:261–265

Chenite A, Chaput C, Wang D, Combes C, Buschmann MD, Hoemann CD, Leroux JC, Atkinson BL, Binette F, Selmani A (2000) Novel injectable neutral solutions of chitosan form biodegradable gels in situ. Biomaterials 21:2155–2161

Chujo Y, Sada K, Saegusa T (1993a) Cobalt (III) bipyridyl-branched polyoxazoline complex as a thermally and redox reversible hydrogel. Macromolecules 26:6320–6323

Chujo Y, Sada K, Saegusa T (1993b) Iron(II) bipyridyl-branched polyoxazoline complex as a thermally reversible hydrogel. Macromolecules 26:6315–6319

Das D, Ghosh P, Dhara S, Panda AB, Pal S (2015) Dextrin and poly(acrylic acid)-based biodegradable, non-cytotoxic, chemically cross-linked hydrogel for sustained release of ornidazole and ciprofloxacin. ACS Appl Mater Interfaces 7:4791–4803

Daubresse C, Grandfils C, Jerome R, Teyssie P (1994) Enzyme immobilization in nanoparticles produced by inverse microemulsion polymerization. J Colloid Interf Sci 168:222–229

Deepika P, Avijeet KS, Michael JS (2013) Recyclability of poly (N-isopropylacrylamide) microgel-based assemblies for organic dye removal from water. Colloid Polym Sci 291:1795–1802

Eisenberg SR, Grodzinski AJ (1984) Electrically modulated membrane permeability. J Membr Sci 19:173–194

Erdener K, Omer BU, Dursun S (2000) Swelling equilibria and dye adsorption studies of chemically crosslinked superabsorbent acrylamide/maleic acid hydrogels. Eur Polym J 38:2133–2141

Flory PJ (1941a) Thermodynamics of high polymer solutions. J Chem Phys 9:660–661

Flory PJ (1941b) Molecular size distribution in three dimensional polymers I. gelation. J Am Chem Soc 63:3083–3090

Ganji F, Vasheghani-Farahani S, Vasheghani-Farahani E (2010) Theoretical description of hydrogel swelling: a review. Iran Polym J 19:375–398

Geever LM, Cooney CC, Lyons JG, Kennedy JE, Nugent MJ, Devery S, Higginbotham CL (2008) Characterisation and controlled drug release from novel drugloaded hydrogels. Eur J Pharm Biopharm 69:1147–1159

George M, Abraham TE (2006) Polyionic hydrocolloids for the intestinal delivery of protein drugs. J Control Release 114:1–14

Gerola AP, Silva DC, Matsushita AFY, Borges O, Rubira AF, Muniz EC, Valente AJM (2016) The effect of methacrylation on the behavior of Gum Arabic as pH-responsive matrix for colon-specific drug delivery. Eur Polym J 78:326–339

Ghorai S, Sarkar A, Raoufi M, Panda AB, Schonherr H, Pal S (2014) Enhanced removal of methylene blue and methyl violet dyes from aqueous solution using a nanocomposite of hydrolyzed polyacrylamide grafted xanthan gum and incorporated nanosilica. ACS Appl Mater Interfaces 6:4766–4777

Ghorai S, Sarkar AK, Panda AB, Pal S (2013) Effective removal of Congo red dye from aqueous solution using modified xanthan gum/silica hybrid nanocomposite as adsorbent. Bioresour Technol 144:485–491

Ghorai S, Sinhamahpatra A, Sarkar A, Panda AB, Pal S (2012) Novel biodegradable nanocomposite based on XG-g-PAM/SiO_2: application of an efficient adsorbent for Pb^{2+} ions from aqueous solution. Bioresour Technol 119:181–190

Gils PS, Ray D, Sahoo PK (2010) Designing of silver nanoparticles in gum arabic based semi-IPN hydrogel. Int J Biol Macromol 46:237–244

Gupta VK, Pathania D, Singh P, Kumar A, Rathore BS (2014) Adsorptional removal of methylene blue by gum-cerium (IV) tungstate hybrid cationic exchanger. Carbohydr Polym 101:684–691

He H (2006) Multifunctional medical device based on pH-sensitive hydeogels for controlled drug delivery. PhD Thesis, Chemistry department, Ohio State University, Canton, Ohio

Heller J (1993) Polymers for controlled parenteral delivery of peptides and proteins. Adv Drug Deliver Rev 10:163–204

Hennink WE, Nostrum CV (2002) Novel crosslinking methods to design hydrogels. Advan Drug Deliv Rev 54:13–36

Hiremath JN, Vishalakshi B (2015) Evaluation of a pH-responsive guar gum-based hydrogel as adsorbent for cationic dyes: kinetic and modelling study. Polym Bull 72:3063–3081

Hirsch SG, Spontak RJ (2002) Temperature dependent properties developed in hydrogels derived from hydroxypropyl cellulose. Polymer 43:123–129

Hoffman AS (1987) Applications of thermally reversible polymers and hydrogels in therapeutics and diagnostics. J Control Release 6:297–305
Hoffman AS (2002) Hydrogels for biomedical applications. Adv Drug Deliv Rev 54:3–12
Hosseinzadeh H (2015) Synthesis of carrageenan/multi-walled carbon nanotube hybrid hydrogel nanocomposite for adsorption of crystal violet from aqueous solution. Pol J Chem Technol 17:70–76
Hu XH, Gong X (2016) A new route to fabricate biocompatible hydrogels with controlled drug delivery behavior. J Colloid Interface Sci 470:62–70
Hua X, Tanb H, Hao L (2016) Functional hydrogel contact lens for drug delivery in the application of oculopathy therapy. J Mech Behav Biomed Mater 64:43–52
Huggins ML (1942) Theory of solutions of high polymers. J Am Chem Soc 64:1712–1719
Huggins ML (1943) Properties of rubber solutions and gels. Ind Eng Chem 35:216–220
Ichikawa T, Nakajima T (1996) Superabsorptive polymers (from natural polysaccharides and peptides). In JC Salamone (ed) Polymeric materials encyclopedia. CRC Press, New York, pp 8051–8059
Imeson AP (2000) Carrageenan. In: Phillips GO, Williams PA (eds) Handbook of hydrocolloids. Woodhead Publishing Limited, Cambridge UK, pp 87–102
Jaiswal M, Koul V (2013) Assessment of multicomponent hydrogel scaffolds of poly(acrylic acid-2-hydroxy ethyl methacrylate)/gelatin for tissue engineering applications. J Biomater Appl 27:848–861
Jamnongkan T, Kaewpirom S (2010) Potassium release kinetics and water retention of controlled-release fertilizers based on chitosan hydrogels. J Polym Environ 18:413–421
Jones DS, Andrews GP, Gorman SP (2005) Characterisation of cross-linking effects on the physicochemical and drug diffusional properties of cationic hydrogels designed as bioactive urological biomaterials. J Pharm Pharmacol 57:1251–1259
Kabiri K, Omidian H, Zohuriaan-Mehr MJ, Doroudiani S (2011) Superabsorbent hydrogel composites and nanocomposites: a review. Polym Compos 32:277–289
Kamath KR, Park K (1993) Biodegradable hydrogels in drug delivery. Adv Drug Deliver Rev 11:59–84
Karadag E, Saraydin D, Guven O (1995) Behaviours of acrylamide/itaconic acid hydrogels in uptake of uranyl ions from aqueous solutions. Sep Sci Tech 30:374
Karadag E, Saraydin D, Guven O (1996) Interaction of some cationic dyes with acrylamide/itaconic acid hydrogels. J Appl Polym Sci 61:2367–2372
Karadag E, Saraydin D, Guven O (1997) Interaction of nicotine and its pharmaceutical derivatives with acrylamide/itaconic acid hydrogels. J Appl Polym Sci 66:733
Karadag E, Saraydin D, Oztop HN, Guven O (1994) Adsorption of bovine serum albumin onto acrylamide-itaconic acid hydrogels. Polym Adv Tech 5:664–668
Katayama S, Hirokawa Y, Tanaka T (1984) Reentrant phase transition in acrylamide- derivative copolymer gels. Macromolecules 17:2641–2643
Katono H, Maruyama A, Sanui K, Ogata N, Okano T, Sakurai Y (1991) Thermoresponsive swelling and drug release switching of interpenetrating polymer networks composed of poly (acrylamide-co-butyl methacrylate) and poly(acrylic acid). J Control Release 16:215–227
Kaur H, Banipal TS, Thakur S, Bakshi MS, Kaur G, Singh N (2013) Novel biodegradable films with extraordinary tensile strength and flexibility provided by nanoparticles. ACS Sustain Chem Eng 1:127–136
Khan A, Othman MBH, Razak KA, Akil HM (2013) Synthesis and physicochemical investigation of chitosan-PMAA-based dual-responsive hydrogels. J Polym Res 20:1–8
Khullar P, Singh V, Mahal A, Dave PN, Thakur S, Kaur G, Singh J, Kamboj SS, Bakshi MS (2012) Bovine serum albumin bioconjugated gold nanoparticles: synthesis, hemolysis and cytotoxicity toward cancer cell lines. J Phys Chem C 116:8834–8843
Killion JA, Geever LM, Devine DM, Kennedy JE, Higginbotham CJ (2011) Mechanical properties and thermal behavior of PEGDMA hydrogels for potential bone regeneration applications. J Mech Behav Biomed Mater 4:1219–1227

Klouda L, Mikos AG (2008) Thermoresponsive hydrogels in biomedical applications. Eur J Pharm Biopharm 68:34–45
Kloxin AM, Kloxin CJ, Bowman CN, Anseth KS (2010) Mechanical properties of cellularly responsive hydrogels and their experimental determination. Adv Mater 22:3484–3494
Kulicke WM, Aggour YA, Nottelmann H, Elsabee M (1989) Swelling and rheological studies of some starch hydrogels. Starch-Starke 41:140–146
Kulshrestha VK, Chatterjee AC, Mukherjee SN (1962) Macromol. Chem Phys 54:205
Kumar V (2013) Characterization of conducting polymer composites and non-conducting polymers with exposure of neutrons, gamma rays and charged particles. PhD Thesis, Department of Physics, Sant Longowal Institute of Engineering and Technology, Longowal (Sangrur), Punjab (India)
Kurniawan TA, Chan GYS, Lo WH, Babel S (2006) Physico–chemical treatment techniques for wastewater laden with heavy metals. Chem Eng J 118:83–98
Kyzas GZ, Lazaridis NK (2009) Reactive and basic dyes removal by sorption onto chitosan derivatives. J Colloid Interf Sci 331:32–39
Laftah WA, Hashim S, Ibrahim AN (2011) Polymer hydrogels: a review. Polym Plast Technol Eng 50:1475–1486
Lee KY, Mooney DJ (2012) Alginate: properties and biomedical applications. Prog Polym Sci 37:106–126
Li F, Li S, El Ghzaoui A, Nouailhas H, Zhuo R (2007) Synthesis and gelation properties of PEG-PLA-PEG triblock copolymers obtained by coupling monohydroxylated PEG-PLA with adipoyl chloride. Langmuir 23:2778–2783
Lim J, Chouai A, Lo ST, Liu W, Sun X, Simanek EE (2009) Design, synthesis, characterization, and biological evaluation of triazine dendrimers bearing paclitaxel using ester and ester/disulfide linkages. Bioconjug Chem 20:2154–2161
Ma L, Gao C, Mao Z, Zhou J, Shen J, Hu X, Han C (2003) Collagen/chitosan porous scaffolds with improved biostability for skin tissue engineering. Biomaterials 24:4833–4841
Mahdavinia GR, Aghaie H, Sheykhloie H, Vardini MT, Etemadi H (2013) Synthesis of CarAlg/MMt nanocomposite hydrogels and adsorption of cationic crystal violet. Carbohydr Polym 98:358–365
Maitra J, Shukla VK (2014) Cross-linking in hydrogels—a review. Amer J Polym Sci 4:25–31
Mastropietro DJ, Omidian H, Park K (2012) Drug delivery applications for superporous hydrogels. Expert Opin Drug Delivery 9:71–89
Mathur MA, Shailender KM, Scranton A (1996) Methods for synthesis of hydrogel networks: a review. J Macromol Sci Rev Macromol Chem Phys C 36:405–430
Mitsumata T, Suemitsu Y, Fujii K, Fujii T, Taniguchi T, Koyama K (2003) pH-response of chitosan, κ–carrageenan, carboxymethyl cellulose sodium salt complex hydrogels. Polymer 44:7103–7111
Mittal H, Maity A, Ray SS (2015a) Effective removal of cationic dyes from aqueous solution using gum ghatti-based biodegradable hydrogel. Int J Biol Macromol 79:8–20
Mittal H, Maity A, Ray SS (2015b) Synthesis of co-polymer-grafted gum karaya and silica hybrid organic–inorganic hydrogel nanocomposite for the highly effective removal of methylene blue. Chem Eng J 279:166–179
Mittal H, Mishra SB (2014) Gum ghatti and Fe_3O_4 magnetic nanoparticles based nanocomposites for the effective adsorption of rhodamine B. Carbohydr Polym 101:1255–1264
Mittal H, Fosso-Kankeu E, Mishra SB, Mishra AK (2013a) Biosorption potential of gum ghatti-g-poly(acrylic acid) and susceptibility to biodegradation by B. subtilis. Int J Biol Macromol 62:370–378
Mittal H, Mishra SB, Mishra AK, Kaith BS, Jindal R (2013b) Flocculation characteristics and biodegradation studies of gum ghatti based hydrogels. Int J Biol Macromol 58:37–46
Mittal H, Mishra SB, Mishra AK, Kaith BS, Jindal R, Kalia S (2013c) Preparation of poly (acrylamide-co-acrylic acid)-grafted gum and its flocculation and biodegradation studies. Carbohydr Polym 98:397–404

Mittal H, Parashar V, Mishra SB, Mishra AK (2014) Fe_3O_4 MNPs and gum xanthan based hydrogels nanocomposites for the efficient capture of malachite green from aqueous solution. Chem Eng J 255:471–482

Mittal H, Ray SS, Okamoto M (2016) Recent progress on the design and applications of polysaccharide-based graft copolymer hydrogels as adsorbents for wastewater purification. Macromol Mater Eng 301:496–522

Miyata T (2002) Gels and interpenetrating polymer networks. In: Yui N (ed) Supramolecular design for biological applications. CRC Press, Boca Raton, pp 95–136

Mondal IH (2003) Graft copolymerization of nitrile monomers onto sulfonated jute-cotton blended fabric. J Appl Polym Sci 87:2262–2266

Morris GE, Vincent B, Snowden MJ (1997) Adsorption of lead Ions onto N—isopropylacrylamide and acrylic acid copolymer microgels. J Colloid Interf Sci 190:198–205

Nemours EIP (1936) Methacrylate Resins. Ind Eng Chem 28:1160–1163

Padhi JR, Nayak D, Nanda A, Rauta PR, Ashe S, Nayak B (2016) Development of highly biocompatible Gelatin and i-Carrageenan based composite hydrogels: in depth physiochemical analysis for biomedical applications. Carbohy Polym 153:292–301

Palaniraj A, Jayaraman V (2011) Production, recovery and applications of xanthan gum by Xanthomonas campestris. J Food Eng 106:1–12

Park TG, Hoffman AS (1993) Sodium chloride-induced phase transition in nonionic poly (N-isopropylacrylamide) gel. Macromolecules 26:5045–5048

Peppas NA, Bures P, Leobandung W, Ichikaw H (2000) Hydrogels in pharmaceutical formulations. Eur J Pharm Biopharm 50:27–46

Peppas NA, Langer R (1994) New challenges in biomaterials. Science 263:1715–1720

Pongjanyakul T, Puttipipatkhachorn S (2007) Xanthan-alginate composite gel beads: molecular interaction and in vitro characterization. Int J Pharm 331:61–71

Pooja D, Panyaram S, Kulhari H, Rachamalla SS, Sistla R (2014) Xanthan gum stabilized gold nanoparticles: characterization, biocompatibility, stability and cytotoxicity. Carbohydr Polym 110:1–9

Prajapati VD, Jani GK, Moradiya NG, Randeria NP (2013) Pharmaceutical applications of various natural gums, mucilages and their modified forms. Carbohydr Polym 92:1685–1699

Prettyman JB, Eddington DT (2011) Leveraging stimuli responsive hydrogels for on/off control of mixing. Sensor Actuat B Chem 157:722–726

Pulapura S, Kohn J (1992) Trends in the development of bioresorbable polymers for medical applications. J Biomater Appl 6:216–250

Qiu Y, Park K (2001) Environment-sensitive hydrogels for drug delivery. Advan Drug Deliv Rev 53:321–339

Qu Y, Chu BY, Peng JR, Liao F, Qi TT, Shi K, Qian ZY (2015) A biodegradable thermo-responsive hybrid hydrogel: therapeutic applications in preventing the post-operative recurrence of breast cancer. NPG Asia Materials 7:20

Rani P, Sen G, Mishra S, Jha U (2012) Microwave assisted synthesis of polyacrylamide grafted gum ghatti and its application as flocculant. Carbohydr Polym 89:275–281

Ratner BD, Hoffman AS (1976) Synthetic hydrogels for biomedical applications. In: Andrade JD (ed) Hydrogels for medical and related applications. ACS Symposium Series 31, pp 1–36

Rehab A, Akelah A, Issa R, D'Antone S, Solaro R, Chiellini E (1991) Controlled release of herbicides supported on polysaccharide based hydrogels. J Bioact Compat Polym 6:52–63

Richter A (2010) Hydrogels for actuators. In: Hydrogel sensors and actuators. Springer, Dresden, pp 221–248

Ricka J, Tanaka T (1984) Swelling of ionicgels-quantitative performance of the Donnan theory. Macromolecules 17:2916–2921

Rinaudo M (2006) Chitin and chitosan: properties and applications. Prog Polym Sci 31:603–632

Romero MR, Wolfel A, Igarzabal CIA (2016) Smart valve: polymer actuator to moisture soil control. Sens Actuat B: Chem 234:53–62

Saraydin D, Karadag E, Guven O (1995) Adsorption of some heavy metal ions in aqueous solutions onto acrylamide-maleic acid hydrogels. Separation Sci Tech 30:3291

Saraydin D, Karadag E, Guven O (1996) Adsorption of some basic dyes by acrylamide-maleic acid hydrogels. Sep Sci Tech 31:423–434

Saraydin D, Karadag E, Oztop HN, Guven O (1994) Adsorption of BSA onto acrylamide-maleic acid hydrogels. Biomaterials 15:917–920

Sarika PR, Cinthya K, Jayakrishnan A, Anilkumar PR, James NR (2014) Modified gum arabic cross-linked gelatin scaffold for biomedical applications. Mater Sci Eng, C 43:272–279

Sastry SK, Lakonishok M, Wu S, Truong TQ, Huttenlocher A, Turner CE, Horwitz AF (1999) Quantitative changes in integrin and focal adhesion signaling regulate myoblast cell cycle withdrawal. J Cell Biol 144:1295–1309

Schmidt C (1844) Uber Pflanzenschleim und Bassorin. Annalen 51:29–62

Sharma K, Kaith BS, Kumar V, Kalia S, Kumar V, Swart HC (2014) Water retention and dye adsorption behavior of Gg-cl-poly(acrylic acid-aniline) based conductive hydrogels. Geoderma 232:45–55

Sharma K, Kaith BS, Kumar V, Kumar V, Som S, Kalia S, Swart HC (2013) Synthesis and properties of poly(acrylamide-aniline)-grafted gum ghatti based nanospikes. RSC Adv 3:25830–25839

Sharma R, Kaith BS, Kalia S, Pathania D, Kumar A, Sharma N, Street RM, Schauer C (2015a) Biodegradable and conducting hydrogels based on guar gum polysaccharide for antibacterial and dye removal applications. J Environ Manage 162:37–45

Sharma R, Kalia S, Kaith BS, Pathania D, Kumar A, Thakur P (2015b) Guaran-based biodegradable and conducting interpenetrating polymer network composite hydrogels for adsorptive removal of methylene blue dye. Polym Degrad Stab 122:52–65

Shukla NB, Rattan S, Madras G (2012) Swelling and dye-adsorption characteristics of an amphoteric superabsorbent polymer. Ind Eng Chem Res 51:14941–14948

Silva SS, Mano JF, Reis RL (2010) Potential applications of natural origin polymer-based systems in soft tissue regeneration. Crit Rev Biotechnol 30:200–221

Singhal R, Gupta K (2016) A review tailor-made hydrogel structures (classifications and synthesis parameters). Polym Plast Technol Eng 55:54–70

Stojanovic SI, Nikolic L, Nikolic V, Petrovic S, Stankovic M, Ranisavljevic IM (2011) Stimuli-sensitive hydrogels for pharmaceutical and medical applications. Facta Universitatis Series: Phy Chem Technol 9:37–56

Stortz CA, Cerezo AS (2000) Novel findings in carrageenans, agaroids and "hybrid" red seaweed galactans. Curr Topics Phytochem 4:121–134

Tan X, Feldman SR, Chang JW, Balkrishnan R (2012) Topical drug delivery systems in dermatology: a review of patient adherence issues. Expert opi Drug Deliv 9:1263–1271

Tasdelen B, Cifci DI, Meric S (2015) Thermo and pH-responsive polymers and adsorbents for separation purposes. Jacobs J Civil Eng 1:1–3

Thakur S, Pandey S, Arotiba OA (2016) Development of a sodium alginate-based organic/inorganic superabsorbent composite hydrogel for adsorption of methylene blue. Carbohydr Polym 153:34–46

Tighe BJ (1976) The design of polymers for contact lens applications. Br Polym J 8:71–90

Treloar LRG (1951) The physical properties of amorphous polymer networks. Chem Ind 45:955–958

Treloar LRG (1952) The thermodynamic study of rubber-like elasticity. Proc Royal Soc London Series B-Biolog Sci 139:506–521

Ullah F, Othman MBH, Javed F, Ahmad Z, Akil HM (2015) Classification, processing and application of hydrogels: a review. Mater Sci Eng, C 57:414–433

Usov AI (2011) Polysaccharides of the red algae. Adv Carbohydr Chem Biochem 65:115–217

Wang C, Stewart RJ, Kopecek J (1999) Hybrid hydrogel assembled from synthetic polymers and coiled-coil protein domains. Nature 397:417–420

Wang Y, Chen L, Tan LI, Zhao Q, Feng Luo, Wei Y, Qian Z (2014) PEG-PCL based micelle hydrogels as oral docetaxel delivery systems for breast cancer therapy. Biomaterials 35:6972–6985

Wichterle O, Lim D (1960) Hydrophilic gels for biological use. Nature 185:117–118

Yan Z, Lin Z, Yizhong C (2016) Swelling properties and environmental responsiveness of superabsorbent composite based on starch-G-poly acrylic acid/organo-zeolite. J Macromol Sci B 55:662–679

Yetimoglu EK, Kahraman MV, Ercan O (2007) N-vinylpyrrolidone/acrylic acid/2- acrylamido-2-methylpropane sulfonic acid based hydrogels: synthesis, characterization and their application in the removal of heavy metals. React Funct Polym 67:451–460

Yoshida R, Sakai K, Okano T, Sakurai Y (1993) Pulsatile drug delivery systems using hydrogels. Adv Drug Deliver Rev 11:85–108

Zeng M, Zhang L, Zhou Y (2004) Effects of solid substrate on structure and properties of casting waterborne polyurethane/carboxymethylchitin films. Polym 45:3535–3545

Zhang X, Matteus Goosen FA, Wyss SP, Pichora D (1993) Biodegradable polymers for orthopedic applications. Polym Rev 33:81–102

Zhang Y, Wu F, Li M, Wang E (2005) pH switching on-off semi-IPN hydrogel based on cross-linked poly(acrylamide-co-acrylic acid) and linear polyallyamine. Polymer 46:769–7700

Zhao H, Zhou F, Peng W, Zheng J, Dziugan P, Zhang B (2015) The effects of carrageenan on stability of arachin and the interactions between them. Food Hydrocoll 43:763–768

Zhu J, Marchant RE (2011) Design properties of hydrogel tissue-engineering scaffolds. Expert Rev Med Devic 8:607–626

Zohuriaan-Mehr MJ, Kabiri K (2008) Superabsorbent polymer materials: a review. Iranian Polym J 17:451–477

Chapter 3
Macroporous Hydrogels: Preparation, Properties, and Applications

Maria Valentina Dinu and Ecaterina Stela Dragan

Abstract This chapter will be focused on the latest developments in the preparation and properties of macroporous hydrogels (MHGs), and their potential for biomedical applications and separation processes. A wide variety of synthetic and natural polymers have been used for the fabrication of novel macroporous hydrogels. Their morphology could be tailored by the synthesis strategy and synthesis parameters such as initial monomer concentration, cross-linking degree, and gel preparation temperature.

Keywords Cryogels · Drug delivery · Dyes · Heavy metal ions
Macroporous hydrogels · Mechanical properties · Pore-forming techniques
Porosity · Swelling · Tissue engineering

Abbreviations

AA	Acrylic acid
AAm	Acrylamide
ALG	Alginate
BAAm	*N,N'*-methylenebisacrylamide
CELL	Cellulose
COL	Collagen
CRS	Chondroitin sulfate
CS	Chitosan
DDS	Drug delivery systems
DMAEMA	2-(*N,N*-dimethylamino)ethyl methacrylate
DS	Diclofenac sodium
Dx	Dextran
DxS	Dextran sulfate
DOX	Doxorubicin

M. V. Dinu · E. S. Dragan (✉)
"Petru Poni" Institute of Macromolecular Chemistry, Aleea Grigore Ghica Voda 41A, 700487 Iasi, Romania
e-mail: sdragan@icmpp.ro

V. K. Thakur and M. K. Thakur (eds.), *Hydrogels*, Gels Horizons: From Science to Smart Materials, https://doi.org/10.1007/978-981-10-6077-9_3

ECM	Extracellular matrix
ELA	Elastin
HA	Hyaluronic acid
HAP	Hydroxyapatite
HEMA	2-hydroxyethyl methacrylate
HEP	Heparin
HPMA	Hydroxypropyl methacrylamide
HPMC	Hydroxypropyl methylcellulose
GA	Glutaraldehyde
GEL	Gelatin
IPN	Interpenetrating polymeric networks
LL	Levofloxacin lactate
MHGs	Macroporous hydrogels
MAA	Methacrylic acid
NaCMC	Carboxymethylcellulose sodium salt
NPA	Naproxen
NSAIDs	Nonsteroidal anti-inflammatory drugs
PA	Hydrolyzed PS-*g*-PAN
PAA	Poly(acrylic acid)
PAAm	Poly(acrylamide)
PDMAEM	Poly(*N*,*N*-dimethylaminoethyl methacrylate)
PEG	Poly(ethylene glycol)
PHEMA	Poly(2-hydroxyethyl methacrylate)
PHPMA	Poly(2-hydroxypropyl methacrylamide)
PGA	Poly(glycolic acid)
PCL	Poly(ε-caprolactone)
PLA	Poly(lactic acid)
PLGA	Poly(lactic-co-glycolic acid)
PMMA	Poly(methylmethacrylate)
PNIPAAm	Poly(*N*-isopropylacrylamide)
PPO	Poly(propylene oxide)
PS	Potato starch
Rh	Rhodamine B
PVA	Poly(vinyl alcohol)
PVP	Poly(vinyl pyrrolidone)
PU	Polyurethanes
$scCO_2$	Supercritical carbon dioxide
VPTT	Volume phase transition temperature

1 Introduction

Hydrogels are cross-linked macromolecular networks, which are capable to retain large quantities of water or biological fluids without dissolving due to the presence of physical or chemical cross-links (Thakur and Thakur 2014a, b; 2015). Hydrogels having pore sizes in the range of 10 nm–10 μm are called microporous, while those with pores above 10 μm are usually termed macroporous or superporous (Bencherif et al. 2013). Macroporous hydrogels (MHGs) have large and/or interconnected pores, which make them able to absorb water at a very fast rate. MHGs are suitable for biomedical and biotechnological applications, as drug delivery systems (DDS), for example, for gastric/intestinal release of drugs, proteins, or peptides.

The high water content, biocompatibility, and in vivo structural resemblance to extracellular matrix (ECM) make them an excellent substrate for tissue engineering (TE) applications. By controlling the chemical composition and cross-linking density, the polymeric network and the pore sizes could be conveniently tuned and structured into various geometries depending on the application requirements.

Not only the chemical composition of the hydrogel could be modified, but some physical properties of hydrogels could be also modulated in a similar manner. The mechanical properties of the most MHGs are depending on the content of cross-linker, co-monomer composition, polymerization conditions, or the attachements of additional binding sites. Another advantage of MHGs is that a stimuli-responsive behavior could be induced by using sensitive polymers as polymer matrix component, e.g., poly(*N*-isopropylacrylamide) (PNIPAAm). The specific response to an external stimulus is very useful in loading the polymer matrices with drugs or growth factors, which could be subsequently released using an appropriate stimulus (physical, mechanical or chemical). Various "smart" hydrogels have been reported, to be able to change their shape and/or properties when are exposed to an external stimulus such as temperature, pH, ionic strength, light, electric and magnetic field, or a combination of them (Zhao et al. 2015). These stimuli-responsive hydrogels contribute to the design of new generation of biomaterials used as self-regulated and targeted DDSs, specialized sorbents for separations, or bioreactors in various biological and biomedical applications.

A broad range of synthetic polymers have been involved in the synthesis of MHGs, such as poly(acrylamide) (PAAm), poly(acrylic acid) (PAA), PNIPAAm, poly(2-hydroxyethyl methacrylate) (PHEMA), poly(ethylene glycol) (PEG), poly (vinyl alcohol) (PVA), poly(2-hydroxypropyl methacrylamide) (PHPMA), polyurethanes (PU), poly(vinyl pyrrolidone) (PVP), poly(*N*,*N*-dimethylaminoethyl methacrylate) (PDMAEM), as well as polypeptides and polyesters. Synthetic polymers lead to hydrogels with controllable chemical structure, architecture, physico-chemical properties, degradation rate or mechanical strength (Bajaj et al. 2014; Nasim Annabi et al. 2010; Zhu and Marchant 2011). On the other hand, natural hydrogels based on chitosan (CS), cellulose (CELL), alginate (ALG), gelatin (GEL), collagen (COL), chondroitin sulfate (CRS), elastin (ELA), hyaluronic acid (HA), or heparin (HEP) are attractive for biomedical field due to their excellent biocompatibility and biodegradability (Croisier and Jérôme 2013; Giri et al. 2012; Hsieh et al. 2007; Shen et al. 2016). Nevertheless, the major limitations in using natural polymers are associated with their low mechanical strength, and uncontrollable structure. By combination of natural and synthetic polymers, well-defined and tailored hydrogel materials with improved properties are obtained, such as poly(PEG-co-peptides), ALG-*g*-(PEG-PPO-PEG), poly (PLGA-co-serine), COL-acrylate, poly(HPMA-*g*-peptide), HA-*g*-PNIPAAm, and poly(MAA-co-AAm)/CS (Aminabhavi et al. 2015; Dragan 2014a, b; Dragan et al. 2016; Sun et al. 2015a; Zhao et al. 2015).

In this context, our chapter will summarize the most important achievements in the field of MHGs, especially highlighting the techniques currently used for their preparation, as well as their properties and applications in TE, drug delivery, chromatography, and environmental protection.

2 Classification of Macroporous Hydrogels

Based on their cross-linking mechanism, MHGs could be first classified into physical and chemical hydrogels. Physical hydrogels involve non-covalently cross-linked networks, and their most relevant benefits are reversibility and the absence of chemical reactions, and, consequently, the lack of side products (Suhag et al. 2015). However, the weak physical interaction that leads to polymer network formation also reveals their major disadvantages, which include a lower stability and weaker mechanical properties compared to the chemically cross-linked hydrogels. Control over the network stability and its mechanical properties could be achieved by preparation of chemically cross-linked hydrogels via covalent bonds formed using a variety of coupling reactions, such as Michael-type addition, "click" chemistry, Schiff base coupling, disulfide bond formation, photo- and enzyme-mediated cross-linking (Guarino et al. 2015; Lau and Kiick 2015). However, the biocompatibility and safety issues caused by the presence in traces of chemical reagents and side products after the hydrogel synthesis and purification are the major limitations of using chemically cross-linked hydrogels in biomedical applications. Since the presence of cross-linking agents could induce cytotoxicity or disorders at the systemic level, many studies have been focused on developing of new approaches, such as 3D printing and photolithography, to create MHGs for a wide range of biomedical applications (Bajaj et al. 2014; Park and Park 2016).

According to the physical status of polymeric matrix, the MHGs could be classified as beads, monoliths, membranes, rods, disks, films, flakes, palettes, and capsules. When the MHGs are stimuli-sensitive, they could be distinguished by the nature of the stimulus responsible for swelling behavior; thus, they could be grouped in temperature-, pH-, ionic strength-, light-, and magnetic-responsive hydrogels (Sood et al. 2016). Furthermore, taking into account the polymer source, MHGs could be also classified as natural, synthetic, and synthetic/natural hybrid (composite) hydrogels. The decision regarding the selection of hydrogels derived from natural sources or synthetic polymers is intensively debated in the literature, but it generally depends on the required properties and desired application (Bajaj et al. 2014; Croisier and Jérôme 2013; Giri et al. 2012; Hsieh et al. 2007; Nasim Annabi et al. 2010; Zhu and Marchant 2011).

3 Preparation Techniques of Macroporous Hydrogels

Several techniques are usually selected for preparation of MHGs, such as cross-linking polymerization in the presence of pore-forming agents (Okay 2000), porogen leaching (Capes et al. 2005; Kuo and Wang 2013; Přádný et al. 2010, 2014; Sato et al. 2015; Sokic et al. 2014; Studenovska et al. 2008; Zhang et al. 2012), lyophilization of the hydrogel swollen in water (Dash et al. 2013; Dragan et al. 2009, 2012c; Fernandes et al. 2013; Thein-Han and Misra 2009; Wei et al. 2015; Zhong et al. 2011), cryogelation (Dragan 2014a; Dragan et al. 2016; Gunko et al. 2013;

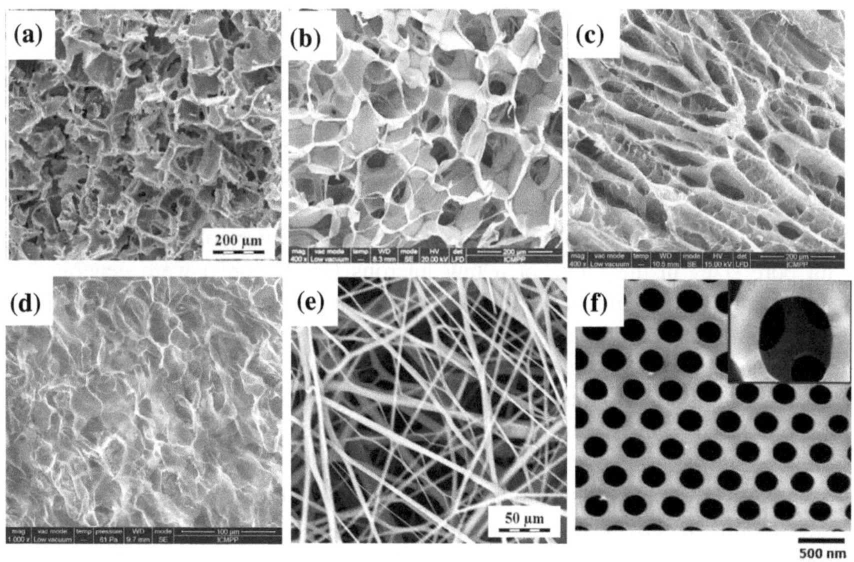

Fig. 1 SEM micrographs of various MHGs with porosity induced by: **a** free radical polymerization of 2-hydroxyethyl methacrylate (HEMA) in the presence of NaCl crystals (Prádný et al. 2014); **b** cross-linking of CS using ice-templating process (Dinu et al. 2013c); **c** cross-linking of CS using unidirectional freezing in liquid nitrogen combined with porogen leaching technique (Dinu et al. 2013b); **d** lyophilization of semi-IPN PAAm/CS composite hydrogels (Dragan et al. 2012c); **e** electrospun fibers of 7 wt% CS (Rogina 2014); and **f** combination of photo-patterning and colloidal templating method to obtain PHEMA hydrogel micropatterns (Lee et al. 2012)

Henderson et al. 2013; Lozinsky 2014), cross-linking polymerization in the presence of certain inorganic compounds able to release porogen gases (Croisier and Jérôme 2013; Ji et al. 2011; Nasim Annabi et al. 2010), electrospinning (Rogina 2014), colloidal particle templating, 3D printing, and photolithography (Bajaj et al. 2014; He et al. 2015; Park and Park 2016). In this section, some of these methods are presented in more detail, in Fig. 1 being presented the SEM micrographs of several MHGs synthesized by different pore-forming techniques showing the differences in their internal morphology.

3.1 Porogen Leaching

According to this method, water-soluble substances (sugars, salts, polymers) are present during the free-radical polymerization process, and are removed from the matrices after polymerization (Přádný et al. 2010). The major advantage of porogen leaching technique consists in the facile control of pore size and overall porosity only by selecting the required particle size and concentration within the prepolymer. Various porogens, such as sodium chloride crystals (Přádný et al. 2010, 2014), sugar crystals (Capes et al. 2005), poly(L-lactide) fibers (Studenovska et al. 2008), polystyrene microspheres (Kuo and Wang 2013), GEL microspheres (Sokic et al. 2014), or ALG microcapsules (Phull et al. 2013; Sato et al. 2015), have been used to obtain MHGs especially for tissue engineering applications. The scaffolds based on HEMA were used for cell culture experiments, and it was reported that the cell growth was reduced and even stopped after 7 days, because of limited nutrient supply since the large pores became blocked with cells (Lesny et al. 2006). Therefore, the same group of researchers reported the synthesis of MHGs based on cross-linked PHEMA with dual porosity (Přádný et al. 2014). MHGs with pore diameters in the range of 10 and 100 μm (large pores) were obtained by addition of NaCl crystals to the reaction mixture and their removal after polymerization. MHGs with pore sizes in the range of a few microns (small pores) were generated by 1-dodecanol (a poor solvent for PHEMA), when a phase separation occurs (Přádný et al. 2014). The preliminary cell experiments showed that the presence of two generations of pores (large and small pores, i.e. dual porosity) have significantly influenced proliferation, and adhesion of cells (Přádný et al. 2014).

The reduced control over the orientation and degree of pore interconnectivity, as well as the difficulties concerning the removal of soluble particles from the interior of 3D matrices are the major limitations associated with porogen leaching strategy. Moreover, it is necessary to mention that in the most of the particle leaching methods, different organic solvents are involved, which are highly toxic and require long drying times to assure their complete removal. To address these issues and to optimize the synthesis of MHGs with controlled pore architecture, different strategies have been further developed.

3.2 *Freeze-Drying Technique (Lyophilization)*

Freeze-drying (lyophilization) is also a simple and efficient process to create porous materials and consists in rapid freezing of polymeric system followed by the sublimation of solvent (water) under reduced pressure. For example, cross-linked pullulan/dextran, cross-linked CS, CS/COL, and CS–ZnO composite sponges were prepared using this technique (Autissier et al. 2010; Croisier and Jérôme 2013). It was shown that the degree of porosity could be modulated by the freeze-drying pressure. Thus, scaffolds based on cross-linked pullulan/dextran containing pores with sizes of 55 ± 4 μm and a porosity of 33 ± 12% were prepared at high freeze-drying pressure (6.5 mbar), whereas low freeze-drying pressure (0.1 mbar) led to scaffolds consisting of bigger pores with sizes of 243 ± 14 μm and a porosity of 68 ± 3% (Autissier et al. 2010). Polymer concentration was also found to have a strong influence on the thickness of pore walls; a lower polymer concentration resulted in thinner walls and high pore interconnectivity (Přádný et al. 2010). Lyophilization offers MHGs with higher specific pore area and thinner pore walls compared to porogen leaching technique (Bencherif et al. 2013). The porous structure of these hydrogels is formed during the first freeze/thaw cycle and consists not only in the formation of ice crystals, but also in the phase separation into two phases, i.e. one polymer-rich phase and a polymer-poor phase. The phase separation which occurs in the initial freezing step is resulting in the formation of a polymer-poor phase that still contains important amounts of polymer (Ricciardi et al. 2005). To overcome this problem, an alternative method that involves the repetition of freeze–thaw cycles was used (Ricciardi et al. 2005; Qi et al. 2015). This approach was used to obtain PVA hydrogels using cycles consisting of a 20-h freezing step at −22 °C, followed by a 4-h thawing step at 25 °C (Ricciardi et al. 2005).

Freeze-drying technique has been also applied to create pores within the non-porous hydrogels after their synthesis; thus, the hydrogel is first swollen in water, then lyophilized (Dash et al. 2013; Dragan et al. 2009, 2012c; Fernandes et al. 2013; Thein-Han and Misra 2009; Wei et al. 2015; Zhong et al. 2011). In this context, architectures with interconnected pores of various sizes were reported for a wide range of hydrogel materials, including natural and synthetic hydrophilic polymers. The internal morphology and the pore size were mainly affected by the increase of cross-linking degree (Dash et al. 2013; Wei et al. 2015), which leads to hydrogels with a more compact structure and smaller pore sizes. However, the porous materials generated by lyophilization are stiff and inelastic, while the swollen ones are soft, flexible, and elastic. Polymer concentration affects the thickness of pore walls as it was shown for CS–hydroxyapatite (HAP) hydrogels, where thicker pore walls were obtained with the increase of CS concentration (Thein-Han and Misra 2009).

Nevertheless, the architecture of hydrogels formed using this method is extremely sensitive to the kinetics of thermal quenching process; therefore, the precise tuning of their pore size encounters many difficulties. Other major drawback associated with the MHGs prepared using this technique are the low mechanical strength. Moreover, it was found that the freeze-drying process often results in the formation of a surface skin because of collapse of the hydrogel matrix at the scaffold–air interface due to the interfacial tension caused by solvent evaporation (Nasim Annabi et al. 2010). In addition, freeze-drying is a cost-effective technique which asks longer processing time for a complete removal of the solvent.

3.3 Ice-Templating (Cryogelation)

In ice-templating technique or cryogelation, the ice crystals constitute the solid porogen and derives from the porogen leaching (Gunko et al. 2013; Henderson et al. 2013; Lozinsky 2014). Cryogelation is a versatile route which has been widely applied for preparation of porous materials based on either natural and synthetic polymers, or their combination (Apopei and Dragan 2013; Berillo et al. 2014; Dinu et al. 2007; Dragan and Cocarta 2016; Dragan et al. 2016; Gunko et al. 2013; Gyarmati et al. 2015; Henderson et al. 2013; Lozinsky 2014; Oztoprak et al. 2014). In the ice-templating process, the precursors (initiators, activators, monomers, polymers) are concentrated within the unfrozen liquid micro-phase (ULMP) between the ice crystals. Porogen removal is achieved by thawing, which leads to cross-linked networks containing large and highly interconnected pores, and are well-known as cryogels. Furthermore, cryogels are characterized by superfast responsiveness at water absorption, possess a tissue-like elasticity, and are able to resist to different degrees of deformations (Berillo et al. 2014; Dinu et al. 2007, 2011b; Orakdogen et al. 2011). To create 3D networks with permeable walls, which could allow the unrestricted diffusion of low-molar mass substances when the large pores are filled with cells, a strategy combining porogen leaching and ice-templating process was developed (Dinu et al. 2013b, c). The schematic representation of this strategy is depicted in Fig. 2.

As Fig. 2 shows, glutaraldehyde (GA) interacts with the amine groups of CS, within the ULMP, and imine bonds are generated. The fractionated poly(methylmethacrylate) (PMMA) particles were enclosed within the pore walls of CS network during cross-linking of CS by GA at −18 °C. After reaction, the PMMA particles were washed out from the cryogel by acetone. MHGs based on CS or PHEMA with interconnected pores and less compact walls were successfully obtained when cryogelation was performed in the presence of PMMA particles. The mesh of fractionated PMMA particles and the crystallization rate were found to be the key parameters, having a strong influence on hydrogel morphology.

The ice-templating process also allows a better control on the morphology of resulting macroporous structures using, for example, the unidirectional freezing approach at a controlled immersion rate (Bai et al. 2013; Gutiérrez et al. 2008; Nieto-Suárez et al. 2016). Thus, various systems consisting of water soluble polymers, inorganic colloidal dispersions or hybrid composites were unidirectional frozen in different cooling agents such as frozen acetic acid/ethylether mixture (−80 °C),

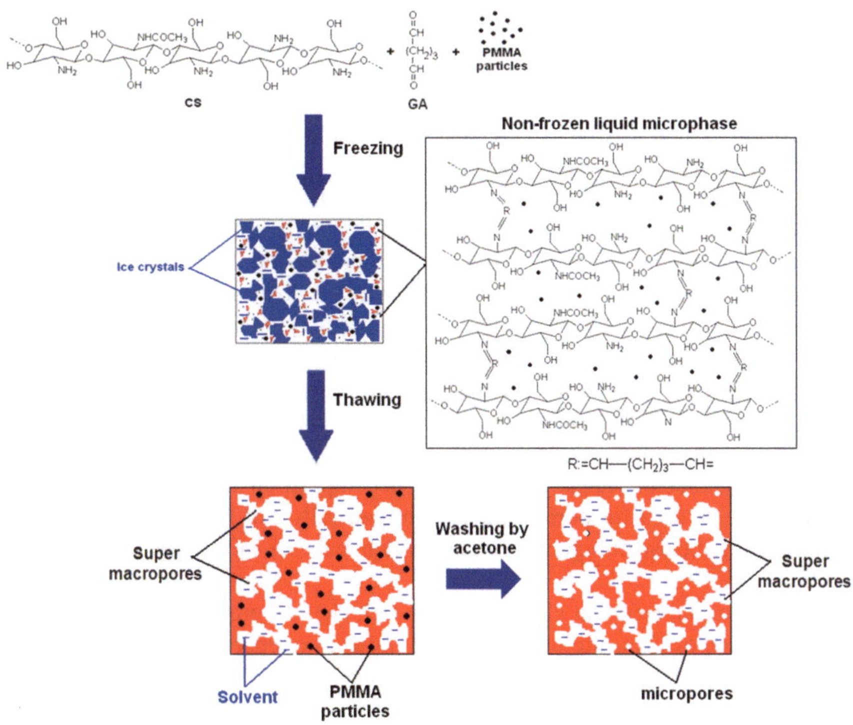

Fig. 2 Schematic representation of CS network formation using the combination of cryogelation process and porogen leaching technique (Dinu et al. 2013b)

frozen ethanol (−110 °C) or liquid nitrogen (−196 °C). In this approach, micro-channeled structures aligned along the freezing direction were formed, leading to macroporous materials with well-patterned channel morphologies (e.g., micro-honeycomb or lamellar). By unidirectional freezing approach various MHGs with aligned porous structures, such as xylan (Köhnke et al. 2014), aligned porous structures from water-soluble polymers such as PVA (Gutiérrez et al. 2008), poly(L-lactic acid) (Kim et al. 2009), PEG (Wu et al. 2012), PLGA (Hu et al. 2008), poly(NIPAAm) (Bai et al. 2013) based hydrogels and recently aligned porous zeolite/CS cryogels (Zhang et al. 2015) and aligned CS/GEL scaffolds (Nieto-Suárez et al. 2016) have been prepared. The size of tubular pores and the pore wall thickness was adjusted by varying the initial monomer or polymer concentration, the molecular weight of polymer as well as the crystallization speed (Bai et al. 2013; Dinu et al. 2013b; Nieto-Suárez et al. 2016; Wu et al. 2010).

3.4 Gas Foaming

The gas-foaming technique uses the nucleation and growth of gas bubbles distributed throughout a polymer to create pores within the 3D network (Bajaj et al. 2014; Lips et al. 2005; Nasim Annabi et al. 2010). The gas bubbles are formed by a foaming/blowing agent via chemical reactions or are expelled from a presaturated gas/polymer mixture prepared at high pressure. The foaming/blowing agent is a chemical compound which is mixed into the prepolymer and generates a gas in special conditions or when it chemically decomposes. The most commonly foaming agents used to prepare hydrogels with interconnected pores are sodium or ammonium bicarbonate, which are able to generate CO_2 in mild acidic solutions. For example, MHGs based on poly(acrylic acid/acrylamide) (AA/AAm) or PEG were prepared using sodium bicarbonate as blowing agent (Nasim Annabi et al. 2010). Ammonium bicarbonate is another gas blowing agent that was used to produce MHGs with interconnected pores via its decomposition to CO_2 and NH_3. The blowing agents used for pore formation are generally cell friendly, this technique not involving the use of organic solvents, which makes it well-suited for biomedical applications.

The use of supercritical carbon dioxide ($scCO_2$) as "green" medium to induce porosity inside of CS scaffolds (Ji et al. 2011) was also reported. The $scCO_2$ method allows the preparation of porous scaffolds suitable for cell cultures directly upon depressurization, based on hydrophilic polymers such as PLA, PLGA, and PCL (Bencherif et al. 2013; Nasim Annabi et al. 2010). To increase the solubility of CO_2 in hydrophilic polymers, various techniques, such as CO_2–water emulsion templating or the use of co-solvent systems, have been developed (Bajaj et al. 2014). Using dense gas CO_2 to generate porosity eliminates the use of surfactants or foam stabilizers that are required in conventional gas-foaming techniques.

Gas-foaming techniques can be consequently carried out with a wide variety of materials using commonly available gas blowing agents such as sodium or ammonium bicarbonate. However, it is still not clear why, in some cases, proper pore interconnectivity is achieved without any problems, while in others, more sophisticated improvements of the process are necessary.

3.5 Electrospinning

Electrospinning is another technique frequently used to prepare interconnected porous matrices for tissue engineering applications (Rogina 2014). A polymer solution, held at the end of a capillary tube by surface tension, is charged by applying a high voltage (Rogina 2014; Xu et al. 2016). Almost simultaneously, a thin jet of solution is drawn from the tube by electrostatic forces toward an oppositely charged collector. Along the jet flight toward the collecting electrode, the solvent evaporates and fibers with a lower diameter compared to the jet are finally formed (Reneker and

Chun 1996). Generally, a stationary ground plate is used to collect the fibers but, when a preferred fiber orientation is required, a rotating drum is employed as collector. The properties of resulting materials (morphology, porosity, fiber diameter) are controlled by modulating the temperature, viscosity, solution conductivity, and applied voltage (Xu et al. 2016). Porous fibers based on PHEMA or PVA and PAA in the micro- and nanometer range have been prepared by electrospinning (Přádný et al. 2010). Using a combination of electrospinning and salt-leaching, macroporous HA/COL hydrogels as nanofibers were formed (Kim et al. 2008). Nevertheless, the inability to prepare various 3D hydrogel shapes, the poor mechanical properties, and the low control over porosity and pore size are still the most significant limitations of using this technique in development of macroporous materials.

3.6 Other Preparation Techniques (Colloidal Crystal Templating, 3D Printing, Photolithography)

Within the last two decades, an increased interest was focused on using colloidal particles as templates to design 3D-ordered macroporous materials (Bencherif et al. 2013; He et al. 2015). In this context, stunning breakthroughs were achieved in the development of 3D-ordered architectures with interconnected pores, well-controlled pore size, having various chemical compositions and desired properties. The colloidal crystal templating is a facile and effective method adaptable to obtain porous materials with different architectures, such as films, monoliths, rods, and even hydrogels, depending on material composition and preparation method. The resulted MHGs consist of 3D ordered close-packed spherical cavities with sizes varying from 1 μm to mm range, surrounded by thin walls. The structural architecture has been described as the inverse replica of 3D assemblies of spherical colloidal particles closely packed in the ordered arrangement of a face-centered-cubic crystal, similar to that of natural opal gemstones. The 3D replica is obtained by back-filling the interparticle spaces with the material that will constitute the polymer matrix and the subsequent removal of colloids. A wide range of potential applications have been investigated for these porous materials containing inverse opal structures (He et al. 2015), starting from applications where the interconnected pores allow an effortless and unrestricted access to large active surface areas (e.g., tissue regeneration and immunotherapy) to those that require the presence of periodic 3D structures (e.g., photonic crystals and optical sensors).

3D printing and photolithography processes have been also applied to design MHGs (Bajaj et al. 2014; Lee et al. 2012; Park and Park 2016; Visser et al. 2015). These techniques allow the full control over the pore size and distribution, as well as on pore wall composition and geometry, holding thus a high potential in terms of organ-specific 3D design and development. Nevertheless, complex devices are required for the preparation of porous matrices, and several limitations may be imposed by the nature of materials that can be structured using these techniques.

3D inkjet printing is a technique in which a liquid precursor solution is dropwise expelled from a nozzle and accumulated on a support, where it solidifies (Bajaj et al. 2014). Thus, using a technique similar to layer-by-layer deposition, complex 3D porous structures could be build with a resolution controlled by the droplet size.

Photolithography is also a top-down technology, which allows designing of nearly arbitrary scaffold geometries. Preparation of hydrogel scaffolds has been mainly based on photo-polymerization of acrylic monomers or acrylate-functionalized prepolymers, such as PEG diacrylate (Wu et al. 2015), acrylated GEL, ALG, or other vinyl-functionalized macromolecules (Lee et al. 2012).

4 Properties of Macroporous Hydrogels

4.1 Swelling

Swelling behavior could be considered one of the most important properties of hydrogels, which determines their wide range of applications. When the macroporous material is in contact with solvent molecules, usually water, the polymer chains tend to expand to their fully solvated state, while the cross-linked structure applies an oppositely oriented force that pulls back the chains (Bilici et al. 2010; Caykara et al. 2007), until the expanding and retracting forces counterbalance each other and the swelling equilibrium is achieved. Swelling studies are usually carried out by immersing the dried hydrogel into the swelling medium, and then the hydrogel is removed from the solution and weighed after the excess solution from the surface is blotted (Dragan and Dinu 2015a). The relative values of weight and volume swelling ratios of hydrogels provide information about their internal structure in swollen state (Dinu et al. 2007, 2011a, c, 2013a). During the swelling process, the network pores are rapidly filled with solvent and, in the same time, the polymeric region of hydrogel takes up solvent molecules from the environment, the water absorption depending on the affinity of the solvent molecules for the polymer chains. Thus, two different processes are considered to govern the amount of solvent taken by the hydrogel at equilibrium: (a) solvation of network polymer chains, and (b) filling the pores with solvent. If the swelling is assumed to be isotropic, which means the pore volume remains constant upon swelling, the volume swelling ratio of porous networks is only caused by the solvation of network polymer chains. Accordingly, the higher the difference between equilibrium weight and volume swelling ratio, the higher the network porosity is. For example, the porosity of PAAm-based cryogels (Dinu et al. 2007), PAAm/Dx s-IPN cryogels (Dinu et al. 2011a), and CS-based cryogels (Dinu et al. 2013b) has been estimated taking into account both the values of weight and volume swelling ratios at equilibrium.

The water content of swollen hydrogels plays a key role in the biomedical and pharmaceutical applications because the equilibrium swelling ratio influences the surface mobility, the solute diffusion coefficient, and optical and mechanical properties of the hydrogel (Chhatri et al. 2011; Dragan et al. 2016; Sood et al. 2016; Wei et al. 2015). The swelling properties of MHGs are determined by various factors, including the monomer nature and concentration, the cross-linking density, and other environmental parameters, such as temperature, pH, and ionic strength (Chhatri et al. 2011; Dinu et al. 2013c; Dragan and Apopei 2011; Dragan et al. 2012b; Kennedy et al. 2014; Kuang et al. 2011; Li et al. 2015a, b; Park and Kim 2006; Salerno et al. 2011; Topuz and Okay 2009; Truong et al. 2013; Wei et al. 2015).

The swelling ratio (SR) is generally calculated with Eq. (1):

$$SR = \frac{W_t}{W_d} \tag{1}$$

where W_t is the weight (g) of swollen hydrogel at time t and W_d is the weight (g) of dried hydrogel.

For example, the swelling behavior of semi-IPN and IPN PAAm/CS composite hydrogels with macroporous structure generated by lyophilization was studied as a function of contact time by measuring the solvent uptake at different time intervals (Dragan et al. 2012c). It was found that the SR values were significantly influenced by the cross-linker ratio, the increase of this ratio decreasing the SR values. At lower cross-linking density, the network has a higher hydrodynamic free volume to accommodate more solvent molecules, thereby increasing matrix swelling. The mobility and relaxation of polymer chains are hindered with the increase of the cross-linker ratio, which in turn impedes the mobility of water, finally decreasing the SR values (Dragan et al. 2012c).

The rapid response rate to external stimuli of "smart" hydrogels is the key parameter deciding their further applications, and consequently, various methods have been used to increase the response kinetics (Bilici et al. 2010; Li et al. 2015a, b; Topuz and Okay 2009; Truong et al. 2013; Welzel et al. 2012). Furthermore, the pore size and architecture have also a strong influence on swelling rate, the MHGs showing a higher swelling–deswelling rate than the conventional hydrogels (Berillo and Volkova 2014; Bilici et al. 2010; Caykara et al. 2007; Dinu et al. 2011a; Dragan and Apopei 2013a; Kennedy et al. 2014; Rivero et al. 2015; Salerno et al. 2011). Even if their deswelling rates have been increased by several techniques, their reswelling from collapsed or dried state was still not improved; for example, the equilibrium swollen state in water was attained in hundreds of minutes by conventional hydrogels (Dragan et al. 2012c; Park and Kim 2006; Salerno et al. 2011). The previously reported slow reswelling rate of hydrogels is mainly caused by the collapse of their porous structure due to the cohesional forces between the solvated polymer chains. However, all hydrogels formed by cryogelation have been swollen within seconds (Bilici et al. 2010; Dinu et al. 2011a; Dragan and Apopei 2013a; Topuz and Okay 2009; Welzel et al. 2012). This different behavior is generated by the particular structure of cryogels, which does not collapse during drying or deswelling that enables the diffusion of water within the cryogel network.

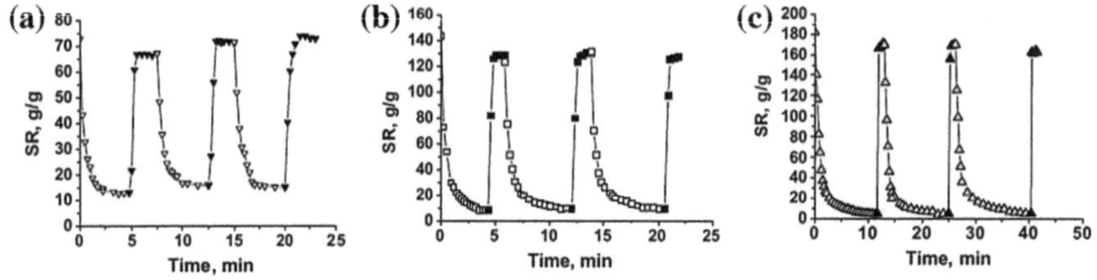

Fig. 3 Deswelling (empty symbols)–reswelling (full symbols) kinetics of MHGs in ethanol/water: **a** PAAm/PS80.3H (down triangle); **b** PAAm/PA80.3H (square); **c** PAAm/PA80.5H (up triangle) (Dragan and Apopei 2013a)

The deswelling/reswelling kinetics in a good/poor solvent pair of MHGs with the same cross-linker ratio, $X = 1/80$, but different by the nature of entrapped polymer and the initial concentration of monomers, are presented as an example in Fig. 3 (Dragan and Apopei 2013a). The MHGs entrapping gelatinized potato starch (PS) or hydrolyzed PS-*g*-PAN (PA) have been prepared by cross-linking polymerization of AAm with *N,N′*-methylenebisacrylamide (BAAm), at low temperature (−18 °C).

A fast-swelling response of collapsed cross-linked structures is evident in Fig. 3 for all gels, while the deswelling response of these gels from their equilibrium swollen state was dependent on the gel structure. Thus, the longer time required for deswelling of PA-containing MHGs (PAAm/PA80.3H, Fig. 3b) was assigned to the presence of ionic groups within the structure of PA as polyelectrolyte, which have a higher affinity toward water molecules than gelatinized PS (PAAm/PS80.3H, Fig. 3a). Moreover, a significant influence on deswelling rate was observed by modulating the concentration of AAm monomer, a faster deswelling response being shown by the cryogel containing a lower content of monomer (PAAm/PA80.3H, 3 wt% monomer concentration, Fig. 3b), compared to PAAm/PA80.5H (5 wt% monomer concentration, Fig. 3c).

The water uptake (WU, Eq. 2), the equilibrium water content (EWC, Eq. 2), and the water retention (WR, Eq. 4) are also used to describe the swelling behavior of MHGs (Dragan and Cocarta 2016).

$$\mathrm{WU} = \frac{W_t - W_d}{W_d} \tag{2}$$

$$\mathrm{EWC} = \frac{W_e - W_d}{W_e} \times 100 \tag{3}$$

$$\mathrm{WR} = \frac{W_t - W_d}{W_e - W_d} \times 100 \tag{4}$$

where W_t represents the mass of gel in swollen state at time t, W_d is the mass of gel in dry state, W_e represents the mass of swollen gel at equilibrium, and W_d is the mass of gel in dry state.

4.2 Porosity

Another remarkable characteristic of hydrogels is their highly porous structure. The porosity, pore size, and pore interconnectivity play an important role in direct function of biomaterial scaffolds (e.g., in growth and penetration of cells into the 3D structure, or in diffusion of nutrients and oxygen when the vascular system is not functional) (Lesny et al. 2006; Zhu and Marchant 2011; Zhao et al. 2015; Oh et al. 2015). The porosity and pore size of hydrogels could be tuned by changing different parameters, such as temperature, concentration of polymers or monomers, cross-linking degree, rate of freezing, addition of organic solvents or salts (Berillo and Volkova 2014; Dinu et al. 2007; Dinu et al. 2013a, b, c; Dragan and Apopei 2013a; Dragan et al. 2012b, Gunko et al. 2013; Lozinsky 2014; Savina et al. 2016).

The swollen state porosity, P_s (%), and total porosity, P (%), of MHGs have been generally calculated with Eqs. (5) and (6), respectively (Dinu et al. 2007, 2011a, b; Dragan and Dinu 2015a; Ozmen et al. 2007).

$$P_s = \left[1 - q_v\left(1 + \frac{(q_w - 1)d_2}{d_1}\right)^{-1}\right] \times 100 \tag{5}$$

$$P = \left(1 - \frac{d_0}{d_2}\right) \times 100 \tag{6}$$

where $q_v = (D_w/D_{dry})^3$ is the equilibrium volume swelling ratio, $q_w = (m_w/m_{dry})$ is the equilibrium weight swelling ratio, D_w and D_{dry} are the diameters of equilibrium swollen and dry gels, respectively, m_w and m_{dry} are the weight of gels after reaching the swelling equilibrium in water and after drying, d_1 is the density of solvent (water), d_2 is the density of polymer (Dinu et al. 2007; Ozmen et al. 2007), $d_0 = m_{dry}/(\pi D^2_{dry} l_{dry}/4)$ is the density of porous network, where l_{dry} is the length of cylindrical gel in the dry state.

The pore volume and the communicating pore fraction were also evaluated for several hydrogel systems (Dinu et al. 2013a, b; Přádný et al. 2010, 2014).

4.3 Mechanical Properties

The mechanical strength of macroporous materials mainly depends on the hydrogel composition and structure (Sood et al. 2016). The swelling degree, co-monomer composition, polymerization conditions, and cross-linking density are the most important parameters influencing the mechanical properties of hydrogels (Argun et al. 2014; Berillo and Volkova 2014; Chatterjee et al. 2015; Chiu et al. 2013; Dash et al. 2013; Dinu et al. 2011b, 2012, 2013a; Karpushkin et al. 2013; Lazaridou et al.

2015; Oztoprak et al. 2014; Park and Kim 2006; Přádný et al. 2014; Rivero et al. 2015; Strom et al. 2015; Tavsanli et al. 2015; Topuz and Okay 2009; Truong et al. 2013). The most common methods employed in the investigation of mechanical properties of hydrogels include extensiometry and compression (Sood et al. 2016). In strip extensiometry technique, a tensile force is applied to strips of porous material held between two grips. When the tensile force is applied to a ring instead of a single strip, the method is known as ring extensiometry. Shear-strain charts are thus obtained between the applied force and the elongation of material. The shear-strain charts are very useful to obtain different specific parameters describing the mechanical properties of hydrogels including Young's modulus, tensile strength, and yield strength. The viscoelastic nature of hydrogels by elongating the material strip to a precise length and examining the stress relaxation response over time at a constant strain is also given by extensiometry. Compression is the other technique used to check the mechanical properties of hydrogels (Dinu et al. 2007, 2011a, 2013a; Ozmen et al. 2007; Topuz and Okay 2009). The porous material is placed between two plates and compressed, the variable pressure applied to the hydrogel surface allowing the calculation of different mechanical parameters using various theoretical models. The main advantage of compression test over extensiometry is that it does not limit the material geometry to strips or rings.

As it was mentioned above, the mechanical properties of MHGs could be influenced by the type and degree of cross-linking (Argun et al. 2014; Přádný et al. 2014; Suhag et al. 2015; Topuz and Okay 2009). Thus, the compressive elastic modulus and compressive strength of physically cross-linked pH-responsive hydrogels in swollen state at pH 7 were found to vary with the gel composition from 3 to 11 kPa and 178 to 206 kPa, respectively (Suhag et al. 2015). The swollen gels had a viscoelastic behavior and encountered deformation from 70 to 85% before failure, indicating the formation of robust 3D hydrogel structures. The mechanical properties of MHGs could also be modulated by selecting the co-monomers and synthesis temperatures (RT or cryogelation) (Dinu et al. 2007, 2011a, 2012; Ozmen et al. 2007; Rivero et al. 2015).

Poly(α-amino acid)s-based cryogels with relatively high mechanical strength able to withstand large mechanical loading, even though the water content was more than 95% and the porosities higher than 90%, were reported (Sedlacík et al. 2015). The study regarding their rheological properties indicated that the storage moduli are within the range of 1–6 kPa that falls within the relevant stiffness range of artificial scaffolds.

Mechanical properties of MHGs are also influenced by architectural parameters, such as porosity, pore size, pore shape, fiber diameter, and fiber spacing (Izadifar et al. 2012; Lazaridou et al. 2015; Nasim Annabi et. 2010; Yamane et al. 2007). The effect of pore size was studied using woven CS/HA copolymeric fibrous scaffolds (Yamane et al. 2007), being observed a larger compression modulus for the scaffolds with the smallest pore sizes (100 μm) than for the scaffolds with 200 or 400 μm pore sizes.

Unidirectional freeze-drying method has been exploited for preparation of GEL scaffolds with high compressive strength (Wu et al. 2010). Highly porous HAP materials exhibited unusually high compressive strength up to 145 MPa for only 47% porosity suggesting these HAP-based materials could be used for load-bearing applications such as artificial bones.

To improve the mechanical properties of hydrogels, various strategies have been applied, including the addition of other molecules or polymers (Argun et al. 2014; Chhatri et al. 2011; Dinu et al. 2012; Park and Kim 2006; Tavsanli et al. 2015), the use of various cross-linking agents and different cross-linking ratios (Dash et al. 2013; Přádný et al. 2014), changes in monomer chemistry, variation of molecular weight and polymer concentration (Chiu et al. 2013; Přádný et al. 2014; Rivero et al. 2015). The compression modulus of hydrogels prepared using these strategies ranged from 0.005 to 20 MPa. However, these modifications could also induce chemical toxicity and impaired nutrient diffusion within the polymer matrix (Sood et al. 2016).

4.4 Thermal Stability

The thermal stability of macroporous semi-IPN hydrogels strongly depends on the polymer matrix structure and the trapping mode, being higher (Liu and Ciu 2011) or lower than that of polymer itself. For example, the thermal behavior of PAAm/Dx semi-IPN gels was compared to that corresponding to PAAm gels and Dx alone (Dinu et al. 2013a). Dx decreased the PAAm/Dx gel thermal stability, indicating the interdiffusion and physical entanglement of Dx chains within the 3D PAAm network (no covalent bonds were formed between Dx and PAAm). However, the hydrogel thermal stability did not depend on gel preparation temperature.

Thermal degradation of the gels was enhanced by chemical cross-linking for GEL above its melting temperature as showed by thermal scanning experiments (Dash et al. 2013); the prepared gels were not degraded until 50 °C. Hydrogels with high thermal stability were also obtained by grafting methyl methacrylate on starch or hydroxypropyl starch (Kumar et al. 2015).

4.5 Biodegradability and Biocompatibility

Hydrogel biodegradability is a required feature in biomedical applications demanding controlled in vivo resorption or local dissolution to support cell activities and promote tissue regeneration (Rogina 2014; Zhu and Marchant 2011). In this respect, temperature or pH could act as local environmental parameters, which can accelerate/decelerate the hydrogel degradation by chain disentanglement, hydrolysis or even proteolysis (Zhao et al. 2015).

Several strategies were involved in controlling the degradation rate of naturally derived hydrogels, including the change in deacetylation degree of material, modification of molecular weight, concentration, combination with other polymers, and incorporation of additives (Park and Park 2016). Local environmental parameters, such as temperature and pH, are also important parameters that affect the hydrogel degradation by accelerating/decelerating the hydrolysis processes.

Many natural polymers such as proteins, cellulose, starch, and chitin are biodegradable, but they have limited applications in preparation of hydrogels with tailored mechanical properties. On the other hand, the synthetic biodegradable polymers including PLA, PCL, poly(glycolic acid) (PGA), and their copolymers, have been also involved in the preparation of biodegradable MHGs (Zhu and Marchant 2011). Hydrolytically degradable hydrogels were also synthesized by photo-polymerization of polyester-containing macromers. In addition, when PEG cross-linkers containing functional groups, such as acetal, ketal, or disulfide, have been used, biodegradable PEG-containing hydrogels were prepared (Truong et al. 2013; Zhao et al. 2015).

It was reported that the hydrogels with lower porosity or larger pore sizes degraded more rapidly than those with higher porosity or smaller pore sizes (Chiu et al. 2013; Izadifar et al. 2012; Sedlacík et al. 2015; Su et al. 2012). This behavior is attributed to the thicker pore walls, associated self-catalyzed hydrolysis, and domination of bulk degradation over surface degradation. The hydrogel structure (the mesh size of cross-linked network) also affects the degradation profile, a highly cross-linked hydrogel (smaller mesh size) exhibiting longer degradation times (Chiu et al. 2013).

The major limitation of using synthetic biodegradable hydrogels as tissue engineering scaffolds is the lack of cell-specific bioactivities. To overcome this drawback, bioactive molecules have been incorporated into the synthetic hydrogels to mediate specific cell functions (Lau and Kiick 2015). Thus, using the molecular design, the physical properties (e.g., network parameters, diffusive profile and mechanical strength) and bioactivities of these hydrogels (e.g., cell attachment, migration and scaffold biodegradation) could be tailored.

Preliminary studies on biocompatibility of hydrogels based on AA and 2-(dimethylamino)ethyl methacrylate (DMAEMA) (poly(AA-co-DMAEMA)) investigated using HeLa and McCoy mouse fibroblast cell lines have showed their non-cytotoxicity (Suhag et al. 2015). PNIPAm-based hydrogels were also used without any significant cytotoxic effect to support fibroblast cell adhesion and proliferation for extended periods of time (up to 20 days) (Rivero et al. 2015). However, the scaffold biocompatibility could be improved by modification with natural polymers and natural particles such as HAP (Park and Park 2016; Sood et al. 2016; Thein-Han and Misra 2009; Wu et al. 2010; Zhong et al. 2011).

5 Applications of Macroporous Hydrogels

5.1 Drug Delivery

The MHGs used in drug delivery act as transport vehicles of the drug or of the therapeutically active component (Aminabhavi et al. 2015; Dragan 2014a, b; Elviri et al. 2015; Kumar et al. 2015; Mastropietro et al. 2012; Rogina 2014; Sood et al. 2016; Vishal Gupta and Shivakumar 2010; Wu et al. 2016). Different methods were developed to combine the hydrogels and drug, and finally to deliver the drug at the desired site. In this respect, the MHGs could be designed as reservoirs where the drug microparticles are placed (Fernandes et al. 2013; Sood et al. 2016). Alternatively, MHGs can be immersed in an aqueous solution containing the dissolved drug, allowing the absorption of solution until the hydrogel is fully swollen. The saturated MHGs are then dried, leaving behind the drug throughout the hydrogel open-pore structures (Mastropietro et al. 2012; Suhag et al. 2015; Varaprasad et al. 2011; Zhang et al. 2015). According to the last method, the drug is incorporated into the reaction mixture during hydrogel synthesis, being the least attractive option, due to the impurities, purification, and drug-loading issues. Frequently, for gastric delivery systems, a solid delivery system is prepared by entrapping various ratios of slow-dissolving hydroxypropyl methylcellulose (HPMC) and fast-dissolving PVP polymers (Mastropietro et al. 2012). The higher the content of HPMC the more prolonged releasing time was and the release kinetics followed better the zero-order release profile. However, the fast-dissolving polymer showed just a little effect on drug release, probably due to its quick dissolution into the testing medium.

The medication strategy involved in protein therapy usually requires frequent dose administration due to the short blood circulation time of proteins. Therefore, the development of novel sustained drug release systems that would overcome this major drawback was mandatory (Cheng et al. 2003; Dragan et al. 2016; Lazaridou et al. 2015; Suhag et al. 2015). In this respect, temperature-sensitive macroporous PNIPAAm hydrogels were prepared by leaching technique, using NaCl crystals as pore-forming agent (Cheng et al. 2003). When these MHGs were investigated as bovine serum albumin (BSA) delivery systems, the kinetics of release was dependent on changes in temperature; the lower temperatures led to a better BSA release, whereas higher temperatures resulted in a collapse of hydrogel, closing the pores and, consequently, slowing the release. The BSA release was also tested using pH-responsive hydrogels based on AA and 2-(dimethylamino)ethyl methacrylate (DMAEMA) (poly(AA-co-DMAEMA) (PAD) hydrogels) in various buffer solutions that simulate the pHs of gastrointestinal tract (Suhag et al. 2015). Thus, the hydrogel composition and the equilibrium swelling ratio had a significant influence on the release of BSA. Increasing the pH from 1.2 up to 7.4, the cumulative release of BSA significantly increased, ranging from 24 to 73%.

Single-network cryogels (SNCs) and IPN cryogels were recently investigated for lysozyme loading and release (LYS) (Dragan et al. 2016). The IPN hydrogels were obtained following a sequential strategy, which involves in the first step the formation of the first network by cryopolymerization of MAA sodium salt, with either AAm or HEMA as co-monomers, followed by the cross-linking of CS with PEGDGE, which generates the second network. A decrease in monomer concentration from 10 to 5 wt/v%, as well as the use of HEMA instead of AAm, favored the release of LYS, these results being attributed to the higher SR values and to the sponge-type morphology of IPNs.

Nonsteroidal anti-inflammatory drugs (NSAIDs) commonly cause nausea, heartburn, stomach pains, and ulceration when they are orally administered for the treatment of chronic inflammatory diseases. Therefore, various strategies were previously reported to protect and deliver these drugs (Dragan and Cocarta 2016; Mastropietro et al. 2012; Kumar et al. 2015). Thus, the ibuprofen release from pectin-based MHGs was investigated as a function of various parameters including pH, temperature, and porosity (Mastropietro et al. 2012). These biodegradable hydrogels showed a low release of ibuprofen at pH 1.2 (14%), which considerably increases with the increase of pH (79% at pH 7.4) revealing their high potential for intestinal drug delivery.

Macroporous pH-responsive hydrogels obtained by grafting of methyl methacrylate on starch and hydroxypropylated starch were also found to preferential deliver esomeprazole magnesium into pH 6.8 (Kumar et al. 2015). A gamma scintigraphy, using ^{99m}Tc as marker, was used to follow the gastrointestinal transit behavior of the optimized hydrogel, indicating a minimal release in the stomach and an increase in intestine.

A fast release of levofloxacin lactate (LL) and rhodamine B (Rh) was demonstrated for temperature-responsive poly[(NIPAAm)-co-(aminoethyl methacrylate β-cyclodextrin)] MHGs (Li et al. 2015a). Thus, about 30% of LL and Rh were released during the first 30 min, and about 80% of drugs after 300 min, which was much faster than the drug release from PNIPAAm hydrogels. The release of LL and Naproxen (NPA) was also studied from temperature-sensitive macroporous IPN hydrogels based on NIPAAm and AAm (Jiang et al. 2015). Poly(ethylene glycol diacrylate) and poly(ε-caprolactone diacrylate) were used as cross-linker agents due to their excellent biocompatibilities. A release of about 20% LL and NPA was observed during the first 2.5 h, while 80% of drugs was released after 20 h, indicating the entrappement of LL and NPA within the inner part of hydrogels.

Curcumin, a food ingredient, was successfully encapsulated in cryogels composed of CS, *kappa*-carrageenan, and carboxymethylcellulose sodium salt (NaCMC) (Nakagawa et al. 2013). The encapsulation efficiencies were in the range of 89–99%, the values depending on the cooling procedure used for freezing. The release profiles indicated a burst release which corresponds to a first-order release kinetic.

IPN superporous hydrogels based on CS and PVA prepared by gas foaming using sodium bicarbonate as pore-forming agent were investigated for the release of rosiglitazone maleate (Vishal Gupta and Shivakumar 2010). The sustained drug release from these superporous hydrogels suggests their potential application as gastro-retentive DDS for rosiglitazone maleate deliver.

pH–temperature dual-responsive semi-IPN hydrogels based on salecan and poly (*N*,*N*-diethylacrylamide-co-MAA) have been recently reported and investigated as doxorubicin (DOX)-releasing systems (Wei et al. 2015). The results showed that both the increase of temperature (37 °C) and the decrease of pH (4.0) significantly accelerated the DOX release rate. Moreover, increasing the salecan content and decreasing the amount of cross-linker led to faster release rate and higher release quantity. The colorimetric assay with tetrazolium dye 3-(4,5-dimethylthiazol-2-yl)-2,5-diphenyltetrazolium bromide (MTT) showed that the drug-free hydrogels were non-toxic for A549 cells, while the cytotoxicity of released DOX was comparable with that of free DOX. Fast multi-responsive macroporous IPN hydrogels were recently synthesized and investigated for the release of diclofenac sodium (DS) (Dragan and Cocarta 2016). The first network based on PDMAEM cross-linked with BAAm was prepared at −18 °C, and then swollen in a solution mixture consisting of initiators, AAm and BAAm, the second network being sequentially obtained using the same reaction conditions. The DS release from the IPN cryogels was modulated by DMAEM concentration in the first network, as well as by temperature and pH values of the release medium. Thus, a pulsatile DS release was observed, which was dependent on temperature changes: The drug release rate was lower below the volume phase transition temperature (VPTT) (34 °C) and faster above the VPTT (37 and 40 °C). The DS cumulative release was almost negligible in pH 1.3, whereas in pH 7.4 it was significant and increased with the increase of temperature from 34 up to 40 °C. The Korsmeyer–Peppas and Higuchi mathematical models described very well the DS release, which occurred by an anomalous mechanism, both diffusion and chain relaxation being involved in the release of drug from IPN hydrogels. The ability of PDMAEM/PAAm IPN cryogels to adsorb and release drugs in a controlled manner upon changing the pH and temperature recommends them as potential candidates for preparation of novel controlled drug delivery systems.

The design of composite materials by combination of two or more components into a single system would overcome the individual drawbacks, leading to synergetic functions that are not present in individual components. The entrapment of nanoparticles into 3D polymeric hydrogels started to gain more interest in the last years as an innovative strategy for development of new drug delivery platforms (Zhao et al. 2015).

5.2 Tissue Engineering and Regenerative Medicine

Lately, a considerable progress has been accomplished in transplantation of various organs, surgical reconstruction, or use of artificial prostheses. However, the drawbacks associated with these procedures have led to the consideration of tissue regeneration by cell proliferation on various scaffolds (Bajaj et al. 2014; Park and Park 2016).

Many attempts were being made to engineer virtually every human tissue. Currently, the TE involving cells and biomaterial-based therapy to restore bones, cartilages, or intervertebral disks is being explored. The TE approach is considered a promising technique to repair or regenerate the damaged tissue, restoring the functions during regeneration and being subsequently integrated with the host tissue. Thus, significant attention has been given to 3D polymer scaffolds (Bajaj et al. 2014; Bencherif et al. 2013; Chung and Park 2007; Croisier and Jérôme 2013; Guarino et al. 2013; Kuo and Wang 2013; Nasim Annabi et al. 2010; Oh et al. 2015; Owen et al. 2013; Park and Park 2016; Thein-Han and Misra 2009; Yue et al. 2010; Zhu and Marchant 2011). A scaffold designed to be used in TE should be endowed with high biocompatibility, appropriate porosity, controllable biodegradability, as well as enhanced mechanical stability (Nasim Annabi et al. 2010; Oh et al. 2015; Park and Park 2016; Sood et al. 2016). Furthermore, these scaffolds should provide the required support as artificial ECM, allowing the adhesion and proliferation of cells and maintaining their differentiated functions (Park and Park 2016; Přádný et al. 2014; Studenovska et al. 2008, Tripathi et al. 2013). For example, the presence of both macro- and micropores is essential in 3D scaffolds, the macropores (>50 μm) promoting the cell migration, while micropores favor the cell–cell interaction and the transport of nutrients and metabolites, which improves the tissue formation and growth (Henderson et al. 2013; Lesny et al. 2006; Nasim Annabi et al. 2010; Přádný et al. 2014). The porous sponges and nonwoven fibrous materials are the most commonly used 3D scaffolds in cartilage TE (Bajaj et al. 2014; Bencherif et al. 2013; Owen et al. 2013; Park and Park 2016). Highly porous networks provide larger surface areas for more cell attachment (about four times) compared to the less porous substrates, resulting in more cartilage tissue formation (Welzel et al. 2012). Pore interconnectivity has also a major influence on the ability of a 3D scaffold to support chondrogenesis as well as on the quality of formed tissue, even in 3D scaffolds with comparable porosities (Bajaj et al. 2014; Nasim Annabi et al. 2010; Park and Park 2016). Thus, the interconnected pores facilitate the homogeneous cell seeding and a better nutrient transport throughout the 3D scaffold, while the lack of pore interconnectivity induces an inhomogeneous chondrogenesis. In bone TE, the scaffold acts as a temporary biodegradable template introduced at the damaged site or lost bone to guide the regeneration of bone tissue, being gradually replaced by the regenerated tissue.

Scaffolds from natural polymers are usually prepared by freeze-drying (Hsieh et al. 2007), and a comprehensive review of CS-based hydrogel scaffolds for TE is given in Croisier and Jérôme (2013). Synthetic polymers including PGA, PLA, PLGA, PCL, PEG, and PHEMA are also frequently used in TE, all of them being FDA approved (Hu et al. 2008; Lesny et al. 2006; Nasim Annabi et al. 2010; Přádný et al. 2014; Park and Park 2016). Moreover, different techniques have been used to improve the bioactivity and cell adhesion on synthetic polymers, such as grafting or blending with hydrophilic/hydrophobic natural polymers, such as CS, HA, COL, or CRS (Fathi et al. 2013; Giri et al. 2012; Guarino et al. 2015; Izadifar et al. 2012; Khoshakhlagh and Moore 2015; Park and Park 2016; Sood et al. 2016; Zhong et al. 2011).

Scaffolds consisting of nanocomposite materials with controllable biodegradability and bioactivity, as well as suitable mechanical properties, have been also explored and developed for potential application in bone TE (Bajaj et al. 2014; Park and Park 2016; Thein-Han and Misra 2009). In this respect, highly porous 3D scaffolds containing high and medium molecular weight CS and various contents of HAP (0.5, 1 and 2 wt%) were obtained by freezing and lyophilization. In composite scaffolds, the proliferation of pre-osteoblasts (MC 3T3-E1) was about 1.5 times higher than CS scaffolds after seven days of culture. The biological response on 3D nanocomposite scaffolds was also improved in terms of better cell adhesion and well-spread cellular morphology compared to CS scaffold (Thein-Han and Misra 2009).

Cryogel scaffolds based on various synthetic or natural polymers (Henderson et al. 2013; Lozinsky 2014), for example, PAAm, PDMAm, PHEMA, PVA, PNIPAM, poly(*N*-vinylcaprolactam), PEO, CS (Croisier and Jérôme 2013), GEL, dextran (Dx), agarose, ALG, starPEG-HEP (Welzel et al. 2012), as well as blends like agarose/ALG (Tripathi and Kumar 2011; Tripathi et al. 2013), agarose/GEL (Tripathi et al. 2009), CS/GEL (Giri et al. 2012), CS/agarose/GEL, poly(AN)/GEL (Jain et al. 2009), poly(AN-co-NVP) and PNIPAAm/CS IPN (Jain et al. 2015) have been prepared and evaluated as scaffolds for TE and for other biotechnological applications. For instance, the proper pore size and interconnectivity within starPEG-HEP cryogel matrix enabled human umbilical vein endothelial cells (HUVECs) to migrate into the material (Welzel et al. 2012). These starPEG-heparin cryogels allow a secondary biofunctionalization including the binding of adhesion ligands, or incorporation of heparin-binding growth factors. In another study it was demonstrated that the nature of functional groups which are in contact with the cells has a major influence on the final cell adhesion onto the cryogel matrix (Tripathi et al. 2013). The surface properties and thus the adhesion and proliferation of cells onto ALG/GEL cryogel matrices were affected by changing the nature of the cross-linker.

Cryogels composed of AN and NVP and IPNs of PNIPAAm and CS which support in vitro growth and function of fibroblasts (COS-7) and human liver hepatocarcinoma cells (HepG2) were recently reported (Jain et al. 2015). The low albumin adsorption and platelet adherence on these cryogels demonstrated their hemocompatibility. Furthermore, the in vivo implantation of PNIPAAm–CS cryogels in mice also showed their biocompatibility with the surrounding tissue. A closed continuous bioreactor based on these cryogels maintained the availability of HepG2 cells at high density for seven days. Moreover, the off-line clinical evaluation showed the detoxification of circulating plasma obtained from patients with acute or chronic liver deficiency using the cells attached on PNIPAAm–CS cryogels.

5.3 Chromatography

Various diagnostic, biotechnological, and biomedical applications require the preparative separation of cells (Gunko et al. 2013; Lamprou et al. 2015; Lozinski 2014; Savina et al. 2016). In this respect, affinity chromatography which is based on the interaction between a receptor from the cell surface and an immobilized ligand, was reported as a promising method for cell separation. However, most of the currently available porous systems have pore sizes smaller than the cell size, causing the column clogging. Another limitation of this technique is the difficulty to elute the bound cells. Application of novel adsorbents, such as supermacroporous monolithic cryogels, allowed the overcoming of these drawbacks (Savina et al. 2016). Cryogels are characterized by highly interconnected large (10–100 μm) pores, sponge-like morphology, high elasticity and offer a plethora of possibilities for chemical modification with a wide range of ligands. During the transport of cells through the matrix by convective migration, the target cells bind to the affinity ligands, while the other cells are non-retained and are easily removed from the cryogel column during washing. The elastic, sponge-like structure of cryogel matrices allows a unique and efficient way of cell detachment by mechanical compression. The recent developments in affinity based cryogels and molecularly imprinted cryogels in protein purification were comprehensively reviewed (Andaç et al. 2016; Ertürk and Mattiasson 2014).

5.4 Environmental Protection

5.4.1 Heavy Metal Ions Sorption

Protection of groundwater and marine waters is a complex issue of our modern life that should stand as an imperative responsibility of humankind. Once contaminated, the aquatic environment would be practically very difficult and extremely expensive to clean it up. Amongst all pollutants, heavy metal ions (i.e., ions of Cr(VI), As(V), Cu(II), Pb(II), Cd(II), Ni(II), Hg(II)) and their complexes are considered as the most problematic ones because of their high persistence into the ecosystem, bioaccumulation and high toxicity. Consequently, intense efforts have been undertaken to minimize the harmful impact of heavy metal ions. Adsorption on MHGs is a process that proved to be suitable and remarkably efficient for the removal of heavy metal ions from aqueous systems, such as Cu^{2+} (Apopei et al. 2012; Deze et al. 2012; Dragan 2014a, b; Dragan and Dinu 2013, 2015b, Wang and Li 2015; Yi et al. 2016; Zhang et al. 2015, Zhu and Li 2015), Ni^{2+} (Apopei et al. 2012; Liu et al. 2016; Zhu and Li 2015); Cd^{2+} (Apopei et al. 2012; Deze et al. 2012; Zhu and Li 2015), Zn^{2+} (Apopei et al. 2012), Pb^{2+} (Zhu and Li 2015), CrO_4^{2-} (Zhu et al. 2013), AsO_4^{3-} (Sahiner et al. 2015), and uranyl ions (Karadağ and Kundakcı 2015). For example, poly((3-acrylamidopropyl)trimethylammonium chloride) (p(APTMACl)) cryogels were used as a superporous sorbents for the removal of toxic AsO_4^{3-} anions from an aqueous medium. The fast swelling of p(APTMACl) cryogels in water (about 7 s) led to a faster AsO_4^{3-} sorption (less than 30 min) compared to 12 h for conventional p(APTMACl) hydrogels. The p(APTMACl) cryogels exhibited a maximum adsorption capacity of about 120 mg AsO_4^{3-}/g cryogel. Furthermore, Fe_2O_3/p(APTMACl) magnetic cryogels were synthesized and investigated for sorption of AsO_4^{3-} anions. The Fe_2O_3/p(APTMACl) cryogels containing AsO_4^{3-} were easily removed by applying an external magnetic field. These composite cryogels have been reused in the arsenate removal from aqueous environments up to five times without significant loss of sorption capacity. An increasing interest was focused on polysaccharides-based sorbents because of their biodegradability and low-cost. Biopolymers, such as CS (Dragan and Dinu 2013; Zhang et al. 2016) and PS (Apopei et al. 2012), were used as components in preparation of MHGs with promising sorption properties for various pollutants (dyes, heavy metal ions). In our studies we have been interested in the evaluation of sorption capacity for various metal ions (Cu^{2+}, Cd^{2+}, Ni^{2+}, and Zn^{2+}) of semi-IPN cryogels based on PAAm as a matrix and anionically modified PS as entrapped polysaccharide (Apopei et al. 2012). The experimental data have been analyzed by fitting various mathematical models (Langmuir, Freundlich, Sips, and Temkin isotherms). Sips isotherm model was found to be the best-fitted model on all experimental data, the theoretical sorption capacity being of 40.72 mg Cu^{2+}/g, 19.72 mg Cd^{2+}/g, 9.31 mg Ni^{2+}/g, and 7.48 mg Zn^{2+}/g. Sorption of Cu^{2+} ions was also investigated on PAAm/CS IPN hydrogels (Dragan and Dinu 2013, 2015b). The semi-IPN hydrogels showed no sorption capacity for Cu^{2+}, while IPN hydrogels had a high Cu^{2+} sorption capacity. The different behavior was attributed to the structural changes occurring within the PAAm matrix during the generation of the second network at high pH, when carboxylate groups are formed by partial hydrolysis of amide groups.

Sorption of Cu^{2+} ions was also investigated on the ordered porous zeolite/CS (Zel/CS) monoliths, and it was reported that the sorption capacity of the monoliths depended on the contact time, the Zel/CS ratio, and the metal ion concentration (Zhang et al. 2015). A maximum sorption capacity of 89 mg Cu^{2+}/g was found for these monoliths. Moreover, the Cu^{2+}-containing Zel/CS monoliths showed catalytic activity, reducing 4-nitrophenol to 4-aminophenol.

Ion-imprinted MHGs were also synthesized and evaluated for removal of heavy metal ions. Thus, Wang and Li synthesized Cu(II)-imprinted poly(polyethyleneimine/hydroxyethyl acrylate) (Cu(II)-p(PEI/HEA)) hydrogels by ^{60}Co-γ-induced copolymerization (Wang and Li 2015). The Cu(II)-p(PEI/HEA) hydrogels exhibited a maximum sorption capacity for Cu^{2+} of 40 mg/g.

A novel and facile strategy using graphene oxide (GO)-stabilized high internal phase emulsion (HIPE) templates for preparation of interconnected macroporous PAA hydrogels was recently developed (Yi et al. 2016). The interconnectivity of the obtained polymer foams was tailored by the wettability and amount of GO. The highly permeable macroporous PAA hydrogels showed an excellent sorption capacity of Cu^{2+} ions of 280 mg/g.

5.4.2 Dyes Sorption

The pollution of water with dyes represents a major environmental issue due to the wide diversity of dyes employed in industrial applications. The presence of dyes affects the photosynthetic activity in aquatic life due to the decrease of light penetration through water, and thus, their decontamination is demanding. Various techniques, such as chemical precipitation, solvent extraction, coagulation/flocculation, oxidation, and adsorption, have been employed for dye removal from wastewaters. Among them, adsorption is considered as a straightforward and efficient technique for water purification, a plenitude of sorbents, synthetic or coming from natural sources, being investigated for the removal of dyes. In the last decade, an increased interest was addressed in finding novel sorbents with enhanced separation properties, high adsorption capacities, rapid sorption/desorption rates, and mild conditions for their regeneration (Dragan 2014a, b; Dragan and Dinu 2015a, b). In this context, MHGs are gaining more and more interest for application as novel sorbents (Dragan and Apopei 2011; Dragan and Apopei-Loghin 2013b; Dragan et al. 2012a, b, c; Perju et al. 2012; Savina et al. 2016; Sun et al. 2015b; Yi et al. 2016).

As it was already shown (Sect. 3.3), cryogelation is one of the most accessible and versatile techniques to generate permanent macropores in hydrogels. Ionic multicomponent cryogels, having enhanced mechanical and chemical resistance, have been studied as sorbents in the removal of ionic dyes (Dragan et al. 2012b; Perju et al. 2012). The sorption of two ionic dyes, the cationic dye Methylene Blue (MB) and the anionic dye Direct Blue 1 (DB1), was reported on PAAm/CS semi- or IPN hydrogels (Dragan et al. 2012b, c). The semi-IPN hydrogels sorbed the anionic dye DB1 due to the positively charged groups of CS, while the IPN hydrogels exhibited a higher affinity for the cationic dye MB owing to the presence of anionic COO^- groups formed during the generation of the second network. Three kinetic models, i.e., the pseudo-first-order kinetic model, PFO, the pseudo-second-order kinetic model, PSO, and the intra-particle diffusion model, were used to investigate the mechanism controlling the dye sorption. The theoretical sorption capacities obtained using PFO model were very close to the experimental values both for semi-IPNs and IPNs, and this suggested a physical sorption as the controlling mechanism of dye sorption (Dragan et al. 2012b). The PAAm/CS IPN cryogels were also used for the selective separation of MB from its mixture with methyl orange (MO).

A high sorption capacity of MB was also reported for semi-IPN composite cryogels containing anionically modified PS as entrapped polymer into a matrix of PAAm (Dragan et al. 2012a; Dragan and Apopei-Loghin 2013b). The enhancement of sorption capacity was obtained when a controlled hydrolysis of PAAm matrix was performed, due to the generation of a high density of COO^- groups able to interact with the cationic dye. The sorption kinetics was well fitted by PFO model, the overall sorption process being thus controlled by physisorption. The semi-IPN cryogels have also shown a high degree of reusability and stability, their sorption capacities being almost similar after six cycles of sorption/desorption.

MB removal was also studied on xylan/poly(acrylic acid)/Fe_2O_3 magnetic nanocomposite hydrogels (Sun et al. 2015b). The sorbent had a semi-IPN network structure and exhibited a structure with interconnected pores. A MB removal of 90% was obtained at pH 8 when a sorbent dosage of 3 g/L and an initial MB concentration of 400 mg/L were used. Macroporous PAA hydrogels with an open-cell structure having a sorption capacity of MB of 130 mg/g were recently reported (Yi et al. 2016). The recycling performance of the macroporous PAA hydrogel was also investigated. It was found that 75% of the initial sorption capacity of the hydrogel was preserved after five recovery cycles.

6 Conclusions

In the last decades, an outstanding improvement has been done in the development of macroporous materials with biomedical and technological applications. With the advancement of novel drug delivery platforms and the raised demanding for efficient and low-cost sorbents, macroporous systems based on hydrogels started to gain a considerable research interest. However, the most significant drawbacks of conventional hydrogels are their low mechanical strengths and swelling response rates. To overcome these issues, several strategies were developed, including synthesis of macroporous hydrogels by porogen leaching, gas foaming, and ice-templating, as well as the design of composite materials by combination of two or more components into a single system, such as organic/inorganic composites or IPN hydrogels. Moreover, by tuning the chemical composition, the cross-linking density, and the conditions of preparation, the polymeric network and the pore sizes could be conveniently controlled and structured into various geometries depending on the application requirements.

Nevertheless, macroporous hydrogels with fast responsive properties to external stimuli are necessary in a wide range of applications, and the easiest way to achieve this aim is to include sensitive polymers as polymer matrix components, such as the well-known temperature-responsive poly(*N*-isopropylacrylamide) (PNIPAAm). These stimuli-responsive properties are very useful in loading the polymer matrices with various payloads, which could be subsequently released "on demand" upon stimulation with an appropriate stimulus. In this respect, the stimuli-sensitive hydrogels are expected to have a huge potential in the development of next generation of biomaterials used as drug delivery systems, specialized sorbents for various separations, scaffolds for tissue engineering, or as bioreactors in a wide range of biomedical and technological applications.

Furthermore, the design of cross-linked networks containing large and highly interconnected pores, also known as cryogels, brought into attention some of their remarkable properties such as tissue-like elasticity, high mechanical strength, and superfast swelling rate, which can address specific biological and medical challenges. Multicomponent macroporous materials based on synthetic or natural polymers, characterized by enhanced mechanical and chemical resistance, showed high efficiency in wastewater decontamination.

All the examples mentioned in this chapter represent only some of the approaches currently used for macroporous hydrogels preparation, as well as the techniques involved in the improvement of their properties and thus to extend their application range.

Acknowledgements This work was supported by Romanian National Authority for Scientific Research, CNCSIS-UEFISCDI, grant number PN-II-ID-PCE-2011-3-0300.

References

Aminabhavi TM, Nadagouda MN, More UA, Joshi SD, Kulkarni VH, Noolvi MN, Kulkarni PV (2015) Controlled release of therapeutics using interpenetrating polymeric networks. Expert Opin Drug Delivery 12:669–688

Andaç M, Galaev IY, Denizli A (2016) Affinity based and molecularly imprinted cryogels: applications in biomacromolecule purification. J Chromatogr B 1021:69–80

Apopei DF, Dragan ES (2013) Semi-interpenetrating polymer networks based on polyacrylamide and starch or modified starch. J Nanostruct Polym Nanocompos 9:16–20

Apopei DF, Dinu MV, Trochimczuk A, Dragan ES (2012) Sorption isotherms of heavy metal ions onto semi-IPN cryogels based on polyacrylamide and anionically modified potato starch. Ind Eng Chem Res 51:10462–10471

Argun A, Can V, Altun U, Okay O (2014) Nonionic double and triple network hydrogels of high mechanical strength. Macromolecules 47:6430–6440

Autissier A, Visage CL, Pouzet C, Chaubet F, Letourneur D (2010) Fabrication of porous polysaccharide-based scaffolds using a combined freeze-drying/cross-linking process. Acta Biomater 6:3640–3648

Bai H, Polini A, Delattre B, Tomsia AP (2013) Thermoresponsive composite hydrogels with aligned macroporous structure by ice-templated assembly. Chem Mater 25:4551–4556

Bajaj P, Schweller RM, Khademhosseini A, West JL, Bashir R (2014) 3D Biofabrication strategies for TE and regenerative medicine. Annu Rev Biomed Eng 16:247–276

Bencherif SA, Braschler TM, Renaud P (2013) Advances in the design of macroporous polymer scaffolds for potential applications in dentistry. J Periodontal Implant Sci 43:251–261

Berillo D, Volkova N (2014) Preparation and physicochemical characteristics of cryogel based on gelatin and oxidised dextran. J Mater Sci 49:4855–4868

Berillo D, Mattiasson B, Kirsebom H (2014) Cryogelation of chitosan using noble-metal ions: in situ formation of nanoparticles. Biomacromolecules 15:2246–2255

Bilici C, Karayel S, Demir TT, Okay O (2010) Self-oscillating pH-responsive cryogels as possible candidates of soft materials for generating mechanical energy. J Appl Polym Sci 118:2981–2988

Capes JS, Ando HY, Cameron RE (2005) Fabrication of polymeric scaffolds with a controlled distribution of pores. J Mater Sci Mater Med 16:1069–1075

Caykara T, Kucuktepe S, Turan E (2007) Swelling characteristics of thermo-sensitive poly [(2-diethylaminoethyl methacrylate)-co-(N, N-dimethylacrylamide)] porous hydrogels. Polym Int 56:532–537

Chatterjee P, Dai A, Yu H, Jiang H, Dai LL (2015) Thermal and mechanical properties of poly (N-isopropylacrylamide)-based hydrogels as a function of porosity and medium change. J Appl Polym Sci 132:42776–42768

Cheng SX, Zhang JT, Zhuo RX (2003) Macroporous poly(Nisopropylacrylamide) hydrogels with fast response rates and improved protein release properties. J Biomed Mater Res A 67A:96–103
Chhatri A, Bajpai J, Bajpai AK, Sandhu SS, Jain N, Biswas J (2011) Cryogenic fabrication of savlon loaded macroporous blends of alginate and polyvinyl alcohol (PVA). Swelling, deswelling and antibacterial behaviors. Carbohydr Polym 83:876–882
Chiu YC, Kocagoz S, Larson JC, Brey EM (2013) Evaluation of physical and mechanical properties of porous poly (ethylene glycol)-co-(L-lactic acid) hydrogels during degradation. PLoS ONE 9:607–628
Chung HJ, Park TG (2007) Surface engineered and drug releasing pre-fabricated scaffolds for tissue engineering. Adv Drug Delivery Rev 59:249–262
Croisier F, Jérôme C (2013) Chitosan-based biomaterials for tissue engineering. Eur Polym J 49:780–792
Dash R, Foston M, Ragauskas AJ (2013) Improving the mechanical and thermal properties of gelatin hydrogels cross-linked by cellulose nanowhiskers. Carbohydr Polym 91:638–645
Deze EG, Papageorgiou SK, Favva EP, Katsaros FK (2012) Porous alginate aerogel beads for effective and rapid heavy metal sorption from aqueous solutions: effect of porosity in Cu^{2+} and Cd^{2+} ion sorption. Chem Eng J 209:537–546
Dinu MV, Ozmen MM, Dragan ES, Okay O (2007) Freezing as a path to build macroporous structures: superfast responsive polyacrylamide hydrogels. Polymer 48:195–204
Dinu MV, Perju MM, Dragan ES (2011a) Porous semi-interpenetrating hydrogel networks based on dextran and polyacrylamide with superfast responsiveness. Macromol Chem Phys 212:240–251
Dinu MV, Perju MM, Dragan ES (2011b) Composite IPN ionic hydrogels based on polyacrylamide and dextran sulfate. React Funct Polym 71:881–890
Dinu MV, Perju MM, Cazacu M, Dragan ES (2011c) Polyacrylamide-dextran polymeric networks: effect of gel preparation temperature on their morphology and swelling properties. Cell Chem Technol 45:197–203
Dinu MV, Schwarz S, Dinu IA, Dragan ES (2012) Comparative rheological study of ionic semi-IPN composite hydrogels based on polyacrylamide and dextran sulphate and of polyacrylamide hydrogels. Colloid Polym Sci 290:1647–1657
Dinu MV, Cazacu M, Dragan ES (2013a) Mechanical, thermal and surface properties of polyacrylamide/dextran semi-interpenetrating network hydrogels tuned by the synthesis temperature. Cent Eur J Chem 11:248–258
Dinu MV, Prádny M, Dragan ES, Michálek J (2013b) Ice-templated hydrogels based on chitosan with tailored porous morphology. Carbohydr Polym 94:170–178
Dinu MV, Prádny M, Dragan ES, Michálek J (2013c) Morphogical and swelling properties of porous hydrogels based on poly(hydroxyethyl methacrylate) and chitosan modulated by ice-templating process and porogen leaching. J Polym Res 20:275–285
Dragan ES (2014a) Design and applications of interpenetrating polymer network hydrogels. A review. Chem Eng J 243:572–590
Dragan ES (2014b) Advances in interpenetrating polymer network hydrogels and their applications. Pure Appl Chem 86:1707–1721
Dragan ES, Apopei DF (2011) Synthesis and swelling behavior of pH-sensitive semi-interpenetrating polymer network composite hydrogels based on native and modified potatoes starch as potential sorbent for cationic dyes. Chem Eng J 178:252–263
Dragan ES, Apopei DF (2013a) Multiresponsive macroporous semi-IPN composite hydrogels based on native or anionically modified potato starch. Carbohydr Polym 92:23–32
Dragan ES, Apopei-Loghin DF (2013b) Enhanced sorption of methylene blue from aqueous solutions by semi-IPN composite cryogels with anionically modified potato starch entrapped in PAAm matrix. Chem Eng J 234:211–222
Dragan ES, Cocarta AI (2016) Smart macroporous IPN hydrogels responsive to pH, temperature, and ionic strength: synthesis, characterization, and evaluation of controlled release of drugs. ACS Appl Mater Interfaces 8:12018–12030

Dragan ES, Dinu MV (2013) Design, synthesis and interaction with Cu^{2+} of ice templated composite hydrogels. Res J Chem Environ 17:4–10
Dragan ES, Dinu MV (2015a) Interpenetrating polymer network composite cryogels with tailored porous morphology and sorption properties. Methods Mol Biol 1286:239–252
Dragan ES, Dinu MV (2015b) Progress in polysaccharide/zeolites and polysaccharide hydrogel composite sorbents and their applications in removal of heavy metal ions and dyes. Curr Green Chem 2:342–353
Dragan ES, Cazacu M, Nistor A (2009) Ionic organic/inorganic materials. III. stimuli responsive hybrid hydrogels based on oligo(N, Ndimethylaminoethylmethacrylate) and chloroalkyl-functionalized siloxanes. J Polym Sci A Polym Chem 47:6801–6813
Dragan ES, Dinu MV, Apopei DF (2012a) Macroporous anionic interpenetrating polymer networks composite hydrogels and their interaction with methylene blue. Int J Chem 1:548–569
Dragan S, Lazar MM, Dinu MV, Doroftei F (2012b) Macroporous composite IPN hydrogels based on poly(acrylamide) and chitosan with tuned swelling and sorption of cationic dyes. Chem Eng J 204–206:198–209
Dragan S, Lazar MM, Dinu MV (2012c) Preparation and characterization of IPN composite hydrogels based on polyacrylamide and chitosan and their interaction with ionic dyes. Carbohydr Polym 88:270–281
Dragan ES, Cocarta AI, Gierszewska M (2016) Designing novel macroporous composite hydrogels based on methacrylic acid copolymers and chitosan and in vitro assessment of lysozyme controlled delivery. Colloids Surf B: Biointerfaces 139:33–41
Elviri L, Asadzadeh M, Cucinelli R, Bianchera A, Bettini R (2015) Macroporous chitosan hydrogels: effects of sulfur on the loading and release behaviour of amino acid-based compounds. Carbohydr Polym 132:50–58
Ertürk G, Mattiasson B (2014) Cryogels-versatile tools in bioseparation. J Chromatogr A 1357:24–35
Fathi A, Lee S, Zhong X, Hon N, Valtchev P, Dehghani F (2013) Fabrication of interpenetrating polymer network to enhance the biological activity of synthetic hydrogels. Polymer 54:5534–5542
Fernandes M, Gonalves IC, Nardecchia S, Amaral IF, Barbosa MA, Martins MCL (2013) Modulation of stability and mucoadhesive properties of chitosan microspheres for therapeutic gastric application. Int J Pharm 454:116–124
Giri TK, Thakur A, Alexander A, Badwaik H, Tripathi DK (2012) Modified chitosan hydrogels as drug delivery and tissue engineering systems: present status and applications. Acta Pharm Sin B 2:439–449
Guarino V, Alvarez-Perez MA, Borriello A, Napolitano T, Ambrosio L (2013) Conductive PANi/PEGDA macroporous hydrogels for nerve regeneration. Adv Healthc Mater 2:218–227
Guarino V, Galizia M, Alvarez-Perez M, Mensitieri G, Ambrosio L (2015) Improving surface and transport properties of macroporous hydrogels for bone regeneration. J Biomed Mater Res, Part A 103A:1095–1105
Gunko VM, Savina IN, Mikhalovsky SV (2013) Cryogels: morphological, structural and adsorption characterization. Adv Colloid Interf Sci 187–188:1–46
Gutiérrez MC, Ferrer ML, del Monte F (2008) Ice-templated materials: sophisticated structures exhibiting enhanced functionalities obtained after unidirectional freezing and ice-segregation-induced self-assembly. Chem Mater 20:634–648
Gyarmati B, Mészár EZ, Kiss L, Deli MA, László K, Szilágyi A (2015) Supermacroporous chemically cross-linked poly(aspartic acid) hydrogels. Acta Biomater 22:32–38
He H, Averick S, Mandal P, Ding H, Li S, Gelb J, Kotwal N, Merkle A, Litster S, Matyjaszewski K (2015) Multifunctional hydrogels with reversible 3D ordered macroporous structures. Adv Sci 2:1500069–1500075
Henderson TM, Ladewig K, Haylock DN, McLean KM, O'Connor AJ (2013) Cryogels for biomedical applications. J Mater Chem B 1:2682–2695
Hsieh WC, Chang CP, Lin SM (2007) Morphology and characterization of 3D micro-porous structured chitosan scaffolds for tissue engineering. Colloids Surf B: Biointerfaces 57:250–255

Hu XX, Shen H, Yang F, Bei JZ, Wang SG (2008) Preparation and cell affinity of microtubular orientation-structured PLGA(70/30) blood vessel scaffold. Biomaterials 29:3128–3136

Izadifar Z, Chen X, Kulyk W (2012) Strategic design and fabrication of engineered scaffolds for articular cartilage repair. J Funct Biomater 3:799–838

Jain E, Srivastava A, Kumar A (2009) Macroporous interpenetrating cryogel network of poly (acrylonitrile) and gelatin for biomedical applications. J Mater Sci Mater Med 20:S173–S179

Jain E, Damania A, Kumar Shakya A, Kumar A, Sarin SK, Kumar A (2015) Fabrication of macroporous cryogels as potential hepatocyte carriers for bioartificial liver support. Colloids Surf B: Biointerfaces 136:761–771

Ji C, Annabi N, Khademhosseini A, Dehghani F (2011) Fabrication of porous chitosan scaffolds for soft tissue engineering using dense gas CO_2. Acta Biomater 7:1653–1664

Jiang Y, Wu Y, Huo Y (2015) Thermo-responsive hydrogels with N-isopropylacrylamide/ acrylamide interpenetrating networks for controlled drug release. J Biomater Sci Polym Ed 26:917–930

Karadağ E, Kundakcı S (2015) Application of highly swollen novel biosorbent hydrogels in uptake of uranyl ions from aqueous solutions. Fibers Polym 16:2165–2176

Karpushkin E, Dusková-Smrcková M, Slouf M, Dusek K (2013) Rheology and porosity control of poly(2-hydroxyethyl methacrylate) hydrogels. Polymer 54:661–672

Kennedy S, Bencherif S, Norton D, Weinstock L, Mehta M, Mooney DJ (2014) Rapid and extensive collapse from electrically responsive macroporous hydrogels. Adv Healthc Mater 3:500–507

Khoshakhlagh P, Moore MJ (2015) Photoreactive interpenetrating network of hyaluronic acid and Puramatrix as a selectively tunable scaffold for neurite growth. Acta Biomater 16:23–34

Kim TG, Chung HJ, Park TG (2008) Macroporous and nanofibrous hyaluronic acid/collagen hybrid scaffold fabricated by concurrent electrospinning and deposition/leaching of salt particles. Acta Biomater 4:1611–1619

Kim JW, Taki K, Nagamine S, Ohshima M (2009) Preparation of porous poly(l-lactic acid) honeycomb monolith structure by phase separation and unidirectional freezing. Langmuir 25:5304–5312

Köhnke T, Elder T, Theliander H, Ragauskas AJ (2014) Ice templated and cross-linked xylan/ nanocrystalline cellulose hydrogels. Carbohydr Polym 100:24–30

Kuang J, Yuk KY, Huh KM (2011) Polysaccharide-based superporous hydrogels with fast swelling and superabsorbent properties. Carbohydr Polym 83:284–290

Kumar P, Ganure AL, Subudhi BB, Shukla S (2015) Preparation and characterization of pH-sensitive methylmethacrylate-g-starch/hydroxypropylated starch hydrogels: in vitro and in vivo study on release of esomeprazole magnesium. Drug Delivery Transl Res 5:243–256

Kuo Y-C, Wang C-C (2013) Guided differentiation of induced pluripotent stem cells into neuronal lineage in alginate–chitosan–gelatin hydrogels with surface neuron growth factor. Colloids Surf B: Biointerfaces 104:194–199

Lamprou A, Gavriilidoua A-F-M, Storti G, Soosa M, Morbidelli M (2015) Application of polymeric macroporous supports for temperature-responsive chromatography of pharmaceuticals. J Chromatogr A 1407:90–99

Lau HK, Kiick KL (2015) Opportunities for multicomponent hybrid hydrogels in biomedical applications. Biomacromolecules 16:28–42

Lazaridou A, Kritikopoulou K, Biliaderis CG (2015) Barley β-glucan cryogels as encapsulation carriers of proteins: impact of molecular size on thermo-mechanical and release properties. Bioact Carbohydr Dietary Fibre 6:99–108

Lee Y, Park S, Han SW, Lim TG, Koh W-G (2012) Preparation of photolithographically patterned inverse opal hydrogel microstructures and its application to protein patterning. Biosens Bioelectron 35:243–250

Lesny P, Prádny M, Jendelová P, Michálek J, Vacík J, Syková E (2006) Macroporous hydrogels based on 2-hydroxyethyl methacrylate. Part 4. growth of rat bone arrow stromal cells in three-dimensional hydrogels with positive and negative surface charges and in polyelectrolyte complexes. J Mater Sci Mater Med 17:829–833

Li Y, Guo H, Gan J, Zheng J, Zhang Y, Wu K, Lu M (2015a) Novel fast thermal-responsive poly (N-isopropylacrylamide) hydrogels with functional cyclodextrin interpenetrating polymer networks for controlled drug release. J Polym Res 22:78–91
Li Z-X, Lu M-G, Wu K, Zhang Y-F, Miao L, Li Y-W, Guo H-L, Zheng J (2015b) Temperature-responsiveness and sustained delivery properties of macroporous PEG-co-PNIPAAm-co-PCL hydrogels. Polym Eng Sci 55:223–230
Lips PAM, Velthoen IW, Dijkstra PJ, Wessling M, Feijen J (2005) Gas foaming of segmented poly(ester amide) films. Polymer 46:9396–9403
Liu Y, Cui Y (2011) Preparation and properties of temperature-sensitive soy protein/poly (N-isopropylacrylamide) interpenetrating polymer network hydrogels. Polym Int 60:1117–1122
Liu L, Xie JP, Li YJ, Zhang Q, Yao JM (2016) Three-dimensional macroporous cellulose-based bioadsorbents for efficient removal of nickel ions from aqueous solution. Cellulose 23:723–736
Lozinsky VI (2014) A brief history of polymeric cryogels. In: Okay O (ed) Advances in polymer science. Springer International Publishing, New York, pp 1–48
Mastropietro DJ, Omidian H, Park K (2012) Drug delivery applications for superporous hydrogels. Expert Opin Drug Deliv 9:71–89
Nakagawa K, Sowasod N, Tanthapanichakoon W, Charinpanitkul T (2013) Hydrogel based oil encapsulation for controlled release of curcumin by using a ternary system of chitosan, kappa-carrageenan, and carboxymethylcellulose sodium salt. Food Sci Technol 54:600–605
Nasim Annabi MS, Nichol JW, Zhong X, Ji C, Sandeep Koshy MBE, Khademhosseini A, Dehghani F (2010) Controlling the porosity and microarchitecture of hydrogels for tissue engineering. Tissue Eng B 16:371–383
Nieto-Suárez M, López-Quintel MA, Lazzari M (2016) Preparation and characterization of crosslinked chitosan/gelatin scaffolds by ice segregation induced self-assembly. Carbohydr Polym 141:175–183
Oh BHL, Bismarck Alexander, Chan-Park Mary B (2015) Injectable, interconnected, high-porosity macroporous biocompatible gelatin scaffolds made by surfactant-free emulsion templating. Macromol Rapid Commun 36:364–372
Okay O (2000) Macroporous copolymer networks. Prog Polym Sci 25:711–779
Orakdogen N, Karacan P, Okay O (2011) Macroporous, responsive DNA cryogel beads. React Funct Polym 71:782–790
Owen SC, Fisher SA, Tam RY, Nimmo CM, Shoichet MS (2013) Hyaluronic acid click hydrogels emulate the extracellular matrix. Langmuir 29:7393–7400
Ozmen MM, Dinu MV, Dragan ES, Okay O (2007) Preparation of macroporous acrylamide-based hydrogels: cryogelation under isothermal conditions. J Macromol Sci Part A Pure Appl Chem 44:1195–1202
Oztoprak Z, Hekimoglu T, Karakutuk I, Tuncaboylu DC, Okay O (2014) Porous rubber cryogels: effect of the gel preparation temperature. Polym Bull 71:1983–1999
Park H, Kim D (2006) Swelling and mechanical properties of glycol chitosan/poly(vinyl alcohol) IPN-type superporous hydrogels. J Biomed Mater Res 78A:662–667
Park S, Park KM (2016) Engineered polymeric hydrogels for 3D tissue models. Polymers 8:23–36
Perju MM, Dinu MV, Dragan ES (2012) Sorption of methylene blue onto ionic composite hydrogels based on polyacrylamide and dextran sulfate: kinetics, isotherms, and thermodynamics. Sep Sci Technol 47:1322–1333
Phull MK, Eydmann T, Roxburgh J, Sharpe JR, Lawrence-Watt DJ, Phillips G, Martin Y (2013) Novel macro-microporous gelatin scaffold fabricated by particulate leaching for soft tissue reconstruction with adipose-derived stem cells. J Mater Sci Mater Med 24:461–467
Přádný M, Šlouf M, Martinová L, Michálek J (2010) Macroporous hydrogels based on 2-hydroxyethyl methacrylate. Part 7: methods of preparation and comparison of resulting physical properties. e-Polymers 043:1–12
Přádný M, Dušková-Smrčková M, Dušek K, Janoušková O, Sadakbayeva Z, Šlouf M, Michálek J (2014) Macroporous 2-hydroxyethyl methacrylate hydrogels of dual porosity for cell

cultivation: morphology, swelling, permeability, and mechanical behavior. J Polym Res 21:567–579
Qi X, Hu X, Wei W, Yu H, Li J, Zhang J, Dong W (2015) Investigation of Salecan/poly(vinyl alcohol) hydrogels prepared by freeze/thaw method. Carbohydr Polym 118:60–69
Reneker DH, Chun I (1996) Nanometer diameter fibers of polymer, produced by electrospinning. Nanotechnology 7:216–223
Ricciardi R, D'Errico G, Auriemma F, Ducouret G, Tedeschi AM, De Rosa C, Laupretre F, Lafuma F (2005) Short time dynamics of solvent molecules and supramolecular organization of poly (vinyl alcohol) hydrogels obtained by freeze-thaw techniques. Macromolecules 38:6629–6639
Rivero RE, Alustiza F, Rodríguez N, Bosch P, Miras MC, Rivarola CR, Barbero CA (2015) Effect of functional groups on physicochemical and mechanical behavior of biocompatible macroporous hydrogels. React Funct Polym 97:77–85
Rogina A (2014) Electrospinning process: versatile preparation method for biodegradable and natural polymers and biocomposite systems applied in tissue engineering and drug delivery. Appl Surf Sci 296:221–230
Sahiner N, Demirci S, Sahiner M, Yilmaz S, Al-Lohedan H (2015) The use of superporous p (3-acrylamidopropyl)trimethyl ammonium chloride cryogels for removal of toxic arsenate anions. J Environ Manag 152:66–74
Salerno A, Borzacchiello R, Netti PA (2011) Pore structure and swelling behavior of porous hydrogels prepared via a thermal reverse-casting technique. J Appl Polym Sci 122:3651–3660
Sato R, Noma R, Tokuyama H (2015) Preparation of macroporous poly(N-isopropylacrylamide) hydrogels using a suspension–gelation method. Eur Polym J 66:91–97
Savina IN, Ingavle GC, Cundy AB, Mikhalovsky SV (2016) A simple method for the production of large volume 3D macroporous hydrogels for advanced biotechnological, medical and environmental applications. Sci Rep 6:21154–21162
Sedlačík T, Proks V, Šlouf M, Dušková-Smrčková M, Ha Studenovská, Rypáček F (2015) Macroporous biodegradable cryogels of synthetic poly(α-amino acids). Biomacromolecules 16:3455–3465
Shen X, Shamshina JL, Berton P, Gurauc G, Rogers RD (2016) Hydrogels based on cellulose and chitin: fabrication, properties, and applications. Green Chem 18:53–75
Sokic S, Christenson M, Larson J, Papavasiliou G (2014) In situ generation of cell-laden porous MMP-sensitive PEGDA hydrogels by gelatin leaching. Macromol Biosci 14:731–739
Sood N, Bhardwaj A, Mehta S, Mehta A (2016) Stimuli-responsive hydrogels in drug delivery and tissue engineering. Drug Deliv 23:748–770
Strom A, Larsson A, Okay O (2015) Preparation and physical properties of hyaluronic acid-based cryogels. J Appl Polym Sci 132:42194–42204
Studenovska H, Slouf M, Rypacek F (2008) Poly(HEMA) hydrogels with controlled pore architecture for tissue regeneration applications. J Mater Sci Mater Med 19:615–621
Su Y, Zhao Q, Sun J, Wu J (2012) Synthesis and characterization of biodegradable macroporous cryogels crosslinked by chitosan oligosaccharide-graft-acrylic acid. Soft Matter 8:4382–4389
Suhag D, Bhatia R, Das S, Shakeel A, Ghosh A, Singh A, Sinha OP, Chakrabarti S, Mukherjee M (2015) Physically cross-linked pH-responsive hydrogels with tunable formulations for controlled drug delivery. RSC Adv 5:53963–53972
Sun X-F, Jing Z, Wang H, Liu Y (2015a) Physical–chemical properties of xylan/PAAc magnetic semi-interpenetrating network hydrogel. Polym Compos 36:2317–2325
Sun X-F, Liu B, Jing Z, Wang H (2015b) Preparation and adsorption property of xylan/poly (acrylic acid) magnetic nanocomposite hydrogel adsorbent. Carbohydr Polym 118:16–23
Tavsanli B, Can V, Okay O (2015) Mechanically strong triple network hydrogels based on hyaluronan and poly(N, N-dimethylacrylamide). Soft Matter 11:8517–8524
Thakur VK, Thakur MK (2014a) Recent trends in hydrogels based on psyllium polysaccharide: a review. J Clean Prod 82:1–15
Thakur VK, Thakur MK (2014b) Recent advances in graft copolymerization and applications of chitosan: a review. ACS Sustain Chem Eng 2(12):2637–2652

Thakur VK, Thakur MK (2015) Recent advances in green hydrogels from lignin: a review. Int J Biol Macromol 72:834–847

Thein-Han WW, Misra RDK (2009) Biomimetic chitosan–nanohydroxyapatite composite scaffolds for bone tissue engineering. Acta Biomater 5:1182–1197

Topuz F, Okay O (2009) Macroporous hydrogel beads of high toughness and superfast responsivity. React Funct Polym 69:273–280

Tripathi A, Kumar A (2011) Multi-featured macroporous agarose–alginate cryogel: synthesis and characterization for bioengineering applications. Macromol Biosci 11:22–35

Tripathi A, Kathuria N, Kumar A (2009) Elastic and macroporous agarose–gelatin cryogels with isotropic and anisotropic porosity for tissue engineering. J Biomed Mater Res 90A:680–694

Tripathi A, Vishnoi T, Singh D, Kumar A (2013) Modulated crosslinking of macroporous polymeric cryogel affects in vitro cell adhesion and growth. Macromol Biosci 13:838–850

Truong VX, Barker IA, Tan M, Mespouille L, Dubois P, Dove AP (2013) Preparation of in situ-forming poly(5-methyl-5-allyloxycarbonyl-1,3-dioxan-2-one)-poly(ethylene glycol) hydrogels with tuneable swelling, mechanical strength and degradability. J Mater Chem B 1:221–229

Varaprasad K, Narayana Reddy N, Ravindra S, Vimala K, Mohana Raju K (2011) Synthesis and characterizations of macroporous poly(acrylamide-2-acrylamido-2-methyl-1-propanesulfonic acid) hydrogels for in vitro drug release of ranitidine hydrochloride. Int J Polym Mater Polym Biomater 60:490–503

Vishal Gupta N, Shivakumar HG (2010) Preparation and characterization of superporous hydrogels as gastroretentive drug delivery system for rosiglitazone maleate. DARU 18:200–210

Visser J, Melchels FPW, Jeon JE, van Bussel EM, Kimpton LS, Byrne HM, Dhert WJA, Dalton PD, Hutmacher DW, Malda J (2015) Reinforcement of hydrogels using three-dimensionally printed microfibers. Nature Commun 6:6933–6943

Wang J, Li Z (2015) Enhanced selective removal of Cu(II) from aqueous solution by novel polyethylenimine-functionalized ion imprinted hydrogel: behaviors and mechanisms. J Hazard Mater 300:18–28

Wei W, Qi X, Liu Y, Li J, Hu X, Zuo G, Zhang J, Dong W (2015) Synthesis and characterization of a novel pH-thermo dual responsive hydrogel based on salecan and poly(N, N-diethylacrylamide-co-methacrylic acid). Colloids Surf B: Biointerfaces 136:1182–1192

Welzel PB, Grimmer M, Renneberg C, Naujox L, Zschoche S, Freudenberg U, Werner C (2012) Macroporous starPEG-heparin cryogels. Biomacromolecules 13:2349–2358

Wu X, Liu Y, Li X, Wen P, Zhang Y, Long Y, Wang X, Guo Y, Xing F, Gao J (2010) Preparation of aligned porous gelatin scaffolds by unidirectional freeze-drying method. Acta Biomater 6:1167–1177

Wu J, Zhao Q, Sun J, Zhou Q (2012) Preparation of poly(ethylene glycol) aligned porous cryogels using a unidirectional freezing technique. Soft Matter 8:3620–3626

Wu Y, Chen YX, Yan J, Yang S, Dong P, Soman P (2015) Fabrication of conductive polyaniline hydrogel using porogen leaching and projection microstereolithography. J Mater Chem B 3:5352–5360

Wu C, Wang D, Wu H, Dan Y (2016) Synthesis and characterization of macroporous sodium alginate-g-poly(AA-co-DMAPMA) hydrogel. Polym Bull. https://doi.org/10.1007/s00289-016-1653-z

Xu S, Deng L, Zhang J, Yin L, Dong A (2016) Composites of electrospun-fibers and hydrogels: a potential solution to current challenges in biological and biomedical field. J Biomed Mater Res Part B: Appl Biomater 104B:640–656

Yamane S, Iwasaki N, Kasahara Y, Harada K, Majima T, Monde K, Nishimura S, Minami A (2007) Effect of pore size on in vitro cartilage formation using chitosan-based hyaluronic acid hybrid polymer fibers. J Biomed Mater Res Part A 81A:586–593

Yi W, Wu H, Wang H, Du Q (2016) Interconnectivity of macroporous hydrogels prepared via grapheme oxide-stabilized Pickering high internal phase emulsions. Langmuir 32:982–990

Yue Z, Wen F, Gao S, Ang MY, Pallathadka PK, Liu L, Yu H (2010) Preparation of three-dimensional interconnected macroporous cellulosic hydrogels for soft tissue engineering. Biomaterials 31:8141–8152

Zhang X, Cao C, Ma X, Li Y (2012) Optimization of macroporous 3-D silk fibroin scaffolds by salt-leaching procedure in organic solvent-free conditions. J Mater Sci Mater Med 23:315–324

Zhang Y, Yana W, Sun Z, Pan C, Mi X, Zhao G, Gao J (2015) Fabrication of porous zeolite/chitosan monoliths and their applications for drug release and metal ions adsorption. Carbohydr Polym 117:657–665

Zhang L, Zeng Y, Cheng Z (2016) Removal of heavy metal ions using chitosan and modified chitosan: a review. J Mol Liq 214:175–191

Zhao F, Yao D, Guo R, Deng L, Dong A, Zhang J (2015) Composites of polymer hydrogels and nanoparticulate systems for biomedical and pharmaceutical applications. Nanomaterials 5:2054–2130

Zhong X, Ji C, Chan AKL, Kazarian SG, Ruys A, Dehghani F (2011) Fabrication of chitosan/poly (e-caprolactone) composite hydrogels for tissue engineering applications. J Mater Sci Mater Med 22:279–288

Zhu Q, Li Z (2015) Hydrogel-supported nanosized hydrous manganese dioxide: synthesis, characterization, and adsorption behavior study for Pb^{2+}, Cu^{2+}, Cd^{2+} and Ni^{2+} removal from water. Chem Eng J 281:69–80

Zhu J, Marchant RE (2011) Design properties of hydrogel tissue-engineering scaffolds. Expert Rev Med Dev 8:607–626

Zhu W, Song H, Du K, Zeng H, Yao S (2013) Rapid removal of Cr(VI) ions from aqueous solutions by the macroporous poly(N, N-dimethylamino ethyl methacrylate) hydrogels. J Appl Polym Sci 128:2729–2735

Chapter 4
Hydrogel-Based Strategies for Stem Cell Therapy

Shuaiqiang Zhang, Yan Nie, Hongyan Tao and Zongjin Li

Abstract Stem cells from different sources provide considerable expectation for applications in tissue engineering and regenerative medicine because of their ability to proliferate and differentiate into functional cells. However, poor cell engraftment and survival after transplantation are key factors limiting the current stem cell-based therapy. The utilization of engineered microenvironments with synthetic biomaterials has been progressively successful in controlling the transplanted cells fate by imitating the native stem cell niche. Recently, synthetic 3D extracellular matrices (ECMs) have been extensively explored as scaffolds for tissue regeneration by emulating the components of natural stem cell niche to minimize implanted cell death. For their tunable tissue-like properties, biodegradability, and biocompatibility in all different forms of biomaterials, hydrogels are most normally used as substrates and scaffolds, which serve as a promising cell delivery vehicle to illustrate stem cell biology. Here, we will focus on recent advances in generating hydrogel, as well as the application of hydrogel for stem cell-based therapy. Finally, hydrogel-based controlled-release strategies for improving therapeutic efficiency of stem cells will be discussed.

Keywords Hydrogel · Stem cell · Stem cell therapy · Extracellular matrix
Microenvironments

S. Zhang · Y. Nie · H. Tao · Z. Li (✉)
Department of Pathophysiology, Nankai University School of Medicine,
94 Weijin Road, Tianjin 300071, China
e-mail: zongjinli@nankai.edu.cn

Z. Li
The Key Laboratory of Bioactive Materials, Ministry of Education,
Nankai University, College of Life Sciences, Tianjin 300071, China

V. K. Thakur and M. K. Thakur (eds.), *Hydrogels*, Gels Horizons: From Science to Smart Materials, https://doi.org/10.1007/978-981-10-6077-9_4

1 Introduction

Stem cells represent a population of cells with the potential applications including regenerative medicine and tissue engineering owing to their proliferation and differentiation ability (Tabar and Studer 2014). Nowadays, transplantation of embryonic stem cells or adult stem cell, such as mesenchymal stem cells (MSCs), has constituted a revolution in regenerative medicine. However, poor cell engraftment and survival after transplantation in injured area limit the therapeutic efficacy of cellular therapy for tissue regeneration (He et al. 2015; Li et al. 2009b). Recent studies demonstrated that co-transplantation of stem cells with bioactive molecules that mimic the natural stem cell niches could prevent apoptosis of engrafted cell triggered by anoikis (He et al. 2015; Li et al. 2009a; Yao et al. 2015).

The most versatile carrier system for cellular delivery in stem cell therapy, hydrogel is polymers cross-linked (Tian et al. 2012) via ionic interactions, chemical bonds, hydrogen bonds, hydrophobic interactions, or physical bonds (Phadke et al. 2012). These materials are three-dimensional networks (Dragan 2014) with a strong power of liquid absorption (Laftah et al. 2011), up to several thousand percent in volume while swell readily without dissolving (Srivastava et al. 2016). The high hydrophilicity is mainly associated with the hydrophilic moieties of hydrogels (Qiu and Park 2012) include the amino, amide, carboxyl, and hydroxyl groups distributed along the backbone of polymer chains (Komeri et al. 2015). They are soft and elastic with desirable biocompatibility in the swollen state (Das 2013), which are resembling the living tissues (Elluru et al. 2013).

Hydrogel can be made from either synthetic polymers or natural polymers, which was first introduced as useful biomaterials by Wichterle and Lim (1960) due to their good biological compatibilities, suitable degradation profiles, and tissue-like properties (Wichterle and Lim 1960). In the past decades, the research in the field of hydrogels has progressed dramatically, particularly in tissue engineering, as they could provide a conductive environment to support the transplanted cells (Guvendiren and Burdick 2013). In this chapter, we will provide a systematic review of hydrogels with varied sources, structures, functions, as well as current knowledge and involvement of hydrogels in stem cell therapy. Moreover, we will provide insights into the enhancement of therapeutic efficacy via co-transplantation of stem cells and hydrogel for tissue regeneration.

2 Classifications of Hydrogels

Hydrogels possess some degrees of flexibility for their large water content, which are very similar to natural tissue (Zhu and Marchant 2011). The water absorption capacity of the hydrogels is facilitated by the hydrophilic functional groups attached to the polymer backbone (Bardajee and Hooshyar 2013), while their resistance to

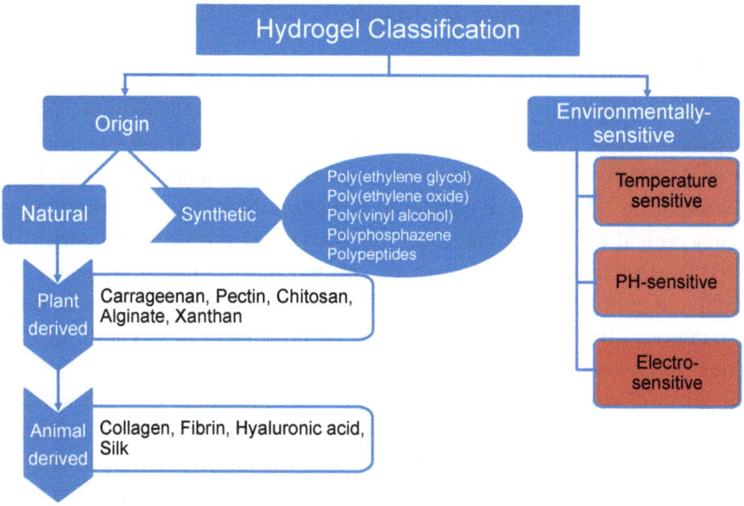

Fig. 1 Principle scheme shows the most common classification of hydrogels

dissolution arises from cross-links between network chains. The classification of hydrogels could be based on their sources and properties (Laftah et al. 2011), and detailed classification based on their origin, the environment response (Fig. 1).

2.1 *According to the Origin of Hydrogel*

On the basis of their compositions, hydrogels can be divided into two groups, natural polymer hydrogels and synthetic polymer hydrogels (Cho et al. 2015; Thakur and Thakur 2015). Natural hydrogels consist of natural polymers including polypeptides (Bakota et al. 2011), polynucleotides (Thakur and Thakur 2014a; Ye et al. 2011), and polysaccharides (Anderson et al. 2011). Both animal and plant products can be the sources of natural polymer hydrogels (Huebsch et al. 2010; Liu et al. 2015; Mano et al. 2007). Majorities of synthetic hydrogels are formed by polymerization of ethylene or ethylene-activated monomers (Khetan et al. 2013) and the different hydrophilicity of monomers as well as cross-linking density in synthetic hydrogels show varied equilibrium swelling values (Wang et al. 2012). During the process of hydrogels synthesis, monomers with specific functions can be introduced to the backbone of hydrogel (Thakur et al. 2016; Xia et al. 2015).

2.1.1 Plant-Derived Polymer Hydrogels

Carrageenan

Carrageenan is a natural family of linear polysaccharides (Sipahigil and Dortunc 2001). It is a large, highly flexible molecule with 25,000 galactose derivatives (Roger et al. 2012). The carrageenan has a good biodegradability and high capacity of water retention. Therefore, it can serve as a good thickening and stabilizing agent in pharmaceutical (Sipahigil and Dortunc 2001) and cosmetic industries (Khetan et al. 2013). The relatively low cost of carrageenan promotes its usage in environmental and commercial applications (Popa et al. 2011). On the basis of different sulfation degrees (between 15 and 40%), the carrageenan can be classified into three categories as λ-carrageenan (three-sulfate), ι-carrageenan (mono-sulfate), and κ-carrageenan(di-sulfate) (Ye et al. 2011). The first one is a thickener agent, whereas the others are gel-forming systems (Coviello et al. 2007). In addition, recent studies have shown the potential of carrageenan to emulate the charged proteins in native ECM due to its abundant sulfate groups and biocompatibility (Liu et al. 2014b).

Pectin

Pectin is a structural heteropolysaccharide constitutes cell walls of plants, which could be isolated from several types of fruits, such as citrus peel and apple pomace. Pectin has a molecular weight of typically 60–130,000 g/mol, varying with origin and extraction conditions while rhamnogalacturonan-I (RG-I), homogalacturonan (HGA), and rhamnogalacturonan-II (RG-II) are the main polysaccharide domains of it (Anderson et al. 2011; Leipzig et al. 2011). Currently, pectin has been intensively used as gelling or thickening agent in food industry and excipient in pharmaceutical industry due to its non-toxicity, low cost of production, and gelling activity (Mogosanu and Grumezescu 2014). Consequently, besides its applications in food industry and pharmacy, the therapeutic potential of hydrogels based on pectin has been explored, such as their antibacterial abilities in wound healing and their mineral nucleation effects in bone cells culture.

Chitosan

Chitosan is a linear polysaccharide of *N*-acetylglucosamine and glucosamine, which isolated from natural chitin by partial deacetylation (Vermonden et al. 2012; Thakur and Thakur 2014b). The structure of chitosan is similar to glycosaminoglycans (GAGs) (Liu et al. 2012), which is a bioactive material with great biocompatibility and biodegradability but no allergenicity (Liu et al. 2012). Previous research has already demonstrated that chitosan hydrogel was an effective cell carrier for the delivery of stem cells into target tissues. Currently, chitosan has been diffusely used as the biological scaffold materials in tissue engineering. The application of chitosan hydrogels could enhance the stem cells' engraftment and survival and promote neovascularization (Busilacchi et al. 2013).

Alginate

Alginates are a natural anionic polysaccharide distributed widely in the cell walls of brown seaweed, which is consisting of (1-4)-linked β-D-mannuronic acid (M) and α-L-guluronic acid (G) (Glotzbach et al. 2012). Alginates have been widely used in tissue regeneration for its low toxicity, favorable biocompatibility, suitable microstructure, and growth factors releasing (Lutolf et al. 2009; Lee and Mooney 2012). Besides, alginate has a relatively mild gelation process by addition of divalent cations such as Ca^{2+} (Gombotz and Wee 2012). Several studies have been carried out using alginate for molecules delivery (Vermonden et al. 2012) and other biomedical applications. For example, the simple sulfonation of uronic acids in alginate that specifically binds heparin-binding proteins (HBPs) with several growth factors has been developed to obtain a sustained release of target proteins and enhanced therapeutic efficiency in a mouse hindlimb ischemia model (Ruvinov et al. 2010).

Xanthan

Xanthan is a natural polysaccharide produced predominantly by *Xanthomonas campestris* (Snippert and Clevers 2011). As an important industrial biopolymer, xanthan is largely used as food thickening agent, cosmetics, and drilling fluids (Bueno et al. 2013). It has a cellulosic backbone formed by successive additions of D-glucose-1-phosphate and D-glucose from 2 mol of UDP D-glucose (Hager and Arendt 2013). For every alternate glucose, there is a side chain consisting of β-D-mannose-(1,4)-β-D-glucuronic acid–(1,2)-α-D-mannose (Harding et al. 2011). Since the discovery of xanthan, it has been widely investigated and utilized to increase the drug delivery rate as tablet excipient (Mikac et al. 2013), but not much results have been reported concerning its use in a sustained drug release (Snippert and Clevers 2011).

2.1.2 Animal-Derived Polymer Hydrogels

Collagen

Collagen is a family of around 20 related proteins and is the main structural protein in the animal body (Trombino et al. 2012). Type I, II, and III are majority of the internal structure of fibrillary collagen, which is the subfamily of collagen (Macaya and Spector 2012). Type I collagen is made up of an uninterrupted triple helix of approximately 300 nm in length and 1.5 nm in diameter flanked by short extra-helical telopeptides. Type II collagen contains the amino acids found in the framework of human cartilage, which is the primary form of collagen. Besides, the length of triple helix in type III collagen is slightly longer with slightly short telopeptides compared to type I (Cameron et al. 2002). Because the collagen derived from animal origin is biocompatible, biodegradable, and nontoxic, it is suitable for tissue engineering utilization (Fullana and Wnek 2012). It has been confirmed that collagen is very useful as a dressing for treatment of many types of wounds, such as pressure sores, donor sites, and decubitus ulcers as well as for in vitro test systems (Mitra et al. 2011). In brief, collagen has been applied in multiple biomedical applications including drug delivery, controlled release of therapeutic agents (Murali et al. 2011), substrate for cell culture (Van Vlierberghe et al. 2011), and gene delivery (Kempf et al. 2011).

Fibrin

Fibrin is a natural polymer formed in the living body and degraded in a highly controlled manner (Zhou and Xu 2011). It is a fibrous, non-globular protein (Feng et al. 2013). Meanwhile, fibrin is biodegradable and biocompatible (Moloney et al. 2015), making it a suitable temporary scaffold for tissue engineering. Thus, many researchers have paid attention to using fibrin hydrogels as cell-supporting scaffolds for tissue engineering in wound healing and tissues repair (Kulkarni et al. 2014). As delivery systems, preclinical studies (Weisel and Litvinov 2013) have illustrated the capacity of fibrin hydrogel loaded human umbilical cord mesenchymal stem cells (hUC-MSCs) to create macropores for enhancing cell viability (Liu et al. 2013).

Hyaluronic Acid

Hyaluronic acid (HA) is a high molecular weight, nonsulfated glycosaminoglycan (Choi et al. 2011), which consists of alternating units of a repeating disaccharide, β-1,4-D-glucuronic acid–β-1, 3-*N*-acetyl-D-glucosamine (Van Vlierberghe et al. 2011). It is a widely and highly hydrated polyanionic, which occurs in molecular weights ranging from 100 kDa in serum to 8000 kDa in the vitreous (Burdick and Prestwich 2011). It can be made into hydrogels via covalent cross-linking with different kinds of hydrazide derivatives and radical polymerization of glycidyl methacrylate (Rayahin et al. 2015). HA is distributed widely throughout epithelial, connective, cartilage, and neural tissues. Moreover, its structural and biological properties mediate its contribution significantly to cell proliferation, migration, and matrix organization (Prestwich 2011), which made HA as potential candidate for tissue engineering applications for such as wound healing, soft tissue augmentation, artificial skin, and facial intradermal implants (Kim et al. 2011). In clinical, HA has been utilized in several applications, including as a drug delivery system (Oh et al. 2011) in osteoarthritis for viscosupplementation, and as a dermal filler (Yang et al. 2011).

Silk

Silk fibroin (SF) is a fibrous protein synthesized by silkworms and spiders (Neffe et al. 2013). SF fibers are about 10–25 μm in diameter (de Moraes and Beppu 2013), of which primary structure consists of glycine, serine, and alanine (Van Vlierberghe et al. 2011). SF exhibits excellent environmental stability, biocompatibility, biodegradability, and mechanical properties (Lu et al. 2011). SF has been produced in various forms such as films, hydrogels, regenerated fiber, nanofiber, tubes, and three-dimensional scaffolds in the recent years (Teramoto and Kojima 2014). Meanwhile, it has triggered comprehensive studies of silk-based materials for biomedical applications as well (Liu et al. 2014c). For example, SF is used in skin grafts, blood vessels, artificial ligaments, and regenerative medicine (Mogosanu and Grumezescu 2014).

2.1.3 Hydrogels Based on Synthetic Polymers

Poly (Ethylene Glycol)

Poly (ethylene glycol) (PEG) is only a simple molecule, with the general structures: HO–(CH_2CH_2On)–CH_2CH_2–OH (Roberts et al. 2012). The most common form of PEG is a linear or branched polyether terminated with hydroxyl groups, neutral polyether (Banerjee et al. 2012). In addition, there are different forms of PEG with different molecular weight. For example, PEGs are colorless, viscous liquids when the molecular weights are less than 1000, but PEGs are waxy, white solids on higher molecular weight. PEG is synthesized by anionic ring opening polymerization of ethylene oxide initiated by nucleophilic attack of a hydroxide ion on the epoxide ring (Jaiswal et al. 2012). The commonly used molecular weights of PEG in regenerative applications range from a few hundreds to approximately 20,000 (Obermeier et al. 2011). Although PEG polymers with low molecular weight have been shown to induce toxicity after degradation in vivo, the nontoxic PEGs of higher molecular weight have been widely used in foods, cosmetics, and pharmaceuticals for many years (Hoffman 2012).

Poly (Ethylene Oxide)

Poly (ethylene oxide, PEO) is a crystalline, thermoplastic polymer (Ibrahim and Johan 2012) with the repeat units represented by –(C_2H_4O)–. PEO is nonionic and water soluble, and has been widely utilized for stabilizing colloids in formulating pharmaceuticals and cosmetics. Laurenco first reported it in 1860 by condensation of ethylene glycol and ethylene dibromide. Owing to its unexceptionable biodegradability and low toxicity, the potential application of PEO in regeneration medicine has drawn attention of scientific fields (Ma et al. 2012). PEO has many interesting properties as will be seen in the following examples. PEO can dissolve in quantities of common organic solvents such as acetonitrile, alcohols, chloroform, ethylene dichloride, and amides (Weber et al. 2012). It also can lead to an elastic water compound production, which will reduce frictional drag of the flowing water markedly. Usually, commercial polymers are supplied in one or a few molecular weight grades. However, PEO is commercially available in a greatly wide range of molecular weights from diethylene glycol, ethylene glycol to polymers and its molecular weights can be many times greater than a million (Calderón and Zhao 2014).

Poly (Vinyl Alcohol)

Poly (vinyl alcohol, PVA) is a semicrystalline, biodegradable, and water-soluble polymer (Mallakpour and Barati 2011), which is universally produced from poly (vinyl acetate) by alcoholysis, hydrolysis, or aminolysis (Li et al. 2012). The extent of hydrolysis and molecular weight can affect the hydrophilicity and solubility of this polymer (Jannesari et al. 2011). PVA hydrogel is a renowned and biodegradable polymer, which has been used in different biomedical areas (Morimune et al. 2011). Owing to the flexibility, biocompatibility, and swelling capability of PVA, many studies have been widely reported as a wound dressing (Wang et al. 2011; Zhang et al. 2012).

Polyphosphazene

Polyphosphazenes are a kind of tailorable polymers including a wide range of hybrid with controllable physicochemical properties and synthetic flexibility (Xu et al. 2014). Structurally, they are hybrid inorganic–organic polymers with a backbone of alternating nitrogen and phosphorus atoms, usually two organic side groups attached to each phosphorus atom (Deng et al. 2010). Polyphosphazenes are synthesized through two-step reaction processes, which begin with cyclic trimer hexachlorocyclotriphosphazene ($[NPCl_2]_3$) (Peach et al. 2012). Recently, polyphosphazenes have been a promising class of polymers and relatively novel in biomaterials field (Wang et al. 2013). They open up broad prospects for biological, biochemical, and biomedical applications (Huang et al. 2011) for their prominent biodegradability and biocompatibility (Zheng et al. 2011). Their especially flexible inorganic backbone is formed by alternated nitrogen atoms and phosphorus (Allcock 2012), which are the two key elements for biocompatibility. Heretofore, more than 700 polyphosphazenes polymers have been reported for tissue engineering applications and drug delivery (Silva Nykänen et al. 2011).

Polypeptides

A polypeptide consists of at least ten amino acids that are joined together by peptide bonds (Huang and Heise 2013). Besides proteinogenic amino acids, numerous classes of other amino acids are naturally incorporated in proteins (Chen et al. 2011). Therefore, polypeptides can be obtained from natural sources as well from synthetic ways (Cheng and Deming 2012). Polypeptides are generally prepared by using a starting monomer *N*-carboxyanhydride, and large amounts of polypeptides or co-polypeptides can be synthesized from various combinations of amino acids. Short polypeptides are synthesized by adding amino acids stepwise, whereas longer polypeptides are obtained from linking short ones. Owing to their various biomedical applications, polypeptides have received increasing attention in stem cell therapy, antibiotics, drug delivery, and tissue engineering (Deming 2002). Furthermore, polypeptides are unique in their ability to self-assemble into ordered structures, which is especially valuable when complex functions or hierarchical architectures are required for a special application (Lu et al. 2014).

2.2 *Environmentally-Sensitive Polymer Hydrogels*

Environment-sensitive hydrogels are also called “smart” or “intelligent” hydrogels for their response to environmental changes (Laftah et al. 2011). The smart hydrogels can be induced lots of responses by many physical and chemical stimuli. The chemical stimuli include pH, ions, and specific molecular, while the physical stimuli include electric fields, temperature, light, sound, and pressure (Kost and Langer 2012). Therefore, intelligent hydrogels are classified according to their sensitivity to temperature, electric, light, sound, and pH. Considering the importance of thermo-sensitive, pH response, and electric smart hydrogels, they will be discussed in detail below.

2.2.1 Temperature-Sensitive Hydrogels

In the wake of the temperature increases, most polymer hydrogels can increase their water solubility. Polymer hydrogels with lower critical solution temperature (LCST), however, decrease their water solubility as the temperature increases. All these thermo-sensitive hydrogels can be classified into three categories, as follows.

Positively Thermo-sensitive

Interpenetrating polymer network (IPN) is a kind of polymer comprising two or more networks based on polyacrylic acid and polyacrylamide or poly (AAm-co-BMA) (Ichikawa and Fukumori 2000). Hydrogels synthesized by IPN swell at high temperature but shrink at low temperature, which called positively thermo-sensitive hydrogels. The swelling of those hydrogels was reversible, responding to stepwise temperature changes (Priya James et al. 2014). The formation of complex structure by hydrogen bonds in these types of hydrogel leads to this character of hydrogels synthesized by IPN (Liu et al. 2014a).

Negatively Thermo-sensitive

Negatively thermo-sensitive hydrogel includes the hydrogel that based on interpenetrating polymer network (IPN) on the basis of poly (*N*-isopropylacrylamide, PNIPAAm) and poly (tetramethylene glycol, PTMEG) and poly (NIPAAm-co-BMA) (Korotych et al. 2013). They have a critical parameter called low critical solution temperature (LCST), which means the hydrogels will show swelling behavior at lower LCST and shrink when the temperature increases above LCST (Chu et al. 2013). When the percentage ratio of hydrophobic to hydrophilic contents is changed, the low critical solution temperature will be also changed (Samchenko et al. 2011).

Thermo-reversible

This kind of hydrogel has the similar structure and contents as the negative and positive temperature hydrogels (Thakur and Kessler 2015). However, compared with this former two types of thermo-sensitive hydrogels, there exist differences in their types of bonds as the polymer chains are not covalently cross-linked (Chen and Cheng 2009). Due to its ease of control and thermo-reversible behavior, thermo-sensitive hydrogels have been widely used in a variety of areas. For example, thermo-reversible gelling systems have been used in drug delivery, specifically for instillation into the body cavities such as nose, eyes, and ears. Currently, FDA and EPA have approved some of them for applications in pharmaceutical ingredients, additives, and agricultural products (Sharma et al. 2014).

2.2.2 pH-Sensitive Hydrogels

pH-sensitive hydrogels are the most widely used hydrogels among the various environment-sensitive hydrogels for their practical applications. pH-sensitive hydrogels are materials that respond to pH changes in the surrounding medium because they exhibit swelling and de-swelling according to pH of the environment (Hu et al. 2011). The swelling behavior occurs due to changes in the hydrophobic–hydrophilic nature of chains or due to hydrogen bonds and the complexation of inter- and intramolecular or electrostatic repulsion (Milosavljevic et al. 2011). The acceptation or releasing of protons of all the pH-sensitive hydrogels contains pendant basic (e.g., ammonium salts) or acidic (e.g., carboxylic and sulfonic acids) groups are in response to the environmental pH changes. They can be classified into either anionic or cationic hydrogels (Vaghani and Patel 2011), depending on the nature of pendant groups in the network (Gao et al. 2013a). Currently, pH-sensitive hydrogels have been frequently applied to develop the controlled-release formulations for oral administration.

2.2.3 Electro-Sensitive Hydrogels

Electro-sensitive hydrogels are sensitive to electric current which generally made of polyelectrolytes (He et al. 2011). Electro-sensitive hydrogels undergo swelling or shrinking depending on the applied electric field (Wu et al. 2013). In some cases, the hydrogels show shrinking on the other side but swelling on one side owing to the bending of hydrogels. The shape of hydrogels has three types that include swelling, shrinking, and bending hydrogels (Li et al. 2007). These phases depend on various conditions. In general, these polyelectrolyte hydrogels are ion-conductive, so the electro-responsive behavior is dramatically affected by the distribution and concentration of the ions in the system (Jin et al. 2013).

3 Current Status of Stem Cell Survival After Transplantation

Research advances indicated that stem cell therapy has begun to fulfill its long-held promise for functional regeneration of damaged tissues. However, several technological parameters must be optimized before utilizing. One severe problem is that implanted stem cells will encounter a harsh environment after transplanted into damaged tissues. Because of apoptosis, the >90% of implanted cells will be lost within hours to days after transplantation. Considering the poor cell survival rate, there have been little systematic investigations on if the implanted cells can directly engage in regeneration processes or can functionally integrate with the injured areas after transplantation. Therefore, a hydrogel-based environment of transplanted stem cells is considered to be a vital strategy for improving the survival rate.

3.1 Stem Cell Niches

The self-renewal and differentiation of stem cells are controlled by their surrounding microenvironment, which is named as stem cell niches. This concept was first proposed in 1978 as a hypothesis to describe the physiological microenvironment supporting the stem cell (Walker et al. 2009). When stem cells are transplanted into the injured area, the majority of living cells rapidly decreases. This type of cell death may be owing to insufficient signaling molecules, such as ECM and growth factors in the microenvironments surrounding the grafted cells.

The niche functions are the regulatory agency of stem cell differentiation and self-renewal. Niches including niche cells, soluble factors derived from niche cells and ECM, which play the role of nourishing stem cells and keep them in steady state. The interaction of stem cells with other cells or growth factors within the niche will promote the stem cells self-renewal and directional differentiation, and maintain cell or tissue dynamic balance. The "spatiotemporal dialog" between niche and niche can not only protect stem cells from being affected by a variety of signal, but also prevent the excessive proliferation of stem cells leading to cancer.

Extracellular Matrix

As the constitutive part of stem cell niche, ECM plays the vital role for stem cell homeostasis in vivo. In general, these extracellular macromolecules can provide a microenvironment in which the signals derive from ECM–cell interactions by their three-dimensional and assembly organization. For example, neural stem cell (NSC) niche, the lacking of tenascin C will influence the function and number of NSCs in the sub-ventricular zone. Tenascin C tends stem cell to generate glial offspring, seemingly to modulate stem cell sensitivity to fibroblast growth factor-2 (FGF-2) and bone morphogenetic protein-4 (BMP-4). Moreover, in the hematopoietic system, the absence of tenascin C shows to affect stem cell populations (Kazanis and ffrench-Constant 2011). Thus, the specific behaviors that the ECM imposes on stem cell niche should be considered when the stem cell transplanted (Yi et al. 2013).

Soluble Factors

The necessary role of soluble factors including hormones, growth factors, and cytokines in stem cell differentiation and self-renewal has contributed extensively toward the development and validation of regenerative medicine (Hwang et al. 2011). In NSC niche, the proliferation and differentiation of NSC require FGF-2 and EGF. In vivo, exogenous EGF could induce the adult NSC to differentiate into glial cells. On the contrary, compared to the young mice, FGF-2 stimulates NSC proliferation and restores neurogenesis in older mice. The above information shows that different stem cell niches possess plasticity and can respond to signaling cues that are once present in the younger niches to initiate cell differentiation and tissue regeneration (Discher et al. 2009).

Cell–Cell Interactions

In vivo, stem cells are in close neighborhood with other adult cell types. Therefore, the cell–cell interacts through direct contacts or cell-secreted soluble morphogens play necessary roles in phenotypical identity and determining the functions of stem cells (Hwang et al. 2011). For instance, adhering junctions and tight junctions between NSC can allow the establishment of basal–apical polarity (Walker et al. 2009).

The native stem cell niches and their interactions formed during ontogeny are the fundamental problems in the biology of stem cells. In addition to particular conditions, however, the self-renewal of stem cell cannot be maintained for long periods without niche in vitro. So, a vacant niche could be occupied by the transplanted stem cells and can promote stem cells engraftment. Currently, most of the focus of research on stem cells is to figure out the underlying molecular mechanisms and developmental pathways in stem cell differentiation into particular cell lineage for regenerative therapies (Heng et al. 2004). In order to illustrate the mechanism of controlling stem cell fate and enhance survival of implanted stem cells, stem cell engineered microenvironments should be compounded by means of biomaterials in vitro to mimic the stem cell niche in vivo (Guvendiren and Burdick 2013).

3.2 Hydrogel Niches and Their Application in Stem Cell Therapy

Stem cells are resident in a complex microenvironment in vivo (Discher et al. 2009) with components of many biological cues, such as cell–cell interactions, growth factors, biomechanical forces, and ECM molecules. The stem cell microenvironment plays a vital role in regulating the spatial distribution of soluble signaling molecules, nutrients, and oxygen. Fortunately, hydrogels are 3D networks consisting of cross-linked hydrophilic polymers. Hydrogels are soft and rubbery in the swollen state, resembling natural living tissues. Currently, Matrigel matrix has been utilized in numerous researches, which imitates mechanical and biochemical properties of ECM in order to facilitate cellular self-organization. For example, the anoikis of human embryonic stem cell-derived endothelial cells (hESC-ECs) stored at 4 °C and suspended in PBS can be initiated, whereas Matrigel will block the processes and further increase cell survival after transplantation (He et al. 2015). To date, various hydrogels including alginate-based and chitosan hydrogels have shown desirable biocompatibility. Therefore, their physical and chemical properties will provide stability, structural support, carriers, or immunoisolation microenvironment for grafted cells to survive and enhance stem cell survival for tissue regeneration (Fig. 2).

3.2.1 Hydrogels as ECM

The natural ECM is a 3D network made up of matrix proteins, which offered as supporting framework to cells or soluble factors. Hydrogels with highly hydrated

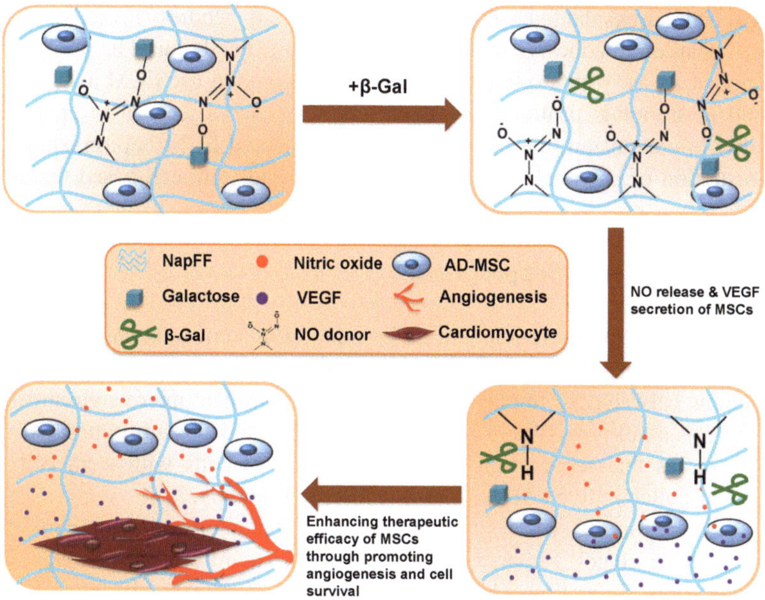

Fig. 2 Nitric oxide (NO) releasing hydrogel could enhance the therapeutic effect of MSCs. Encapsulated MSCs with hydrogel could prevent transplanted cells effusing from injection positions. Moreover, NO molecule released from hydrogel catalyzed by β-galactosidase is able to promote angiogenic cytokines secretion of MSCs, resulting in improved tissue angiogenesis, MSCs survival, and heart function (Yao et al. 2015)

network structural are synthetic polymer or natural which are similar to the native ECM. These natural or synthetic polymers are ideal scaffold for stem cells culture and transplantation (Guvendiren and Burdick 2013).

Many studies have shown that the importance of synthetic matrices functionalized with ECM-derived ligands could promote the differentiation of transplanted stem cells. For instance, the effects of the Arg-Gly-Asp (RGD) peptides on the osteogenic and chondrogenic differentiation of mesenchymal stem cells (MSCs) have been vastly studied. Polylactide and PCL surfaces conjugated with heparin not only support high-affinity adsorption of BMP-2 but also enhance adhesion and proliferation of MSCs. Besides, PEG hydrogels functionalized with heparin-induced osteogenic differentiation and sustained human MSCs viability (Dellatore et al. 2008).

In experimental researches or clinical applications, stem cells are commonly digested into single cells before transplantation. Therefore, the loss of interactions between ECM and stem cells could initiate cell apoptosis and cause a decrease in cell viability. Fortunately, various researches have employed ECM in stem cell therapy to increase the therapeutic efficiency for the following two aspects. On the one hand, the co-transplantation of stem cells with ECM can minimize the major obstacle of extremely poor survival and engraftment of stem cell in damage area. On the other hand, engineered ECM can mimic the natural stem cell niches thus provide numbers of subtle and instructive cues to control the behavior of both resident and transplanted cells. Moreover, recent studies revealed that the down-regulation of adhesive molecules of transplanted cells could be restored in the presence of ECM, which offered a theoretical rationale to use ECM as a protective scaffold to stimulate proliferation of the grafted cells as well as to enhance cell viability. In conclusion, the development of engineered ECM is promising for the administration of stem cells in tissue engineering and stem cell-based therapy.

3.2.2 Hydrogel with Support Cells

The mobilization, proliferation, and differentiation of stem cells are closely related with support cells. Hydrogels not only do duty for a platform to modulate the interactions between the implanted cells and their physical microenvironment, but also serve as a delivery system for regeneration of sophisticated tissues. Currently, certain studies that the co-transplantation of support cells with stem cells have resulted in collaborative effects in stem cell-based therapy.

As a carrier for embryonic stem cells (ESCs), a biodegradable polymers of oligo (poly (ethylene glycol) fumarate, OPF) hydrogel (Kinard et al. 2012) was used in the myocardial infarction (MI) therapy (Jo et al. 2001). In vitro cultures, co-cultivation with OPF, the differentiation of ESCs can be detected after developing 14 days (Wang et al. 2010). After co-transplanted into the mice of MI model for 4 weeks, myocardial differentiation of transplanted stem cells can be detected.

And both 24-h cell retention and 4-week graft size were significantly greater than that of control group. Several weeks after the test, OPF hydrogel remarkable improved the heart function with reduced infarct size and collagen deposition (Temenoff et al. 2002).

3.2.3 Hydrogel with Controlled-Release Strategy

Hydrogels can also be utilized in growth factors releasing or molecule secretion, which could promote cell migration into the injured area and tissue regeneration, or cell differentiation in vivo.

Nitric Oxide (NO) Molecule

Several researches have revealed that self-assembling peptides and subtle designed synthetic polymers could form hydrogel under certain circumstance. These polymers were developed for cells delivery system to implanted into a MI model thus improved the cell retention rate (Liu et al. 2012; Reis et al. 2012). So, a peptide hydrogel combined with naphthalene conjugated short peptide, FFGGG, and β-galactose caged nitric oxide (NO) donor (NapFF-NO) (Fig. 3) (Yao et al. 2015) was developed, which can release NO molecule in response to β-galactosidase (Gao et al. 2013b; Zhao et al. 2013). This synthesized hydrogel, NapFF-NO, together with adipose MSCs which were used for MI treatment. It proved that NO hydrogel could not only increase the proliferation of MSCs, but also increase the MSCs engraftment and improve heart function after MI (Fig. 4) (Yao et al. 2015). Moreover, the transgenic mice with VEGFR2 promoter could be directly visualized angiogenesis in living animal with the advanced in vivo bioluminescence imaging (BLI) technique (Yao et al. 2015).

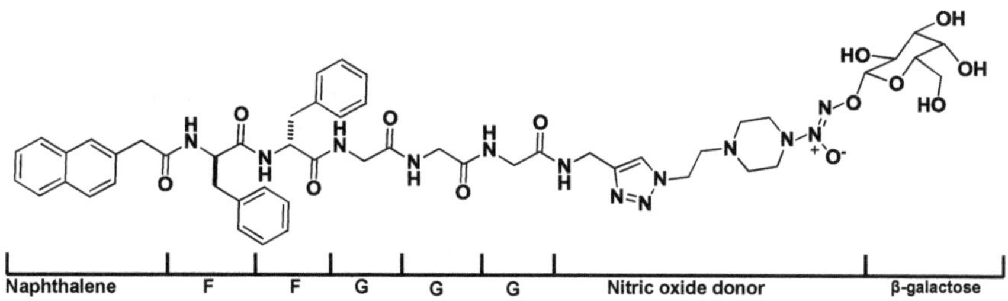

Fig. 3 Structural formula of NapFF-NO hydrogel. It consists of 5 amino acids peptide FFGGG modified by a naphthalene molecule (Nap) and the β-galactose caged nitric oxide (NO) donor (Yao et al. 2015)

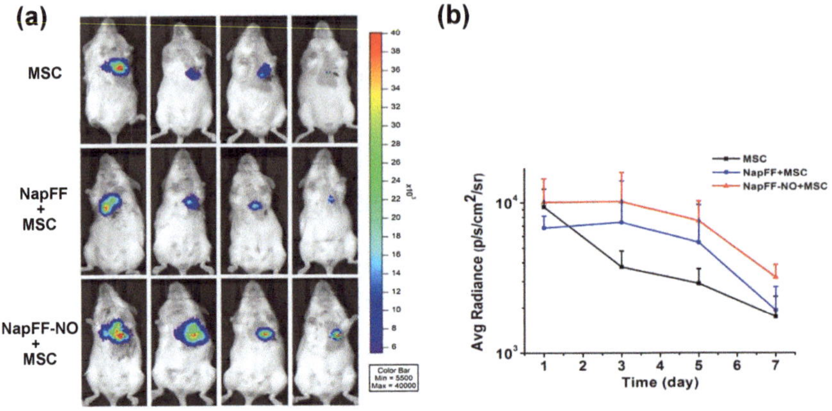

Fig. 4 Bioluminescence imaging of MSCs survival in heart. **a** Representative bioluminescence imaging of mice transplanted with MSC, MSC and control hydrogel, or MSC with NO release hydrogel. Compared with transplantation cells without biomaterials, co-transplantation of cells with NO release hydrogel could remarkably enhanced cell survival. **b** Detailed quantitative analysis of signals (Yao et al. 2015)

Hydrogel with Growth Factors

Growth factors have a certain influence on the growth of tissue, such as insulin-like growth factor (IGF), transforming growth factor beta (TGF-β), and bone morphogenetic protein (BMP). TGF-β family plays a vital role in tissue development. TGF-β1 can increase the interaction between cells while at the same time TGF-β2 involves in the process of MSC differentiation. IGF-1 can increase the secretion of collagen type II and mucopolysaccharides. BMPs can adjust the cell differentiation status and increase the synthesis of extracellular matrix.

Growth factor as a third component can be applied to regulate the stem cell behaviors, such as cell proliferation, differentiation, migration, self-renewal, and accelerate the formation of tissue. For example, the hydrogel combined with growth factors can control tissue regeneration and stem cell differentiation. Co-transplantation of growth factors and stem cells has been broadly investigated in many different studies. The incorporation of growth factors has been utilized to motivate grafted cell differentiation and recruit undifferentiated progenitor cells into carrier system. Plenty of researches have proved the regenerative potential of growth factor and stem cells co-delivery system in vivo and in vitro (Kim et al. 2007, 2010; Terella et al. 2010).

Basic fibroblast growth factor (bFGF) is very effective in promoting stem cell differentiation, proliferation, and survival. However, considering the form of bFGF, the efficiency is limited due to the rapid clearance rate. Controlled-release bFGF systems could be obtained by using biodegradable hydrogels in order to overcome this defect. Biodegradable hydrogels present remarkable biocompatibility and achieve a gradually release of bFGF in vivo because of hydrogel degradation within several weeks. But above all, altering the water content of hydrogel can control the release rate. For example, the controlled release of bFGF has been shown to effectively enhance donor cell engraftment (Takehara et al. 2008)

Hansoo Park et al. also synthesized a biodegradable hydrogel that was composed of OPF with gelatin microparticles (MPs) and embedded MSCs loaded the TGF-β1 for tissue engineering (Park et al. 2007). TGF-β1-loaded MPs and MSCs were encapsulated in OPF, which consists of the *N*, *N*, *N*′, *N*′-tetramethylethylenediamine and the radical initiator's ammonium persulfate, and then cross-linked at 37 °C to form a poly (ethylene glycol)-diacrylate cross-linker hydrogel composites (Park et al. 2005). This sustained growth factor release not only provides a good matrix, but also remarkably promotes cell proliferation, differentiation.

Hydrogel with Immobilized Growth Factors

To acquire higher specific bioactivity after cell transplantation, the growth factors can be covalent immobilized with biomaterial matrices, which can hinder growth factors diffusion and receptor-mediated endocytosis at injured area (Discher et al. 2009; Eshghi and Schaffer 2008; Masters 2011). Several immobilized growth factors including bFGF, VEGF, EGF, IGF-1, BMP have been applied in tissue regeneration and repair, which have achieved greater therapeutic effects (Masters 2011).

Synthetic peptides designed to emulate the biological properties of parent proteins have the ability to provide clinical benefits similar to full-length growth factors in the meantime reduce cost (Madl et al. 2014). Mounting evidence supports the use of bioactive mimicking peptide (e.g., RGD, QK, QHREDGS) as surrogate for growth factors (fibronectin, VEGF, angiopoietin-1) to increase the bioactivity of engineered matrices (Dang et al. 2014; Ingavle et al. 2014; Leslie-Barbick et al. 2011). A 12 amino acids peptide sequence GYGSSSRRAPQT, the C domain of IGF-1 (IGF-1C), has been identified as the active region of IGF-1 (Jansen et al. 1983) and shown to be relevant to promotion of corneal epithelial wound healing (Yamada et al. 2004). Recent study (Feng et al. 2016) has developed chitosan (CS)-based injectable hydrogel with immobilized IGF-1C (CS-IGF-1C) to strengthen the supportive niches function and favor the survival and therapeutic benefits of transplanted MSCs in a mouse model of acute kidney injury (AKI). This hydrogel is accompanied by its well-documented innocuous, biocompatibility, and thermo-sensitive. CS-IGF-1C exerted pro-proliferative, proangiogenic, and anti-apoptotic effects on cultured MSCs. These results attributed to the favorable niches produced by cell–hydrogel interaction, which gives rise to cell survival enhancement (Fig. 5).

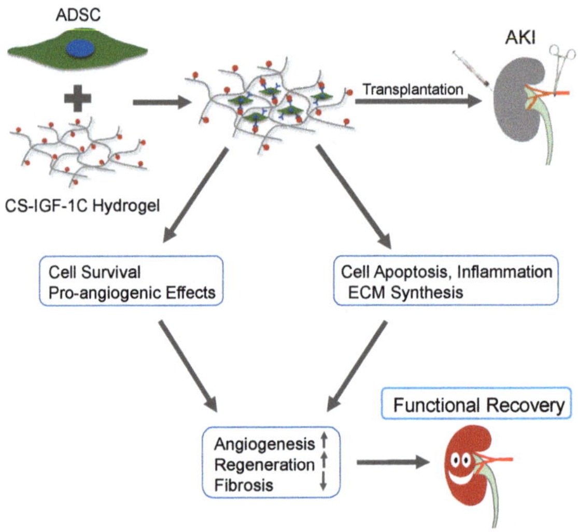

Fig. 5 Co-transplanted ADSCs into acute kidney injury model, the bioactive hydrogel by immobilizing the C domain peptide of IGF-1 (CS-IGF-1C) could protect delivered ADSCs. The hydrogel could facilitate their paracrine and anti-inflammatory effects of ADSCs, but also can inhibit ECM synthesis in the kidney, resulting in function and structure recovery (Feng et al. 2016)

4 Conclusions and Future Perspective

Hydrogel-based culture platforms have a vast potential to construct complex synthetic stem cell microenvironments. The particular bioactive constituents can be merged into the stem cell microenvironment in vivo for the aim of controlling stem cell function and fate. Such components, including biomolecules or support cells, can manipulate stem cell fates, such as cell differentiation, proliferation, and survival. And hydrogels can be applied for regulating the in vivo cell microenvironment for enhancing the engraftment and survival of transplanted stem cells, or even recruit resident stem cells to the injured sites. In conclusion, the strategy of stem cells co-transplantation with hydrogels and controlled-release molecules will afford new insights into the mechanistic modulation of stem cell regeneration.

Acknowledgements This work was partially supported by grants from the National Natural Science Foundation of China (81671734, 81371620, 81320108014), National Key R & D Program of China (2017YFA0103200), and Key Projects of Tianjin Science and Technology Support Program.

References

Allcock HR (2012) Polyphosphazene elastomers, gels, and other soft materials. Soft Matter 8:7521–7532

Anderson SB, Lin CC, Kuntzler DV, Anseth KS (2011) The performance of human mesenchymal stem cells encapsulated in cell-degradable polymer-peptide hydrogels. Biomaterials 32:3564–3574

Bakota EL, Wang Y, Danesh FR, Hartgerink JD (2011) Injectable multidomain peptide nanofiber hydrogel as a delivery agent for stem cell secretome. Biomacromolecules 12:1651–1657

Banerjee SS, Aher N, Patil R, Khandare J (2012) Poly(ethylene glycol)-prodrug conjugates: concept, design, and applications. J Drug Delivery 2012:103973

Bardajee GR, Hooshyar Z (2013) A novel biocompatible magnetic iron oxide nanoparticles/hydrogel based on poly (acrylic acid) grafted onto starch for controlled drug release. J Polym Res 20:1–13

Bueno VB, Bentini R, Catalani LH, Petri DF (2013) Synthesis and swelling behavior of xanthan-based hydrogels. Carbohydr Polym 92:1091–1099

Burdick JA, Prestwich GD (2011) Hyaluronic acid hydrogels for biomedical applications. Adv Mater 23:H41–H56

Busilacchi A, Gigante A, Mattioli-Belmonte M, Manzotti S, Muzzarelli RA (2013) Chitosan stabilizes platelet growth factors and modulates stem cell differentiation toward tissue regeneration. Carbohydr Polym 98:665–676

Calderón MÁR, Zhao W (2014) Applications of polymer nanofibers in bio-materials, biotechnology and biomedicine: a review. Wiley

Cameron GJ, Alberts IL, Laing JH, Wess TJ (2002) Structure of type I and type III heterotypic collagen fibrils: an X-ray diffraction study. J Struct Biol 137:15–22

Chen JP, Cheng TH (2009) Preparation and evaluation of thermo-reversible copolymer hydrogels containing chitosan and hyaluronic acid as injectable cell carriers. Polymer 50:107–116

Chen C, Wang Z, Li Z (2011) Thermoresponsive polypeptides from pegylated poly-L-glutamates. Biomacromolecules 12:2859–2863

Cheng J, Deming TJ (2012) Synthesis of polypeptides by ring-opening polymerization of alpha-amino acid N-carboxyanhydrides. Top Curr Chem 310:1–26

Cho D-W, Lee J-S, Jang J, Jung JW, Park JH, Pati F (2015) Natural, synthetic and semi-synthetic polymers, Morgan & Claypool Publishers

Choi KY, Min KH, Yoon HY, Kim K, Park JH, Kwon IC, Choi K, Jeong SY (2011) PEGylation of hyaluronic acid nanoparticles improves tumor targetability in vivo. Biomaterials 32:1880–1889

Chu L-Y, Xie R, Ju X-J, Wang W (2013) Functional membranes with thermo-responsive hydrogel gates. Springer, pp. 111–133

Coviello T, Matricardi P, Marianecci C, Alhaique F (2007) Polysaccharide hydrogels for modified release formulations. J Control Release 119:5–24

Dang LTH, Feric NT, Laschinger C, Chang WY, Zhang BY, Wood GA, Stanford WL, Radisic M (2014) Inhibition of apoptosis in human induced pluripotent stem cells during expansion in a defined culture using angiopoietin-1 derived peptide QHREDGS. Biomaterials 35:7786–7799

Das N (2013) Preparation methods and properties of hydrogel: a review. Int J Pharm Pharm Sci 5:112–117

de Moraes MA, Beppu MM (2013) Biocomposite membranes of sodium alginate and silk fibroin fibers for biomedical applications. J Appl Polym Sci 130:3451–3457

Dellatore SM, Garcia AS, Miller WM (2008) Mimicking stem cell niches to increase stem cell expansion. Curr Opin Biotechnol 19:534–540

Deming TJ (2002) Methodologies for preparation of synthetic block copolypeptides: materials with future promise in drug delivery. Adv Drug Delivery Rev 54:1145–1155

Deng M, Kumbar SG, Wan Y, Toti US, Allcock HR, Laurencin CT (2010) Polyphosphazene polymers for tissue engineering: an analysis of material synthesis, characterization and applications. Soft Matter 6:3119
Discher DE, Mooney DJ, Zandstra PW (2009) Growth factors, matrices, and forces combine and control stem cells. Science 324:1673–1677
Dragan ES (2014) Design and applications of interpenetrating polymer network hydrogels. a review. Chem Eng J 243:572–590
Elluru M, Ma HY, Hadjiargyrou M, Hsiao BS, Chu B (2013) Synthesis and characterization of biocompatible hydrogel using pluronics-based block copolymers. Polymer 54:2088–2095
Eshghi S, Schaffer DV (2008) Engineering microenvironments to control stem cell fate and function. The Stem Cell Research Community
Feng X, Tonnesen MG, Mousa SA, Clark RA (2013) Fibrin and collagen differentially but synergistically regulate sprout angiogenesis of human dermal microvascular endothelial cells in 3-dimensional matrix. Int J Cell Biol 2013:231279
Feng G, Zhang J, Li Y, Nie Y, Zhu D, Wang R, Liu J, Gao J, Liu N, He N, Du W, Tao H, Che Y, Xu Y, Kong D, Zhao Q, Li Z (2016) IGF-1 C Domain-modified hydrogel enhances cell therapy for AKI. J Am Soc Nephrol
Fullana MJ, Wnek GE (2012) Electrospun collagen and its applications in regenerative medicine. Drug Delivery Transl Res 2:313–322
Gao H, Wang N, Hu X, Nan W, Han Y, Liu W (2013a) Double hydrogen-bonding pH-sensitive hydrogels retaining high-strengths over a wide pH range. Macromol Rapid Commun 34:63–68
Gao J, Zheng W, Zhang J, Guan D, Yang Z, Kong D, Zhao Q (2013b) Enzyme-controllable delivery of nitric oxide from a molecular hydrogel. Chem Commun (Camb) 49:9173–9175
Glotzbach JP, Wong VW, Levi B, Longaker MT, Gurtner GC (2012) Delivery strategies for stem cell-based therapy. J Healthc Eng 3:1–20
Gombotz WR, Wee SF (2012) Protein release from alginate matrices. Adv Drug Deliver Rev 64:194–205
Guvendiren M, Burdick JA (2013) Engineering synthetic hydrogel microenvironments to instruct stem cells. Curr Opin Biotechnol 24:841–846
Hager AS, Arendt EK (2013) Influence of hydroxypropylmethylcellulose (HPMC), xanthan gum and their combination on loaf specific volume, crumb hardness and crumb grain characteristics of gluten-free breads based on rice, maize, teff and buckwheat. Food Hydrocolloids 32:195–203
Harding SE, Smith IH, Lawson CJ, Gahler RJ, Wood S (2011) Studies on macromolecular interactions in ternary mixtures of konjac glucomannan, xanthan gum and sodium alginate. Carbohyd Polym 83:329–338
He L, Zuo Q, Xie S, Huang Y, Xue W (2011) Intelligent hydrogels for drug delivery system. Recent Pat Drug Delivery Formulation 5:265–274
He N, Xu Y, Du W, Qi X, Liang L, Wang Y, Feng G, Fan Y, Han Z, Kong D, Cheng Z, Wu JC, He Z, Li Z (2015) Extracellular matrix can recover the downregulation of adhesion molecules after cell detachment and enhance endothelial cell engraftment. Sci Rep 5:10902
Heng BC, Cao T, Haider HK, Wang DZ, Sim EK, Ng SC (2004) An overview and synopsis of techniques for directing stem cell differentiation in vitro. Cell Tissue Res 315:291–303
Hoffman AS (2012) Hydrogels for biomedical applications. Adv Drug Deliver Rev 64:18–23
Hu Y, You J-O, Auguste DT, Suo Z, Vlassak JJ (2011) Indentation: a simple, nondestructive method for characterizing the mechanical and transport properties of pH-sensitive hydrogels. J Mater Res 27:152–160
Huang J, Heise A (2013) Stimuli responsive synthetic polypeptides derived from N-carboxyanhydride (NCA) polymerisation. Chem Soc Rev 42:7373–7390
Huang X, Huang X-J, Yu A-G, Wang C, Dai Z-W, Xu Z-K (2011) "Click chemistry" as a facile approach to the synthesis of polyphosphazene glycopolymers. Macromol Chem Phys 212:272–277

Huebsch N, Arany PR, Mao AS, Shvartsman D, Ali OA, Bencherif SA, Rivera-Feliciano J, Mooney DJ (2010) Harnessing traction-mediated manipulation of the cell/matrix interface to control stem-cell fate. Nat Mater 9:518–526

Hwang Y, Phadke A, Varghese S (2011) Engineered microenvironments for self-renewal and musculoskeletal differentiation of stem cells. Regenerative Med 6:505–524

Ibrahim S, Johan MR (2012) Thermolysis and conductivity studies of poly(ethylene oxide) (PEO) based polymer electrolytes doped with carbon nanotube. Int J Electrochem Sci 7:2596–2615

Ichikawa H, Fukumori Y (2000) A novel positively thermosensitive controlled-release microcapsule with membrane of nano-sized poly(N-isopropylacrylamide) gel dispersed in ethylcellulose matrix. J Control Release 63:107–119

Ingavle GC, Gehrke SH, Detamore MS (2014) The bioactivity of agarose PEGDA interpenetrating network hydrogels with covalently immobilized RGD peptides and physically entrapped aggrecan. Biomaterials 35:3558–3570

Jaiswal A, Ghosh SS, Chattopadhyay A (2012) One step synthesis of C-dots by microwave mediated caramelization of poly(ethylene glycol). Chem Commun (Camb) 48:407–409

Jannesari M, Varshosaz J, Morshed M, Zamani M (2011) Composite poly(vinyl alcohol)/poly (vinyl acetate) electrospun nanofibrous mats as a novel wound dressing matrix for controlled release of drugs. Int J Nanomed 6:993–1003

Jansen M, van Schaik FM, Ricker AT, Bullock B, Woods DE, Gabbay KH, Nussbaum AL, Sussenbach JS, Van den Brande JL (1983) Sequence of cDNA encoding human insulin-like growth factor I precursor. Nature 306:609–611

Jin SP, Gu JX, Shi YJ, Shao KR, Yu XH, Yue GR (2013) Preparation and electrical sensitive behavior of poly (N-vinylpyrrolidone-co-acrylic acid) hydrogel with flexible chain nature. Eur Polym J 49:1871–1880

Jo S, Shin H, Mikos AG (2001) Modification of oligo(poly(ethylene glycol) fumarate) macromer with a GRGD peptide for the preparation of functionalized polymer networks. Biomacromolecules 2:255–261

Kazanis I, ffrench-Constant C (2011) Extracellular matrix and the neural stem cell niche. Dev Neurobiol 71:1006–1017

Kempf M, Miyamura Y, Liu PY, Chen AC, Nakamura H, Shimizu H, Tabata Y, Kimble RM, McMillan JR (2011) A denatured collagen microfiber scaffold seeded with human fibroblasts and keratinocytes for skin grafting. Biomaterials 32:4782–4792

Khetan S, Guvendiren M, Legant WR, Cohen DM, Chen CS, Burdick JA (2013) Degradation-mediated cellular traction directs stem cell fate in covalently crosslinked three-dimensional hydrogels. Nat Mater 12:458–465

Kim J, Kim IS, Cho TH, Lee KB, Hwang SJ, Tae G, Noh I, Lee SH, Park Y, Sun K (2007) Bone regeneration using hyaluronic acid-based hydrogel with bone morphogenic protein-2 and human mesenchymal stem cells. Biomaterials 28:1830–1837

Kim J, Kim IS, Cho TH, Kim HC, Yoon SJ, Choi J, Park Y, Sun K, Hwang SJ (2010) In vivo evaluation of MMP sensitive high-molecular weight HA-based hydrogels for bone tissue engineering. J Biomed Mater Res A 95:673–681

Kim IL, Mauck RL, Burdick JA (2011) Hydrogel design for cartilage tissue engineering: a case study with hyaluronic acid. Biomaterials 32:8771–8782

Kinard LA, Kasper FK, Mikos AG (2012) Synthesis of oligo(poly(ethylene glycol) fumarate). Nat Protoc 7:1219–1227

Komeri R, Thankam FG, Muthu J (2015) Influence of matrix and bulk behaviour of an injectable hydrogel on the survival of encapsulated cardiac cells. Rsc Adv 5:31439–31449

Korotych O, Samchenko Y, Boldeskul I, Ulberg Z, Zholobak N, Sukhodub L (2013) N-isopropylacrylamide-based fine-dispersed thermosensitive ferrogels obtained via in-situ technique. Mater Sci Eng C Mater Biol Appl 33:892–900

Kost J, Langer R (2012) Responsive polymeric delivery systems. Adv Drug Delivery Rev 64:327–341

Kulkarni M, O'Loughlin A, Vazquez R, Mashayekhi K, Rooney P, Greiser U, O'Toole E, O'Brien T, Malagon MM, Pandit A (2014) Use of a fibrin-based system for enhancing angiogenesis and modulating inflammation in the treatment of hyperglycemic wounds. Biomaterials 35:2001–2010

Laftah WA, Hashim S, Ibrahim AN (2011) Polymer hydrogels: a Review. Polym-Plast Technol 50:1475–1486

Lee KY, Mooney DJ (2012) Alginate: properties and biomedical applications. Prog Polym Sci 37:106–126

Leipzig ND, Wylie RG, Kim H, Shoichet MS (2011) Differentiation of neural stem cells in three-dimensional growth factor-immobilized chitosan hydrogel scaffolds. Biomaterials 32:57–64

Leslie-Barbick JE, Saik JE, Gould DJ, Dickinson ME, West JL (2011) The promotion of microvasculature formation in poly(ethylene glycol) diacrylate hydrogels by an immobilized VEGF-mimetic peptide. Biomaterials 32:5782–5789

Li H, Chen J, Lam KY (2007) Transient simulation of kinetics of electric-sensitive hydrogels. Biosens Bioelectron 22:1633–1641

Li Z, Han Z, Wu JC (2009a) Transplantation of human embryonic stem cell-derived endothelial cells for vascular diseases. J Cell Biochem 106:194–199

Li Z, Lee A, Huang M, Chun H, Chung J, Chu P, Hoyt G, Yang P, Rosenberg J, Robbins RC, Wu JC (2009b) Imaging survival and function of transplanted cardiac resident stem cells. J Am Coll Cardiol 53:1229–1240

Li W, Yue J, Liu S (2012) Preparation of nanocrystalline cellulose via ultrasound and its reinforcement capability for poly(vinyl alcohol) composites. Ultrason Sonochem 19:479–485

Liu Z, Wang H, Wang Y, Lin Q, Yao A, Cao F, Li D, Zhou J, Duan C, Du Z, Wang Y, Wang C (2012) The influence of chitosan hydrogel on stem cell engraftment, survival and homing in the ischemic myocardial microenvironment. Biomaterials 33:3093–3106

Liu J, Xu HH, Zhou H, Weir MD, Chen Q, Trotman CA (2013) Human umbilical cord stem cell encapsulation in novel macroporous and injectable fibrin for muscle tissue engineering. Acta Biomater 9:4688–4697

Liu FY, Seuring J, Agarwal S (2014a) A non-ionic thermophilic hydrogel with positive thermosensitivity in water and electrolyte solution. Macromol Chem Phys 215:1466–1472

Liu H, Cheng J, Chen F, Hou F, Bai D, Xi P, Zeng Z (2014b) Biomimetic and cell-mediated mineralization of hydroxyapatite by carrageenan functionalized graphene oxide. ACS Appl Mater Interfaces 6:3132–3140

Liu Y, Qi N, Song T, Jia M, Xia Z, Yuan Z, Yuan W, Zhang KQ, Sun B (2014c) Highly flexible and lightweight organic solar cells on biocompatible silk fibroin. ACS Appl Mater Interfaces 6:20670–20675

Liu J, Willför S, Xu C (2015) A review of bioactive plant polysaccharides: biological activities, functionalization, and biomedical applications. Bioact Carbohydr Dietary Fibre 5:31–61

Lu Q, Zhang B, Li M, Zuo B, Kaplan DL, Huang Y, Zhu H (2011) Degradation mechanism and control of silk fibroin. Biomacromolecules 12:1080–1086

Lu H, Wang J, Song Z, Yin L, Zhang Y, Tang H, Tu C, Lin Y, Cheng J (2014) Recent advances in amino acid N-carboxyanhydrides and synthetic polypeptides: chemistry, self-assembly and biological applications. Chem Commun (Camb) 50:139–155

Lutolf MP, Gilbert PM, Blau HM (2009) Designing materials to direct stem-cell fate. Nature 462:433–441

Ma GP, Fang DW, Liu Y, Zhu XD, Nie J (2012) Electrospun sodium alginate/poly(ethylene oxide) core-shell nanofibers scaffolds potential for tissue engineering applications. Carbohyd Polym 87:737–743

Macaya D, Spector M (2012) Injectable hydrogel materials for spinal cord regeneration: a review. Biomed Mater 7:012001

Madl CM, Mehta M, Duda GN, Heilshorn SC, Mooney DJ (2014) Presentation of BMP-2 mimicking peptides in 3D hydrogels directs cell fate commitment in osteoblasts and mesenchymal stem cells. Biomacromolecules 15:445–455

Mallakpour S, Barati A (2011) Efficient preparation of hybrid nanocomposite coatings based on poly(vinyl alcohol) and silane coupling agent modified TiO2 nanoparticles. Prog Org Coat 71:391–398

Mano JF, Silva GA, Azevedo HS, Malafaya PB, Sousa RA, Silva SS, Boesel LF, Oliveira JM, Santos TC, Marques AP, Neves NM, Reis RL (2007) Natural origin biodegradable systems in tissue engineering and regenerative medicine: present status and some moving trends. J R Soc Interface 4:999–1030

Masters KS (2011) Covalent growth factor immobilization strategies for tissue repair and regeneration. Macromol Biosci 11:1149–1163

Mikac U, Baumgartner S, Sepe A, Kristl J (2013) MRI study of hydrophilic xanthan tablets with incorporated model drug. JB Principles Diffus Theor Exp Appl 18:1–5

Milosavljevic NB, Ristic MD, Peric-Grujic AA, Filipovic JM, Strbac SB, Rakocevic Z, Kalagasidis Krusic MT (2011) Sorption of zinc by novel pH-sensitive hydrogels based on chitosan, itaconic acid and methacrylic acid. J Hazard Mater 192:846–854

Mitra T, Sailakshmi G, Gnanamani A, Raja ST, Thiruselvi T, Gowri VM, Selvaraj NV, Ramesh G, Mandal AB (2011) Preparation and characterization of a thermostable and biodegradable biopolymers using natural cross-linker. Int J Biol Macromol 48:276–285

Mogosanu GD, Grumezescu AM (2014) Natural and synthetic polymers for wounds and burns dressing. Int J Pharm 463:127–136

Moloney TC, Fhlathartaigh MN, Kulkarni M, Pandit A, Dowd E (2015) Fibrin as a scaffold for delivery of GDNF overexpressing stem cells to the adult rat brain. ACS Biomater Sci Eng 1:559–566

Morimune S, Kotera M, Nishino T, Goto K, Hata K (2011) Poly(vinyl alcohol) nanocomposites with nanodiamond. Macromolecules 44:4415–4421

Murali R, Anumary A, Ashokkumar M, Thanikaivelan P, Chandrasekaran B (2011) Hybrid biodegradable films from collagenous wastes and natural polymers for biomedical applications. Waste Biomass Valor 2:323–335

Neffe AT, Wischke C, Racheva M, Lendlein A (2013) Progress in biopolymer-based biomaterials and their application in controlled drug delivery. Expert Rev Med Devices 10:813–833

Obermeier B, Wurm F, Mangold C, Frey H (2011) Multifunctional Poly(ethylene glycol)s. Angew Chem Int Ed Engl 50:7988–7997

Oh EJ, Choi JS, Kim H, Joo CK, Hahn SK (2011) Anti-Flt1 peptide—hyaluronate conjugate for the treatment of retinal neovascularization and diabetic retinopathy. Biomaterials 32:3115–3123

Park H, Temenoff JS, Holland TA, Tabata Y, Mikos AG (2005) Delivery of TGF-beta1 and chondrocytes via injectable, biodegradable hydrogels for cartilage tissue engineering applications. Biomaterials 26:7095–7103

Park H, Temenoff JS, Tabata Y, Caplan AI, Mikos AG (2007) Injectable biodegradable hydrogel composites for rabbit marrow mesenchymal stem cell and growth factor delivery for cartilage tissue engineering. Biomaterials 28:3217–3227

Peach MS, Kumbar SG, James R, Toti US, Balasubramaniam D, Deng M, Ulery B, Mazzocca AD, McCarthy MB, Morozowich NL, Allcock HR, Laurencin CT (2012) Design and optimization of polyphosphazene functionalized fiber matrices for soft tissue regeneration. J Biomed Nanotechnol 8:107–124

Phadke A, Zhang C, Arman B, Hsu CC, Mashelkar RA, Lele AK, Tauber MJ, Arya G, Varghese S (2012) Rapid self-healing hydrogels. Proc Natl Acad Sci U S A 109:4383–4388

Popa EG, Gomes ME, Reis RL (2011) Cell delivery systems using alginate–carrageenan hydrogel beads and fibers for regenerative medicine applications. Biomacromolecules 12:3952–3961

Prestwich GD (2011) Hyaluronic acid-based clinical biomaterials derived for cell and molecule delivery in regenerative medicine. J Control Release 155:193–199

Priya James H, John R, Alex A, Anoop KR (2014) Smart polymers for the controlled delivery of drugs—a concise overview. Acta Pharm Sin B 4:120–127

Qiu Y, Park K (2012) Environment-sensitive hydrogels for drug delivery. Adv Drug Deliver Rev 64:49–60

Rayahin JE, Buhrman JS, Zhang Y, Koh TJ, Gemeinhart RA (2015) High and low molecular weight hyaluronic acid differentially influence macrophage activation. ACS Biomater Sci Eng 1:481–493

Reis LA, Chiu LL, Liang Y, Hyunh K, Momen A, Radisic M (2012) A peptide-modified chitosan-collagen hydrogel for cardiac cell culture and delivery. Acta Biomater 8:1022–1036

Roberts MJ, Bentley MD, Harris JM (2012) Chemistry for peptide and protein PEGylation. Adv Drug Deliver Rev 64:116–127

Roger VL, Go AS, Lloyd-Jones DM, Benjamin EJ, Berry JD, Borden WB, Bravata DM, Dai S, Ford ES, Fox CS, Fullerton HJ, Gillespie C, Hailpern SM, Heit JA, Howard VJ, Kissela BM, Kittner SJ, Lackland DT, Lichtman JH, Lisabeth LD, Makuc DM, Marcus GM, Marelli A, Matchar DB, Moy CS, Mozaffarian D, Mussolino ME, Nichol G, Paynter NP, Soliman EZ, Sorlie PD, Sotoodehnia N, Turan TN, Virani SS, Wong ND, Woo D, Turner MB, American Heart Association Statistics C, Stroke Statistics S (2012) Heart disease and stroke statistics—2012 update: a report from the American Heart Association. Circulation 125: e2–e220

Ruvinov E, Leor J, Cohen S (2010) The effects of controlled HGF delivery from an affinity-binding alginate biomaterial on angiogenesis and blood perfusion in a hindlimb ischemia model. Biomaterials 31:4573–4582

Samchenko Y, Ulberg Z, Korotych O (2011) Multipurpose smart hydrogel systems. Adv Colloid Interface Sci 168:247–262

Sharma S, Lohan S, Murthy RS (2014) Formulation and characterization of intranasal mucoadhesive nanoparticulates and thermo-reversible gel of levodopa for brain delivery. Drug Dev Ind Pharm 40:869–878

Silva Nykänen VP, Nykänen A, Puska MA, Silva GG, Ruokolainen J (2011) Dual-responsive and super absorbing thermally cross-linked hydrogel based on methacrylate substituted polyphosphazene. Soft Matter 7:4414

Sipahigil O, Dortunc B (2001) Preparation and in vitro evaluation of verapamil HCl and ibuprofen containing carrageenan beads. Int J Pharm 228:119–128

Snippert HJ, Clevers H (2011) Tracking adult stem cells. EMBO Rep 12:113–122

Srivastava A, Shukla S, Kumar R (2016) Applications and characteristics properties of polymeric hydrogel. J Chem Chem Sci 6(2):166–181

Tabar V, Studer L (2014) Pluripotent stem cells in regenerative medicine: challenges and recent progress. Nat Rev Genet 15:82–92

Takehara N, Tsutsumi Y, Tateishi K, Ogata T, Tanaka H, Ueyama T, Takahashi T, Takamatsu T, Fukushima M, Komeda M, Yamagishi M, Yaku H, Tabata Y, Matsubara H, Oh H (2008) Controlled delivery of basic fibroblast growth factor promotes human cardiosphere-derived cell engraftment to enhance cardiac repair for chronic myocardial infarction. J Am Coll Cardiol 52:1858–1865

Temenoff JS, Athanasiou KA, LeBaron RG, Mikos AG (2002) Effect of poly(ethylene glycol) molecular weight on tensile and swelling properties of oligo(poly(ethylene glycol) fumarate) hydrogels for cartilage tissue engineering. J Biomed Mater Res 59:429–437

Teramoto H, Kojima K (2014) Production of Bombyx mori silk fibroin incorporated with unnatural amino acids. Biomacromolecules 15:2682–2690

Terella A, Mariner P, Brown N, Anseth K, Streubel SO (2010) Repair of a calvarial defect with biofactor and stem cell-embedded polyethylene glycol scaffold. Arch Facial Plast Surg 12:166–171

Thakur VK, Kessler MR (2015) Self-healing polymer nanocomposite materials: a review. Polymer 69:369–383

Thakur VK, Thakur MK (2014a) Recent trends in hydrogels based on psyllium polysaccharide: a review. J Clean Prod 82:1–15

Thakur VK, Thakur MK (2014b) Recent Advances in Graft Copolymerization and Applications of Chitosan: a Review. ACS Sustain Chem Eng 2(12):2637–2652

Thakur VK, Thakur MK (2015) Recent advances in green hydrogels from lignin: a review. Int J Biol Macromol 72:834–847

Thakur MK, Thakur VK, Gupta RK, Pappu A (2016) Synthesis and applications of biodegradable soy based graft copolymers: a review. ACS Sustain Chem Eng 4:1–17

Tian HY, Tang ZH, Zhuang XL, Chen XS, Jing XB (2012) Biodegradable synthetic polymers: preparation, functionalization and biomedical application. Prog Polym Sci 37:237–280

Trombino S, Cassano R, Ferrarelli T, Isacchi B, Bilia AR, Picci N (2012) Collagen α-tocopherulate for topical applications: preparation, characterization, and antioxidant activity evaluation. Macromol Res 20:939–943

Vaghani SS, Patel MM (2011) pH-sensitive hydrogels based on semi-interpenetrating network (semi-IPN) of chitosan and polyvinyl pyrrolidone for clarithromycin release. Drug Dev Ind Pharm 37:1160–1169

Van Vlierberghe S, Dubruel P, Schacht E (2011) Biopolymer-based hydrogels as scaffolds for tissue engineering applications: a review. Biomacromolecules 12:1387–1408

Vermonden T, Censi R, Hennink WE (2012) Hydrogels for protein delivery. Chem Rev 112:2853–2888

Walker MR, Patel KK, Stappenbeck TS (2009) The stem cell niche. J Pathol 217:169–180

Wang H, Zhou J, Liu Z, Wang C (2010) Injectable cardiac tissue engineering for the treatment of myocardial infarction. J Cell Mol Med 14:1044–1055

Wang JC, Wang XB, Xu CH, Zhang M, Shang XP (2011) Preparation of graphene/poly(vinyl alcohol) nanocomposites with enhanced mechanical properties and water resistance. Polym Int 60:816–822

Wang H, Liu Z, Li D, Guo X, Kasper FK, Duan C, Zhou J, Mikos AG, Wang C (2012) Injectable biodegradable hydrogels for embryonic stem cell transplantation: improved cardiac remodelling and function of myocardial infarction. J Cell Mol Med 16:1310–1320

Wang M, Fu J, Huang D, Zhang C, Xu Q (2013) Silver nanoparticles-decorated polyphosphazene nanotubes: synthesis and applications. Nanoscale 5:7913–7919

Weber C, Hoogenboom R, Schubert US (2012) Temperature responsive bio-compatible polymers based on poly(ethylene oxide) and poly(2-oxazoline)s. Prog Polym Sci 37:686–714

Weisel JW, Litvinov RI (2013) Mechanisms of fibrin polymerization and clinical implications. Blood 121:1712–1719

Wichterle O, Lim D (1960) Hydrophilic gels for biological use. Nature 185:117–118

Wu J, Ren Y, Sun J, Feng L (2013) Carbon nanotube-coated macroporous poly (N-isopropylacrylamide) hydrogel and its electrosensitivity. ACS Appl Mater Interfaces 5:3519–3523

Xia Y, Zhu K, Lai H, Lang M, Xiao Y, Lian S, Guo C, Wang C (2015) Enhanced infarct myocardium repair mediated by thermosensitive copolymer hydrogel-based stem cell transplantation. Exp Biol Med (Maywood) 240:593–600

Xu J, Zhao Q, Jin Y, Qiu L (2014) High loading of hydrophilic/hydrophobic doxorubicin into polyphosphazene polymersome for breast cancer therapy. Nanomedicine 10:349–358

Yamada N, Yanai R, Nakamura M, Inui M, Nishida T (2004) Role of the C domain of IGFs in synergistic promotion, with a substance P-derived peptide, of rabbit corneal epithelial wound healing. Invest Ophthalmol Vis Sci 45:1125–1131

Yang JA, Park K, Jung H, Kim H, Hong SW, Yoon SK, Hahn SK (2011) Target specific hyaluronic acid-interferon alpha conjugate for the treatment of hepatitis C virus infection. Biomaterials 32:8722–8729

Yao X, Liu Y, Gao J, Yang L, Mao D, Stefanitsch C, Li Y, Zhang J, Ou L, Kong D, Zhao Q, Li Z (2015) Nitric oxide releasing hydrogel enhances the therapeutic efficacy of mesenchymal stem cells for myocardial infarction. Biomaterials 60:130–140

Ye Z, Zhou Y, Cai H, Tan W (2011) Myocardial regeneration: roles of stem cells and hydrogels. Adv Drug Delivery Rev 63:688–697

Yi SY, Hao YB, Nan KJ, Fan TL (2013) Cancer stem cells niche: a target for novel cancer therapeutics. Cancer Treat Rev 39:290–296

Zhang C, Huang S, Tjiu WW, Fan W, Liu T (2012) Facile preparation of water-dispersible graphene sheets stabilized by acid-treated multi-walled carbon nanotubes and their poly(vinyl alcohol) composites. J Mater Chem 22:2427–2434

Zhao Q, Zhang J, Song L, Ji Q, Yao Y, Cui Y, Shen J, Wang PG, Kong D (2013) Polysaccharide-based biomaterials with on-demand nitric oxide releasing property regulated by enzyme catalysis. Biomaterials 34:8450–8458
Zheng C, Xu J, Yao X, Xu J, Qiu L (2011) Polyphosphazene nanoparticles for cytoplasmic release of doxorubicin with improved cytotoxicity against Dox-resistant tumor cells. J Colloid Interface Sci 355:374–382
Zhou H, Xu HH (2011) The fast release of stem cells from alginate-fibrin microbeads in injectable scaffolds for bone tissue engineering. Biomaterials 32:7503–7513
Zhu J, Marchant RE (2011) Design properties of hydrogel tissue-engineering scaffolds. Expert Rev Med Devices 8:607–626

Chapter 5
Protein- and Nanoparticle-Loaded Hydrogels Studied by Small-Angle Scattering and Rheology Techniques

Aristeidis Papagiannopoulos and Stergios Pispas

Abstract In the last decades, hydrogels have been used for controlled loading and release in pharmaceutical applications. In tissue engineering, protein–hydrogel hybrid systems play a critical role in wound healing and tissue growth (Vermonden et al. in Chem Rev 112:2853–2888, 2012). At the same time, the mechanical and morphological properties of hydrogels have been modified and tuned by addition of nanoparticles (Haraguchi et al. in Macromolecules 36:5732–5741, 2003). The mechanical properties of hydrogels are one of their key characteristics. For example in injectable hydrogels, shear-thinning behavior is a defining factor (Guvendiren et al. in Soft Matter 8:260–272, 2012). Furthermore, the rheological behavior of a protein- or nanoparticle-loaded hydrogels may be influenced by the presence of the added compound, especially when the last acts as a cross-linking agent. The multi-scale hierarchical structures produced by hydrogel nanocomposites can be resolved by small-angle neutron scattering and X-ray scattering (SANS and SAXS) in the relevant length scales from 1 to 1000 nm (combined with ultra-small-angle X-ray and neutron scattering: USAXS and USANS). The study of such systems under deformation (e.g., Rheo-SANS) gives invaluable insight into the structural details that define mechanical properties (Shibayama in Polym J 43:18–34, 2011). In this chapter, the recent developments in the field of hydrogels and nanoparticle-loaded-hydrogel systems, based mainly on SANS/SAXS and rheological techniques, are presented. A wide range of experimental realizations and examples of promising hydrogel–protein combinations is covered, and the analyses used to connect the structure–rheology properties are demonstrated in a unifying way.

A. Papagiannopoulos · S. Pispas (✉)
Theoretical and Physical Chemistry Institute, National Hellenic Research Foundation,
48 Vassileos Constantinou Avenue, 11635 Athens, Greece
e-mail: pispas@eie.gr

V. K. Thakur and M. K. Thakur (eds.), *Hydrogels*, Gels Horizons: From Science to Smart Materials, https://doi.org/10.1007/978-981-10-6077-9_5

1 Introduction

A network of cross-linked hydrophilic macromolecular chains in an aqueous medium is defined as hydrogel (Thakur and Thakur 2014a, b, 2015). There are numerous excellent comprehensive reviews on hydrogel cross-linking mechanisms, fundamental characterization and classification (Gulrez et al. 2011), and hydrogel nanocomposites (Zhao et al. 2015). Their importance is in the fact that they can take up large amounts of water and biological fluids (Jonker et al. 2012) compared to their dry mass as quantified by the swelling ratio $\mathrm{SR} = \frac{W_s - W_d}{W_d} \times 100\%$ (where W_s and W_d are the weights of the swollen and the dried hydrogel, respectively). This way they create a biocompatible environment able to accommodate drugs, proteins, and cells and can be used for drug and protein delivery in tissue engineering (Zustiak et al. 2013) and wound healing (Censi et al. 2012; Jonker et al. 2012). Hydrogels can be stimuli responsive so that their swelling, loading, and release can respond to external conditions (Kim et al. 2013). The internal morphology of hydrogels resembles the one of the semidilute polymer solutions in good solvent (neutral hydrophilic polymers) or even semidilute polyelectrolyte solutions (charged polymers). At the same time, the presence of permanent (chemical gel) or temporary (physical gel) cross-links creates constraints in the density fluctuations at higher length scales. Cross-linking is therefore responsible for the elastic nature of hydrogels. Their soft-solid properties allow them to be mechanically stable and capable for shear-thinning flow (Kim et al. 2013).

The hierarchical nature and self-similarity are inherently connected with soft matter. The statistical character of macromolecules induces self-similarity both in structure and in dynamics. Hence, the chapter begins with the fundamental relation between these morphological features and small-angle scattering data. Then, the principles of viscoelastic response of soft matter are analyzed in terms of the underlying length-scale-dependent structure. Eventually, the application of small-angle scattering, rheology, and their combination to characterize nanocomposite- and protein-loaded hydrogels is presented with the aid of highly illustrative examples from the recent literature.

2 Methods for Investigation of Structural and Viscoelastic Properties in Hydrogels

2.1 Structural Characterization by Small-Angle Scattering

Small-angle scattering methods are widely used to characterize the morphology of soft matter at the mesoscale, i.e., from 1 to 1000 nm. In small-angle neutron scattering (SANS), the incoming radiation is scattered by the atomic nuclei, while in small-angle X-ray scattering (SAXS) the incoming photons detect the presence of the outer electrons of the atoms in the sample. In polymer solutions or colloidal dispersions, the difference in scattering length density between the solvent and the dissolved or dispersed material gives rise to the scattering contrast (Higgins and Benoit 1994; Dmitri and Michel 2003). The great advantage of neutrons is the difference in scattering lengths between hydrogen and deuterium. For example, using D_2O instead of H_2O as a solvent is common practice when studying aqueous solutions with SANS. The scattering contrast between the solvent and the solute is this way is greatly enhanced while the incoherent scattering is diminished. Another possibility is to use the appropriate volume ratios of H_2O and D_2O as a mixed solvent (contrast variation) in order to match the scattering length density of one component of the solute (contrast matching). Then, the scattering pattern of the rest of the dispersed components is extracted.

In dilute solutions or dispersions, the scattered intensity is the sum of the scattered intensities from individual scattering objects since their positions are uncorrelated. For an assembly of identical objects, the scattered intensity is written as in Eq. 1.

$$I(q) = N_p \left\langle \left| \int \mathrm{d}\vec{r} \cdot \Delta\rho(\vec{r}) \cdot \exp(i\vec{q}\,\vec{r}) \right|^2 \right\rangle \quad (1)$$

The integral of Eq. 1 takes into account all the neutron or photon path differences arising from an individual scattering particle, while N_p is the number of particles per unit volume. The neutron scattering length density difference $\Delta\rho(\vec{r})$ between the solvent and the particle is in general a function of position. The average is taken over all possible orientations (which are equivalent in dilute solutions of randomly translating and rotating uncorrelated particles), and hence the resulting scattered intensity depends on the magnitude of the scattering wave vector $q = \frac{4\pi}{\lambda}\sin\left(\frac{\theta}{2}\right)$, where λ is the wavelength and θ the scattering angle.

Alternatively, the scattered intensity from N_p identical independent particles in solution is written as in Eq. 2 where $P(q)$ is the form factor, V_p the volume, and $\Delta\rho_\mathrm{p}$ the volume average scattering contrast of the particle (Pedersen 1997).

$$I(q) = N_\mathrm{p} \cdot V_\mathrm{p}^2 \cdot \Delta\rho_\mathrm{p}^2 \cdot P(q) \quad (2)$$

For objects with well-defined geometry, e.g., homogeneous spheres, core–shell spherical micelles, and homogeneous cylinders, the form factor has clear characteristic features (oscillations) in the high-q region of the spectrum (Fig. 1a). In the case of objects with sharp interfaces as the ones just mentioned, the high-q region shows an overall power-law dependence $I(q) \sim q^{-4}$ (Porod's law).

Scattering objects are often soft particles, for example polymer chains, colloidal aggregates, and microgels. These formulations do not show either a sharp boundary interface or a solid (fixed) internal structure. Their surface can be diffuse, and their interior experiences concentration fluctuations. Their intraparticle spatial correlations are described by the density autocorrelation function $g(\vec{r}) = \int \mathrm{d}\vec{r}\,' \cdot n(\vec{r} - \vec{r}\,') \cdot n(\vec{r}\,')$.

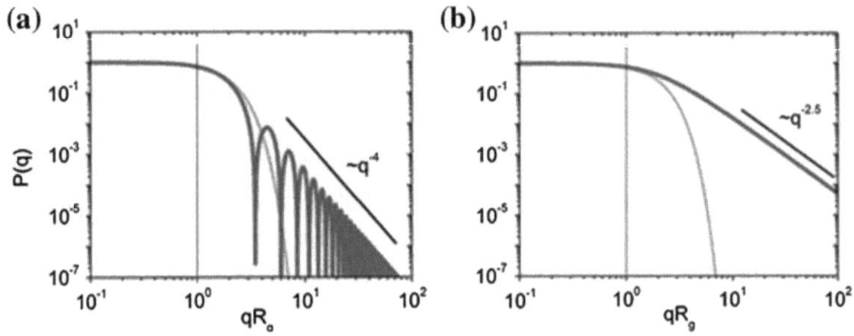

Fig. 1 Form factors from homogeneous spheres (**a**) and fractal aggregates with $D = 2.5$ (**b**). The light gray lines are the Guinier approximation, and the vertical lines highlight the point where $qR_g = 1$. Black lines indicate the power-law slopes

The number density of scatterers $n(\vec{r})$ describes the distribution of scattering length density within the particle (Teixeira 1986; Sorensen 2001). In aggregates of particles or polymers, the internal morphology has fractal nature and its average structure may be described by self-similar structures (Sorensen 2001; Beaucage 1996). The density autocorrelation function has the general form (isotropically averaged) of Eq. 3 where D is the fractal and d the spatial dimension. The term r^{D-d} results in the power-law behavior of the scattering curve as $I(q) \sim q^{-D}$ (Fig. 1b). The cutoff term $h(r/\xi)$ is a function that drops faster than any power law and defines the boundary of the soft particle at a characteristic length ξ which is termed correlation length.

$$g(r) = r^{D-d} \cdot h(r/\xi) \tag{3}$$

The density autocorrelation function and the scattering profile are Fourier transform pairs (Glatter 1977; Koch et al. 2003) since Eqs. 1 and 2 can be rearranged into Eq. 4 under the isotropic assumption.

$$P(q) = \int \mathrm{d}r \cdot \frac{\sin(qr)}{qr} \cdot 4\pi r^2 \cdot g(r) \tag{4}$$

At low q, the scattering profile is defined by the radius of gyration (R_g) of the particles (Eq. 5) which is the well-known Guinier approximation and proves that at low q the exact shape and internal structure of the scattering object do not affect $P(q)$. The distinction between low and high q is defined by R_g at $qR_g \approx 1$. The choice of cutoff function has an effect on the transition of the scattering profile from the Guinier to the power-law region, while the limiting trends are obviously unaffected (Sorensen 2001).

$$P(q) = \exp\left(-q^2 R_g^2\right) \tag{5}$$

Table 1 Fractal dimensions for several internal self-similar morphologies (Beaucage 1996; Schmidt 1991)

Morphology	Fractal dimension
Rigid rod	$D = 1$
Self-avoiding coil	$D = 5/3$
Gaussian coil	$D = 2$
Branched swollen coil	$D = 2$
Thin platelet	$D = 2$
Randomly branched ideal polymer	$D = 2.29$
Mass fractal	$2<D<3$
Surface fractal	$3<D<4$
Sharp interface	$D = 4$

In conclusion, the scattering profile from individual particles is described by two distinct limiting regimes. At $qR_g < 1$, the radius of gyration can be defined with the Guinier approximation. At $qR_g > 1$, the internal morphology is revealed. If the particles are solid with well-defined morphology, characteristic features appear. If the internal structure is soft and self-similar, then the fractal dimension is measured. In Table 1, the characteristic fractal exponents for several morphologies are presented. For a calculation that can be reliably compared to experimental data, polydispersity effects may need to be included (Pedersen 1997), while in SANS the instrument resolution function (uncertainty in q) has to be taken into account (Radulescu et al. 2015; Barker and Pedersen 1995). Both treatments lead to a smearing of the characteristic oscillations and curvature of the scattering profile.

In soft materials, the morphology may very often depend on the length scale. In other words, these materials are structured in multiple levels in a hierarchical manner. A combination of ultra-small-angle neutron scattering (USANS), ultra-small-angle X-ray scattering (USAXS), and SAXS covered almost four orders of magnitude in q (10^{-5}–10^{-1} Å^{-1}) exploring length scales from 5 nm to about 20 μm. The authors have resolved the hierarchical morphology of polyisoprene and poly-(styrene-random-butadiene) copolymer rubbers reinforced by carbon black (Koga et al. 2008). A combination of multiple components from the unified Beaucage model was used to model the experimental data (Beaucage 1995). It was shown that the filler particles comprised the smallest structural level and that their surface was rough with a fractal exponent $\sim$3.5. The carbon black particles (13 nm in radius) assemble in aggregates of fused particles that contain about nine of them. The aggregates are further assembled in agglomerates within the polymeric network with a mass fractal dimension 2.6. The results are confirmed by transmission electron microscopy (TEM) experiments.

A three-level hierarchical model was found for poly(ethylene oxide)-b-poly(*N*-isopropylacrylamide) (PEO-b-PNIPAM) diblock copolymers in aqueous solutions (Papagiannopoulos et al. 2013) below the lower critical solution temperature in PNIPAM. The scattering profile was modeled (Fig. 2a) by a three-term hierarchical unified Beaucage model in the form of Eq. 6.

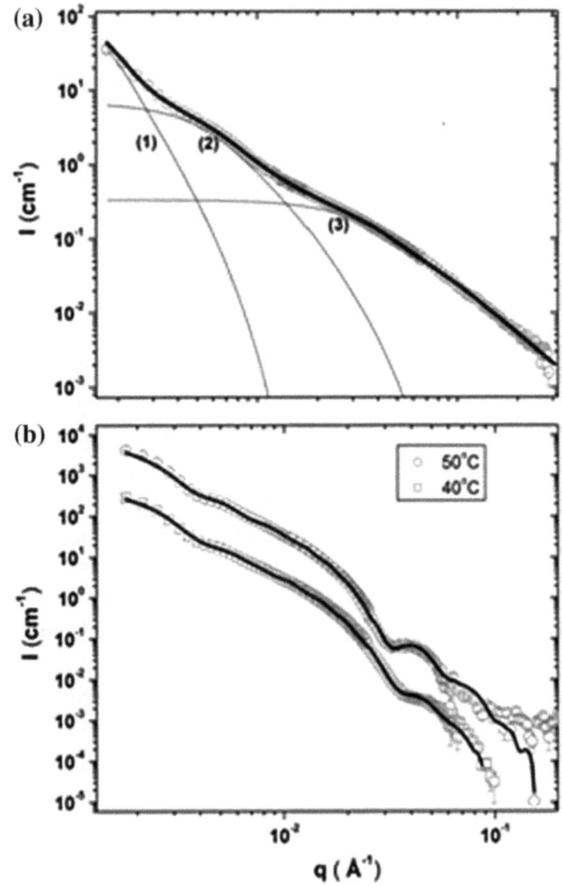

Fig. 2 **a** Modeling SANS data for 6.7 mg/ml PEO-b-PNIPAM in D_2O at 30 °C. Thick line is the fitting curve, while lines (1), (2), and (3) correspond to the terms of Eq. 6. **b** SANS data for 3.3 mg/ml PEO-b-PNIPAM in D_2O at 40 and 50 °C fitted by the three-layer ellipsoidal vesicle model. Data of 40 °C are divided by 10 for clarity. Reprinted from (Papagiannopoulos et al. 2013), Copyright 2013, with permission from Elsevier

$$\begin{aligned} I(q) = {} & G_1 \cdot \exp\left(-q^2 R_{g_1}^2/3\right) + \frac{B_1}{q^{d_1}} \cdot \left[\operatorname{erf}\left(qR_{g_1}/\sqrt{6}\right)\right]^{3d_1} \cdot \exp\left(-q^2 R_{g_2}^2/3\right) \\ & + G_2 \cdot \exp\left(-q^2 R_{g_2}^2/3\right) + \frac{B_2}{q^{d_2}} \cdot \left[\operatorname{erf}\left(qR_{g_2}/\sqrt{6}\right)\right]^{3d_2} \cdot \exp\left(-q^2 R_{g_3}^2/3\right) \\ & + G_3 \cdot \exp\left(-q^2 R_{g_3}^2/3\right) + \frac{B_3}{q^{d_3}} \cdot \left[\operatorname{erf}\left(qR_{g_3}/\sqrt{6}\right)\right]^{3d_3} \end{aligned} \tag{6}$$

where the prefactors B_i are such to provide a smooth transition from the Guinier to the power-law regime. The power laws of the higher levels are cutoff by a Guinier term of their subsequent level (Hammouda 2010; Beaucage 1995).

The small length-scale population was identified as unperturbed coils of diblock copolymer chains while the intermediate length scale as aggregates of coils. It was found that the number density of unperturbed coils was proportional to concentration while the intermediate one was concentration-independent. This led to the conclusion that the intermediate aggregates act as nucleation sites for the virtually unperturbed coils. The aggregates form larger clusters (highest level of hierarchy) that are rough surface fractals that change to mass fractals or polydisperse clusters as concentration increases. Interestingly above the LCST of PNIPAM, the system passes to a concentration-independent conformation of well-defined ellipsoidal vesicles (Fig. 2b). The vesicles are made of three homogeneous layers where the central one consists of PNIPAM and the outer ones of both PNIPAM and PEO segments.

In polymer solutions, the chains start to overlap when the average interchain distance becomes comparable to the size of unperturbed chains. This concentration is termed overlap concentration (c^*) and separates the dilute from the semidilute regime. In semidilute solutions, the correlation length (ξ) is the average distance between segments of different neighboring chains. The small-angle scattering profile from a semidilute solution contains information about the intra- and inter-chain correlations depending on the length scale that is being probed (Daoud et al. 1975). At length scales higher than the ξ, the density autocorrelation function scales as $g(r) \sim 1/r$. At small length scales, it is dominated by the intrachain correlations. Similarly to the self-similar case of Eq. 3 $g(r) \sim r^{-d+1/v}$. In this case, v is the unperturbed chain scaling exponent. The small-angle scattering profile is given in Eq. 7 for different q regimes. The structure factor $S(q)$ is traditionally used in reference to the interference arising from interacting colloids, chains, etc.

$$S(q) \sim \begin{cases} \frac{1}{1+q^2\xi^2} & \text{for } q < 1/\xi \\ q^{-1/v} & \text{for } q > 1/\xi \end{cases} \tag{7}$$

For interacting identical particles with form factor $P(q)$, the resulting scattering pattern is modified by the structure factor as $P(q) \cdot S(q)$ and Eq. 4 in the form $S(q) = \int \mathrm{d}r \cdot \frac{\sin(qr)}{qr} \cdot 4\pi r^2 \cdot g(r)$ provides the interparticle correlations.

In semidilute polyelectrolyte solutions, density fluctuations at $r > \xi$ are suppressed by the high osmotic pressure of the counterions causing a decrease of the scattered intensity at low q. This results in the "polyelectrolyte peak" in the small-angle scattering profiles which is commonly used to define the correlation length in polyelectrolyte solutions (Dobrynin et al. 1995).

Normally, swollen polymer networks are in the semidilute range of concentration. The presence of cross-links introduces spatial inhomogeneities in concentration that are static (Panyukov and Rabin 1996). This is an important difference from solutions where density fluctuations are able to relax by thermal motion. The cross-linking process creates a unique density distribution of monomers with fluctuations that do not average to zero during time. In liquids, the time (and ensemble) average of the density fluctuation vanishes. In a gel that resulted from a semidilute polymer solution with concentration ρ_{av}, the time-averaged concentration $\langle \rho_{\mathrm{t}} \rangle$ is not uniform as in the case of the solution (Fig. 3). The time-independent component in dynamic light scattering intensity correlation functions on gels arises from the appearance of stationary speckle patterns caused by these frozen density

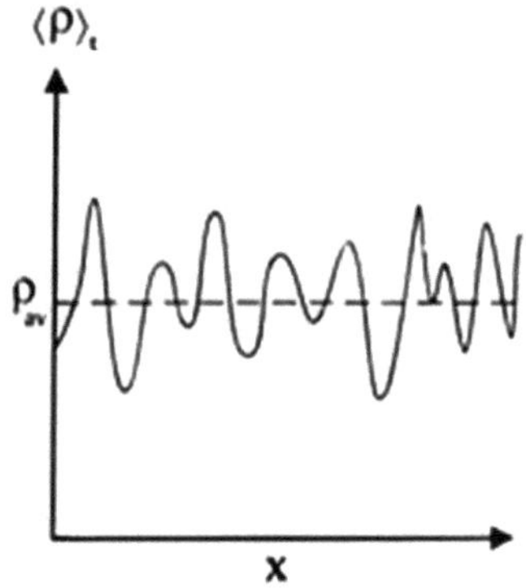

Fig. 3 Illustration of time-averaged concentration of monomers in the gel (solid line) and in the solution (broken line) as a function of position *x*. Reprinted with permission from (Panyukov and Rabin 1996). Copyright 1996 American Chemical Society

fluctuations (Panyukov and Rabin 1996). At length scales shorter than the correlation length, density fluctuations are expected to be similar to the ones in the semidilute solution without cross-links.

Structure factors from swollen polymer networks are often described by the two-component scattering function of Eq. 8. The Ornstein–Zernike $I_{\mathrm{OZ}}(q)$ and the Debye–Bueche (squared Lorentz) $I_{\mathrm{SL}}(q)$ terms describe the thermal concentration fluctuations and static concentration inhomogeneities, respectively (Shibayama 2011; Horkay et al. 2000). The two characteristic length scales are the correlation length (ξ) of the semidilute solution and the characteristic size of the inhomogeneities (Ξ). Since normally $\xi \ll \Xi$, the first term dominates at short and the second at long length scales. There is an illustrative example of applying Eq. 8 in hydrogels used for protein loading and release (Smeets et al. 2014).

$$I(q) = \frac{I_{\mathrm{OZ}}(0)}{1+q^2\xi^2} + \frac{I_{\mathrm{SL}}(0)}{\left(1+q^2\Xi^2\right)^2} \tag{8}$$

$$I(q) = \frac{I_{\mathrm{OZ}}(0)}{1+q^2\xi^2} + \frac{I_{fr}(0)}{\left(1+\frac{D_f+1}{3}q^2\Xi^2\right)^{\frac{1}{2}D_f}} \tag{9}$$

Gels are complex materials that can often have self-similar properties, especially near their gelation threshold. Hence, the observed fractal power-law exponents D_f for the long wavelength behavior of the structure factor may be different than $D_f = 4$. A generalized Zimm function can replace the Debye–Bueche term in Eq. 8 (Eq. 9) so that a characteristic exponent D_f can be extracted from the experimental data (Shibayama et al. 1992). Mesh size (ξ) has been proved to correlate not only with the short-scale dynamics (see next section) of a hydrogel but also with the macroscopic friction coefficient which is a crucial parameter in biological lubrication (Urueña et al. 2015).

2.2 Viscoelasticity of Polymer Fluids Probed by Rheology

Rheology investigates the mechanical strength and flow properties of complex fluids and consequently hydrogels. Hence, it is a basic tool for characterization of industrially relevant properties. At the same time, rheological measurements provide information on the internal structure of soft materials based on their dynamic response. They can detect bulk phase transitions as the solution-to-gel transition (solgel) which can be triggered by concentration, pH, and temperature changes (Pape et al. 2014). Several geometries can provide well-defined conditions of flow for a rheological experiment. Concentric cylinders (Couette), cone-and-plate and parallel disks are traditionally used (Larson 1999).

In a viscoelastic fluid, both viscous and elastic components appear while their relevant strength depends on the timescale of measurement. For example, in polymeric liquids elasticity originates from the entropy of macromolecular chains. The simplest model of a viscoelastic fluid is the combination of a dashpot and a spring in series, i.e., the Maxwell model. The two components introduce the viscosity η_0 and the storage modulus G_0, respectively. The total strain ε is the sum of the individual strains of the elastic (ε_1) and the viscous (ε_2) components (Eq. 10a). The total stress (σ) is transferred to both elements (Eq. 10b).

$$\varepsilon = \varepsilon_1 + \varepsilon_2 \tag{10a}$$

$$\sigma = G_0 \varepsilon_1 = \eta_0 \frac{\mathrm{d}\varepsilon_2}{\mathrm{d}t} \tag{10b}$$

Differentiating 10a and substituting 10b, the constitutive equation for the Maxwell model is obtained.

$$\frac{\mathrm{d}\varepsilon}{\mathrm{d}t} = \frac{1}{G_0}\frac{\mathrm{d}\sigma}{\mathrm{d}t} + \frac{1}{\eta_0}\sigma \tag{11}$$

The stress relaxation modulus $G(t)$ is obtained by imposing a constant step strain and measuring the time dependence of the resulting stress. A Maxwell model reacts like an elastic solid at short timescales while flows like a liquid at long times. Solving 11 for an imposed step strain $\varepsilon(t) = \varepsilon_0 u(t)$ and defining $G(t) = \frac{\sigma(t)}{\varepsilon_0}$, one gets:

$$G(t) = G_0 e^{-t/\tau_0} \tag{12}$$

Stress in Maxwell model relaxes exponentially with a characteristic time $\tau_0 = \frac{\eta_0}{G_0}$. Hence, the relaxation time separates the elastic and viscous behavior in the Maxwell fluid.

In an oscillatory experiment, a sinusoidal stress $\sigma(t) = \sigma_0 e^{i\omega t}$ is imposed and the resulting sinusoidal strain $\varepsilon(t) = \varepsilon_0 e^{i\omega t + \varphi}$ is measured (or the opposite). The storage ($G'(\omega)$) and loss moduli (($G''(\omega)$)) are defined as the real and imaginary part of the complex modulus $G^*(\omega) = \frac{\sigma(t)}{\varepsilon(t)}$ (Eq. 13), respectively. The characteristic frequency $\omega_0 = 1/\tau_0$ separates the viscous ($\omega < \omega_0$) from the elastic behavior ($\omega > \omega_0$). At low frequencies, $G' \sim \omega^2$ and $G'' \sim \omega$ with $G' < G''$. At high frequencies, $G' \sim \omega^0$ and $G'' \sim \omega^{-1}$ with $G' > G''$ (Fig. 4a).

$$G'(\omega) = G_0 \frac{\omega^2 \tau_0^2}{1 + \omega^2 \tau_0^2} \tag{13a}$$

$$G'(\omega) = G_0 \frac{\omega \tau_0}{1 + \omega^2 \tau_0^2} \tag{13b}$$

An isolated Gaussian macromolecular chain in solution can be modeled by the Rouse model (Rubinstein and Colby 2003). A monomer has a friction coefficient $\eta_s b$ and a diffusion coefficient $\frac{kT}{\eta_s b}$ (neglecting numerical prefactors) where kT is the thermal energy, η_s the solvent viscosity, and b the monomer size. The characteristic time of a monomer to diffuse the size of itself is $\tau_0 = \frac{b^2}{D} = \frac{\eta_s b^3}{kT}$. In Rouse model, there are no hydrodynamic interactions between monomers. The friction coefficient of a chain of N monomers is $N\zeta$ and the size of the chain $N^{1/2}b$ so the characteristic time for the whole chain is $\tau_R = \frac{Nb^2}{kT/\eta_s Nb} = \frac{\eta_s b^3}{kT} N^2 = \tau_0 \cdot N^2$. The motion of the chain is decomposed in N modes $p = 1, \ldots, N$ where in the pth mode the chain divided in p subchains. The relaxation time of the pth mode is $\tau_p = \tau_0 \cdot \left(\frac{N}{p}\right)^2$. At a certain frequency ω, a mode with characteristic time faster than $1/\omega$ is unrelaxed and contributes kT to the relaxation modulus. The viscoelastic moduli are eventually written as in Eq. 14.

$$G'(\omega) = kT \sum_{p=1}^{N} \frac{\omega^2 \tau_p^2}{1 + \omega^2 \tau_p^2} \tag{14a}$$

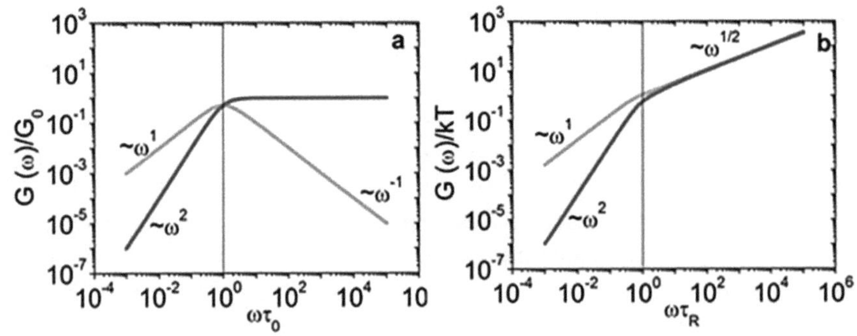

Fig. 4 Viscoelastic moduli (G' ▬ and G'' ▬) of **a** Maxwell (Eq. 13) and **b** Rouse model (Eq. 14)

$$G'(\omega) = kT\sum_{p=1}^{N}\frac{\omega\tau_p}{1+\omega^2\tau_p^2} \tag{14b}$$

As shown in Fig. 4b, the viscoelastic moduli are Maxwell-like up to frequencies near the inverse terminal relaxation time $1/\tau_R$. Above that frequency, power-law dependence is found where $G' = G'' \sim \omega^{1/2}$. This is a signature of self-similarity in dynamics that results from the self-similarity in chain conformation. At low frequencies, rheological measurements probe the slow motion of the whole chain. Increasing the frequency, faster motions of smaller segments of the chain are probed. In dilute solution (isolated chains), the hydrodynamic interaction between monomers has to be taken into account to obtain a realistic result. Zimm model is the proper theory to analyze the viscoelasticity of ideal chains in dilute solution. Approximately, the relaxation time of a chain segment is estimated by assuming that as it diffuses it drags the volume of the solvent it occupies as a whole. Still, the viscoelastic character shows Maxwell-like behavior at times longer than the terminal relaxation time and power-law behavior at shorter times ($G' \cong G''/1.73 \sim \omega^{2/3}$). In semidilute solutions of neutral or charged linear macromolecules (Dobrynin et al. 1995), a chain adopts an ideal conformation at length scales larger than the correlation length. Additionally, the hydrodynamic interactions are screened at these length scales. Consequently, Rouse model adequately describes the dynamics in semidilute solutions in the absence of entanglements.

In a gelling liquid, the correlation length (size of the typical cluster) increases as more bonds are created. As the point of gelation is approached, the size of clusters spans the whole fluid volume and the correlation length diverges. Below the critical point for gelation, the fluid is mostly viscous $G'' > G'$, while in the gelled fluid it is predominantly elastic $G'' < G'$. At the critical point for gelation, the structure is self-similar with no characteristic microscopic length scale (Vilgis and Winter 1988). There the viscoelastic moduli present self-similarity in time as there is no characteristic relaxation time and they follow a power law $G' \sim G'' \sim \omega^a$. Hence in critical gels, the stress relaxation modulus is represented (Winter and Mours 1997) by its critical exponent a and its strength S_g (Eq. 15).

$$G(t) = S_g \cdot t^{-a} \tag{15}$$

In formed gels and networks, the viscoelastic modulus is predominantly elastic and frequency independent for a wide frequency range. There is no terminal relaxation time to liquid-like flow except from the cases where there are physical bonds that eventually relax and flow begins. Under the assumption of an affine network, i.e., a network that follows the bulk deformation of the sample, each elastic chain strand that is fixed between two cross-links contributes kT to the elastic modulus (Larson 1999). Hence, G is proportional to the number density (n) of the effective elastic strands (Eq. 16). In entangled polymer solutions, topological constrains restrict the chains forcing them to reptation inside a tube formed by neighboring chains. These constrains act as temporary cross-links for timescales faster than a terminal relaxation time. This time is much slower than the terminal relaxation time in the unentangled regime.

$$G = nkT \tag{16}$$

In Fig. 5, the frequency dependence of G' and G'' of hydrogels composed from two amphiphilic macromolecules *N*-palmitoyl-Gly-His (PalGH) and sodium palmitate (PalNa) at different molar percentages (Matsumoto et al. 2015). In all cases, G' is higher than G'' for the full frequency range. For 0, 20, and 100% molar ratios, G' is almost frequency independent.

Very soft hydrogels (Shen et al. 2007) have been produced by an artificial triblock protein $AC_{10}A$ made of a middle random coil block (C_{10}) and two associative leucine zipper endblocks (A). The physical bonds created by these chains are not permanent, and hence the hydrogels flow at sufficiently low frequencies in a Maxwell fluid manner (Fig. 6a). Nevertheless, a plateau of the storage modulus appears at the rest of the frequency range, while the loss modulus drops. The

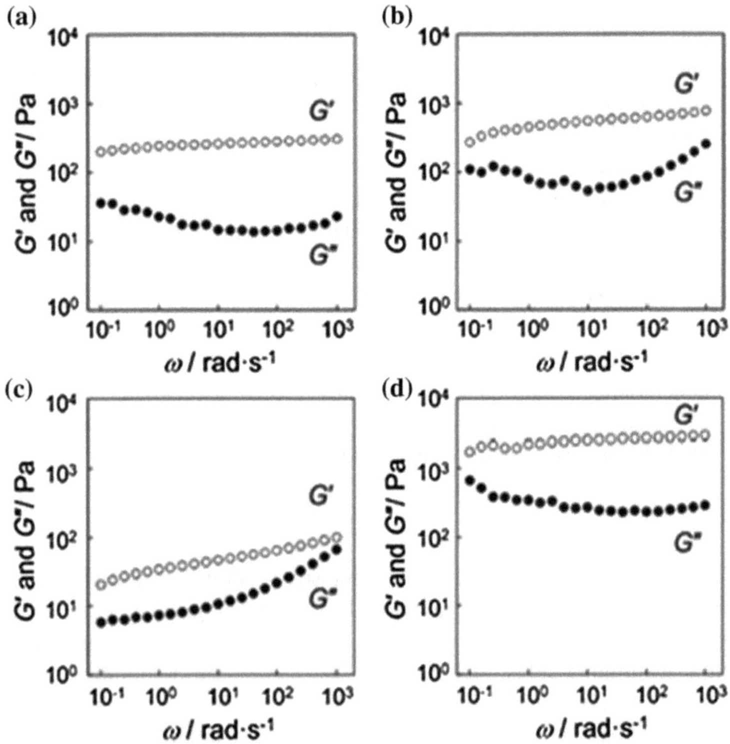

Fig. 5 Viscoelastic moduli from PalGH/PalNa supramolecular hydrogels at several molar ratios. **a** 0, **b** 20, **c** 50, and **d** 100%. Reproduced from (Matsumoto et al. 2015) with permission from the PCCP Owner Societies

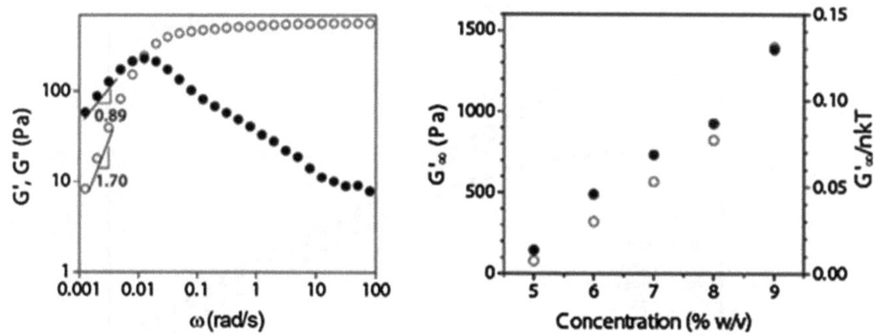

Fig. 6 Viscoelastic moduli (G' ○, G'' ●) at 7% w/v $AC_{10}A$ hydrogel (left). Plateau storage modulus as a function of $AC_{10}A$ concentration (G'_{∞} ○, G'_{∞}/nkT ●). Reproduced from (Shen et al. 2007) with permission from The Royal Society of Chemistry

measured G at the plateau G'_{∞} (Fig. 6b) is compared to the expected one assuming that all chains are elastically effective (Eq. 16). The observed G'_{∞}/nkT ratio does not exceed 15% because the middle flexible group allows individual protein chains to form closed loops instead of interchain bridges (Shen et al. 2007).

A linear oscillatory experiment needs to be performed well within the linear regime of the material response. This is traditionally checked before every experiment by performing an oscillatory test of increasing stress (or strain) amplitude at a preset frequency. Up to a certain value of amplitude, measurements of viscoelastic moduli are amplitude independent (Choi et al. 2015). Nonlinear response may be caused by the disturbance of the fluids internal structure. In soft hydrogels, for example, breaking of physical bonds occurs after a certain strain and the viscoelastic modulus weakens.

Nonlinear rheological measurements provide information on the performance of a fluid during flow. For example, a hydrogel during injection is preferred to have a reduced viscosity, the so-called shear-thinning property (Guvendiren et al. 2012; Choi et al. 2015). After the cessation of shear stresses, the hydrogel has to recover its initial integrity, i.e., to have self-healing properties. A traditional experiment to observe the non-Newtonian shear-thinning (or shear thickening) of a complex fluid is the steady shear viscosity test. In Fig. 7, the shear-thinning behavior of xanthan solutions is observed in a wide range of shear rates. At low shear rates, the Newtonian plateau makes the measurement of zero shear viscosity feasible. Above a critical shear rate, power-law decrease of viscosity takes place (Wyatt and Liberatore 2010). Recently, the power-law viscoelasticity of xanthan soft hydrogels was analyzed by particle tracking microrheology (Papagiannopoulos et al. 2016), and its linear response was connected to the nonlinear steady shear viscosity. The use of probe particles in microrheological techniques allows the determination of viscoelasticity as a function of length scale, and this is a key concept for connecting structure and dynamics in complex fluids (Papagiannopoulos 2016).

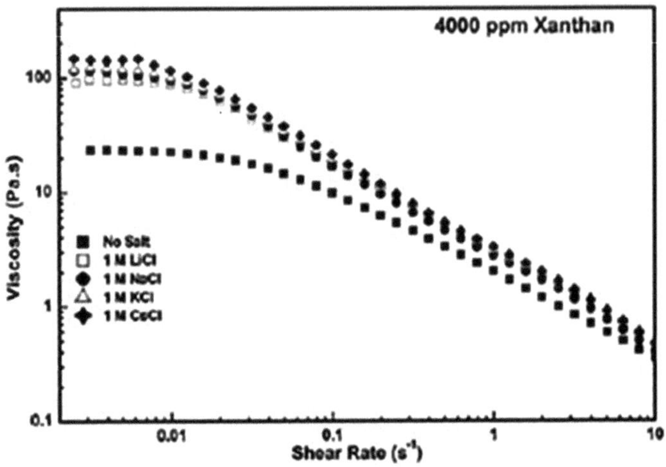

Fig. 7 Steady shear viscosity of xanthan solutions in the absence and presence of several salts. Reproduced from Wyatt and Liberatore 2010) with permission from The Royal Society of Chemistry

3 Investigation of Rheological and Structural Properties of Protein- and Nanoparticle-Loaded Hydrogels

In biotechnology and nanomedicine, hydrogels need to incorporate (loading) and deliver (release) drug molecules and proteins on demand. For example, the release of a protein can be tuned by temperature in a thermoresponsive gel. There the pore size and temperature-induced shrinkage control the burst of water and protein molecules from the hydrogel (Zhang et al. 2010). In order to test loading processes, samples of well-defined geometry (e.g., cylindrical) are introduced to a protein solution in the desired salt and pH conditions (Schillemans et al. 2011) and protein concentration is measured in several time intervals. Protein release is quantified by incubating the loaded hydrogels in buffer solutions and measuring the amount of the released protein.

Bovine serum albumin (BSA) and myoglobin were loaded to oppositely charged and chemically cross-linked dextran networks (Schillemans et al. 2011). Loading and consequently release were found to depend on the distribution of proteins inside the network. BSA unexpectedly showed similar diffusion coefficient for release compared to myoglobin. The size of BSA is comparable to the networks pore size (mesh size ~7 nm), while myoglobin is smaller. BSA globules immobilized by strong electrostatic interaction on the hydrogel pores were found to create steric hindrance to globules that would diffuse deeper into the network. Hence, loading of BSA was not homogeneous but mostly covered the rim of the cylindrical specimens (Fig. 8). The distribution was more homogeneous at high ionic strength proving the electrostatic nature of BSA–hydrogel interaction. In another study, the

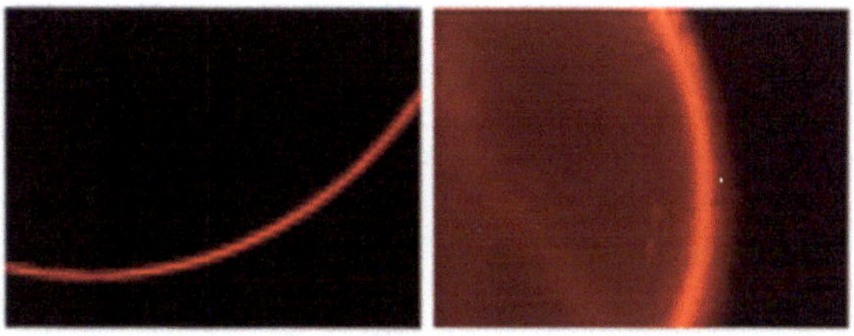

Fig. 8 Fluorescence image FITC-labeled BSA loaded to positively charged dextran hydrogels. No added NaCl (left) and 0.1 M NaCl (right). Reprinted from (Schillemans et al. 2011), Copyright 2011, with permission from Elsevier

retention of immobilized lipase in cationic polyphosphazene hydrogels was tested and release could be tuned by the solution ionic strength. These hydrogels could be used for repeated enzyme loading (Qian et al. 2015).

Interesting connection between the mesh size of hydrogels and the release of a loaded compound was found in eight-armed cross-linked PEG hydrogels (Kirchhof et al. 2015). Gel cylinders were immersed in buffer solutions, and their volume was monitored as a function of time. From the swelling ratio, the average mesh size was estimated. Both volume and mesh size increased exponentially during the first 30 days. Modeling the rheological data with a superposition of four Maxwell models and using network elasticity theory, the mesh size was calculated and was found consistent to the one estimated by swelling studies. Nuclear magnetic resonance was an additional support to the findings. The release of fluorescein isothiocyanate-labeled dextrans from the hydrogels was successfully modeled on the basis on their diffusion coefficient in free solution against the diffusion coefficient in the hydrogels of known mesh size. The authors concluded that a refinement of the method could be done by taking into account a distribution of mesh sizes instead of a fixed average value (Kirchhof et al. 2015).

Disulfide-cross-linked hydrogels were prepared by adding reductants (DTT) in aqueous solutions of poly(acrylamide) chains functionalized with pyridyl disulfide (PDS) side groups (Chen et al. 2015). In situ, rheological measurements followed the kinetics of gelation under several conditions. In Fig. 9, the evolution of viscoelastic moduli (at 0.1 Hz) clearly represents the solgel transition of the system. The crossover point of G' and G'' indicates the gelling point. Gelation time appears to be faster in solutions with a higher DTT to PDS molar ratio. The immobilization of BSA in these hydrogels was tested by confocal microscopy images on FITC-labeled BSA. It was found that due to diffusion limitation BSA would not reach depths much larger than 10 μm.

In Fig. 10, a successful scheme to form (Du et al. 2014) protein-based hydrogels for 3D cell culture is presented. The protein globules are part of the actual scaffold network. The primary structure of tax interactive protein-1 (TIP1) was genetically modified by two or four cysteine residues. The cysteine groups may react with maleimide residues on the ends of four-armed polyethylene glycol (PEG) to create a cross-linked network (Du et al. 2014). Scanning electron microscopy images of the

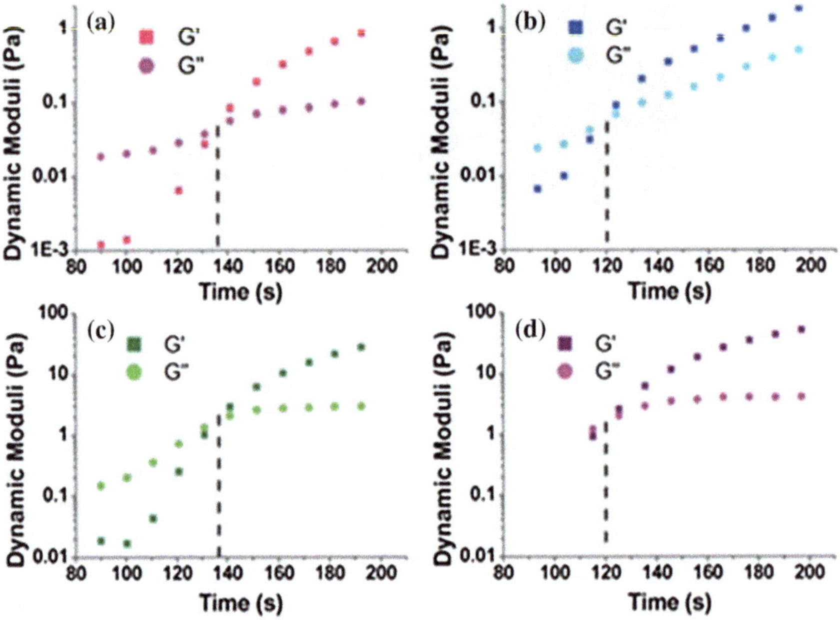

Fig. 9 Viscoelastic moduli as a function of gelation time for different polymer concentrations/DTT to PDS molar ratios. **a** 10%/0.5, **b** 10%/1.0, **c** 15%/0.5, and **d** 15%/1.0. Reproduced from (Chen et al. 2015) with permission from The Royal Society of Chemistry

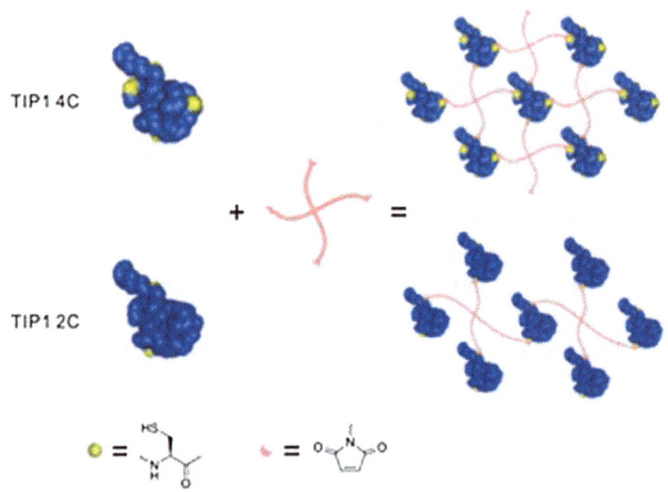

Fig. 10 Schematic illustration of cross-linking four-armed PEG molecules with TIP1 with four (TIP1 4C) or two cysteines (TIP1 2C). Reproduced from (Du et al. 2014)

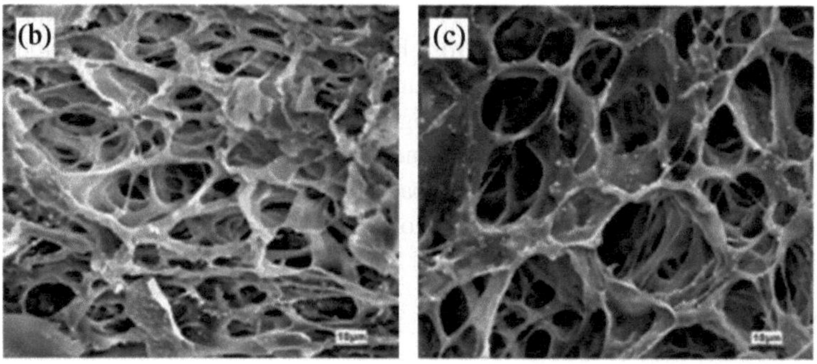

Fig. 11 SEM images from TIP1 2C (**b**) and TIP1 2C RGD (**c**) hydrogels (white bars = 10 μm). Reproduced from (Du et al. 2014)

hydrogels from TIP1 2C without and with the bioactive peptide RGD show the morphological ability for accommodation and culture of cells in 3D (Fig. 11). Their mechanical ability was tested by rheological measurements where $G' \approx 100 - 200$ Pa. Most of the AD293 (Adeno-X 293) cells were alive after a 3-day culture. Additionally, cells could be recovered after dissolving the hydrogel by trypsinase at 37° for 5 min.

Hierarchical hydrogels from genetically engineered elastin-like polypeptides showed self-healing and enhanced mechanical properties (Ghoorchian et al. 2015). The polypeptides form micelles with large hydrophobic cores and hydrophilic coronas. Since their hydrophilic blocks are terminated with metal binding groups, they can be cross-linked with metal ion coordination adding zinc sulfate. The reaction is performed at 45 °C where the micelles are formed. The presence of zinc sulfate appears to stabilize the micelles even at room temperature. SAXS profiles show correlation peaks caused by the interacting micelles. The characteristic intermicellar distances extracted by SAXS analysis show liquid-like scaling, i.e., $d \sim c^{-1/3}$ where c is the micellar concentration. Viscoelastic measurements prove that the presence of hydrophobic domains enhances the modulus which can be tuned by the amount of cross-linking agent.

The structure of cross-linked protein networks can be resolved by small-angle scattering. The form factor $P(q)$ of a single protein can be defined in dilute solution by Eq. 2. In a protein gel, the scattering pattern is modified by the structure factor $S(q) = \int \mathrm{d}r \cdot \frac{\sin(qr)}{qr} \cdot 4\pi r^2 \cdot g(r)$. Hence dividing the concentration-normalized scattering profile in the gel phase by the one in dilute solution, the structure factor is obtained (Eq. 17) assuming that protein conformation is unaltered between the two states.

$$S(q) = \frac{c_{\text{dilute}}}{c_{\text{gel}}} \cdot \frac{I(q)|_{\text{gel}}}{I(q)|_{\text{dilute}}} \tag{17}$$

Bovine pancreatic trypsin (BPTI), myoglobin (Mb), and intestinal fatty acid-binding protein (IFABP) were cross-linked by glutaraldehyde (GA) to produce three hydrogels. The pair correlation functions were extracted from the experimentally determined structure factors (Fig. 12) by Fourier transform (Kaieda et al. 2014). A maximum of the correlation functions at distances of the order of protein diameter indicates the presence of compact clusters. Integration of the pair correlation function over the first coordination cells provides the number of nearest

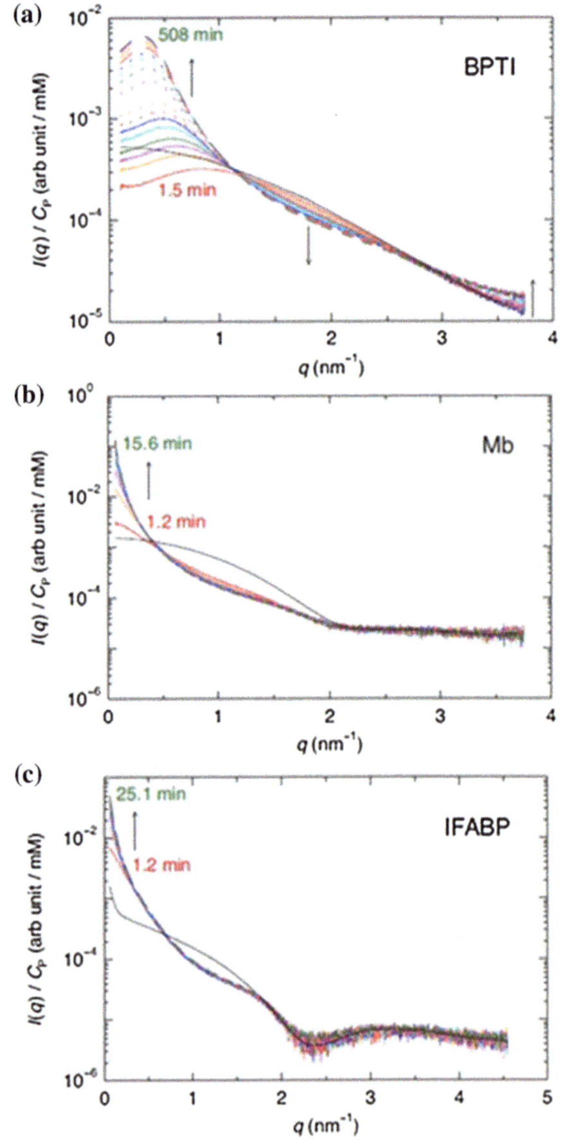

Fig. 12 Time-resolved SAXS profiles from evolution of gels of **a** BPTI, **b** Mb, and **c** IFABP. The times of the first and last profiles are shown. Black lines are the measured form factors of the individual proteins. From (Kaieda et al. 2014) Published by the PCCP Owner Societies

protein neighbors. Additionally, integrating the pair correlation function inside successive coordination shells provides the number of protein globules around another globule as a function of distance, i.e., the coordination number $N(r)$. The power law $N(r) \sim r^3$ for uniform distribution of protein molecules is recovered at distances larger than a correlation length ξ. This is a measure of the cluster–cluster separation distance, and it was found to be bigger for gels of lower protein concentration (Kaieda et al. 2014).

The kinetics of gel formation was followed by time-resolved SAXS. The SAXS profiles as a function of time after mixing a protein with the cross-linker are shown in Fig. 12. The solid black lines are the SAXS before gelation, i.e., the form factors of the proteins. The scattering profiles of MB and IFABP are unaltered at high q which demonstrates the negligible effect of gelation on the form factor. The evolution of the profiles is faster at intermediate q compared to low q indicating that gelation proceeds faster at short than at long length scales (Kaieda et al. 2014).

Hybrid triblock copolymers with a helical self-associating middle block and PNIPAM endblocks that reinforces gelation in a thermoresponsive way have versatile properties (Glassman and Olsen 2013). SANS measurements (Fig. 13a) are sensitive in the temperature response of the system. The scattering curves were analyzed by the $S(q)$ in the Percus–Yevick approximation for hard sphere interaction. The characteristic features of the $S(q)$ are defined by the hard sphere volume fraction (η) and radius (R_{HS}). A core radius (r_0) was also extracted by the fits. The formation of spherical domains is caused by the association of PNIPAM blocks which is enhanced as temperature increases. This is obvious in Fig. 13b where r_0 increases significantly around 30 °C. The trend is similar and even more intense for η. Interaction range of the hard spheres R_{HS} appears less sensitive to temperature. G' and G'' increase abruptly between 20 and 30 °C same range (where the increase of r_0 takes place) and passes through a peak that can be correlated with the maximum in η (Fig. 14c).

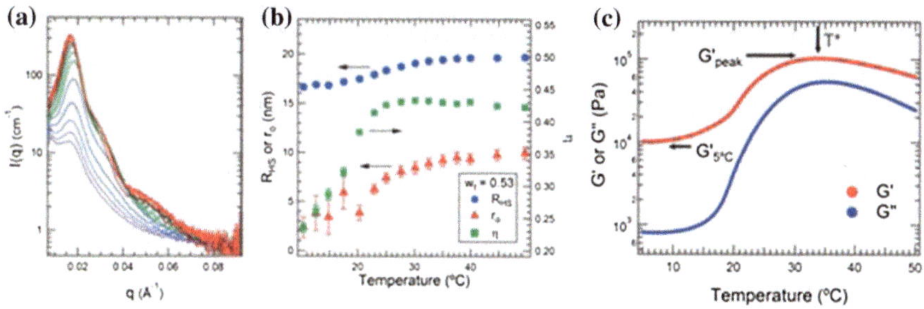

Fig. 13 **a** SANS for different temperatures (increasing from bottom to top from 10 to 50 °C). **b** Extracted parameters from SANS and **c** G' and G'' at 100 rad/s (1% strain amplitude). Results are from 30% (w/v) and triblock weight fraction wt = 0.53. Reproduced from (Glassman and Olsen 2013) with permission from The Royal Society of Chemistry

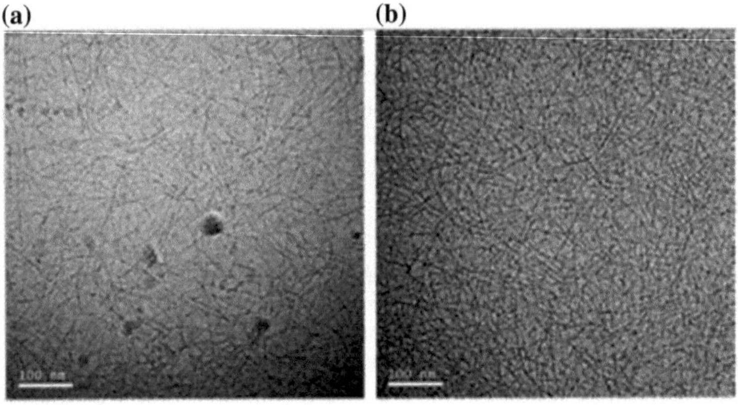

Fig. 14 Cryo-TEM from 0.5 (**a**) and 2 (**b**) wt% β-hairpin networks. Reproduced from (Hule et al. 2008) with permission from The Royal Society of Chemistry

Cytocompatible hydrogels can be formed by the self-assembly of β-hairpin peptide (Hule et al. 2008). Hydrogels elastic modulus was an increasing function of initial concentration, while the formation of the fibrillar networks needed about 1 h to develop. Cryo-TEM images show the network structure at 0.5 and 2 wt% (Fig. 14). It is obvious that at high concentration the fibrillar network is denser. The system was successfully tested as a scaffold for 3D cell culture.

Small-angle scattering quantitatively described the morphology of the β-hairpin fibrillar networks from nanometers to micrometers (Hule et al. 2008). The investigated hydrogels present a power law at the SANS low-q range (3×10^{-3} -10^{-2} Å^{-1}) with fractal dimension (~ 1.7) which is independent of concentration. This indicates an open mass fractal that is homogeneous at these length scales. At high q (10^{-1}–10^{0} Å^{-1}), the correlation length and scaling exponent could be extracted. The correlation length decreased as function of concentration since the density of chains and cross-links increased. The fractal exponent increased from 2.57 to 3.13 (in the concentration range 0.5–2 wt%) which revealed an increasing compactness of the self-assembled fibrils (Hule et al. 2008). At length scales 0.6–20 μm (5×10^{-5}–2×10^{-4} Å^{-1}), USANS measurements showed that sharp interfaces are created as Porod behavior $I(q) \sim q^{-4}$ was found at all concentrations.

Incorporation of hydrophobic magnetite nanoparticles hydrogels based on hydrophobically modified ethoxylated urethanes was studied by neutron spin echo, dielectric spectroscopy, and other techniques (Campanella et al. 2016). The authors used Eq. 18 to model the intermediate scattering function $S(q,t)$ obtained by the dynamic scattering neutron spin echo experiments. The characteristic relaxation rate $\Gamma(q)$ was diffusive at scales larger than 42 Å and revealed Zimm dynamics at short ones. The relaxation rate is connected to the spacing between hydrophobic domains since the terminal relaxation time is proportional to the square of the number of monomers within this distance $\tau_r \sim (d^{D_f})^2$. This distance was extracted by the maximum q of the SANS data as $d = 2\pi/q_{\max}$. Eventually for scaling in good solvent $\Gamma \sim d^{-10/3}$. With this treatment, the increase in distance between hydrophobic domains when the magnetic nanoparticles are embedded is shown to correlate with the dynamics of the largest blobs that connect the hydrophobic regions (Campanella et al. 2016).

$$\frac{S(q,t)}{S(q,0)} = P(q) + (1 - P(q)) \cdot \exp\left(-(\Gamma(q) \cdot t)^{\beta}\right) \tag{18}$$

Equation 18 allowed the decomposition of the scattering from thermal concentration fluctuations and static concentration inhomogeneities (Campanella et al. 2016). In more detail, the non-decaying component $P(q)$ is related to the static concentration inhomogeneities. Finally, the dielectric spectroscopy results that probe segmental dynamics at different timescales are consistent with the decrease of the relaxation rate upon addition of nanoparticles.

Hybrid hydrogels of riboflavin (*R*) and PNIPAM (chemically cross-linked) have been proposed as temperature detection probes (Chakraborty et al. 2014). Riboflavin introduces photoluminescence properties to the hydrogels. The fluorescence intensity of the hydrogels passes through a maximum which is located near the LCST of PNIPAM's coil-to-globule transition. *R* acts as a supramolecular cross-linker in the PNIPAM hydrogels as it is shown by FTIR experiments. The viscoelastic moduli of the hydrogels with and without *R* demonstrated normal gel character with $G' < G''$ and also with both G' and G'' independent of frequency. Increasing *R* concentration resulted in a drop in the modulus which was fairly unexpected as *R* acts as a supramolecular cross-linker. The reduction of H-bond interaction between PNIPAM chains caused by the increase in interchain distances after incorporation of *R* was a possible explanation (Chakraborty et al. 2014). On the other hand, the critical strain for hydrogel breaking increased linearly with the amount of added riboflavin.

The aforementioned hydrogels were used for the production of Ag nanoparticles by placing slashed hydrogel pieces inside $AgNO_3$ solutions (Fig. 15). The linear rheology experiments did not reveal any difference between the nanocomposite hydrogels and the initial hybrid R-PNIPAM ones. Additionally, the critical strain was lower in the nanocomposite hydrogels (Chakraborty et al. 2014).

A two-level Beaucage model (similar to Eq. 6) was employed to describe the data from gelatin hydrogels (Helminger et al. 2014). The model clearly defined the characteristic gyration radii and scaling exponents (Fig. 16). These systems were transformed into ferrogels by in situ growing of magnetite nanoparticles. Contrast variation of the solvent was performed to decompose the scattering into the one of the underlying hydrogel (gelatin) and the one from the NPs. Growth of NPs does not destroy the fractal morphology of the gels, while the size of NPs is limited by the gel mesh size. The experiments were supported by SAXS, TEM, simulations, swelling studies, and magnetic measurements.

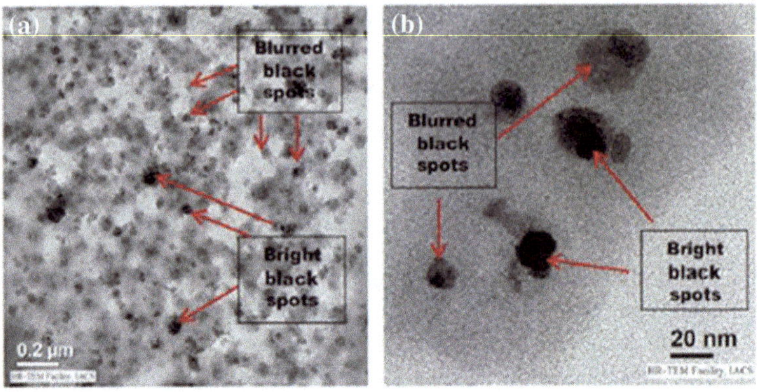

Fig. 15 Transmission electron microscopy images of Ag NPs formed inside R-PNIPAM hybrid hydrogels. Blurred and bright black spots refer to NPs in the interior and the surface of the hydrogel, respectively. Reproduced from (Chakraborty et al. 2014) with permission from The Royal Society of Chemistry

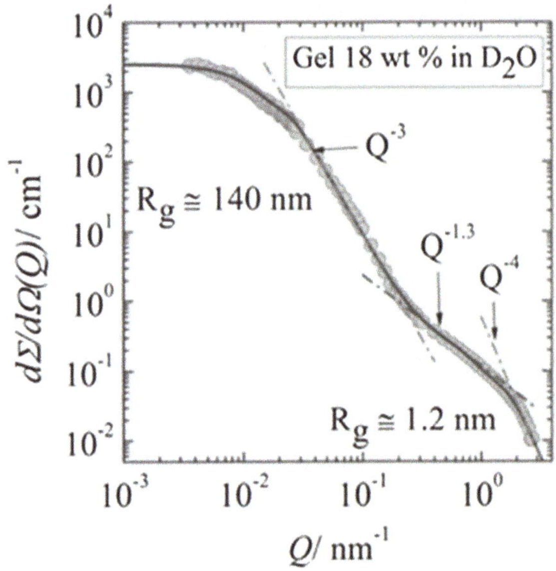

Fig. 16 Fitting SANS and USANS data from 18% gelatin in D_2O with the Beaucage model. Reproduced from (Helminger et al. 2014)

A two-component hydrogel (Berts et al. 2013) was made by mixing aqueous solutions of aldehyde-modified sodium hyaluronan (HAA) with hydrazide-modified poly(vinyl alcohol). It is interesting to discuss the SAXS results from the gel in comparison with the ones from the solution of its main component (HAA). In Fig. 17a, it is clear that the presence of cross-links enhances the scattering at intermediate q range. The authors used an equation that can be thought as a modification of Eq. 9 where the term for semidilute solution scattering $\frac{I_{OZ}(0)}{1+q^2\xi^2}$ has

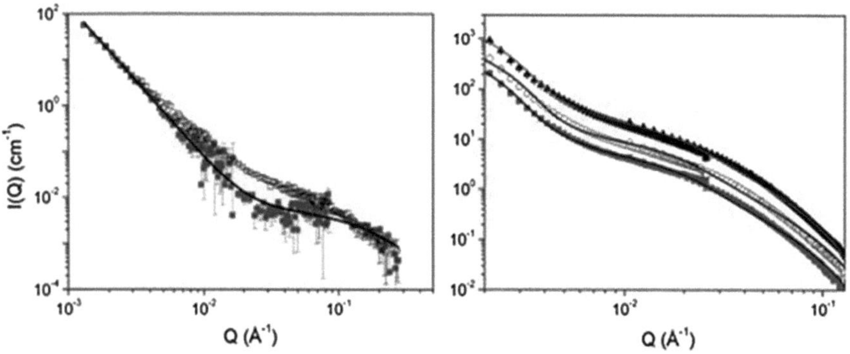

Fig. 17 SANS data from **a** HAA hydrogel (□) and the corresponding solution (■) at 15 mg/ml and **b** HAA hydrogel (15 mg/ml)/HAP nanocomposite with 5 (■), 10 (○) and 20 (▲) % wt NPs. Lines are fits to the experimental data. Reproduced from (Berts et al. 2013). Copyright © 2012 by John Wiley & Sons, Inc. Reprinted by permission of John Wiley & Sons, Inc

been replaced by $\frac{A}{\left(1+q^2R_c^2\right)\cdot(1+qL)}$. This term is used for rodlike polysaccharides with length L and radius of cross section R_c. The large cluster term $\frac{I_{fr}(0)}{\left(1+\frac{D_f+1}{3}q^2\Xi^2\right)^{\frac{1}{2}D_f}}$ has been replaced by a power-law function Bq^{-D_f} because the low-q scattering did not show any Guinier regime (Eq. 19).

$$I(q) = Bq^{-D_f} + \frac{A}{\left(1+q^2R_c^2\right)\cdot(1+qL)} \tag{19}$$

Apparently, the characteristic size of inhomogeneities is bigger than the larger length scale that can be probed ($\Xi > 1/q_{\min}$). The fitting results show that the hydrogel has lower D_f compared to the solution. The reduced fractal dimension is due to the mass fractal nature of the cross-linked system. The authors (Berts et al. 2013) added hydroxyapatite nanoparticles (HAP NPs) to the gels and found that scattered intensity was dominated by the contribution from the NPs. The low-q upturn of the curves (Fig. 17b) was attributed to NP clusters that span the distance between polymer cross-links within the hydrogel nanocomposites.

Associative micellar hydrogels are known to be formed by PLA-b-PEO-b-PLA triblock copolymers due to the presence of the hydrophobic PLA (poly(lactide)) blocks (Agrawal et al. 2008). In the micelles, PLA blocks form the hydrophobic cores and PEO blocks form the hydrophilic coronas. The PEO blocks may either be in loops back-folded onto a single core or be in bridges between two distinct cores (physical cross-links). Addition of laponite clay nanoparticles resulted in reinforcement of the modulus even at low concentrations. The adsorption of PEO on the surface of these model discotic nanoparticles created new junction points within the network (Agrawal et al. 2008). Depending on the type of PLA (r-lactide, i.e., racemic mixture of D- and L- or L-lactide), micellar or lamellae nanocomposites are

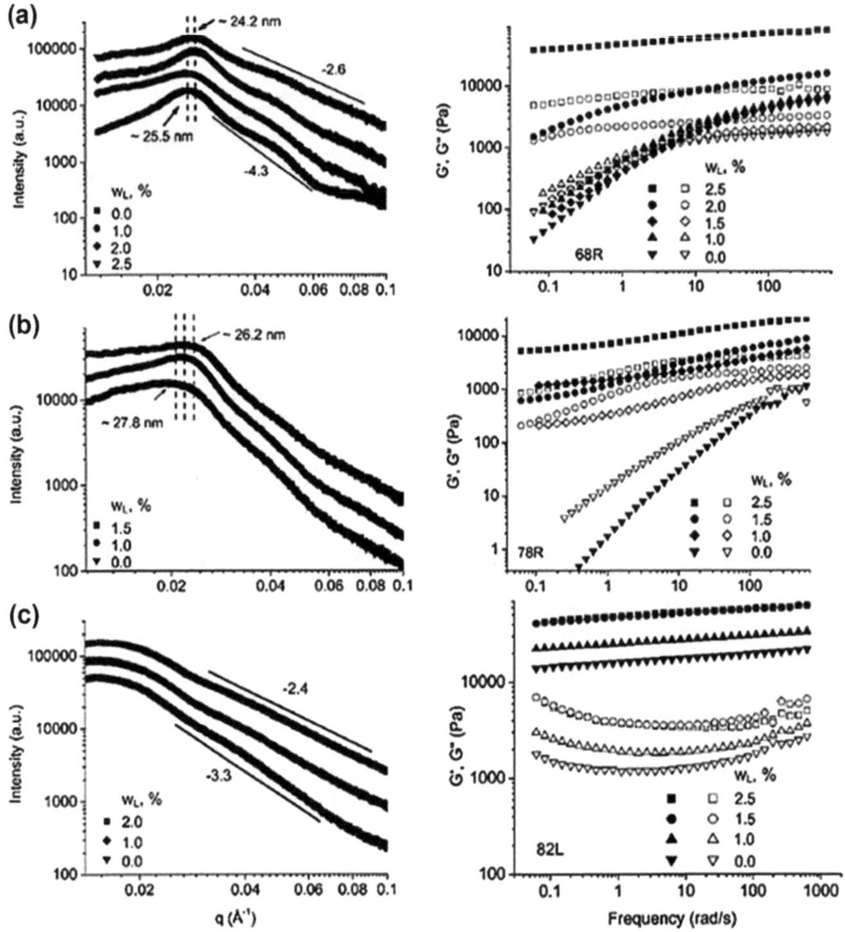

Fig. 18 SANS data (left) and viscoelastic moduli (right) for clay NPs in PLA-b-PEO-b-PLA hydrogels. Reprinted with permission from (Agrawal et al. 2008). Copyright 2008 American Chemical Society

created. In Fig. 18, the effect of clay NPs on the structure and viscoelasticity is presented for macromolecular architectures where the degree of polymerization (DP) of PEO is unaltered ($DP_{PEO} = 202$) and the DP of PLA is varying ($DP_{PLA}(68R) = 68$, $DP_{PLA}(78R) = 78$, $DP_{PLA}(82L) = 82$).

For 68R and 78R (Fig. 18a, b left), the low-q scattering peak has a slight shift to the right as NP concentration increases. The contribution of NPs of this concentration at this q regime is not significant to the micellar one. The peak position (q_{max}) characterizes the average intermicellar distance ($\frac{2\pi}{q_{max}}$). Introducing the NPs, more effective cross-links are created and micelles come closer to each other. In the intermediate q range, the power-law slope decreases gradually from 4 to $\sim$2.5.

The contribution from NPs is not negligible in this regime, and presumably the signature of well-defined interfaces is transformed to scattering from two-dimensional flat surfaces (disks). The situation is similar for 82L (Fig. 18c left) at the intermediate and high q. At low q, a broad peak is caused by the polydisperse disklike randomly oriented lamellar micelles.

The viscoelastic moduli of the PLA-b-PEO-b-PLA hydrogels are greatly affected by the length and type of the PLA block. Adding laponite NPs results in a significant increase in both G' and G'' as a function of NP concentration. Especially for 78R, the liquid-like viscoelasticity changes to predominantly elastic (Fig. 18b right).

In SANS, two-dimensional detectors are useful not only as an effective way of collecting scattered neutrons but also as probes of anisotropy. This is expressed as 2D profiles that are asymmetric in rotation around the axis of the incident beam. Simultaneous SANS and rheological measurements (Rheo-SANS) are used to detect isotropic–anisotropic transitions caused by flow (shear alignment). A good example is the case where polysilicate nanoplatelets (25 nm in diameter and 1 nm in thickness) were incorporated in nanocomposite gels (Schexnailder et al. 2009). PEO chains physically attach on silicate nanoplatelets and hence create hydrogels. The physical bonds break upon shear and apparently are restored after shear cessation. This renders the hydrogels shear-thinning and self-healing (Schexnailder et al. 2009). Up to shear rate 0.1 s^{-1}, the 2D SANS patterns are isotropic (Fig. 19a). At 91 s^{-1}, the 2D patterns appear anisotropic (Fig. 19b–d). Since the flow is on the x-axis, elongation of the 2D patterns along the y-axis suggests alignment of aggregates along the flow. This is explained by the fact that if the size of a particle along a direction is larger than along the other one then $I(q)$ drops steeper along this direction (Guinier approximation is a simple way to the explanation). Additionally, the anisotropy decrease as function of ionic strength implies the importance of electrostatic interactions between nanoparticle platelets.

The deformation and recovery of gel samples are also performed by applying a uniaxial deformation for a certain period of time and then remove the load (Rose et al. 2015). Introduction of silica filler particles to poly(*N,N*-dimethylacrylamide) (PDMA) networks increases both their Young's modulus and strain at failure. The SANS profile is isotropic in the absence of uniaxial strain, and it is recovered after its release. Under 100% strain, 2D SANS is anisotropic. Stretching the sample along the x-direction increases the interparticle distances along this direction and reduces it along the y-direction. The correlation maximum (at low q) shifts to higher q values on the y-direction and lower q values on the x-direction. Additionally, intensity at the maximum drops on the y-direction because of the suppression on fluctuations (Rose et al. 2015). At high q, the scattering patterns are dominated by the form factor of individual spherical silica particles and consequently are isotropic. The characteristic "butterfly" patterns (Shibayama 2011) (at $\varepsilon = 100\%$), i.e., the four-spot shape, arise as the particles avoid each other by localized shear displacements.

As has been already mentioned, the characteristic mesh size of a hydrogel defines the pore radius through which small molecules can diffuse. Release from

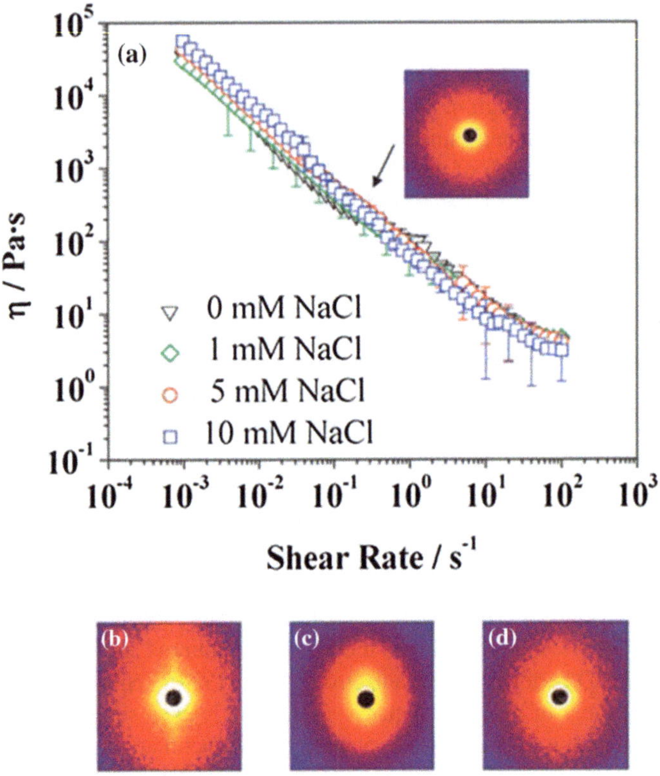

Fig. 19 (**a**) Steady shear viscosity and 2D SANS patterns from PEO cross-linked by silicate nanoparticles in 0 (**b**), 5 (**c**), and 10 (**d**) mM NaCl. Reproduced from (Schexnailder et al. 2009) with permission from the PCCP Owner Societies

poly(organophosphazene) gels was found to depend on the molecular weight of the tested protein (Park et al. 2010). Although increasing the molecular weight of the protein increased the release period, the process was still rapid and further control was needed. Controlling the viscosity of the gels was not sufficient to create a sustained release. The solution was to include proteins in an intermediate level of hierarchy between the network and the single protein. The authors (Park et al. 2010) used oppositely charged polyelectrolytes to form protein/polyelectrolyte complexes. The complexes had sizes larger than the ones of the individual proteins, and especially the BSA/PLA complex had a narrow size distribution. The release rate was significantly suppressed for all complexes in comparison with uncomplexed proteins, while BSA/PLA complex had the slowest release rate.

In complex fluids with multiple relaxation times, viscoelastic response can be analyzed as superposition of dynamic modes. Based on Eq. 12 and the generalization from single Maxwell model to a superposition of characteristic modes (Eqs. 13 and 14), the stress relaxation modulus can be generalized (Eq. 20). Each

Maxwell mode will have a characteristic relaxation time λ and modulus $H(\lambda)$, and hence the viscoelastic response will be described by the stress relaxation spectrum $H(\lambda)$.

$$G(t) = \int_0^{+\infty} \mathrm{d}\ln\lambda \cdot H(\lambda) \cdot \exp(-t/\lambda) \tag{20}$$

It is very interesting to demonstrate an example (Taki et al. 2013) where DNA-modified stress relaxation is quantified by $H(\lambda)$. Synthetic hectorite (SWN) microstructure consists of sheets of $AlO_4(OH)_2$ between SiO4 tetrahedral layers 1 nm in thickness and 50 nm in diameter. The hydrogels formed by SWN have shear-thinning behavior. Their viscoelasticity was modified by interaction with DNA (Taki et al. 2013) where it was found that they became softer. The liquid-like relaxation of DNA/SWN nanocomposite at low strains (linear regime) is changed into solid-like behavior at high strains (nonlinear regime) (Fig. 20).

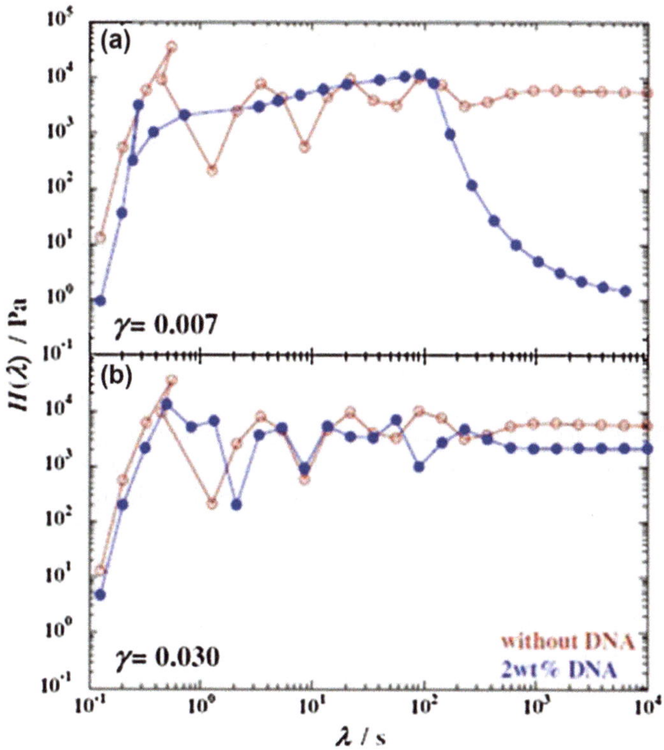

Fig. 20 Relaxation spectra of linear (**a**) and nonlinear (**b**) viscoelastic regime in the absence and presence of DNA in SWN hydrogels. Reprinted from (Taki et al. 2013), Copyright 2013, with permission from Elsevier

Additionally, the modification of the SWN spectrum by DNA is significant only in the linear regime. This observation gave insights into the energy landscapes and jamming processes in the studied complex fluid.

Poly(acrylic acid)/laponite nanocomposites were synthesized by free radical polymerization of acrylic acid in laponite aqueous dispersions (Shen et al. 2014). The polymerization is expected to arise from the surface of the disklike laponite NPs. Hence, the number density of laponite NPs sets an effective cross-linking density. Indeed, the shear-thinning steady shear viscosity measurements showed that the viscosity increases as a function of laponite content. Similarly, the magnitude of viscoelastic moduli increased monotonically as the number of laponite NPs increased. The characteristic time τ^* as the inverse frequency where $G' = G''$ shifted to higher values showing the formation of stronger gels with higher cross-linking densities. The adhesive properties of the hydrogels were tested by peel strength measurements. Peel strength had a non-monotonic dependence on the laponite content. It passed through a maximum revealing that peel strength depends on both the cohesion and flowability of the hydrogels (Shen et al. 2014).

4 Conclusions and Future Perspective

Hydrogels belong to the class of soft, complex, biocompatible materials and therefore are important for practical biotechnology and medical applications as they present mechanical integrity, tunable rheological properties and can accommodate large amounts of aqueous media. They are organized in multiple length scales and have self-similar internal morphology that define their rheological response. Addition or incorporation of components as proteins and nanoparticles alters their internal morphology and mechanical behavior. In this chapter, a detailed account of the methodologies for analyzing data from small-angle scattering and rheology on hydrogel nanocomposites was presented. Recent paradigms from the literature were used to illustrate the application to systems of high interest and provide a unifying framework for these combined investigations. Simultaneous rheological and small-angle scattering experiments are well established, and hence they can support the research on hydrogel nanocomposites. Additionally, a lot of possibilities arise by introducing an intermediate hierarchical level between the hydrogel and the loaded component, for example an amphiphilic core–shell polymeric micelle for carrying a hydrophobic drug within a hydrogel. Finally, the established microrheological techniques are expected to be exploited as a length-scale-sensitive rheology–probe for extracting the bulk and local viscoelastic properties of protein- and nanoparticle-loaded hydrogels.

References

Agrawal SK, Sanabria-DeLong N, Tew GN, Bhatia SR (2008) Nanoparticle-reinforced associative network hydrogels. Langmuir 24(22):13148–13154

Barker JG, Pedersen JS (1995) Instrumental smearing effects in radially symmetric small-angle neutron scattering by numerical and analytical methods. J Appl Crystallogr 28:105–114

Beaucage G (1995) Approximations leading to a unified exponential/power-law approach to small-angle scattering. J Appl Crystallogr 28(6):717–728

Beaucage G (1996) Small-angle scattering from polymeric mass fractals of arbitrary mass-fractal dimension. J Appl Crystallogr 29:134–146

Berts I, Gerelli Y, Hilborn J, Rennie AR (2013) Structure of polymer and particle aggregates in hydrogel compossites. J Polym Sci Part B Polym Phys 51(6):421–429

Campanella A, Holderer O, Raftopoulos KN, Papadakis CM, Staropoli MP, Appavou MS, Muller-Buschbaum P, Frielinghaus H (2016) Multi-stage freezing of HEUR polymer networks with magnetite nanoparticles. Soft Matter 12(13):3214–3225

Censi R, Di Martino P, Vermonden T, Hennink WE (2012) Hydrogels for protein delivery in tissue engineering. J Control. Release 161(2):680–692

Chakraborty P, Bairi P, Roy B, Nandi AK (2014) Rheological and fluorescent properties of riboflavin-poly(*N*-isopropylacrylamide) hybrid hydrogel with a potentiality of forming Ag nanoparticle. RSC Advances 4(97):54684–54693

Chen D, Wu D, Cheng G, Zhao H (2015) Reductant-triggered rapid self-gelation and biological functionalization of hydrogels. Polym Chem 6(48):8275–8283

Choi B, Loh XJ, Tan A, Loh CK, Ye E, Joo MK, Jeong B (2015) Introduction to in situ forming hydrogels for biomedical applications. In: Loh JX (ed) In-Situ gelling polymers: for biomedical applications. Springer Singapore, p 5–35

Daoud M, Cotton JP, Farnoux B, Jannink G, Sarma G, Benoit H, Duplessix C, Picot C, de Gennes PG (1975) Solutions of flexible polymers. neutron experiments and interpretation. Macromolecules 8(6):804–818

Dmitri IS, Michel HJK (2003) Small-angle scattering studies of biological macromolecules in solution. Rep Prog Phys 66(10):1735

Dobrynin AV, Colby RH, Rubinstein M (1995) Scaling theory of polyelectrolyte solutions. Macromolecules 28(6):1859–1871

Du X, Wang J, Diao W, Wang L, Long J, Zhou H (2014) A genetically modified protein-based hydrogel for 3D culture of AD293 cells. PLoS ONE 9(9):e107949

Ghoorchian A, Simon JR, Bharti B, Han W, Zhao X, Chilkoti A, López GP (2015) Bioinspired reversibly cross-linked hydrogels comprising polypeptide micelles exhibit enhanced mechanical properties. Adv Func Mater 25(21):3122–3130

Glassman MJ, Olsen BD (2013) Structure and mechanical response of protein hydrogels reinforced by block copolymer self-assembly. Soft Matter 9(29):6814–6823

Glatter O (1977) A new method for the evaluation of small-angle scattering data. J Appl Crystallogr 10(5):415–421

Gulrez SKH, Al-Assaf S, Phillips GO (2011) Hydrogels: Methods of Preparation, Characterisation and Applications. In: Progress in molecular and environmental bioengineering—From analysis and modeling to technology applications, InTech Open Access Publisher, London

Guvendiren M, Lu HD, Burdick JA (2012) Shear-thinning hydrogels for biomedical applications. Soft Matter 8(2):260–272

Hammouda B (2010) Analysis of the Beaucage model. J Appl Crystallogr 43(6):1474–1478

Haraguchi K, Farnworth R, Ohbayashi A, Takehisa T (2003) compositional effects on mechanical properties of nanocomposite hydrogels composed of poly(*N*, *N*-dimethylacrylamide) and clay. Macromolecules 36(15):5732–5741

Helminger M, Wu B, Kollmann T, Benke D, Schwahn D, Pipich V, Faivre D, Zahn D, Cölfen H (2014) Synthesis and characterization of gelatin-based magnetic hydrogels. Adv Func Mater 24 (21):3187–3196

Higgins JS, Benoit HC (1994). Polymers and neutron scattering
Horkay F, McKenna GB, Deschamps P, Geissler E (2000) Neutron Scattering properties of randomly cross-linked polyisoprene gels. Macromolecules 33(14):5215–5220
Hule RA, Nagarkar RP, Altunbas A, Ramay HR, Branco MC, Schneider JP, Pochan DJ (2008) Correlations between structure, material properties and bioproperties in self-assembled [small beta]-hairpin peptide hydrogels. Faraday Discuss 139:251–264
Jonker AM, Löwik DWPM, van Hest JCM (2012) Peptide- and protein-based hydrogels. Chem Mater 24(5):759–773
Kaieda S, Plivelic TS, Halle B (2014) Structure and kinetics of chemically cross-linked protein gels from small-angle X-ray scattering. Phys Chem Chem Phys 16:4002
Kim M, Tang S, Olsen BD (2013) Physics of engineered protein hydrogels. J Polym Sci Part B Polym Phys 51(7):587–601
Kirchhof S, Abrami M, Messmann V, Hammer N, Goepferich AM, Grassi M, Brandl FP (2015) Diels–Alder hydrogels for controlled antibody release: correlation between mesh size and release rate. Mol Pharm 12(9):3358–3368
Koch MHJ, Vachette P, Svergun DI (2003) Small-angle scattering: a view on the properties, structures and structural changes of biological macromolecules in solution. Q Rev Biophys 36 (02):147–227
Koga T, Hashimoto T, Takenaka M, Aizawa K, Amino N, Nakamura M, Yamaguchi D, Koizumi S (2008) New insight into hierarchical structures of carbon black dispersed in polymer matrices: a combined small-angle scattering study. Macromolecules 41(2):453–464
Larson RG (1999) The structure and rheology of complex fluids. Oxford University Press, New York
Matsumoto K, Shundo A, Ohno M, Fujita S, Saruhashi K, Miyachi N, Miyaji K, Tanaka K (2015) Modulation of physical properties of supramolecular hydrogels based on a hydrophobic core. Phys Chem Chem Phys 17(3):2192–2198
Panyukov S, Rabin Y (1996) Polymer Gels: frozen inhomogeneities and density fluctuations. Macromolecules 29(24):7960–7975
Papagiannopoulos A (2016) Introduction to the most popular microrheology techniques. In: Microrheology with optical tweezers: principles and applications. Pan Stanford, Singapore
Papagiannopoulos A, Sotiropoulos K, Pispas S (2016) Particle tracking microrheology of the power-law viscoelasticity of xanthan solutions. Food Hydrocoll
Papagiannopoulos A, Zhao J, Zhang G, Pispas S, Radulescu A (2013) Thermoresponsive transition of a PEO-b-PNIPAM copolymer: From hierarchical aggregates to well defined ellipsoidal vesicles. Polymer 54(23):6373–6380
Pape ACH, Bastings MMC, Kieltyka RE, Wyss HM, Voets IK, Meijer EW, Dankers PYW (2014) Mesoscale characterization of supramolecular transient networks using SAXS and rheology. Int J Mol Sci 15(1):1096–1111
Park M-R, Chun C, Cho C-S, Song S-C (2010) Enhancement of sustained and controlled protein release using polyelectrolyte complex-loaded injectable and thermosensitive hydrogel. Eur J Pharm Biopharm 76(2):179–188
Pedersen JS (1997) Analysis of small-angle scattering data from colloids and polymer solutions: modeling and least-squares fitting. Adv Coll Interface Sci 70:171–210
Qian Y-C, Chen P-C, Zhu X-Y, Huang X-J (2015) Click synthesis of ionic strength-responsive polyphosphazene hydrogel for reversible binding of enzymes. RSC Advances 5(55):44031–44040
Radulescu A, Szekely NK, Polachowski S, Leyendecker M, Amann M, Buitenhuis J, Drochner M, Engels R, Hanslik R, Kemmerling G, Lindner P, Papagiannopoulos A, Pipich V, Willner L, Frielinghaus H, Richter D (2015) Tuning the instrument resolution using chopper and time of flight at the small-angle neutron scattering diffractometer KWS-2. J Appl Crystallogr 48 (6):1849–1859
Rose S, Marcellan A, Boué NTF, Cousin F, Hourdet D (2015) Structure investigation of nanohybrid PDMA/silica hydrogels at rest and under uniaxial deformation. Soft Matter 11 (29):5905–5917

Rubinstein M, Colby RH (2003) Polymer physics. OUP, Oxford
Schexnailder P, Loizou E, Porcar L, Butler P, Schmidt G (2009) Heterogeneity in nanocomposite hydrogels from poly(ethylene oxide) cross-linked with silicate nanoparticles. Phys Chem Chem Phys 11(15):2760–2766
Schillemans JP, Verheyen E, Barendregt A, Hennink WE, Van Nostrum CF (2011) Anionic and cationic dextran hydrogels for post-loading and release of proteins. J Control Release 150 (3):266–271
Schmidt P (1991) Small-angle scattering studies of disordered, porous and fractal systems. J Appl Crystallogr 24(5):414–435
Shen M, Li L, Sun Y, Xu J, Guo X, Prud'homme RK (2014) Rheology and adhesion of poly (acrylic acid)/laponite nanocomposite hydrogels as biocompatible adhesives. Langmuir 30 (6):1636–1642
Shen W, Kornfield JA, Tirrell DA (2007) Structure and mechanical properties of artificial protein hydrogels assembled through aggregation of leucine zipper peptide domains. Soft Matter 3 (1):99–107
Shibayama M (2011) Small-angle neutron scattering on polymer gels: phase behavior, inhomogeneities and deformation mechanisms. Polym J 43(1):18–34
Shibayama M, Kurokawa H, Nomura S, Muthukumar M, Stein RS, Roy S (1992) Small-angle neutron scattering from poly(vinyl alcohol)-borate gels. Polymer 33(14):2883–2890
Smeets NMB, Patenaude M, Kinio D, Yavitt FM, Bakaic E, Yang F-C, Rheinstadter M, Hoare T (2014) Injectable hydrogels with in situ-forming hydrophobic domains: oligo(d, l-lactide) modified poly(oligoethylene glycol methacrylate) hydrogels. Polym Chem 5(23):6811–6823
Sorensen CM (2001) Light scattering by fractal aggregates: a review. Aerosol Sci Technol 35 (2):648–687
Taki A, John B, Arakawa S, Okamoto M (2013) Structure and rheology of nanocomposite hydrogels composed of DNA and clay. Eur Polymer J 49(4):923–931
Thakur VK, Thakur MK (2014a) Recent trends in hydrogels based on psyllium polysaccharide: a review. J Clean Prod 82:1–15
Thakur VK, Thakur MK (2014b) Recent advances in graft copolymerization and Applications of chitosan: a review. ACS Sustain Chem Eng 2(12):2637–2652
Thakur VK, Thakur MK (2015) Recent advances in green hydrogels from lignin: a review. Int J Biol Macromol 72:834–847
Teixeira J (1986) Experimental methods for studying fractal aggregates. In: Stanley HE, Ostrowsky N (eds) On growth and form: fractal and non-fractal patterns in physics. Springer, Dordrecht, Netherlands, pp 145–162
Urueña JM, Pitenis AA, Nixon RM, Schulze KD, Angelini TE, Gregory Sawyer W (2015) Mesh size control of polymer fluctuation lubrication in gemini hydrogels. Biotribology 1–2:24–29
Vermonden T, Censi R, Hennink WE (2012) Hydrogels for protein delivery. Chem Rev 112 (5):2853–2888
Vilgis TA, Winter HH (1988) Mechanical selfsimilarity of polymers during chemical gelation. Coll Polym Sci 266(6):494–500
Winter H, Mours M (1997) Rheology of polymers near liquid-solid transitions. In: Neutron spin echo spectroscopy viscoelasticity rheology, vol 134. Springer, Heidelberg, pp 165–234
Wyatt NB, Liberatore MW (2010) The effect of counterion size and valency on the increase in viscosity in polyelectrolyte solutions. Soft Matter 6(14):3346–3352
Zhang J-T, Petersen S, Thunga M, Leipold E, Weidisch R, Liu X, Fahr A, Jandt KD (2010) Micro-structured smart hydrogels with enhanced protein loading and release efficiency. Acta Biomater 6(4):1297–1306
Zhao F, Yao D, Guo R, Deng L, Dong A, Zhang J (2015) Composites of polymer hydrogels and nanoparticulate systems for biomedical and pharmaceutical applications. Nanomaterials 5 (4):2054
Zustiak SP, Wei Y, Leach JB (2013) Protein–hydrogel interactions in tissue engineering: mechanisms and applications. Tissue Eng Part B Rev 19(2):160–171

Chapter 6
Preparation, Properties and Application of Hydrogels: A Review

Sumit Mishra, Priti Rani, Gautam Sen and Kartick Prasad Dey

Abstract Hydrogels are primarily synthesized to retain large amounts of aqueous solution. Depending upon modes of synthesis, the hydrogel materials develop various types of network structure. Recently, popular techniques have been developed for synthesis of hydrogels in the presence of crosslinking agents or multifunctional co-monomer which acts as a crosslinker. It can be categorized according to synthesis techniques, bio-degradability, response to environment and their intended applications. These applications may vary from water retention, conditioner to different aspects of biomedical applications and tissue engineering. Hydrogels contain a number of functional groups which may be utilized as such or modified and used to suit our requirements. In this review article, we have focused on the available synthesis techniques of hydrogels along with their inevitable properties and applications.

Keywords Hydrogel · Radical polymerization · Biomedical application
Tissue engineering

1 Introduction

Polymer hydrogels are well-known materials with widespread usage in many fields, particularly in biomedical, pharmaceutical and agricultural sector. The first water-absorbent polymer was synthesized in 1938 after thermal polymerization of acrylic acid and divinylbenzene in an aqueous medium (Zohuriaan-Mehr 2006). Significant advancement in this field was marked in 1960 when these polymeric hydrogels were widely worked upon exploring their hidden potential (Kumar et al. 2006).

S. Mishra (✉) · P. Rani · G. Sen · K. P. Dey
Department of Chemistry, Birla Institute of Technology Mesra, Ranchi 835215, Jharkhand, India
e-mail: sumitmishra1@gmail.com

V. K. Thakur and M. K. Thakur (eds.), *Hydrogels*, Gels Horizons: From Science to Smart Materials, https://doi.org/10.1007/978-981-10-6077-9_6

From the era of 1970s, available literature reports polymeric hydrogels with new chemical structure and resultant better physical properties as well as innovative applications. Advancements in hydrogel properties led to increase its suitability for design and operation of some sophisticated system, like automatic drug delivery systems. Hydrogels need to be smarter to perform necessary function to achieve desired applications. The development of novel hydrogels with their potential applications is in huge demand to suit our requirements.

For example, there are a number of wound-care products available in market today, from injectable foam to hydrocolloid dressings. Recently, hydrogels have found application in wound management systems. Since blood is indispensable for healing process and wounds can damage these blood vessels which may turn fatal, it is important to hold it at the affected areas. For this purpose, the researchers started developing an approach to wound-care management that would rebuild those damaged or dead vessels, creating a matrix of stem cell sponges and hydrogel that develops like skin of a growing foetus. Now, scientists are searching for the best source of such cell sponges, bone marrow and fats seem to be the best options so far. Soon, the clinical trials will be done in which participants will use this advanced hydrogel on diabetic foot ulcers.

More than one million people are suffering from inflammatory bowel disease (IBD) worldwide. The population of patients suffering from this disease is increasing continuously around the world, but its treatment is very limited. People have to depend on daily enemas as a part of therapy which is not comfortable, totally impractical and has great side effects on health. In this regard, clinical development for the effective treatment of chronic and debilitating inflammatory bowel disease (IBD), including Crohn's disease and ulcerative colitis, is necessary. So, scientists have tried to develop an alternative therapy using disease targeted hydrogel that rapidly sticks to the ulcers and slowly releases the drug to the targeted area only. It is a great breakthrough for the patient for more targeted enema-based therapy in future.

The need to develop smarter hydrogels which can simulate the physiological system to address the biomedical issues is one of the primary objectives of contemporary research. Yet, it is important to understand the fundamental similarities and differences between the two. Such understanding will lead to development of multifunctional hydrogels to suit our requirements. Thus, it is important to review the available literature on synthesis, characterization and available applications of hydrogels along with patents and commercial products.

1.1 What Is Hydrogel?

The term 'hydrogel' bears no specific definition as per updated medical and pharmaceutical encyclopaedia. Hydrogels are three-dimensional materials having capacity to hold excess amount of water to maintain the stability of its dimension. It maintains its 3D integrity in its swollen phase by the virtue of crosslinking

(Williams 1990). Polymer hydrogels are either synthetic or of natural origin, homopolymers or co-polymers (Langer and Peppas 2003). It can also be classified on the basis of nature of crosslinked junction.

(a) Chemically crosslinked with stable networks
(b) Physically bonded networks owing to ionic interactions, hydrogen bonds or hydrophobic interactions, or even polymeric chain entanglements (Jen et al. 1996).

Chemically, hydrogels are synthesized by two different techniques: 'three-dimensional polymerization' occurs in the presence of hydrophilic monomer in addition with either a crosslinking agent indirectly or by direct crosslinking shown in (Fig. 1). Alternatively, polymerization is carried out by employing free radical initiators like benzoyl peroxide, ammonium peroxodisulphate, 2,2-azo-isobutyronitrile (AIBN). This may also be achieved by irradiation with UV rays, gamma rays or electron beam shown in (Fig. 2). In three-dimensional polymerization, product is obtained along with considerable amount of residual monomers. These unreacted monomers are often toxic so they need to be separated out; otherwise, there remains a possibility of leaching from the hydrogels (Khutoryanskiy et al. 2013; Montoro et al. 2014; Mathur et al. 1996).

Rosiak used biopolymers like agar and gelatin as well as synthetic polymers like polyvinyl pyrrolidone (PVP) or polyvinyl alcohol (PVA) which have been crosslinked in the presence of gamma radiation and give sterile hydrogels for wound care (Rosiak et al. 1989). Nowadays, these hydrogels are available in market under brand name 'Kikgel' and 'Aqua-gel' used for dressing of wounds (Rosiak et al. 1989).

Another method for hydrogel synthesis is suggested by Khutoryanskiy. His method synthesizes water-soluble polymers using thermal and microwave radiation in aqueous medium (Khutoryanskiy et al. 2013; Cook et al. 2012). In this process, the aqueous solutions of PVA and poly(methyl vinyl ether-alt-maleic anhydride) are homogenized well at room temperature and kept under microwave radiation and thermal treatment under high pressure through autoclave. These synthesis

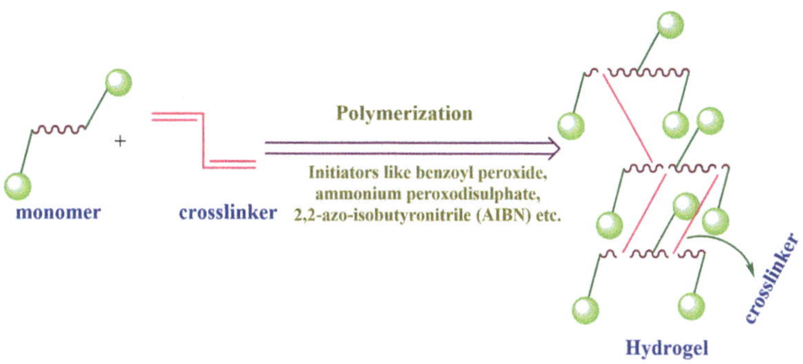

Fig. 1 Schematic representation of hydrogel formation by initiators

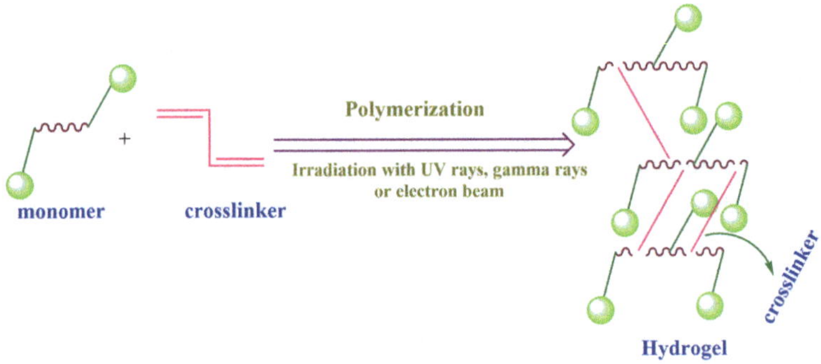

Fig. 2 Schematic representation of hydrogel formation by irradiation

techniques using irradiation as well as thermal procedures are cheap, safe and free from purification steps. They result in formation of hydrogels in the presence of hydrophilic polymers in suitable combination.

1.2 History of Hydrogel

According to Lee, Kwon and Park, the word 'hydrogel' was used in a number of articles since 1894 (Wei and Charlotte 2014). The material ascribed at that time was a colloidal gel made up of inorganic salts rather than a hydrogel. In 1960, the first developed crosslinked network material having hydrogel properties as well as high water affinity was polyhydroxyethylmethacrylate (PHEMA) hydrogel. This was intended to be used in a much-awaited project of permanent contact lens production. Hydrogel was the first soft material developed to be used inside any patient (Nierzwicki and Prins 1975; Wichterle and Lim 1960). Broadly, the history of hydrogels can be categorized (Buwalda et al. 2014) as follows:

- The first era of hydrogel development history involved a wide range of random crosslinking experimental procedures. These involved using initiators for the chemical modifications. The primary objective was to synthesize materials with high swelling index and reasonable mechanical properties.
- The second generation of materials involved optimizing response to specific stimuli like variation in pH, temperature, pressure or even concentration of solvent molecules. These stimuli were used to initiate specific phenomenon like polymerization of monomer, drug delivery. (Buwalda et al. 2014).
- Finally, the third generation of hydrogels led to further advancement. It focused on the development of stereo-complexed materials, e.g. PEG-PLA interaction (Yom-Tov et al. 2014; Abebe and Fujiwara 2012), and hydrogels crosslinked with other physical interaction, e.g. cyclodextrins (Chung et al. 2008; Kirakci et al. 2014).

In 1980, Lim and Sun modified hydrogels and designed calcium alginate microcapsules for cell engineering. Later, Yanna and co-workers (Lim and Sun 1980) worked upon collagen and shark cartilage tissues as natural raw materials for novel dressing which reduced the healing burn. Both natural and synthetic hydrogel generated an interest for encapsulation of cells (Yannas et al. 1989) and have become a popular choice in the field of 'tissue engineering'. Here, they serve as matrices for repairing of tissues and regeneration of tissues and organs (Sefton et al. 2000). It is helpful in the prevention of thrombosis, in drug delivery systems, as biosensor coatings as well as in cell transplants (Nguyen and West 2002; Peppas and Bures 2000; Sawhney and Pathak 1994; Miyata et al. 2002; Chang et al. 2010).

The tunable properties, hydrophilic character, biocompatibility and response to stimuli focus the interest of scientists in the development of the 'smart hydrogels' for many years (Park et al. 1993).

1.3 Types of Hydrogel

Hydrogels can be broadly categorized on the basis of various features. This may include their synthesis procedures and routes, biodegradable properties, applications or their response to various stimuli and some other parameters. Hydrogels are found to exhibit variations with respect to changes in pH, temperature, light, enzyme, electric and other stimuli. Hydrogels showing some environmentally sensitive properties along with their parent polymers are shown in Table 1.

2 Synthesis Procedures of Hydrogels

Therefore, hydrogels may be categorized as homo-polymers, co-polymers, interpenetrating networks and semi-interpenetrating network based on synthesis techniques and routes.

2.1 Homopolymers

Homopolymers are developed from single monomer. It provides a basic structural and functional unit to any polymeric network (Iizawa et al. 2007). Homopolymer crosslinked frame structure depends upon polymerization technique and nature of the monomer. The possible way to prepare homopolymeric hydrogel is by use of PHEMA {poly(2-hydroxyethyl methacrylate)} as monomer, polyethylene glycol dimethacrylate as crosslinking agent and benzoin isobutyl ether as a UV-sensitive initiator. This film was kept in de-ionized water under UV radiation ($\lambda = 253.7$ nm). The film was immersed in water to remove toxic or unreacted

Table 1 Classification of hydrogels with respect to response to stimuli

Response to stimuli	Parent polymer	Source
pH	Poly(methacrylic acid-co-methyl methacrylate); chitosan–carboxymethylcellulose, chitosan–alginate, sodium and chitosan–carbopol Poly(acrlyamide-co-acrylic acid) Chitosan–poly(vinyl alcohol)	Segundo et al. (2008) Saleem et al. (2012) Gemeinhart et al. (2000) Gunasekaran and Chai (2006)
Temperature	Poly(*N*-t butyl acryl amide-co-acrylamide); poly(*N*-isopropylacrylamide) (PNIPAAm)-poly(ethylene glycol) diacrylate	Okay and Ozturk (2002) Derwent and Mieler (2008)
Enzyme	Poly(ethylene glycol)	Aimetti et al. (2009)
pH-thermo	Poly-*N*-isopropylacrylamide	Johannsmann and Bunsow (2008)
IR light	*N*-isopropylacrylamide	Zeng and Jiang (2008)
Electro	Polydimethylaminopropylacrylamide Chondroitin 4-sulphate	Murdan (2003) Masteikova et al. (2003)

substances. PHEMA is used in artificial skin manufacturing and wound dressing besides the contact lenses to provide good healing conditions. PHEMA can also be synthesized by low molecular weight crosslinking agents like trimethylolpropane trimethacrylate. These hydrogels are soft in nature like PMMA and have high oxygen permeability. So, they can also be used for manufacture of contact lenses, as drug delivery matrices and tissue implants. Its application can further be improved upon by increasing its mechanical strength. Cretu has improved the hydrogel properties by introducing hydrophobic compound (caprolactone) into its structures or synthesizing amphiphilic material (Cretu et al. 2004).

The technique for the synthesis of these macromolecular architectures by an elegant utilization of 'click chemistry' was first made by Hilborn and co-workers in 2006 (Ossipov and Hilborn 2006). They synthesized poly(vinyl alcohols) functionalized with acetylene or azide groups. Hydrogel was formed immediately after addition of $CuSO_4$/Na ascorbate. Lin and Anseth also synthesized PEG hydrogel through click chemistry (Benamer et al. 2006). This technique is based on a step-growth synthesis process. In this reaction, monomers having azide and alkyne functional groups remain bonded with each other by the virtue of catalysts to form stable covalent bonds. PEG hydrogel obtained by this method has good mechanical properties and permits modulation of physicochemical characteristics of the PEG hydrogels.

Polyvinylpyrrolidone hydrogels are synthesized by irradiation technique (Benamer et al. 2006). This is done by irradiating the reaction mixture with 60 Co source at a dose rate of 3.2 Gy/min. It can also be used for wound healing.

Polyacrylic acid (PAA) is also a homopolymeric hydrogel (Christensen et al. 2006). It contains 2.5% PAA and 97.5% water in its commercially available grade. It is stable and possesses fairly elastic properties. It is non-toxic, non-inflammatory and simulates the human physiological environment of soft tissue for the application in endoprosthesis.

2.2 Co-polymers

Co-polymers consist of multimonomer systems with one hydrophilic monomer, managed in any standard arrangement like random, block or alternating along with the polymer chain (Kopecek and Yang 2009). Ring opening co-polymerization of a-caprolactone to poly(ethylene glycol)-poly(a-caprolactone)-poly(ethylene glycol) co-polymeric hydrogel was given by Gong for establishment of drug delivery system (Gong et al. 2009; Kim and Peppas 2003). In triblock synthesis mPEG, stannous octoate and hexamethylene diisocyanate were used as initiator, catalyst and coupling agent. When this co-polymeric block is applied in situ, it is capable to form hydrogel. This hydrogel can release both hydrophilic and hydrophobic drugs including proteins.

Carboxymethyl cellulose (CMC) is known to be water soluble and bio-compatible. Wang suggested blending CMC with cellulose or with PVP to form PVP/CMC hydrogel (Wang et al. 2007). The blended PVP/CMC hydrogels possessed good mechanical properties. Its high water retention capacity improved its bio-degradability and also made it suitable as dressing material.

Later on, Thomas developed free radical co-polymerization. This hydrogel has been synthesized using acrylamide and acrylic acid as monomers and *N*,*N*-methylene bisacrylamide and potassium persulfate in aqueous medium (Thomas et al. 2007). This hydrogel was transparent and embedded silver nanoparticles. This silver embedded nano-composite hydrogel possessed antimicrobial activity which could be used further in various applications.

Synthesis of a thermoplastic hydrogel was reported by polymerization of a-benzyl L-glutamate (BLG) *N*-carboxyanhydride. This reaction was carried out using diamine groups placed at the ends of poly(ethylene oxide) chains of the poloxamer. This hydrogel was pH-sensitive and thermosensitive in nature and used for drug delivery applications (Oh et al. 2003).

Multistep gelation process synthesizes hydrogel with multimembrane 'onion-like structure' under controlled physicochemical conditions by Ladet as shown in Fig. 3. He reported synthesis of a multilayered material based on physical hydrogels of amphiphilic polymers. These were synthesized without any external crosslinker by using chitosan and alginate. These novel three-dimensional multimembrane structures are useful in biomedical applications (Ladet et al. 2008).

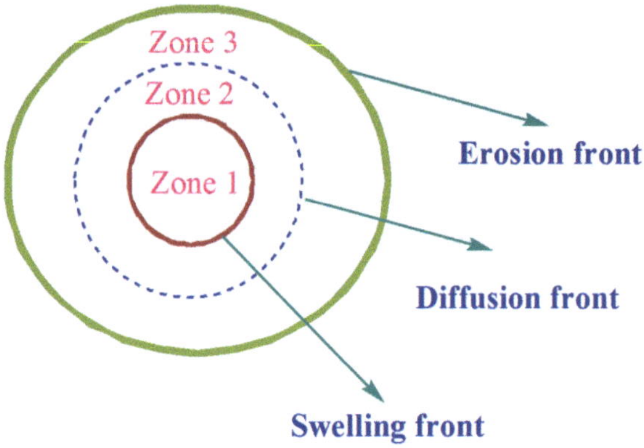

Fig. 3 Schematic representation of swellable matrix tablet during drug release

2.3 *Semi-interPenetrating Networks (Semi-IPNs)*

Semi-interpenetrating networks are formed without any chemical bonds between them. Here, (Semi-IPN) one straight chain polymer penetrates another crosslinked network (Zhang et al. 2009). Semi-IPNs are reported to have fast response rates towards pH and temperature as given by their kinetic studies. The polymerization of cationic polyallylammonium chloride (linear) in acrylamide or acrylic acid is an example of (Semi-IPN) co-polymer hydrogel. It has higher mechanical strength which led to theophylline release by pH variation. This type of semi-IPN was prepared by using *N*,*N*-methylene bisacrylamide as a crosslinker by means of template co-polymerization. (Zhang et al. 2005). The IPN network consists of both covalent and ionic bonds, and 3D structure is formed due to covalent bonds. The ionic bonds were responsible for better mechanical properties and pH reversible property.

Also, a semi-IPN hydrogel network was synthesized for its suitability in nano-level systems. Here, production and stabilization of silver nanoparticles were achieved (3–5 nm size) (Murthy et al. 2008). PVP chains were dispersed physically throughout PAA network. This silver nano-architecture was found to have significant antibacterial effects.

Crosslinked co-polymer of PHEMA and semi-IPN of gum arabic was prepared by the use of ammonium persulfate as initiator and *N*,*N*-methylene bisacrylamide as crosslinker (Gils et al. 2010). The hydrogel was embedded with silver nanoparticles using silver nitrate with trisodium citrate as reducing agent. This led to appreciable antibacterial activity due to the presence of silver nanoparticles inside the hydrogel network.

Some semi-IPN hydrogels of alginate and poly(*N*-isopropylacrylamide) (PNIPAAm) were synthesized using calcium chloride as crosslinker (Ju et al. 2002). Swelling behaviour of semi-IPNs was observed at different pH at higher temperatures, and the formation of a polyelectrolyte complex was reported. It forms by the reaction of carboxyl groups of alginate and amino groups of modified PNIPAAm. It was sensitive towards pH, temperature and ionic strength of solvent.

Semi-IPN hydrogels have also been investigated for delivery in stomach. They show more swelling index in acidic conditions. These studies have been carried out for the delivery of amoxicillin and metronidazole antibiotics. These drugs are used for the treatment of helicobacter pylori infestation in stomach. A guar gum (GG)-based semi-IPN hydrogel and poly(methacrylic acid) are also reported (Li and Liu 2008). The minimum swelling index was observed in acidic pH condition for hydrogel loaded with 5-aminosalicylic acid (5-ASA).

In acidic pH conditions, the model drug 5-ASA formed complex hydrogen-bonded structure with the polymeric network structure. The maximum swelling index is observed for these hydrogels at pH 7.4 due to the electrostatic repulsion of the carboxylic groups. It resulted in minimum release of 5-ASA from the hydrogel network at pH 2.2. In vitro study showed that concentration of crosslinker and amount of GG affected degradation. This enzymatic degradation of hydrogels by cecal bacteria accelerates the release of 5-ASA from hydrogel at pH 7.4.

PVP-based hydrogel (Lu et al. 2010a, b) was synthesized as very promising thermosensitive material. The most vital shortcoming of PVP hydrogel as thermosensitive material is that it did not exhibit thermosensitivity under normal conditions. PVP and CMC resulted in the formation semi-IPN hydrogel having application as volume-phase transition temperature (VPTT). This study explained that the VPTT was significantly dependent on CMC content and pH of the swelling medium. VPTT behaviour is reflected in buffer solution of pH 1.2 but not in alkaline medium. In vitro drug release was studied at various buffer solutions using bovine serum albumin (BSA) as a model drug. These studies suggested that PVP/CMC semi-IPN hydrogels could work as suitable candidates for the delivery of protein drug in the intestine.

2.4 InterPenetrating Networks (IPNs)

An interpenetrating polymer network (IPN) consists of two or more polymeric chains in any network. The synthesis is done by immersing a pre-polymerized hydrogel into a solution of monomers and a polymerization initiator. These are not covalently bonded to each other but partially interconnected with each other on a polymer scale (Lipatov 2002). These networks cannot be distributed unless chemical bonds are broken. In this hydrogel, two or more networks can be arranged in such a manner that they are coupled and cannot be separated from each other. These hydrogels are often referred as 'intelligent polymers' or 'hungry network' (Shivashankar and Mandal 2012). The benefits of using IPNs are its compact matrices, strong mechanical properties, modifiable physical properties and fairly good candidature for drug delivery systems as compared to conventional hydrogels (Mohamadnia et al. 2007). As of now, it is the centre of immense interest for scientific research due to its unexplored potential in a large number of applications, particularly in medicine, industry, biology and environmental areas.

IPN has tailor-made properties. Its pore size and surface chemistry can be tuned for specific drug release kinetics. This is a result of interaction between hydrogel and the surrounding tissues (Li et al. 2007). Chivukula synthesized highly cross-linked IPN hydrogel network that restricts the swelling response of pH-sensitive hydrogel. It allows linear swelling with an abrupt pH change from 7.4 to 2 suited for oral drug delivery application (Chivukula et al. 2006). Chitosan crosslinked/ PNIPAM IPN network has been studied with diclofenac (Alvarez et al. 2005).

Polyurethane (PU) is also a biomaterial. IPN of PU and polyacrylamide controlled water absorption (Abraham et al. 2001). They are mixed together by addition of crosslinking agents vinyl pyrrolidone and methylenebisacrylamide. Then, the reaction mixture is exposed to UV radiation. This type of IPN hydrogels finds application as wound dressing material, artificial muscles, sensor systems and bio-separators.

Series of IPN hydrogels were reported to affect sensitivity towards temperature and pH fluctuations (Liu et al. 2012). This synthesis involved incorporation of polyaspartic acid which is a pH-sensitive polymer, into PNIPAAm hydrogel system. The swelling mechanism confirmed that IPN hydrogels possessed swift shrinking and re-swelling properties with respect to the concentration ratio of two components. These fast responsive properties of hydrogels enhance its applications in biomedical fields.

Next series of hydrogels include in situ polymerization approaches. Calcium alginate (Ca–Alg) and dextran methacrylate derivative (Dex-MA) hydrogels showed potential application in pharmaceutical field (Matricardi et al. 2008). The semi-IPN synthesized by the dispersion of Dex-MA chains into Ca–Alg hydrogel forms a hydrogel having different rheological properties than original Ca–Alg precursor. This enhanced its chances through use in injection of the semi-IPN through hypodermic needle. The IPN synthesized by UV treatment of semi-IPN forms hydrogels which are strong enough for delivery of bioactive molecules like protein.

3 Significant Properties of Hydrogel

3.1 Swelling Properties

The swelling properties of hydrogel matrix are important features that govern its future applications in pharmaceutical, biomedical, ophthalmology and tissue engineering fields. The polymer chains present in hydrogel interact with solvent molecules and start expanding to the fully relaxed and solvated state till they attain equilibrium state. The crosslinked network, on the other hand, applies an opposite force that pulls the chains inside. When these two forces acting in opposite directions balance each other, equilibrium is attained. This equilibrium ratio (Eq. 1.) generally illustrates the swelling behaviour of hydrogels.

$$\text{Equilibrium swelling ratio} = \frac{W_{\text{Swollen}}}{W_{\text{dry}}} \tag{1}$$

where

W_{swollen} weight of the swollen hydrogel
W_{dry} weight of the dry gel

The swelling kinetics of the hydrogels can be determined from the swelling kinetics. First, the weight of the dry hydrogel (W_{dry}) is determined. Then, this dried hydrogel is immersed in water until the swelling equilibrium is reached. This is followed by weighing the swollen hydrogel (W_{swollen}) after removing the excess water. The swelling ratio is given as (Eq. 2).

$$\text{Swelling ratio} = \frac{W_{\text{Swollen}} - W_{\text{dry}}}{W_{\text{dry}}} \tag{2}$$

Many research groups have assessed the swelling/shrinking kinetics of PNIPAAm hydrogel with temperature variations. Yoshida and co-workers have discussed the shrinking kinetics of PNIPAAm hydrogel. They explained that comb-type PNIPAAm hydrogels collapsed fist followed by the hydrogels without grafted side chains. (Yoshida et al. 1995; Kaneko et al. 1995). They also mentioned a comb-type grafted hydrogel synthesized by PEO graft chains in the crosslinked PNIPAAm network (Kaneko et al. 1998). The swelling characteristics are of utmost importance for hydrogels in biomedical and pharmaceutical applications. The equilibrium swelling ratio is an important parameter which affects the solute diffusion coefficient, surface wettability and mechanical properties of the hydrogel. The swelling properties are affected by many factors like nature of monomer, crosslinker concentration and other environmental factors like temperature, pH and ionic strength.

Kiil developed a mathematical model to explain the water-induced swelling, drug dissolution, external and internal mass transport resistance of dissolved drug in an HPMC matrix (Kiil and Dam-Johansen 2003). The main purpose of this model was to determine the position of three distinct moving fronts (Fig. 3).

3.2 *Mechanical Properties of Hydrogel*

The mechanical strength of the hydrogel network depends upon the composition and structure of hydrogels (Shibayama 2012). In general, the polymer hydrogels are very weak, soft, brittle and cannot withstand large deformation. These mechanical properties of hydrogels are very important parameters for pharmaceutical and biomedical applications. The assessment of mechanical property is important for biomedical applications, e.g. ligament and tendon repair, wound dressing material, matrix for drug delivery, tissue engineering and as cartilage replacement material. The mechanical properties are suitable to maintain its physical texture during delivery of therapeutic moieties. Hydrogel is a tailor-made material which can be optimized mechanically by tuning its crosslinking parameters. Higher degree of crosslinking results in formation of a strong hydrogel, but it reduces the percentage elongation of hydrogel and forms more brittle structure. Therefore, optimum degree of crosslinking has been achieved for relatively strong and elastic hydrogel. The mechanical characterization involved relaxation experiments based on generalized Maxwell model. Crosslinking density of hydrogel was determined by Flory's theory and Young modulus. This value was then used to determine the average polymeric mesh size according to the equivalent network theory (Schurz 1991).

Recently, a hydrogel having capacity for the large deformation and development of a slide-ring (SR) gel was given by Okumura and Ito. The process involved the crosslinking of polyrotaxane (Okumura et al. 2001).

3.3 Bio-compatible Properties

Bio-compatibility and non-toxicity is an important property for the hydrogel to make it appreciable in the field of biomedical studies. Most of the hydrogels used for this application have to pass cytotoxicity and in-vivo toxicity tests (Grodzinski 2009). Biocompatibility of the material signifies its suitable response in a specific application. It is divided into two components:

(a) bio-safety: the absence of cytotoxicity, mutagenesis carcinogenesis and related things
(b) bio-functionality: ability of material to perform the specified task for which it is intended (Grodzinski 2009)

This is relevant for tissue engineering applications since the tissue continuously interacts with the body. In case of synthetic hydrogel synthesis, any chemical used in polymerization may cause challenge for in vivo biocompatibility. Further, initiators, organic solvents, stabilizers, emulsifiers, unreacted monomers and crosslinkers may also be toxic to host cells (Bryant et al. 2000). Thus, removal of hazardous chemicals from preformed hydrogels is important. This may involve a number of purification processes like solvent washing or dialysis. In situ gelation of scaffolds presents a unique challenge since reactants which are used to synthesize the gel are injected into the body in the pre-polymer solution form. The utilization of this technique is ideal but needs precaution that all components used in the reaction are safe and non-toxic. Earlier, natural polymers were believed to be better than synthetic ones in terms of biocompatibility; still the presence of synthetic crosslinkers and initiators seems to have the same toxicity concerns as purely synthetic hydrogels.

4 Characterization of Hydrogels

Different types of characterization techniques have been developed for understanding the hydrogel network structure and physical or chemical properties. The physical properties of hydrogels network structure depend upon the equilibrium and dynamic swelling ratios. It depends on volume fraction of polymer, effective molecular weight of the polymer chain in between two crosslinking points and the correlation between two adjacent crosslinks (Lin and Metters 2006; Peppas et al. 2006). Flory (1942) and Huggins (1942) were the first to develop independently the theoretical base 70 years ago for understanding the polymer solutions.

Hydrogels have numerous properties, like absorption capacity, swelling behaviour, permeability, surface properties, optical properties and mechanical properties. All these properties make hydrogel a promising material for a wide variety of applications. The properties of the polymer chains and the crosslinking structures in these aqueous solutions play a vital role in the outcome of the properties of the hydrogel. Hydrogels are characterized by following methods/tests.

4.1 Fourier Transform Infrared Spectroscopy

FTIR analysis provides a reliable information about the crosslinking which is confirmed by appearance of IR bands near 1648 cm^{-1} region. IR absorption spectra give an idea about morphology of hydrogels.

4.2 Atomic Force Microscopy (AFM)

A multimode atomic force microscope helps to examine the surface morphology of the hydrogels.

4.3 Network Pore Size

Pore size determination is an important technique for hydrogel characterization. Different techniques like Quasi-elastic laser light scattering, electron microscopy, mercury porosimetry, rubber elasticity measurements, and equilibrium swelling experiments are used to determine the network pore size of hydrogel.

4.4 X-ray Diffraction

X-ray diffraction studies provide useful insight into the crystalline nature of hydrogel, whether the hydrogels retain their crystallinity or they get deformed during the synthesis procedures

4.5 Swelling Behaviour

Data of specific swelling studies is a must to determine its potential use as a hydrogel, and many researchers have successfully worked upon it.

4.6 Crosslinking and Mechanical Strength

The mechanical strength of the hydrogel depends upon the crosslinker density inside the network structure. Generally, the mechanical strength of the hydrogel increases with increase in the crosslinker concentration.

4.7 Rheology

It depends on the type of interactions based on structural properties (i.e. association, entanglement and crosslinks) present in the system. Polymer solutions are essentially viscous at low frequencies and tend to fit the scaling laws: $G' \sim \omega^2$ and $G'' \sim \omega$. At high frequencies, elasticity dominates ($G' > G''$). This corresponds to Maxwell-type behaviour with a single relaxation time, which may be determined from the crossover point, and this relaxation time increases with concentration. Crosslinked microgel dispersions exhibit G' and G'' that are almost independent of oscillation frequency.

All these characterization techniques give an account for the confirmation of intended crosslinking results, formation of hydrogel, used further for different applications.

5 Application of Smart Hydrogel

In the last few decades due to its hydrophilic character, biocompatibility and initiate stimuli hydrogels received considerable attention towards several applications. These applications are discussed in the following subsections.

5.1 Biomaterial Application

5.1.1 Soft Contact Lenses

Soft contact lenses remain one of the most widely used applications of hydrogels. Hydrogels have the adaptability towards the global ocular curvature and permit the atmospheric oxygen to reach up to cornea by dissolving the water of the lens (Lum et al. 2013). PHEMA was the first ever synthetic hydrogel prepared by DuPont scientists in 1936 (Strain et al. 1939). It was established as a promising and wonderful candidate for manufacture of contact lens by Wichterle and Lim (1960) due to their biocompatibility and mechanical properties. Nowadays, a number of hydrogel contact lens materials have been developed containing various monomers such as N-VP, MAA, MMA and glyceryl methacrylate. These are incorporated to increase the water content of hydrogel contact lens and also to enhance its mechanical properties to allow them to hold the force of the eyelid along with an elevated permeability to oxygen (Lum et al. 2013) (Fig. 4).

5.1.2 Tissue Regeneration and Tissue Engineering Applications

Globally, many people suffer from the loss of an organ or chronic failure of any organ function as a result of some severe disease, accident, etc., every year. This necessitates the need of tissue and organ transplantations, but they are difficult to be carried out due to lesser availability of donors, legal norms, social norms, etc. (Lee and Mooney 2001).

The term 'tissue engineering' came in practice in 1988 as an engineered application. The beautiful combination of fundamental principles of basic

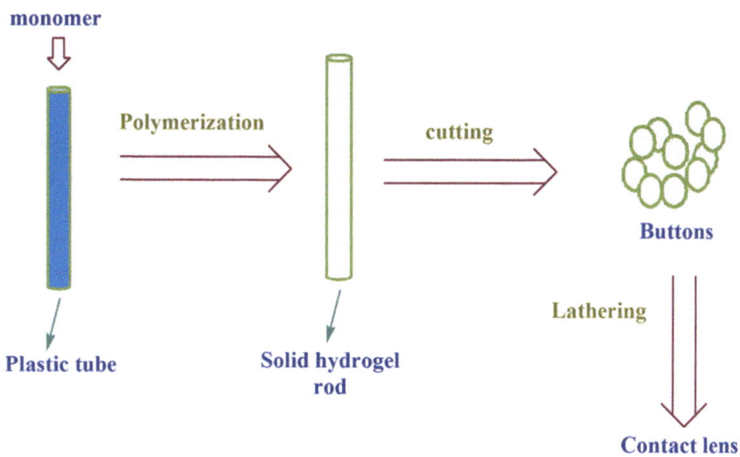

Fig. 4 Schematic representation of lathe-cutting technique

engineering and life sciences towards understanding human physiology and solving its problems by developing substitutes for tissues or organ system was a major breakthrough in the history of mankind. This helped in reviving a hope towards extension of life by regenerating specific tissues and organ systems using engineered materials and suitable synthetic strategies (Chapekar 2000).

Thus, tissue engineering has developed a hope for design of an ideal living substitute which mimics the properties of living tissues in human system (Langer and Vacanti 1993). Scaffolds act as three-dimensional artificial templates in which the tissue targeted for reconstruction is cultured to grow. The highly porous nature of hydrogel allows it for the diffusion of cells during migration. It also transfers the nutrients and excludes the waste products from cellular membranes (Loh and Choong 2013).

Now, both synthetic as well as natural hydrogels are used as scaffolds for various tissue engineering applications. These include repair of tendon, ligament, cartilage, blood vessels, skin and even heart valves (Ma 2004). The synthetic hydrogels targeted to be used as scaffolds are polyurethanes (PU), PVA, PEO, PNIPAAm, PAAc and poly(propylene fumarate-*co*-ethylene glycol) [P(PFco-EG)]. The natural hydrogels to be used in these applications include agarose, alginate, collagen, chitosan, gelatin, fibrin and hyaluronic acid (HA).

The micron-sized hydrogels (microgels) are also used to deliver macromolecules like phagosomes into cytoplasm of antigen-presenting cells. This release is triggered on due to acidic conditions within the tissues. These hydrogels mould themselves according to the pattern of membranes, tissues and possess considerable mechanical strength. This property of hydrogels is also used in cartilage repair (Bindu et al. 2012). Presently, the following hydrogels have been successfully used in tissue engineering applications:

(a) Collagen-coated tissue culture: They are used for implant of cornea, trachea gland cells, etc. (Pal et al. 2009).
(b) Poly(lactic-co-glycolic acid) (PLGA) polymer: They are used in conjunction with preadipocytes for epithelial cell culture of breast cells (Pal et al. 2009).
(c) Porous scaffolding: These are coated with fibrillar collagen, used for the culture of liver cells, to be used in liver implants (Pal et al. 2009).

Nowadays, scientists are expanding the application of hydrogels in the regeneration of the central nervous system. For this application, chemically crosslinked PHEMA tubes have been developed by synthesizing centrifugal force; the outer diameter of these tubes is 2.4 mm, and wall thickness is 40–400 μm, which could be used for guided regeneration in the nervous system (Lum et al. 2013).

5.1.3 Wound Dressing

From early 1980s investigations are going on to promote the skin healing potential of hydrogels and its application in clinical setting. At first, the hydrogels absorb and retain wound exudates. Afterwards, fibroblast proliferation and keratinocyte migration are done for completion of epithelialization of the wound or wound healing. [Bullock et al. 2010; Ribeiro et al. 2009) The dense matrices of hydrogels (100 nm in swollen state) not only check the entry of bacteria but also permit transport of bioactive molecules (e.g. antimicrobial agents or drugs) to the targeted wound site to be healed (Drury and Mooney 2003). Such molecules can be easily entrapped in the polymeric network during the gel formation process, which is gradually released on the wound site as hydrogel absorbs its exudate and swells (Burd 2007; Boonkaew et al. 2014; Chakavala et al. 2012; Kumar et al. 2012a, b; Cui et al. 2011). The unique and tunable mechanical properties of hydrogels increase its suitability towards elasticity and flexibility to adapt with wounds caused in different body sites.

Hydrogels bring immediate relief to patients in distress, as compared to conventional bandages, pads or gauzes. Even, in case of burns, hydrogels are good alternatives for running water. It acts as a coolant to localized wound also reduces the pain and recovers the extent of resultant damage and reduces pain (Cuttle et al. 2009; Coats et al. 2002). The high water retention ability of hydrogels makes them particularly soothing on the wound. Non-adhesive nature of hydrogels causes less pain and discomfort to patient, as cells do not attach firmly to hydrophilic surfaces. Transparency of hydrogel provides an advantage over traditional bandages causing less pain during peeling it off.

A wide variety of hydrogels for wound dressings is commercially available in the market for treatment of minor burns and other skin wounds. They are available in numerous forms like amorphous gels, gel-impregnated gauzes, sheets or plasters (Burd 2007; Grippaudo et al. 2010). Amorphous gels are generally prescribed for superficial burn like cavity wounds, sheets and gel-impregnated gauzes (Winter et al. 1962). Plaster-like hydrogel dressings (e.g. MySkin®) are very user-friendly and attractive, as it can be well positioned on the wound without the use of adhesives and bandages. The development of hydrogel formulations is attaining new heights (Table 2), to address different aspects of wound healing and management (e.g. reduction in infection control, easy dressing).

5.1.4 Drug Delivery

Hydrogels have unique properties that make it useful in drug delivery application. Due to its hydrophilicity, it can hold excessive amounts of water. Hence, it can be used for design of drug delivery systems that control release of solute over a given time period. Many biomaterials have been explored for this purpose, which act by two mechanisms: (1) controlled release may be achieved by varying the crosslinker concentration and controlling the ratio of hydrophilic to hydrophobic monomers. (2) The interaction of hydrogels with drugs is very less, so it is better to release large fraction of active molecules drugs (protein and peptides) through hydrogel carriers. Controlled and targeted drug delivery would decrease the unwanted side effects and aid the recovery aspects. Several mechanisms describe the release of

Table 2 List of various hydrogels used in wound dressing

Burn depth	Hydrogel precursors	Additional components	Source
Partial-thickness	AMPS/ PEGDA	–	Nalampang et al. (2013)
	AMPS	Silver nanoparticles	Boonkaew et al. (2014a), Boonkaew et al. (2014b)
	Chitin	ZnO nanoparticles	Kumar et al. (2012a, b)
	Chitosan	–	Lu et al. (2010a, b)
	PVP/PEG	Honey	Mohd et al. (2012)
	Laponite®/ alginate	Mafenide	Ghadiri et al. (2014)
	PVA/ chitosan	Silver sulfadiazine	Chakavala et al. (2012)
	PVA	Silver nanoparticles	Oliveira et al. (2014)
Full-thickness	Chitosan/ collagen	Lysostaphin	Cui et al. (2011)
	Collagen/ PEG/fibrin	–	Natesan et al. (2013)
	Dextran/ PEGDA	–	Sun et al. (2011)
	Dextran	Chitosan microparticles with EGF and VEGF	Ribeiro et al. (2013)

drug from hydrogel: (a) diffusion, (b) chemical control, (c) swelling and environmentally responsive release.

The diffusion-controlled release systems can be represented by matrix devices. This may be available as vacant chambers in form of a capsule, cylinders or sphere. This system carries drugs covered with a hydrogel membrane as shown in Fig. 5.

In this system, a continuous drug release system is maintained due to high concentration of drug in the centre of system (Peppas and Lowman 1999). Other matrix systems include dispersion or uniform dissolution of drug throughout the

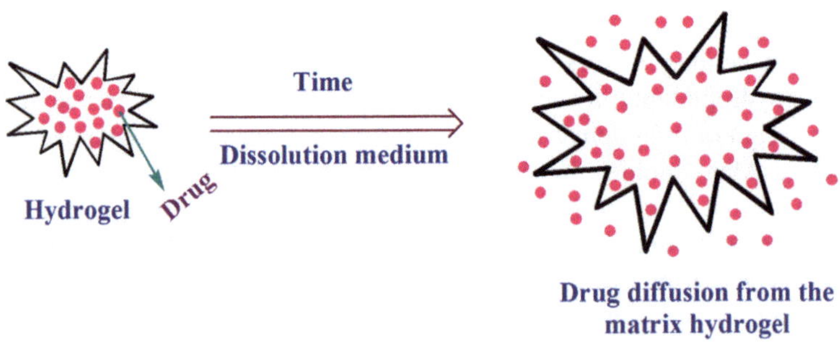

Fig. 5 Schematic representation of drug release from hydrogel matrix

three-dimensional space of hydrogel as shown in Fig. 5. Here, drug release is carried out by the macromolecular pores. In this system, the initial rate of drug release is proportional to the square root of time (Peppas and Lowman 1999).

In the swelling-controlled release devices, the drug is loaded in a glassy polymer and it starts swelling on contact with a bio-fluid. Afterwards, hydrogel starts expanding beyond its boundary and allows the drug to diffuse with the relaxation of polymeric chains. This is called Case II transport mechanism, and it shows constant, time-independent release kinetics. It is called as 'anomalous transport', combining swelling-controlled release with diffusion (Peppas and Lowman 1999).

'Smart' hydrogels are promising materials for controlled release of drug, since they change their properties in response to specific stimuli. Right after the discovery of hydrogels, it has been used as anticancer and antibiotic delivery. Various research experiments have focus on the variety of drugs that can be delivered efficiently through the hydrogel based on the delivery systems shown in Table 3.

5.2 *Industrial Applications*

Hydrogels are significantly used for the adsorption of methylene blue dye from the industrial effluent. Hydrogels beads are used for the adsorption of dioxins.

Table 3 Drug delivery via different hydrogel-based system

S. No.	Carrier medium	Drug	Therapeutic category	Analysis	Ref.
1	Hydrogel	Insulin	Hypogylcemic	Sustain release of insulin	Brown et al. (1996)
2	Hydrogel	Riboflavin water	Soluble vitamin	pH sensitive to localize drug delivery	Amiji (1997)
3	Hydrogel	Salicylic acid	Antiseborrheics	pH-sensitive drug delivery system	Ferreira (2000)
4	Hydrogel	Terbinafine hydrochloride	Antifungal	Controlled drug delivery system	Şen et al. (2000)
5	Hydrogel	5-fluorouracil and diclofenac sodium	Antimetabolite and antiinflammatory	Localized drug delivery	Zhang et al. (2014)
6	IPN hydrogel beads	Simvastatin	Lipid lowering drug	Controlled drug delivery system	Boppana et al. (2010)

A numbers of researchers are trying to develop techniques to sieze the metal ions. Irani and the co-workers synthesized a polyethylene-*g*-poly(acrylic acid)-co-starch/OMMT (LLDPE-g-PAA-co-starch/OMMT) hydrogel composite for PbI (II) removal (Joint et al. 2010; Irani et al. 2015).

Yan and co-workers performed etherification and functionalization of chitosan beads resulting in carboxymethylated chitosan having adsorption capacity of metal ions. These beads selectively adsorb specific ions like Cu(II), Pu(II) and Mg(II) (Yan et al. 2011). This property increases its potential towards dye removal after magnetic doping of hydrogel microsphere with IPN structures (Yan et al. 2011).

Novel and porous bio-absorbent xylan-based hydrogel has been developed by Xin after graft co-polymerization of acrylic acid (AA) and xylan-rich hemicelluloses. It has many applications towards adsorption of heavy metal ions (Pd^{2+}, Cd^{2+} and Zn^{2+}) from aqueous solutions.

The maximum adsorption capacities of Pd^{2+}, Cd^{2+} and Zn^{2+} were reported to be 859, 495 and 274 mg/g, respectively (Peng et al. 2012).

5.3 *Environmental Applications*

Hydrogels are polymeric chains of repeating units with high water absorbing capability. Some hydrogels have capability to absorb water 500 times more than their weight. These superabsorbent properties of hydrogel enhance its potential to conserve water and to solve other environmental issues.

5.3.1 Hydrogels for the Prevention of Soil Erosion

Soil erosion has been controlled since a decade by reducing erosion and increasing water retention and permeability of finely textured agricultural soils. Water-soluble hydrogels have been used to reduce erosion and improve water infiltration among fine-textured agricultural soils. The water-soluble PAM hydrogels play an important role in preventing soil erosion by forming a thin film that covers the soil surface. During irrigation, this film protects the soil from washing away and maintains the optimum moisture conditions within the soil system, so that irrigation water can permeate easily. Several studies have proved that PAM hydrogels are efficient agents used to combat soil erosion.

5.3.2 Use of Hydrogels for Agricultural Purposes

Hydrogel uniqueness enhances its potential towards the field of agriculture also. It is reported that anionic herbicide 2,4-D encapsulated in carboxymethylcellulose (CMC) gel is suitable for the controlled release of herbicide. More perfection in this process has been attained by the addition of bentonite in the gel formulation by insinuating inorganic or organic cations in between of Na^+ saturated bentonite for the slow release rate of herbicide in water and soil (Li et al. 2009).

Nowadays, clay-loaded hydrogels are also gaining more interest because of its unique properties and broad applications. Nano-composite-based hydrogels have been developed to release macro- and micronutrients in slow and controlled manner into independent or concurrent systems. Its physicochemical properties and water uptake capacity improved 5000 times more than its weight by producing nanocomposite by underwent hydrolysis treatment. These nanocomposites had great swelling degree, so they were formulated with high calcium montmorillonite (MMt) contents for agricultural applications (e.g. carriers for nutrient release) (Bortolin et al. 2016).

6 Patents on Hydrogels

As a result of high potential and tailor-made properties of hydrogels, researchers have obtained various patents in its various spheres of utilization. The recent available patents are listed in Table 4.

Table 4 Patents on application of hydrogels

Application	Patent no	Title of patent	Year	Ref.
Biomaterial application	United State Patent US 6861123	Silicone hydrogel contact lens	2005	Turner et al. (2005)
	United State Patent US 3679504	Method of forming colour effects in hydrogel contact lenses and ophthalmic prostheses	2005	Ketelson et al. (2005)
	United State Patent US 4472327	Method of making hydrogel cosmetic contact lenses	1984	Neefe (1984)
	United State Patent US 3808178	Oxygen-permeable contact lens composition, methods and article of manufacture	1974	Gaylord (1974)
	United State Patent US 5423737	Transparent hydrogel wound dressing with release tab	1995	Cartmell et al. (1995)
	United State Patent US 8431151	Antimicrobial nano-structured hydrogel web containing silver	2013	Mather et al. (2013)
	United State Patent US 8409606	Drug delivery through hydrogel plugs	2013	Sawhney et al. (2013)

(continued)

Table 4 (continued)

Application	Patent no	Title of patent	Year	Ref.
	United State Patent US 5514380	Biodegradable hydrogel co-polymer as drug delivery matrix		Song et al. (1996)
	United State Patent US 8383153	Poly(amidoamine) oligomer hydrogel for drug delivery and drug carrier using the same	2013	Lee et al. (2013)
	United State Patent WO 1998043615	Method for oral delivery of proteins	1998	Lowman et al. (1998)
	United Sate Patent US 7066904	Triggered release hydrogel drug delivery system	2006	Arthur et al. (2006)
	United State Patent US 9255178	Photocrosslinkable poly (caprolactone fumarate)	2016	Wang et al. (2016)
	United State Patent US 4654039	Hydrogel-forming polymer compositions for use in absorbent structures	1987	Brandt et al. (1987)
	United State Patent US 5147343	Absorbent products containing hydrogels with ability to swell against pressure	1992	Kellenberger (1992)
	European Patent EP 2646002	Hydrogels microcapsules	2013	Mistry et al. (2013)
Industrial application	European Patent EP 2172475	Lipid peptide for use as hydrogels	2015	Miyaji et al. (2015)
	United State Patent US 6960617	Hydrogels having enhanced elasticity and mechanical strength properties	2005	Omidian et al. (2005)
	United State Patent US 5847089	Carboxyl-modified superabsorbent protein hydrogel	1998	Damodaran and Hwang (1998)
	United State Patent US 7304098	Hydrogel for use in down hole seal applications	2007	Li and Zhou (2007)
	United State Patent US 8244078	Hydrogel-based sensor probe for detecting an environmental state	2012	Hendriks et al. (2012)

(continued)

Table 4 (continued)

Application	Patent no	Title of patent	Year	Ref.
Environmental application	United State Patent 8840839	Environmentally responsive hydrogel	2012	Iordanov et al. (2014)
	United State Patent US 8840839	Hydrogel-based device for detecting an environmental state	2014	Sidney et al. (1993)
	United Sate Patent US 5185024	Application of agricultural polyammonium acrylate or polyacrylamide hydrogels	1993	Aqua source Inc. (1993)
	European Patent EP 2535359	Controlled release agricultural products	2012	Sannino et al. (2012)

7 Conclusion

The hydrogel materials have a great ability to serve as response to stimuli materials. Hydrogels have immense hidden potential in various applications. They are successfully being used in biomedical field in adverse environmental conditions like low pH and high temperatures, in human metabolism. A large numbers of hydrogel materials have been investigated for controlled drug release for drug formulation purpose. This bio-compatible nature of hydrogels makes them a promising candidate for future applications as well as for the development of next generation of materials for biomedical application also environmental field.

A major weakness of all these hydrogels is the need for optimization of their response time, which is either too slow or too fast. These optimized hydrogels as per their response time are in huge demand for various emerging applications. Future hydrogels are in the process of development with perfect control over their structure–property relationships even at micro- and nano-levels. So, practically they are used as smart materials. This relationship is further expected to result in a new range of hydrogels which are specifically suited for an application and unlock a new era of human imagination.

References

Abebe DG, Fujiwara T (2012) Controlled thermoresponsive hydrogels by stereocomplexed PLA-PEG-PLA prepared via hybrid micelles of premixed copolymers with different PEG lengths. Biomacromol 13:1828–1836

Abraham GA, de Queiroz AAA, San RJS (2001) Hydrophilic hybrid IPNs of segmented polyurethanes and copolymers of vinyl pyrrolidone for applications in medicine. Biomaterials 22:1971–1985

Aimetti AA, Machen AJ, Anseth KS (2009) Poly (ethylene glycol) hydrogels formed by thiolene photopolymerization for enzyme-responsive protein delivery. Biomaterials 30(30):6048–6054

Alvarez LC, Concheiro A, Dubovik AS, Grinberg NV, Burova TV, Grinberg VY (2005) Temperature-sensitive chitosan-poly *N*-isopropylacrylamide interpenetrated networks with enhanced loading capacity and controlled release properties. J Control Release 102:629–641

Amiji M (1997) Gelatin-poly(ethylene oxide) semi interpenetrating polymer network with pH-sensitive swelling and enzyme-degradable properties for oral drug delivery. Drug Dev Ind Pharm 23:575–582

Aqua Source Inc (1993) United State Patent US 5185024

Arthur R, Barry JJ, Sahatjian R (2006) US Patent US706690

Benamer S, Mahlous M, Boukrif A, Mansouri B, Youcef SL (2006) Synthesis and characterisation of hydrogels based on poly(vinyl pyrrolidone). Nucl Instrum Methods Phy Res Section B: Beam Interact Mater Atoms 248(2):284–290

Bindu SM, Ashok V, Chatterjee A (2012) Review article as a review on hydrogels as drug delivery in the pharmaceutical field. Int J Pharm Chem Sci 1:642–661

Boonkaew B, Kempf M, Kimble R, Supaphol P, Cuttle L (2014a) Antimicrobial efficacy of a novel silver hydrogel dressing compared to two common silver burn wound dressings: Acticoat™ and PolyMem Silver®. Burns 40:89–96

Boonkaew B, Barber PM, Rengpipat S, Supaphol P, Kempf M, He J et al (2014b) Development and characterization of a novel antimicrobial, sterile hydrogel dressing for burn wounds: single-step production with gamma irradiation creates silver nanoparticles and radical polymerization. J Pharm Sci 103:3244–3253

Boppana R, Kulkarni RV, Mutalik SS, Biswanath S (2010) Interpenetrating network hydrogel beads of carboxymethylcellulose and egg albumin for controlled release of lipid lowering drug. J Microencap 27:337–344

Bortolin A, Serafim AR, Aouada FA, Mattoso LH, Ribeiro C (2016) Macro- and micronutrient simultaneous slow release from highly swellable nanocomposite hydrogels. J Agric Food Chem 64(16):3133–3140

Brandt KA, Goldman SA, Inglin TA, Procter and Gamble Co (1987) Hydrogel-forming polymer compositions for use in absorbent structures. U.S. Patent 4,654,039

Brown LR, Edelman ER, Fischel-Ghodsian F, Langer R (1996) Characterization of glucose-mediated insulin release from implantable polymers. J Pharm Sci 85 (12):1341–1345

Bryant SJ, Nuttelman CR, Anseth KS (2000) Cytocompatibility of UV and visible light photo initiating systems on cultured NIH/3T3 fibroblasts in vitro. J Biomater Sci Polym Ed 11:439–457

Bullock AJ, Pickavance P, Haddow DB, Rimmer S, MacNeil S (2010) Development of a calcium-chelating hydrogel for treatment of superficial burns and scalds. Regen Med 5:55–64

Burd A (2007) Evaluating the use of hydrogel sheet dressings in comprehensive burn wound care. Ostomy Wound Manage 53:52–62

Buwalda SJ, Boere KW, Dijkstra PJ, Feijen J, Vermonden T (2014) Hydrogels in a historical perspective: from simple networks to smart materials. J Control Release 190:254–273

Cartmell JV, Sturtevant WR, Bausmith WE, Wolf ML (1995) US Patent US5423737

Chakavala SR, Patel NG, Pate NV, Thakkar VT, Patel KV, Gandhi TR (2012) Development and *in vivo* evaluation of silver sulfadiazine loaded hydrogel consisting of polyvinyl alcohol and chitosan for severe burns. J Pharm Bioallied Sci 4:S54–S56

Chang C, Duan B, Cai J, Zhang L (2010) Hydrogels based on cellulose for smart swelling and controllable delivery. Eur Polym J 46:92–100

Chapekar MS (2000) J Biomed Mater Res 53:617–620

Chivukula P, Dusek K, Wang D, Duskova SM, Kopeckova P, Kopecek J (2006) Synthesis and characterization of novel aromatic azo bond-containing pH-sensitive and hydrolytically cleavable IPN hydrogels. Biomaterials 27:1140–1151

Christensen L, Breiting V, Vuust J, Hogdall E (2006) Adverse reactions following injection with a permanent facial filler polyacrylamide hydrogel (aquamid): causes and treatment. Eur J Plast Surg 28:464–471

Chung HJ, Lee Y, Park TG (2008) Thermo-sensitive and biodegradable hydrogels based on stereocomplexed Pluronic multi-block copolymers for controlled protein delivery. J Control Release 127:22–30

Coats TJ, Edwards C, Newton R, Staun E (2002) The effect of gel burn dressings on skin temperature. Emerg Med J 19:224–225

Cook JP, Goodall GW, Khutoryanskaya OV, Khutoryanskiy VV (2012) Microwave-assisted hydrogel synthesis: a new method for crosslinking polymers in aqueous solutions. Macromol Rapid Commun 33:332–336

Cretu A, Gattin R, Brachais L, Barbier-Baudry D (2004) Synthesis and degradation of poly (2-hydroxyethyl methacrylate)-graft-poly (ε-caprolactone) copolymers. Polym Degrad Stab 83:399–403

Cui F, Li G, Huang J, Zhang J, Lu M, Lu W (2011) Development of chitosan-collagen hydrogel incorporated with lysostaphin (CCHL) burn dressing with anti-methicillin-resistant *Staphylococcus aureus* and promotion wound healing properties. Drug Deliv 18:173–180

Cuttle L, Pearn J, McMillan JR, Kimble RM (2009) A review of first aid treatments for burn injuries. Burns 35:768–775

Damodaran S, Hwang DC (1998) United State Patent US 5847089

Derwent JJK, Mieler WF (2008) Thermoresponsive hydrogels as a new ocular drug delivery platform to the posterior segment of the eye. Trans Am Ophthalmol Soc 106:206–214

Drury JL, Mooney DJ (2003) Hydrogels for tissue engineering: scaffold design variables and applications. Biomaterials 24(24):4337–4351

Ferreira L (2000) Evaluation of poly(2-hydroxyethyl methacrylate) gels as drug delivery systems at different pH values. Int J Pharm 194:169–180

Flory PJ (1942) Thermodynamics of high polymer solutions. J Chem Phys 10:51–61

Gaylord NG (1974) US Patent US3808178

Gemeinhart AR, Chen J, Park H, Park K (2000) pH sensitive of fast responsive superporous hydrogels. J Biomater Sci Plymer Edn 11(12):1371–1380

Ghadiri M, Chrzanowski W, Rohanizadeh R (2014) Antibiotic eluting clay mineral (Laponite®) for wound healing application: an *in vitro* study. J Mater Sci Mater Med 25(11):2513–2526

Gils PS, Ray D, Sahoo PK (2010) Designing of silver nanoparticles in gum arabic based semi-IPN hydrogel. Int J Biol Macromol 46:237–244

Gong CY, Shi S, Dong PW, Kan B, Gou ML, Wang XH (2009) Synthesis and characterization of PEG-PCL-PEG thermosensitive hydrogel. Int J Pharm 365:89–99

Grippaudo FR, Carini L, Baldini R (2010) Procutase versus 1% silver sulphadiazine in the treatment of minor burns. Burns 36:871–875

Grodzinski J (2009) Polymeric gels and hydrogels for biomedical and pharmaceutical application. Polym Adv Technol 21:27–47

Gunasekaran S, Chai WTC (2006) Swelling of pH-sensitive chitosan-poly(vinyl alcohol) hydrogels. J Appl Polym Sci 102:4665–4671

Hendriks BHW, Deladi S, Kurt R, Suijver JF (2012) United State Patent US 8244078

Huggins ML (1942) Some properties of solutions of long-chain compounds. J Phys Chem 46:151–158

Iizawa T, Taketa H, Maruta M, Ishido T, Gotoh T, Sakohara S (2007) Synthesis of porous poly (N-isopropylacrylamide) gel beads by sedimentation polymerization and their morphology. J Appl Polym Sci 104:842–850

Iordanov VP, Krijnsen HC, Van Bruggen MPB, Janner AM, Kurt R, Koninklijke Philips NV (2014) Hydrogel based device for detecting an environmental state. United State Patent 8,840,839

Irani M, Ismail H, Ahmad Z, Fan M (2015) Synthesis of linear low density polyethylene-g-poly (acrylic acid)-co starch/organomontmorillonite hydrogel composite as an adsorbent for removal of Pb(II) from aqueous solutions. J Environ Sci 27:9–20

Jen AC, Wake MC, Mikos AG (1996) Review: hydrogels for cell immobilization. Biotechnol Bioeng 50:357–364

Johannsmann D, Bunsow J (2008) Electrochemically produced responsive hydrogel films: influence of added salt on thickness and morphology. J Colloid Interface Sci 326:61–65

Joint I, Mühling M, Querellou J (2010) Culturing marine bacteria—an essential prerequisite for biodiscovery. Microb Biotechnol 3:564–575

Ju HK, Kim SY, Kim SJ, Lee YM (2002) pH/temperature-responsive semi-IPN hydrogels composed of alginate and poly(*N*-isopropylacrylamide). J Appl Polym Sci 83:1128–1139

Kaneko Y, Yoshida R, Sakai K, Sakurai Y, Okano T (1995) Temperature-responsive shrinking kinetics of poly(N-isopropylacrylamide) copolymer gels with hydrophilic and hydrophobic comonomers. J Membr Sci 101:13–22

Kaneko Y, Nakamura S, Sakai K, Aoyagi T, Kikuchi A, Sakurai Y, Okano T (1998) Rapid deswelling response of poly(N-isopropylacrylamide) hydrogels by the formation of water release channels using poly(ethylene oxide) graft chains. Macromolecules 31:6099–6105

Kellenberger SR (1992) Absorbent products containing hydrogels with ability to swell against pressure. U.S. Patent 5,147,343 issued September 15, 1992

Ketelson HA, Meadows DL, Stone RP (2005) Dynamic wettability properties of a soft contact lens hydrogel. Colloid Surf B. 40:1–9

Khutoryanskiy VV, Khutoryanskaya OV, Cook JP, Goodall GW (2013) US Patent 0018110

Kiil S, Dam-Johansen K (2003) Controlled drug delivery from swellable hydroxypropylmethyl cellulose matrices: model-based analysis of observed radial front movements. J Control Rel 90:1–21

Kim B, Peppas NA (2003) Poly(ethylene glycol)-containing hydrogels for oral protein delivery applications. Biomed Microdevices 5:333–341

Kirakci K, Šícha V, Holub J, KubÃ P, Lang K (2014) Luminescent hydrogel particles prepared by self-assembly of Î^2-cyclodextrin polymer and octahedral molybdenum cluster complexes. Inorg Chem 53:13012–13018

Kopecek J, Yang J (2009) Peptide directed self assembly of hygrogels. Acta Biomater 5:805–816

Kumar HGS, Satish CS, Satish KP (2006) Hydrogels as controlled drug delivery systems synthesis, crosslinking, water and drug transport mechanism. Ind J Pharm Sci 68:133–280

Kumar PT, Lakshmanan VK, Biswas R, Nair SV, Jayakumar R (2012a) Synthesis and biological evaluation of chitin hydrogel/nano ZnO composite bandage as antibacterial wound dressing. J Biomed Nanotechnol 8:891–900

Kumar PT, Lakshmanan VK, Anilkumar TV, Ramya C, Reshmi P, Unnikrishnan AG (2012b) Flexible and microporous chitosan hydrogel/nano ZnO composite bandages for wound dressing: in vitro and in vivo evaluation. ACS Appl Mater Interfaces 4:2618–2629

Ladet S, David L, Domard A (2008) Multi-membrane hydrogels. Nature 452(7183):76–79

Langer R, Vacanti J (1993) Tissue engineering. Science 260:920–926

Langer R, Peppas NA (2003) Advances in biomaterials, drug delivery, and bionanotechnology. AIChE J 49(12):2990–3006

Lee KY, Mooney DJ (2001) Chem Rev 101:1869–1880

Lee SD, Kim BS, Nguyen MK (2013) US Patent US8383153

Li S, Liu X (2008) Synthesis, characterization and evaluation of semi-IPN hydrogels consisted of poly(methacrylic acid) and guar gum for colon-specific drug delivery. Polym Adv Technol 19:371–376

Li Y, Zhou J (2007) United State Patent US 7304098

Li J, Jiang M, Wu H, Li Y (2009) Addition of modified bentonites in polymer gel formulation of 2,4-D for its controlled release in water and soil. J Agric Food Chem 57:2868–2874

Li S, Yajiang Y, Haibing Li, Xiangliang Y, Huibi Xu (2007) pH-responsive semi-interpenetrating networks hydrogels of poly(acrylic acid-acrylamide-methacrylate) and amylose. I. Synthesis and characterization. J Appl Polym Sci 106(6):3792–3799

Lim F, Sun AM (1980) Microencapsulated islets as bioartificial endocrine pancreas. Science 210:908–910

Lin CC, Metters AT (2006) Hydrogels in controlled release formulations: network design and mathematical modeling. Adv Drug Del Rev 58:1379–1408

Lipatov YS (2002) Polymer blends and interpenetrating polymer networks at the interface with solids. Prog Polym Sci 27:1721–1801
Liu M, Su H, Tan T (2012) Synthesis and properties of thermo and pH-sensitive poly (N-isopropylacrylamide)/polyaspartic acid IPN hydrogels. Carbohydr Polym 87:2425–2431
Loh QL, Choong C (2013) Three-dimensional scaffolds for tissue engineering applications: role of porosity and pore size. Tissue Eng Part B: Rev 19(6):485–502
Lowman AM, Morishita M, Nagai T, Peppas NA (1998) US Patent WO1998043615
Lu G, Ling K, Zhao P, Xu Z, Deng C, Zheng H (2010a) A novel in situ-formed hydrogel wound dressing by the photocross-linking of a chitosan derivative. Wound Repair Regen 18:70–79
Lu S, Liu M, Ni B, Gao C (2010b) A novel pH- and thermo-sensitive PVP/CMC semi-IPN hydrogel: swelling, phase behavior, and drug release study. J Polym Sci Part B Polym Phys 48:1749–1756
Lum E, Golebiowski B, Gunn R, Babhoota M, Swarbrick H (2013) Corneal sensitivity with contact lenses of different mechanical properties. Optom Vis Sci 90(9):954–960
Ma PX (2004) Scaffolds for tissue fabrication. Mater Today 7:30–40
Masteikova R, Chalupova Z, Sklubalova Z (2003) Stimuli-sensitive hydrogels in controlled and sustained drug delivery. Medicina 39:19–24
Mather P, Wu J, Ren D, Hou S (2013) US Patent US8431151
Mathur AM, Moorjani SK, Scranton AB (1996) Methods for synthesis of hydrogel networks: a review. J Macromolecular Sci Part C: Polym Rev 36(2):405–430
Matricardi P, Pontoriero M, Coviello T, Casadei MA, Alhaique F (2008) In Situ Cross-Linkable Novel Alginate-Dextran Methacrylate IPN Hydrogels for Biomedical Applications: Mechanical Characterization and Drug Delivery Properties. Biomacromolecules 9(7):2014–2020
Mistry KK, Hussain AW, Beck PH, Palmer DV, Sales BR, Mint A (2013) European Patent EP2646002
Miyaji N, Iwama T, Gotoh M, Maruyama T, Kohda D (2015) European Patent EP2172475
Miyata T, Uragami T, Nakamae K (2002) Biomolecules sensitive hydrogels. Adv Drug Deliv Rev 54:79–98
Mohamadnia Z, Zohuriaan-Mehr MJ, Kabiri K, Jamshidi A, Mobedi H (2007) pH-sensitive IPN hydrogel beads of carrageenan-alginate for controlled drug delivery. J Bioact Compat Polym 22:342–356
Mohd ZR, Abu BZZ, Yusof N, Mohamed MN, Abdullah MN (2012) Gelam (*Melaleuca* spp.) honey based hydrogel as burn wound dressing. Evid Based Complement Alternat Med 2012:843025
Montoro SR, Medeiros SDF, Alves GM (2014) Chapter 10—nanostructured hydrogels. In: Nanostructured polymer blends. William Andrew, Elsevier, Oxford, pp 325–355
Murdan S (2003) Electro-responsive drug delivery from hydrogels. J Control Release 92:1–17
Murthy PSK, Murali Mohan Y, Varaprasad K, Sreedhar B, Mohana RK (2008) First successful design of semi-IPN hydrogel-silver nanocomposites: a facile approach for antibacterial application. J Colloid Interface Sci 318:217–224
Nalampang K, Panjakha R, Molloy R, Tighe BJ (2013) Structural effects in photopolymerized sodium AMPS hydrogels cross-linked with poly(ethylene glycol) diacrylate for use as burn dressings. J Biomater Sci Polym 24:1291–1304
Natesan S, Zamora DO, Wrice NL, Baer DG, Christy RJ (2013) Bilayer hydrogel with autologous stem cells derived from debrided human burn skin for improved skin regeneration. J Burn Care Res 34:18–30
Neefe CW (1984) US Patent US4472327
Nguyen K, West J (2002) Biomaterials. Photopolymerizable hydrogels for tissue engineering applications 23:4307
Nierzwicki W, Prins W (1975) Hydrogels of crosslinked poly(1-glyceryl methacrylate) and poly (2-hydroxypropyl methacrylamide). J Appl Poly Sci 19(7):1885–1892
Oh SB, Choi YK, Cho CS (2003) Thermoplastic hydrogel based on pentablock copolymer consisting of poly(γ-benzyl L-glutamate) and poloxamer. J Appl Polym Sci 88:2649–2656

Okay O, Ozturk V (2002) Temperature sensitive poly(N-t-butylacrylamide-co-acrylamide) hydrogels: synthesis and swelling behavior. Polymer 43:5017–5026

Okumura Y, Ito K (2001) The polyrotaxane gel: a topological gel by figure-of-eight cross-links. Adv Mater 13:485–487

Oliveira RN, Rouzé R, Quilty B, Alves GG, Soares GD, Thiré RM (2014) Mechanical properties and *in vitro* characterization of polyvinyl alcohol-nano-silver hydrogel wound dressings. Interface Focus 4:20130049

Omidian H, Qiu Y, Yang S, Kim D, Park H, Park K (2005) Purdue Research Foundation. Hydrogels having enhanced elasticity and mechanical strength properties. U.S. Patent No. 6,960,617

Ossipov DA, Hilborn J (2006) Macromolecules 39:1709

Pal K, Banthia AK, Majumdar DK (2009) Polymeric hydrogels: characterization and biomedical applications. A mini review. Des Monom Polym 12:197–220

Park K, Shalaby WSW, Park H (1993). Biodegradable hydrogels for drug delivery. Technomic

Peng XW, Zhong LX, Ren JL, Sun RC (2012) Highly effective adsorption of heavy metal ions from aqueous solutions by macroporous xylan-rich hemicelluloses-based hydrogel. J Agric Food Chem 60(15):3909–3916

Peppas N, Bures P (2000) Hydrogels in pharmaceutical formulations. Eur J Pharm Biopharm 50:27–46

Peppas NA, Lowman AM (1999) Hydrogels. In: Mathiowitz E (ed) Encyclopedia of controlled drug delivery. Wiley, New York, pp 397–418

Peppas NA, Hilt JZ, Khademhosseini A, Langer R (2006) Hydrogels in biology and medicine: from molecular principles to bio nanotechnology. Adv Mater 18:1345–1360

Ribeiro MP, Ana E, Daniela S, Patrícia B, Joaquim H, Catarina F, Jorge CS, João PB, Eduardo P, Paula C, Ilídio JC (2009) Development of a new chitosan hydrogel for wound dressing. Wound Repair Regeneration 17(6):817–824

Ribeiro MP, Morgado PI, Miguel SP, Coutinho P, Correia IJ (2013) Dextran-based hydrogel containing chitosan microparticles loaded with growth factors to be used in wound healing. Mater Sci Eng C Mater Biol Appl 33:2958–2966

Rosiak J, Ruciska-Rybus A, Pekala W (1989) United State Patent US Patent 4871490

Saleem MA, Azharuddin SK, Ali S, Patil CC (2012) Studies on different chitosan polyelectrolyte complex hydrogel for modified release of diltiazem hydrochlorid. Int J Pharm and Pharm Sci 2 (4):64–67

Sannino A, Ambrosio L, Nicolasis L (2012) European Patent EP 2535359

Sawhney A, Pathak C (1994) Adhesion prevention. J Biomed Mat Res 28:831–838

Sawhney AS, Jarrett P, Bassett M, Blizzard C (2013) US Patent US8409606

Schurz J (1991) Rheology of polymer solutions of the network type. Prog Polym Sci 16:1–53

Sefton MV, May MH, Lahooti S, Babensee JE (2000) Making microencapsulation work: conformal coating, immobilization gels and *in vivo* performance. J Control Release 65:173–186

Şen M, Uzun C, Güven O (2000) Controlled release of terbinafine hydrochloride from pH sensitive poly(acrylamide/maleic acid) hydrogels. Int J Pharm 203(1–2):149–157

Segundo EP, Guerrero DQ, Cornejo BNZ, Rondero AG, Arzaluz MGN, Jose Manuel Cornejo-Bravo JMC (2008) Controlled release of model substances from pH-sensitive hydrogels. J Mex Chem Soc 52(4):272–278

Shibayama M (2012) Structure-mechanical property relationship of tough hydrogels. Soft Matter 8:8030–8038

Shivashankar M, Mandal BK (2012). A review on interpenetrating polymer network. Int J Pharm Pharm Sci 4(5):1–7

Sidney RS, Lousis LW, Gray JC (1993) US Patent US5185024

Song SS, Kim HH, Yi YW (1996) US Patent US5514380

Strain DE, Kennelly RG, Dittmar HR (1939) Methacrylate resins. Ind Eng Chem 31:382–387

Sun G, Zhang X, Shen YI, Sebastian R, Dickinson LE, Fox-Talbot K (2011) Dextran hydrogel scaffolds enhance angiogenic responses and promote complete skin regeneration during burn wound healing. Proc Natl Acad Sci U S A 108:20976–20981
Thomas V, Yallapu MM, Sreedhar B, Bajpai SK (2007) A versatile strategy to fabricate hydrogel-silver nanocomposites and investigation of their antimicrobial activity. J Colloid Interface Sci 315:389–395
Turner DC, Steffen RB, Wildsmith C, Matiacio TA (2005) US patent 6861123
Wang S, Lu L, Yaszemski MJ (2016) Mayo Foundation for Medical Education. Photocrosslinkable poly (caprolactone fumarate). U.S. Patent 9,255,178
Wang M, Xu L, Hu H, Zhai M, Peng J, Nho Y, Li J, Wei G (2007) Radiation synthesis of PVP/CMC hydrogels as wound dressing. Nucl Instr Meth Phys Res B 265:385–389
Wei YS, Charlotte AEH (2014) Short to ultrashort peptide hydrogels for biomedical uses. Mater Today 17:381–388
Wichterle O, Lim D (1960) Hydrophilic gels for biological use. Nature 185:117–118
Williams DF (1990) Concise encyclopedia of medical and dental materials. Pergamon Press, Oxford, England (1990)
Winter GD (1962) Formation of the scab and the rate of re-epithelialisation in the skin of the young domestic pig. Nature 193:293–294
Yan H, Dai J, Yang Z, Yang H, Cheng R (2011) Enhanced and selective adsorption of copper(II) ions on surface carboxymethylated chitosan hydrogel beads. Chem Eng J 174:586–594
Yannas IV, Lee E, Orgill DP, Skrabut EM, Murphy GF (1989) Synthesis and characterization of a model extracellular matrix that induces partial regeneration of adult mammalian skin. Proc Natl Acad Sci U S A 86:933–937
Yom-Tov O, Neufeld L, Seliktar D, Bianco-Peled H (2014) A novel design of injectable porous hydrogels with in situ pore formation. Acta Biomater 10:4236–4246
Yoshida R, Uchida K, Kaneko Y, Sakai K, Kikuchi A, Sakurai Y, Okano T (1995) Comb-type grafted hydrogels with rapid deswelling response to temperature changes. Nature 374:240–242
Zeng X, Jiang H (2008) Tunable liquid microlens actuated by infrared light-responsive hydrogel. Appl Phys Lett 93:151101–151103
Zhang YX, Wu FP, Li MZ, Wang EJ (2005) pH switching on-off semi-IPN hydrogel based on cross-linked poly (acrylamide-*co*-acrylic acid) and linear polyallyamine. Polymer 46:7695–7700
Zhang JT, Bhat R, Jandt KD (2009) Temperature-sensitive PVA/PNIPAAm semi-IPN hydrogels with enhanced responsive properties. Acta Biomater 5:488–497
Zhang L, Wang L, Guo B, Ma PX (2014) Cytocompatible injectable carboxymethyl chitosan/N-isopropylacrylamide hydrogels for localized drug delivery. Carbohydr Polym 103:110–118
Zohuriaan-Mehr MJ (2006) Super-absorbents. Iran Polym Soc, 2–4

Chapter 7
Hydrogel-Based Stimuli-Responsive Functionalized Graft Copolymers for the Controlled Delivery of 5-Fluorouracil, an Anticancer Drug

T. S. Anirudhan, P. L. Divya and J. Nima

Abstract 5-Fluorouracil (5-FU) is a widely used anticancer drug. To minimize the toxic side effects of 5-FU, a suitable drug delivery system (DDS) is needed for its controlled release. In the present chapter, we conducted a study that compared two different controlled-release systems into which 5-FU had been incorporated. The two delivery systems compared were 3-methacryloxy propyl trimethoxy silane-coated magnetic nanoparticles polymerized with glycidyl methacrylate-grafted-maleated cyclodextrin [MPTMS-MNP-poly-(GMA-g-MACD)] and aminated-glycidyl methacrylate-grafted cellulose-grafted polymethacrylic acid-succinyl cyclodextrin [Cell-g-(GMA/en)-PMA-SCD]. The successful formation of DDS and 5-FU loaded DDS was confirmed from FT-IR, XRD, and SEM analyses. Studies including swelling, in vitro release kinetics, drug loading efficiency, encapsulation efficiency, and cytotoxicity were performed to compare the efficiency of the drug carriers. The results suggest that both MPTMS-MNP-poly-(GMA-g-MACD) and Cell-g-(GMA/en)-PMA-SCD may be useful delivery vehicle for the controlled release of 5-FU.

1 Introduction

5-Fluorouracil (5-FU or 5-fluoro-2,4-pyrimidinedione) first synthesized in 1957 (Heidelberger et al. 1957) is one of the anticancer agents frequently used for treating solid cancers such as breast, colorectal, and gastric cancers (Cunningham and James 2001). Due to its short biological half-life (10–20 min in blood plasma) and non-specificity (Van Kuilenburg 2004), 5-FU is often given in high dose which induces systemic toxicity. The serious side effects, such as leukoencephalopathy, stroke (Kinno et al. 2014), severe diarrhea (Benson et al. 2004), and gastrointestinal mucositis (Logan et al. 2009) often necessitate a decrease in drug dose or a

T. S. Anirudhan (✉) · P. L. Divya · J. Nima
Department of Chemistry, University of Kerala, Kariavattom, Trivandrum 695581, India
e-mail: tsani@rediffmail.com

V. K. Thakur and M. K. Thakur (eds.), *Hydrogels*, Gels Horizons: From Science to Smart Materials, https://doi.org/10.1007/978-981-10-6077-9_7

discontinuation of treatment, which may cause failure of chemotherapy. 5-FU therapy can be improved and its toxicity may be diminished by facilitating the specific accumulation of this anticancer agent in the tumor infected regions with prolonged exposure of the cells to this agent. In order to selectively deliver these active agents and reducing their toxicity, the association of drugs to delivery systems has become an interesting approach.

Drug delivery systems (DDSs) can ameliorate the problems of conventional administration by enhancing drug solubility, prolonging duration time, reducing side effects, retaining drug bioactivity, and so on (Mok et al. 2008). Recently, stimuli-responsive hydrogels have sparked particular interest in their use in drug delivery applications. Hydrogels are three-dimensional networks of chemical or physically cross-linked polymers that swell in water. Hydrogels are generally biocompatible as reflected in their applications as contact lenses and as an exogenous barrier in the peritoneum (Sakai et al. 2015). Due to their unique physicochemical properties, hydrogels have emerged as a safe and effective depot-based DDS in cancer therapy (Simoes et al. 2012).

In the present chapter, we have evaluated two novel biodegradable polymeric delivery systems based on β-cyclodextrin (β-CD) and Cellulose (Cell), viz., 3-methacryloxy propyl trimethoxy silane-coated magnetic nanoparticles polymerized with glycidyl methacrylate-grafted-maleated cyclodextrin [MPTMS-MNP-poly-(GMA-g-MACD)] (Anirudhan et al. 2015a) and aminated-glycidyl methacrylate-grafted cellulose-grafted polymethacrylic acid-succinyl cyclodextrin [Cell-g-(GMA/en)-PMA-SCD] (Anirudhan et al. 2015b) that could be useful in releasing 5-FU in a controlled manner. Magnetic nanoparticles (MNPs) are widely accepted materials to revolutionize the present therapeutic techniques due to their biocompatibility, low biotoxicity, ease of surface modification, and magnetic properties. The high chemical activity and oxidizing power of naked metallic nanoparticles can be minimized by grafting or coating with silanes. In the magnetic nanocarrier system, the encapsulation of drugs provides higher encapsulation efficiency, protection of drug against in vivo degradation, the reduction of potentially toxic side effects, and the achievement of better drug pharmacokinetics (Gref et al. 1995). Deepthy et al. (2014) studied 3-aminopropyltriethoxy silane-coated magnetic nanoparticles (MNPs) coupled with glutaraldehyde for the immobilization of esterases from psedozyma sp. NII 08165 and found that the MNP immobilized esterases had prolonged shelf life, and there was no loss in enzyme activity. Cyclodextrins (CDs) have a hydrophilic exterior, which is useful for maintaining the bulk hydrophilicity and swelling state of the hydrogel, and their hydrophobic interior can facilitate the entrapment and controlled release of hydrophobic drugs. CDs can be modified to smart hydrogels using different polymerization techniques for various applications in controlled drug delivery and gene therapy. Chemical modification has been applied to CDs to alter the solubility, complex formation capacities, and introduce new functional groups (Nakahata et al. 2011). Stefano et al. (2007) prepared cyclodextrin/PEG-based hydrogels for multiple drug delivery. Rodriguez et al. (2007) prepared cyclodextrin hydrogels for the sustained release of estradiol. Blanco et al. (2011) developed cyclodextrin–agar hydrogels for the delivery of ciprofloxacin. Yuan et al. (2013) prepared chitosan-grafted cyclodextrin for the controlled release of poor water-soluble drug, Ketoprofen. Recently, Kaiwen et al. (2016) fabricated dual pH and temperature responsive hydrogels based on β-cyclodextrin derivatives for atorvastatin delivery. In MPTMS-MNP-poly-(GMA-g-MACD), monomers were grafted using a, a′-azobisisobutyronitrile (AIBN) as initiator and ethyleneglycoldimethacrylate (EGDMA) as crosslinker.

Cellulose (Cell) is the most abundant organic compound in the world which is mostly produced by plants (Trache et al. 2017; Corobea et al. 2016; Voicu et al. 2016). Cell and its derivatives are recommended for different biomedical applications such as drug delivery and tissue engineering (Singh and Rana 2010; Miculescu et al. 2016). Cell-based hydrogels are important due to their biocompatibility, non-toxicity and natural origin. Shukla et al. (2013) prepared cellulose from rice husk for application in drug delivery. Soon et al. (2015) synthesized pH-sensitive hydrogels based on hydroxyethylcellulose-hyaluronic acid for transdermal delivery. Sujan et al. (2016) synthesized temperature, pH, and redox-responsive cellulose-based hydrogels for protein delivery. The properties of Cell can be modified by graft copolymerization which provides efficient adsorbent sites for small molecules and ions (Coskun et al. 2005). Combination of β-CD and hydrogels in a single material enhances the biocompatibility, stability of the inclusion complexes and provides affinity-based regulation of drug loading and release. Polymethacrylic acid (PMA) is a nontoxic, low molecular weight polymer having numerous carboxylic acid moieties and find extensive use in drug delivery applications. Hydrogels made of PMA swells least in the stomach and the drug release is minimal. In Cell-g-(GMA/en)-PMA-SCD, glycidyl methacrylate (GMA)-grafted Cell was functionalized with ethylenediamine (en) and further grafted with PMA and succinyl cyclodextrin (SCD) using N-hydroxyl-succinimide (NHS) as an activating reagent and 1-ethyl-(3-3-dimethylaminopropyl) carbodiimide hydrochloride (EDC) as a coupling agent.

2 Experimental

2.1 Materials

β-CD, 3-methacryloxy propyl trimethoxy silane (MPTMS) (98.0% purity), AIBN, EGDMA, 5-FU, dimethylformamide (DMF), GMA, NHS, EDC, polyvinyl alcohol (PVA), benzoyl peroxide (BPO), and isopropyl alcohol were obtained from Sigma-Aldrich, Milwaukee, WI, USA. Cellulose (Cell) was procured from Central Drug House Ltd, New Delhi. Cyclohexane, methanol, en, methacrylic acid, potassium peroxodisulfate ($K_2S_2O_8$), sodium bicarbonate, hydrochloric acid, sodium chloride, $FeCl_3.6H_2O$, and $FeCl_2.4H_2O$ were purchased from E-Merck, Worli, Mumbai, India. Succinic anhydride (SA), acetone, N,N′-dimethylamino-pyridine (DMAP), and methylene bisacrylamide (MBA) were purchased from Sisco Research Laboratories Pvt Ltd., Mumbai, India. Pyridine was obtained from NICE Chemicals Pvt Ltd., Cochin.

All the reagents were of analytical grade and used without further purification. Double distilled water with specific conductivity less than 1 μΩ/cm was used throughout the study.

2.2 *Preparation of the Drug Carriers*

2.2.1 3-Methacryloxy Propyl Trimethoxy Silane-Coated Magnetic Nanoparticles Polymerized with Glycidyl Methacrylate-Grafted-Maleated Cyclodextrin [MPTMS-MNP-poly-(GMA-g-MACD)]

The preparation of MPTMS-MNP-poly-(GMA-g-MACD) involves the following steps as represented in Fig. 1. MNPs were prepared by complete precipitation of Fe_3O_4 under alkaline condition (Balaji et al. 2009). 0.35 g of MNPs prepared was dispersed in ethanol (50 mL). After treating the solution with ultrasonic wave (30 min), 140 μL MPTMS was added and stirred for 7 h to obtain MPTMS-MNP (Ma et al. 2003). Then MACD was prepared by stirring 4.90 g of maleic anhydride (MAH) (0.05 M) and β-CD (5.682 g) in 30 mL DMF at 80 °C for 10 h. Then, the solution was cooled and poured into chloroform (30 mL). The obtained MACD (2.0 g) was treated with 0.29 g of GMA at 50 °C for 24 h and precipitated using acetone (10 mL) to obtain GMA-g-MACD (Reis et al. 2009). The final DDS was prepared by treating 0.50 g of GMA-g-MACD and 0.50 g of MPTMS-MNP in 50 mL methanol in presence of 0.02 g AIBN and 50 wt% EGDMA (60–70 °C, 3 h).

2.2.2 Aminated-Glycidyl Methacrylate-Grafted Cellulose-Grafted-Polymethacrylic Acid-Succinyl Cyclodextrin [Cell-g-(GMA/en)-PMA-SCD]

The reaction steps followed for the synthesis of Cell-g-(GMA/en)-PMA-SCD are shown in Fig. 2. Definite proportions of grafting agent (9.9 g GMA), crosslinker (0.1 g MBA), and initiator (1.0 g BPO) were added to a suspension of 10.0 g Cell in 100 mL distilled water followed by the addition of 1 mL isopropyl alcohol and 12 mL cyclohexane mixture which was then poured into 75 mL 1.0% PVA and heated at 70–80 °C in a water bath for 3 h to obtain GMA-grafted Cell (Cell-g-GMA) (Anirudhan and Jalajamony 2010). 10.0 g of the obtained Cell-g-GMA was heated with 50 mL of en at 80 °C, at pH 11.0 for 4 h. (Anirudhan and Suchithra 2009). The obtained Cell-g-(GMA/en) (1.0 g) was treated with succinyl CD (SCD, 4.0 g) and PMA hydrogel (3.0 g) in 50 mL of ethanol/water mixture. A mixture of EDC (0.5 g) and NHS (0.1 g) prepared in 5 mL of distilled water was added to the above mixture and stirred at 50 °C for 4 h. The obtained drug carrier, Cell-g-(GMA/en)-PMA-SCD was dried in an air oven at 50 °C for 6 h and stored in vacuum for further use.

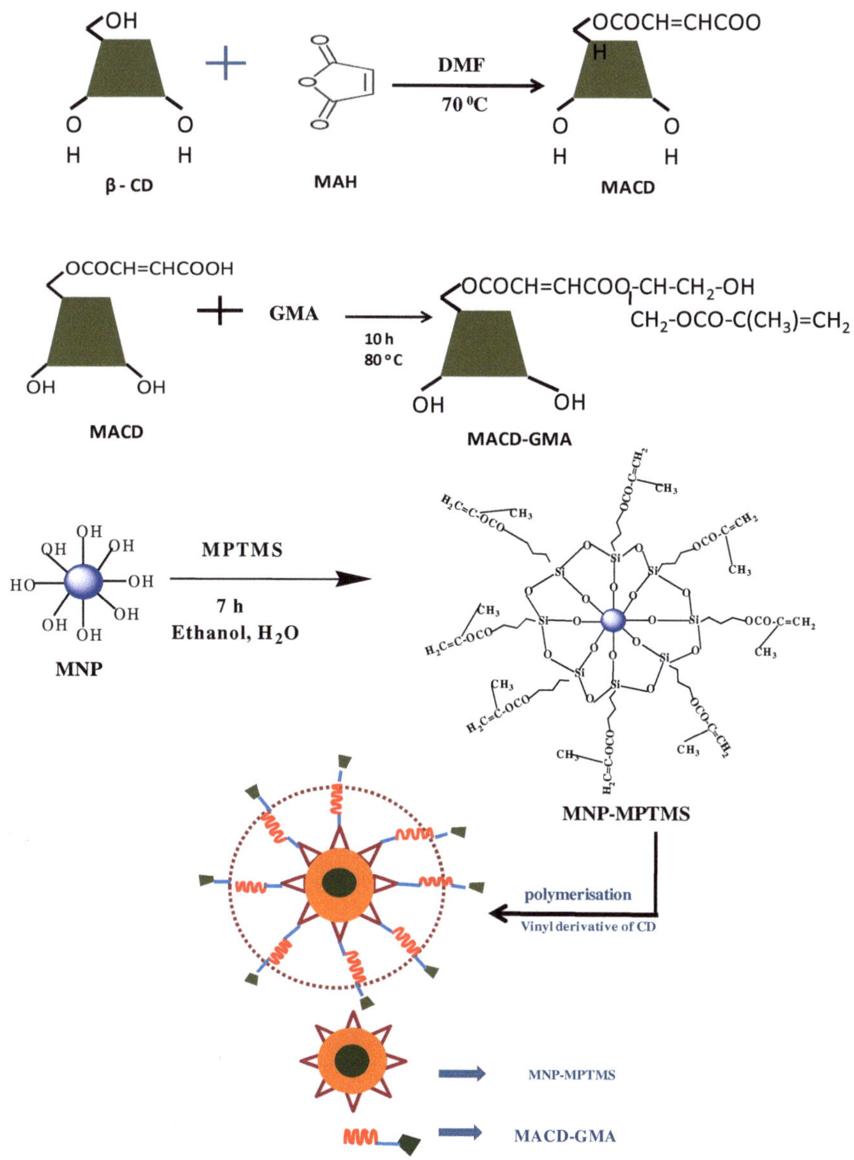

Fig. 1 Synthesis of MPTMS-MNP-poly-(GMA-g-MACD): proposed reaction mechanism

2.3 *Instruments and Methods of Characterization*

Infrared spectra were recorded with a Shimadzu FT-IR spectrometer in the wavelength range 400–4000 cm^{-1} using a KBr window at a resolution of 4 cm^{-1}. The XRD patterns of all the samples were recorded using an X'Pert Pro X-ray

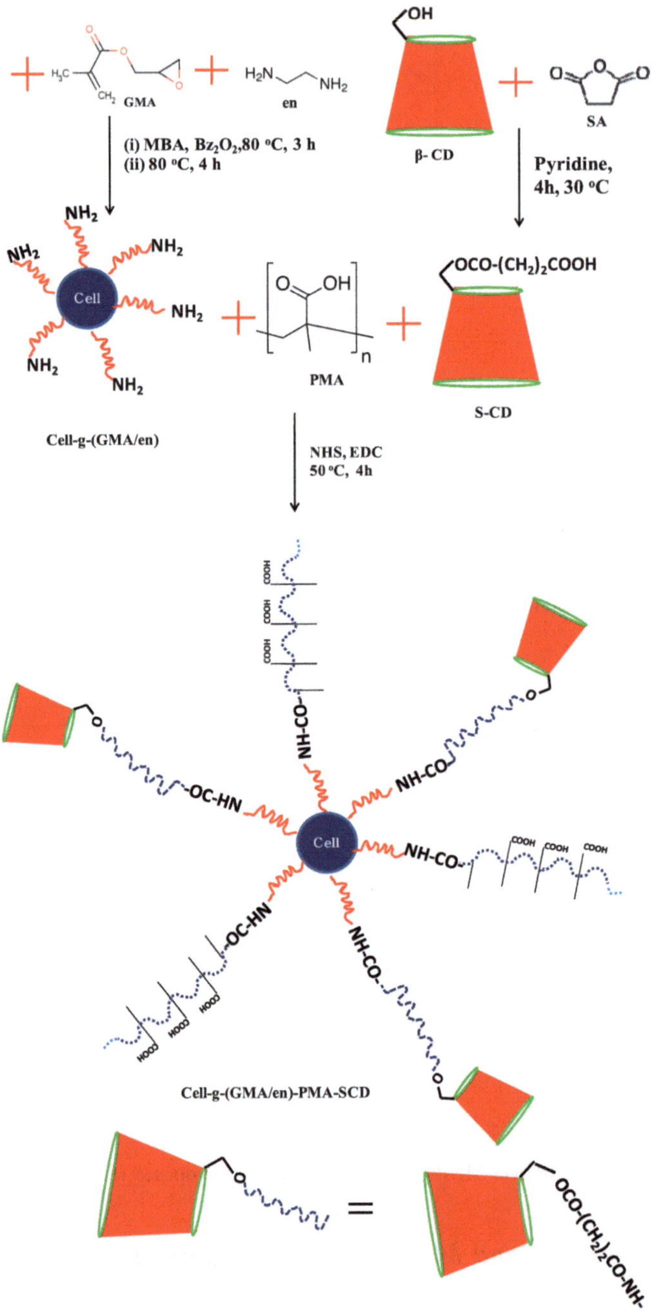

Fig. 2 Proposed reaction mechanism for the synthesis of Cell-g-(GMA/en)-PMA-SCD

diffractometer using Cu Kα radiation (λ = 1.5406 Å). SEM analysis was done using JEOL JSM 6390 LA scanning electron microscope. The absorbance of 5-FU solution was performed on a JASCO-V-530-UV-visible spectrophotometer at the λ_{max} value of 266 nm (Li et al. 2011). All pH measurements were made on a μ processor Systronic pH meter (model 361). A temperature controlled water bath shaker (Labline, India) with a temperature tolerance of ±1 °C was used for equilibrium studies. Borosil glasswares were used throughout the experiments.

2.4 Experimental

2.4.1 Swelling Measurements

Swelling experiments play an important role in determining the drug release rate. The experiments were conducted as a function of time at two different pH conditions 7.4 and 1.2 in which 0.1 g of the dry polymer taken in a weighed tea bag was allowed to swell in 100 mL solutions of desired pH for different intervals of time at 30 °C. The sample containing tea bag was weighed after definite time periods by removing excess water by blotting onto a filter paper. The swelling of polymer was monitored continuously until equilibrium conditions were reached. The degree of swelling is calculated as follows:

$$\text{Swelling } (\%) = \frac{(W_s - W_d)}{W_d} \times 100 \quad (1)$$

where W_s (mg) represents weight of swollen sample and W_d (mg) weight of dry sample.

2.4.2 Drug Encapsulation and Delivery Studies

Drug Loading Efficiency (DLE) and Encapsulation Efficiency (EE)

Drug encapsulation was carried out using equilibrium swelling method. Here, the DDS was allowed to swell in the drug solution of known concentration for 24 h at 37 °C. The solubility of 5-FU in water is low (13 mg/mL), but its sodium salt has higher solubility, 65 mg/mL (Garcia et al. 1994). For maximum encapsulation of drug, the drug solution was neutralized with NaOH and used in the feed mixture. The drug-loaded DDS was allowed to incubate in 10 mL 0.02 M phosphate buffer solution (PBS) at room temperature for 24 h (Rao et al. 2013). Then, the filtrate was separated and absorbance was measured at 266 nm using UV–Visible spectrometer. From the absorbance value, the amount of drug in the DDS can be calculated. The values of DLE and EE were calculated using the equations:

$$DLE\ (\%) = \frac{\text{Total amount of drug} - \text{Amount of free drug}}{\text{Total mass of composite}} \times 100 \qquad (2)$$

$$\text{EE}\ (\%) = \frac{\text{Total drug} - \text{Free drug}}{\text{Total drug}} \times 100 \qquad (3)$$

In Vitro Release of the Drug

In vitro release studies were carried out by placing 0.1 g of drug-loaded DDS into a stoppered bottle containing 50 mL of KCl–HCl solution with pH 1.2 (simulative gastric pH) and gently shaking it in a water bath shaker at 37 °C. Definite aliquots of the supernatant were withdrawn at predetermined time intervals from the release medium replacing it with the same volume of fresh KCl–HCl solutions. The supernatant was analyzed spectrometrically at 266 nm to determine the amount of 5-FU released. The experiments were continued until no more change in the concentration of drug in the solution occurred. The release experiments were repeated using $NaHPO_4$–KH_2PO_4 solution at pH 7.4 (simulated intestinal pH) and 37 °C. The drug release percentage in both cases was calculated by the following equation:

$$\text{Drug released}\ (\%) = \frac{C_t}{C_\infty} \times 100 \qquad (4)$$

where C_t and C_∞ are the concentration of drug released at time t and that of drug released completely, respectively.

Cytotoxicity (3-[4,5-Dimethylthiazol-2-yl]-2,5-Diphenyl Tetrazolium Bromide) (MTT Assay)

MTT assay was carried out as per previously published protocol (Wilson 2000). Briefly, 5×10^3 cells were seeded on 96-well plate and incubated in Dulbecco's modified minimal essential media (DMEM) supplemented with 10.0% fetal calf serum (FCS) for 24 h inside CO_2 incubator set at 5.0% CO_2 and 37 °C. Growth media was replaced with incubation media containing samples at different concentrations in DMEM. Treated cells were incubated overnight and media were replaced with DMEM containing MTT (5 mg/mL). Cells were incubated for 3–4 h for purple colored formazan crystal development. These water-insoluble crystals were dissolved in DMSO with 10.0% sodium dodecyl sulfate (SDS). Absorbance was measured at 570 nm with a microplate reader (Model 680 Biorad, USA) in endpoint mode. Results were expressed as mean value ± standard deviation. Percentage of cytotoxicity was calculated based on the formula:

$$\text{Toxicity (\%)} = \frac{(A_0 - A_1)}{A_0} \times 100 \quad (5)$$

where A_0 is absorbance of control well and A_1 is the absorbance of sample treated well.

All the experiments were carried out in triplicates with standard deviation of 3.0%. The data were discarded, when the relative error exceeded 3.0%, and a new experiment was conducted until the relative error falls within an acceptable range.

3 Results and Discussion

3.1 Development of DDS and Its Characteristics

3.1.1 MPTMS-MNP-poly-(GMA-g-MACD)

For the controlled delivery of 5-FU, a novel DDS was developed by combining the advantages of MNP for targeted delivery so that drug is transported directly to the center of the disease under various conditions and host–guest interaction of cyclodextrin moiety (Fig. 1). The magnetic core is surrounded by silane (MPTMS) with vinylic functionality (MPTMS-MNP) so that further polymerization is possible. To exploit cyclodextrins as suitable DDS's for targeted delivery, it is properly functionalized with suitable monomers. Thus, grafting with MAH followed by GMA provides a new material with hydrophilicity and higher biocompatibility. The grafting was carried out in such a manner that free –COOH groups are available on the surface of the final polymer from MACD. When this GMA-g-MACD is polymerized with MPTMS-MNP using AIBN initiator and EGDMA as crosslinker, a hydrophilic polymeric network is obtained having a dendrimer-like appearance with –OH and –COOH functionality on the surface of the DDS, combining the advantages of both magnetic core and host-guest interactions of β-CD.

3.1.2 Cell-g-(GMA/en)-PMA-SCD

For preparing the DDS, GMA was first graft polymerized with Cell in presence of MBA and BPO (Fig. 2). Amine functionality was then introduced using en by the breaking of epoxide ring of GMA to obtain Cell-g-(GMA/en). SCD was prepared by reacting SA and β-CD in presence of suitable solvents. The final DDS, Cell-g-(GMA/en)-PMA-SCD was prepared by grafting Cell-g-(GMA/en) with PMA and SCD using NHS and EDC so that a hydrogel is obtained with –COOH and –NH functional groups and CD cavity on its surface which ensures the better encapsulation of drug. Introduction of PMA and SCD also increases the biocompatibility of the final material.

3.2 *Characterization*

3.2.1 FT-IR Spectra

The FT-IR spectra of MNP, MPTMS-MNP-poly-(GMA-g-MACD), and 5-FU-MPTMS-MNP-poly-(GMA-g-MACD) are shown in Fig. 3a. The FT-IR of MNP showed a peak at 587 cm^{-1} indicating the presence of iron species. In the DDS, the peaks corresponding to Si–O–Si and >C=O group were obtained at 1050–1250 and 1700 cm^{-1}, but peaks corresponding to >C=C< stretching (1644 cm^{-1}) were absent indicating the polymerization between vinylic functionalization of macromonomers (MPTMS-MNP and GMA-g-MACD) in presence of crosslinker and initiator. In the 5-FU-loaded DDS, characteristic peaks of 5-FU at 3387 cm^{-1} (N–H stretching) and 1705 and 1578 cm^{-1} (due to C=O stretching) were retained with little modification indicating the hydrogen bonding interaction between the drug and DDS (Anirudhan et al. 2014). Due to the interaction between drug and DDS, bond length increases, causing decrease in stretching frequency, which resulted in wave number shift from higher to lower frequency region (Rejinold et al. 2010). Figure 3b presents the FT-IR spectra of Cell, Cell-g-(GMA/en)-PMA-SCD and 5-FU-loaded Cell-g-(GMA/en)-PMA-SCD; (5-FU-Cell-g-(GMA/en)-PMA-SCD). The FT-IR spectra of Cell showed the characteristic broad peak related to hydrogen bonded O–H stretching vibration at 3211 cm^{-1}, and the peaks at 2900 and 1028 cm^{-1} could be attributed to the C–H stretching and C–H bending vibrations of the $-CH_2$ group. Signals corresponding to C=O stretching of hemicelluloses and β-glycosidic linkages were observed at 1790 and 670 cm^{-1}, respectively. The FT-IR spectrum of Cell-g-(GMA/en)-PMA-SCD retains the peaks of its constituents with slight modifications. The FT-IR spectrum of 5-FU shows a broad band between 3000 and

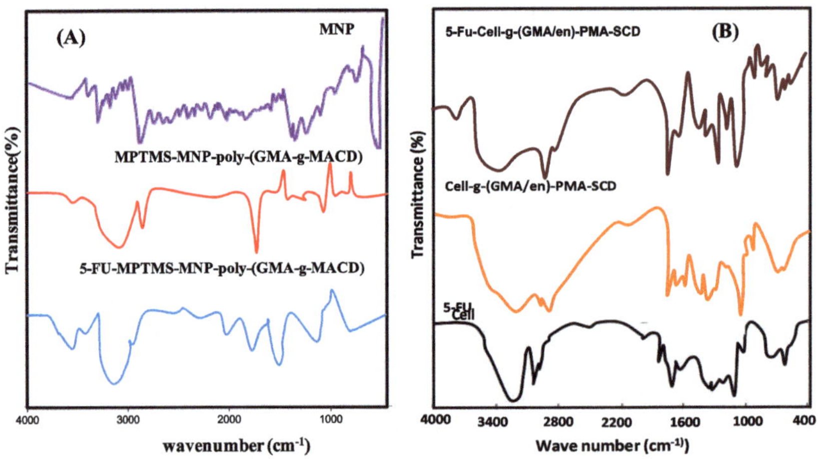

Fig. 3 **a** FT-IR spectra of MNP, MPTMS-MNP-poly-(GMA-g-MACD) and 5-FU-MPTMS-MNP-poly-(GMA-g-MACD). **b** Cell, Cell-g-(GMA/en)-PMA-SCD, and 5-FU-Cell-g-(GMA/en)-PMA-SCD

3500 cm^{-1} which is due to N–H stretching vibrations. This band was observed approximately at 3500 cm^{-1} in the spectrum of 5-FU-Cell-g-(GMA/en)-PMA-SCD due to the overlapping of –OH bond and –NH bond of 5-FU. In 5-FU-Cell-g-(GMA/en)-PMA-SCD, the C=O stretching vibration is shifted to 1625 cm^{-1}, indicating the successful interaction between the DDS and 5-FU. The peak at 1275 cm^{-1} (C–F stretching) in the spectrum of 5-FU is shifted to 1280 cm^{-1} in 5-FU-Cell-g-(GMA/en)-PMA-SCD confirming the successful encapsulation of drug.

3.2.2 XRD Analysis

The XRD patterns of MNP, MPTMS-MNP-poly-(GMA-g-MACD), and 5-FU-MPTMS-MNP-poly-(GMA-g-MACD) are given in Fig. 4a. In the XRD pattern of MNP, characteristic peaks were found at $2\theta = 30.1, 35.4, 43.0, 53.5, 57.0$ and $62.8°$ with their indices (2 2 0), (3 1 1), (4 0 0), (4 2 2), (5 1 1), and (4 4 0), respectively, indicating the spinel structure of Fe_3O_4 (Pan et al. 2011). In the XRD pattern of DDS, the peak corresponding to β-CD cavity was present at $2\theta = 12.5°$. It retains the characteristic peaks of MNP. The characteristic peak of silane was found at $2\theta = 29.5°$. In the 5-FU loaded DDS, peaks at $2\theta = 12.5°$ were absent, indicating the interaction between β-CD cavity of the DDS and the drug. Here, sharp crystalline peak at $2\theta = 28.0°$ corresponding to 5-FU was also absent, indicating the encapsulation of drug by the polymer matrix through electrostatic and host-guest interactions (Lee et al. 2003). The XRD patterns of Cell, Cell-g-(GMA/en)-PMA-SCD and 5-FU-Cell-g-(GMA/en)-PMA-SCD were recorded to determine the possible differences in the crystallinity of the structures (Fig. 4b). The XRD pattern of Cell showed peaks at 2θ values 22.8° and 34.5° corresponding to its crystalline domain nature and a broad hump at $2\theta = 15.5°$ revealed its amorphous nature. The DDS retained almost all the

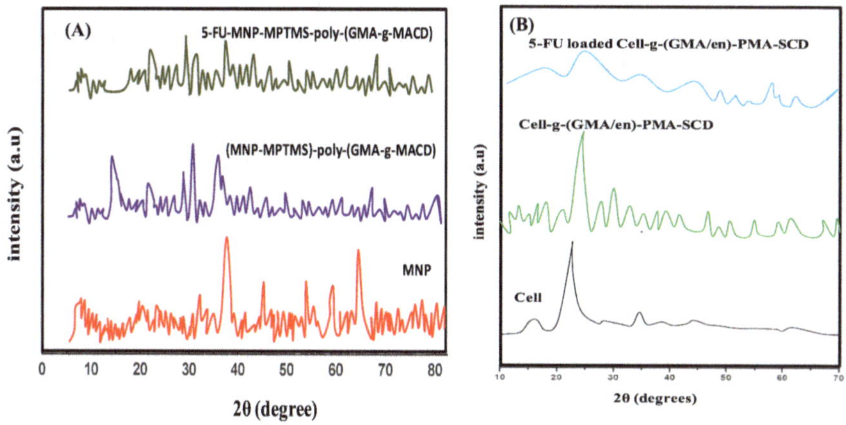

Fig. 4 a XRD spectra of MNP, MPTMS-MNP-poly-(GMA-g-MACD) and 5-FU-MPTMS-MNP-poly(GMA-g-MACD). **b** Cell, Cell-g-(GMA/en)-PMA-SCD, and 5-FU-Cell-g-(GMA/en)-PMA-SCD

peaks of its constituents with slight modifications. After each modification, the crystallinity of the samples was found to decrease. The XRD pattern of 5-FU-Cell-g-(GMA/en)-PMA-SCD was found to be amorphous in nature indicating the successful encapsulation of drug. Here, also the peak corresponding to 5-Fu was absent ($2\theta = 28.0^{o}$).

3.2.3 SEM Analysis

Figure 5a–f illustrates the SEM images of MNP, MPTMS-MNP-poly-(GMA-g-MACD), 5-FU-MPTMS-MNP-poly-(GMA-g-MACD), Cell, Cell-g-(GMA/en)-PMA-SCD, and 5-FU-Cell-g-(GMA/en)-PMA-SCD, respectively. The morphology of each sample was found to be entirely different from one other, indicating successful modification after each step. The surface of MNP showed sphere like agglomerated structure. MPTMS-MNP-poly-(GMA-g-MACD) showed fluffy and porous texture throughout the surface with dark appearance. 5-FU-MPTMS-MNP-poly-(GMA-g-MACD) showed nonporous structure, showing the successful encapsulation of drug. The surface of Cell seemed to have cylindrical shape which is due to the intermolecular forces associated within the structure. Cell-g-(GMA/en)-PMA-SCD showed a porous structure with large number of adsorption sites which facilitate 5-FU loading. The SEM image of 5-FU-Cell-g-(GMA/en)-PMA-SCD showed smooth and uniform morphology indicating the successful encapsulation of drug.

3.3 Swelling Behavior

The swelling behavior of the material plays an important role in determining the rate of release of drug from it. Figure 6a and b represents the pH dependence of the equilibrium swelling for MPTMS-MNP-poly-(GMA-g-MACD) and Cell-(GMA/en)-PMA-SCD, respectively, at 37 °C. To understand the effect of pH on swelling, experiment was conducted under two pH conditions, 7.4 and 1.2. In both carriers, the maximum swelling was obtained at pH 7.4. In both cases, as time increased swelling percentage also increased and found to be gradual and continuous up to 24 h. For MPTMS-MNP-poly-(GMA-g-MACD), maximum swelling percentage at pH 1.2 was found to be 164.0% and that at pH 7.4 was found to be 381.2%, respectively, and for Cell-(GMA/en)-PMA-SCD maximum swelling percentage at pH 1.2 and 7.4 was found to be 266.8 and 378.7% respectively. The maximum swelling percentage obtained at pH 7.4 can be explained on the basis of ionization of –COOH groups present in the drug carrier. In acidic medium, –COOH group is in the unionized state. With increase in the pH of the solution, ionization of –COOH

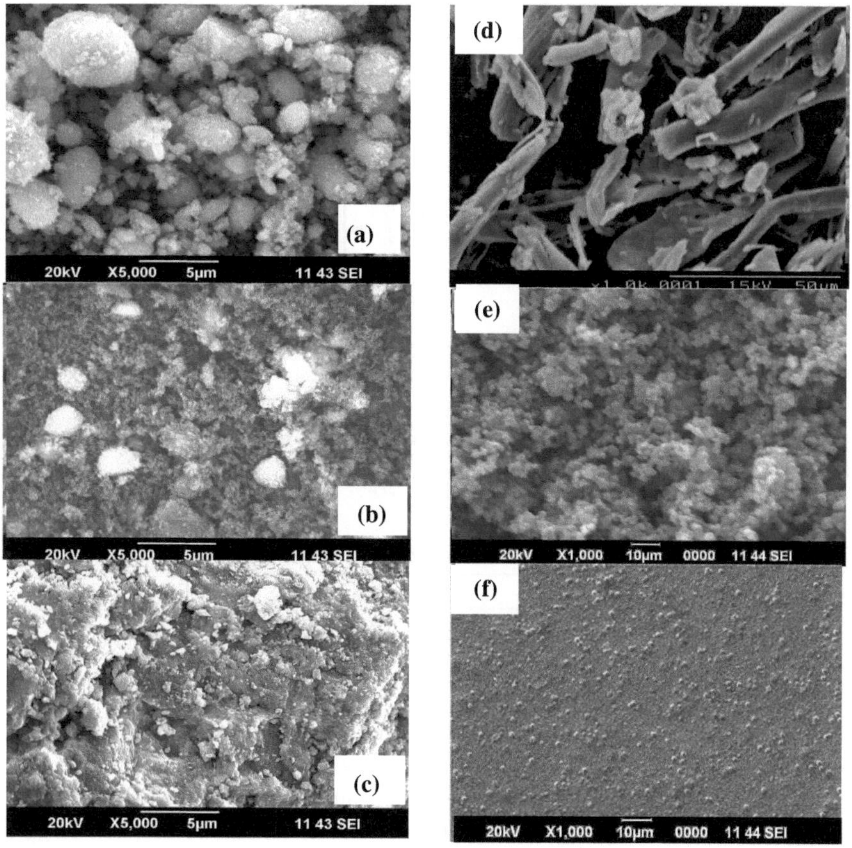

Fig. 5 SEM images of **a** MNP, **b** MPTMS-MNP-poly-(GMA-g-MACD), **c** 5-FU-MPTMS-MNP-poly-(GMA-g-MACD), **d** Cell, **e** Cell-g-(GMA/en)-PMA-SCD, and **f** 5-FU-Cell-g-(GMA/en)-PMA-SCD

group increases which leads to the electrostatic repulsion between the polymer chains due to its similar charge facilitating the swelling property of hydrogel. Presence of β-CD cavity also enhances the swelling capacity of the materials.

3.4 Drug Loading Efficiency (DLE) and Encapsulation Efficiency (EE)

DLE and EE were determined using the two drug carriers and 75 mg/L of the drug solution. The results are given in Table 1. DLE of MPTMS-MNP-poly-(GMA-g-MACD) was found to be 69.0%, but that of Cell-g-(GMA/en)-PMA-SCD was found to be 79.2%. This difference may be due to the availability of more

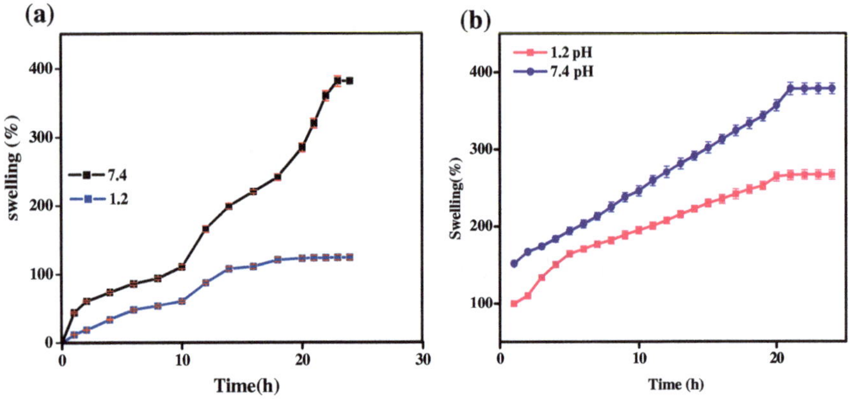

Fig. 6 Swelling behavior of **a** MPTMS-MNP-poly-(GMA-g-MACD) and **b** Cell-g-(GMA/en)-PMA-SCD at pH 1.2 and 7.4

Table 1 Comparison of DLE (%) and EE (%) of the drug carriers using 75 mg/L of drug

Material	Amount of DDS (g)	DLE (%)	EE (%)
MPTMS-MNP-poly-(GMA-g-MACD)	0.1	69.0	90.0
Cell-g-(GMA/en)-PMA-SCD	0.1	79.2	90.3

numbers of –COOH groups and –NH groups from Cell-g-(GMA/en)-PMA-SCD. But EE was found to be same (~90.0%) in both cases which may be due to the presence of β-CD cavities enabling host–guest interactions between DDS and drug and also due to the well-formed hydrogel polymeric network. Hence, drug can easily diffuse through the polymer network and can occupy the active sites present on the polymer including the –COOH, –OH, and –NH functionalities.

3.5 *In Vitro Drug Release*

The in vitro drug release profile of 5-FU from MPTMS-MNP-poly-(GMA-g-MACD) and Cell-g-(GMA/en)-PMA-SCD hydrogels were carried out in two different pH conditions, 1.2 and 7.4 at 37 °C, and the results are presented in Fig. 7a, b, respectively. The rate of release was found to be slow and continuous up to 24 h. The amount of 5-FU released at pH 7.4 was found to be higher than those delivered at pH 1.2. The drug delivery of hydrogels is generally explained using two mechanisms depending on their chemical structure. The first mechanism depends on the swelling capacity of hydrogel which is directly related to the nature and number of functional groups in it. The second mechanism is explained on the basis of interaction between drug and functional groups present in the hydrogel. In both cases, the drug release rate was in agreement with the swelling capacity of the

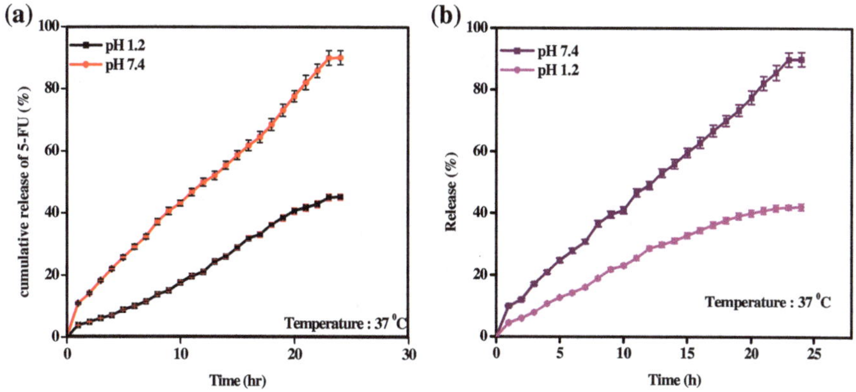

Fig. 7 In vitro release rate of 5-FU from **a** MPTMS-MNP-poly-(GMA-g-MACD) and **b** Cell-g-(GMA/en)-PMA-SCD at different pH

material and encapsulation efficiency of drug. As swelling was maximum at pH 7.4, drug release rate was also maximum. About 90.0% of the drug was released within 24 h from both the carriers. The delayed and controlled release of drug up to 24 h from the DDS can be explained on the basis of interaction between drug and COOH and OH groups (of GMA and MACD moieties) present in MPTMS-MNP-poly-(GMA-g-MACD) polymer network which decreased as swelling increases. Hence, polymer relaxation occurs followed by drug diffusion from polymer network in a controlled manner. Moreover, the drug molecules present within the cavity of β-CD shows delayed release due to host–guest interaction between drug and CD cavity as a result of van der Waals interaction between hydrophobic moiety of drug and that of cavity. Hence, the drug release from β-CD is a time-consuming process and can be sustained up to 24 h. The DDS is made in such a way to protect the drug from external environment till it reaches the intestine (target site) where the pH is 7.4. Since the DDS is in shrinken stage at acidic pHs, the encapsulated drug will be safe till it reaches the interface. But after reaching intestine, the pH of the surrounding medium is 7.4 thereby swelling of the material starts causing controlled delivery of the drug from DDS. Cell-g-(GMA/en)-PMA-SCD also showed a similar mechanism.

3.6 Kinetic Modeling of Drug Release

To investigate more precisely the effect of interpolymer complex formation on the release of 5-FU the results were analyzed according to the Peppa's potential equation (Korsmeyer et al. 1983):

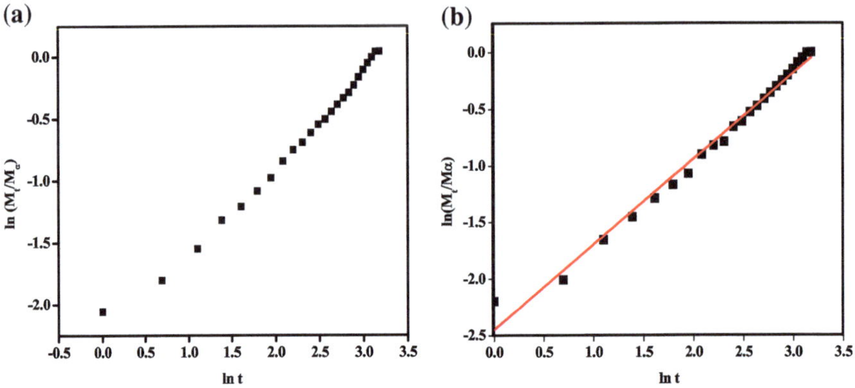

Fig. 8 ln t versus $\ln(M_t/M_\infty)$ plot for **a** MPTMS-MNP-poly-(GMA-g-MACD), **b** Cell-g-(GMA/en)-PMA-SCD

$$\frac{M_t}{M_\infty} = Kt^n \tag{6}$$

where $\frac{M_t}{M_\infty}$ is a fraction of drug released at time t, K is the release rate constant, and n is the release exponent. The drug release parameters, n and K are obtained from the slope and intercept of plot of $\ln(M_t/M_\infty)$ versus ln t. A value of $n = 0.5$ corresponds to Fickian diffusion (Falk et al. 2004), while a value of n between 0.43 and 0.85 represents non-Fickian diffusion (Jin et al. 2010). The value of n for MPTMS-MNP-poly-(GMA-g-MACD) (Fig. 8a) and Cell-g-(GMA/en)-PMA-SCD (Fig. 8b) was found to be 0.709 and 0.750; corresponding R^2 values were found to be 0.990 and 0.993, respectively. Since the value of n lies between 0.43 and 0.85 for both hydrogels, the diffusion of drug and swelling of the polymer play an important role in the release mechanism. These findings suggest that the developed drug carriers exhibit tremendous potential as controlled DDS.

3.7 *Cytotoxicity Assay*

MPTMS-MNP-poly-(GMA-g-MACD) and Cell-g-(GMA/en)-PMA-SCD were tested against human breast cancer (MCF-7) cells at different concentrations using 5-FU (50 and 75 mg/L concentration). In all cases, percentage of cell viability was found to be high for pure DDS. Since these have cell viability more than 80.0%, the materials can be considered as biocompatible (Ying and Hwang 2010). In case of MPTMS-MNP-poly-(GMA-g-MACD), the MNP's were coated with polymers. The polymeric layer improves the biocompatibility of core-shell MNPs and thus lessens toxicity. Similarly in the case of Cell-g-(GMA/en)-PMA-SCD, the material is composed of biodegradable components like Cell, GMA, and CD. So it shows less

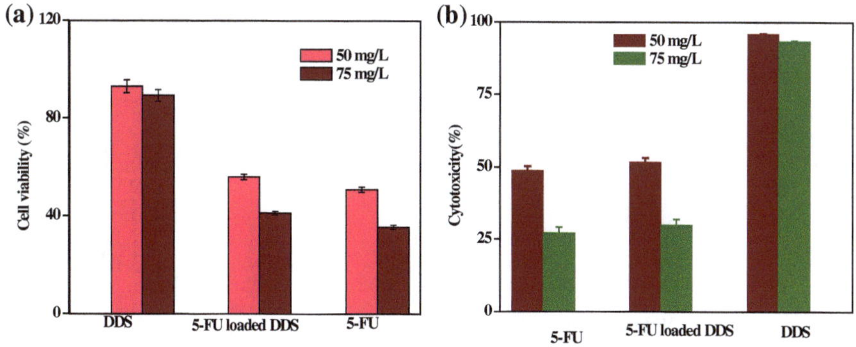

Fig. 9 Change in cytotoxicity of 5-FU, 5-FU-loaded DDS and DDS at different concentrations **a** MPTMS-MNP-poly-(GMA-g-MACD) and **b** Cell-g-(GMA/en)-PMA-SCD

cytotoxicity. In the case of drug alone assay, cell viability was found to be low and decreases from 50.8 to 35.4% with increase in concentration of drug from 50 to 75 mg/L which may be due to the immediate availability of the drug to the cancer cells that caused higher toxicity. But in the assay of drug-loaded carriers, the cell viability was found to be higher than the corresponding drug alone assay (55.8 and 41.2% for MPTMS-MNP-poly-(GMA-g-MACD) and 51.7 and 29.9% Cell-g-(GMA/en)-PMA-SCD at concentrations 50 and 75 mg/L, respectively). This indicates the delayed, controlled, and sustained release of drug from the carrier, thereby decreasing the side effects of drug to the normal cells. This type of cytotoxic assay was reported by earlier workers also (Dutta and Sahu 2012; Jain et al. 2005). The lower cell viability of Cell-g-(GMA/en)-PMA-SCD may be due to the slightly higher release rate of 5-FU from it (Fig. 9).

The morphological image of MCF-7 cells (Fig. 10) treated with the drug carriers and drug-loaded carriers confirmed the above said statements. From the images, it is clear that 75 mg/L drug-loaded DDS caused maximum cell death compared to that of 50 mg/L drug-loaded DDS which is consistent with the drug release data at pH 7.4.

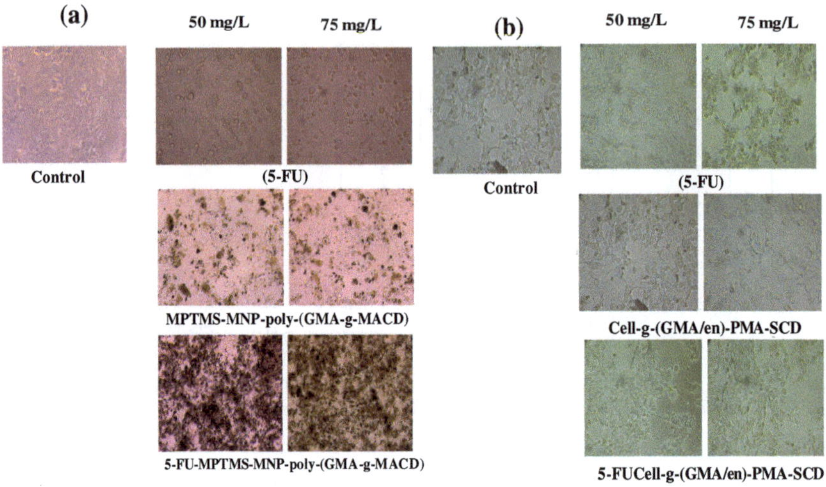

Fig. 10 Images showing the morphological effect of 5-FU, drug carriers and drug-loaded carriers at different concentrations against human breast carcinoma cells (MCF-7) **a** MPTMS-MNP-poly-(GMA-g-MACD) and **b** Cell-g-(GMA/en)-PMA-SCD

4 Conclusions

The present chapter evaluates two stimuli-responsive drug carriers, 3-methacryloxy propyl trimethoxy silane coated magnetic nanoparticles polymerized with glycidyl methacrylate-grafted-maleated cyclodextrin [MPTMS-MNP-poly-(GMA-g-MACD)] and aminated-glycidyl methacrylate-grafted cellulose-grafted poly-methacrylic acid-succinyl cyclodextrin [Cell-g-(GMA/en)-PMA-SCD] for the controlled delivery of 5-Fluorouracil (5-FU), an anticancer drug. The physicochemical properties of the MPTMS-MNP-poly-(GMA-g-MACD) and Cell-g-(GMA/en)-PMA-SCD were investigated. The synthesized drug carriers were characterized using FT-IR, XRD, and SEM techniques. Swelling studies of the hydrogels were carried out with respect to time at pH 7.4 and 1.2 and found maximum at 7.4 in both cases. The maximum swelling percentage at pH 7.4 was found to be 381.2% for MPTMS-MNP-poly-(GMA-g-MACD) and 378.7% for Cell-g-(GMA/en)-PMA-SCD, respectively. Peppas model was used to study the release mechanism, and the results indicate a delayed, swelling controlled, and non-Fickian mechanism for the release of the drugs from polymer networks. Cytotoxicity study revealed that the drug carriers did not show any cytotoxicity, but the drug-loaded carriers showed significant cytotoxicity lesser than that of pure 5-FU confirming the controlled and sustained release of 5-FU from the carrier to the targeted site.

By comparing the two different drug carriers for the same anticancer drug, it can be concluded that although the basic materials are different for two systems the presence of common functional groups and CD moiety gives them almost similar performance in all experiments. Some deviations are obtained in case of DLE and MTT assay study. The higher loading efficiency of Cell-g-(GMA/en)-PMA-SCD may be due to the presence of –NH groups and the availability of higher number of –COOH groups (from PMA) compared to MPTMS-MNP-poly-(GMA-g-MACD). Similarly, the marginal difference in cytotoxic effect (less cell viability) showed by Cell-g-(GMA/en)-PMA-SCD is due to the slightly higher releasing rate of drug from it. In case of MPTMS-MNP-poly-(GMA-g-MACD), the presence of MNPs may cause somewhat delayed release after certain interval. But it can be confirmed that the newly developed drug carriers, MPTMS-MNP-poly-(GMA-g-MACD) and Cell-g-(GMA/en)-PMA-SCD are potential materials for the drug loading and sustained release of 5-FU.

Acknowledgements The authors would like to acknowledge the support of the Head, Department of Chemistry, School of Physical and Mathematical Sciences, University of Kerala, Trivandrum, for providing laboratory facilities for this work.

References

Anirudhan TS, Jalajamony S (2010) Cellulose-based anion exchanger with tertiary amine functionality for the extraction of arsenic(V) from aqueous media. J Environ Manage 91:2201–2207

Anirudhan TS, Suchithra PS (2009) Adsorption characteristics of humic acid immobilized amine modified polyacrylamide/bentonite composite for cationic dyes in aqueous solutions. J Environ Sci 21:884–891

Anirudhan TS, Binusreejayan, Rejeena SR (2014) Synthesis and characterization of chitosan based multilayer and pH sensitive co-polymeric system for the targeted delivery of 5-fluorouracil, an in vitro study. Int J Polym Mater Polym Biomater 63:539–548

Anirudhan TS, Divya PL, Nima J (2015a) Synthesis and characterization of silane coated magnetic nanoparticles/glycidylmethacrylate-grafted-maleated cyclodextrin composite hydrogel as a drug carrier for the controlled delivery of 5-fluorouracil. Mater Sci Eng C 55:471–481

Anirudhan TS, Nima J, Divya PL (2015b) Synthesis, characterization and in vitro cytotoxicity analysis of a novel cellulose based drug carrier for the controlled delivery of 5-fluorouracil, an anticancer drug. Appl Surf Sci 355:64–73

Balaji N, Anantharaman N, Uddin MS, Beegums MNKM (2009) Adsorption and desorption of t-phenylalanine on nano-sized magnetic particles. J Eng Appl Sci 8:1819–1823

Benson AB, Ajani JA, Catalano RB, Engelking C et al (2004) Recommended guidelines for the treatment of cancer treatment induced diarrhea. J Clin Oncol 22:2918–2926

Blanco FB, Lopez VM, Concheiro A, Alvarez LC (2011) Synergistic performance of cyclodextrin-agar hydrogels for ciprofloxacin delivery and antimicrobial effect. Carbohyd Polym 85:765–774

Corobea MC, Muhulet O, Miculescu F, Antoniac IV, Vuluga Z, Florea D et al (2016) Novel nanocomposite membranes from cellulose acetate and clay-silica nanowires. Polym Adv Technol 27(12):1586–1595

Coskun R, Sacak M, Karakisia M (2005) Graft copolymerization of an itaconic acid/acrylamide monomer mixture onto poly(ethylene terephthalate) fibers with benzoylperoxide. J Appl Polym Sci 97:1795–1803

Cunningham D, James RD (2001) Integrating the oral fluoropyrimidines into the management of advanced colorectal cancer. Eur J Cancer 37:826–834

Deepthy A, Abraham M, Rajeev KS (2014) Esterases immobilized on amino silane modified magnetic nanoparticles as a catalyst for biotransformation reactions. Biores Technol 167:547–550

Dutta RK, Sahu S (2012) Development of a novel probe sonication assisted enhanced loading of 5-FU in SPION encapsulated pectin nanocarriers for magnetic targeted drug delivery system. Eur J Pharm Biopharm 82:58–65

Falk B, Garramone S, Shivkumar S (2004) Diffusion coefficient of paracetamol in a chitosan hydrogel. Mater Lett 58:3261–3265

Garcia O, Trigo Blanco RMMD, Teijon JM (1994) Influence of degree of cross-linking on 5-fluorouracil release from poly-(2-hydroxyethyl methacrylate) hydrogels. Biomaterials 15:689–694

Gref R, Domb A, Quellec P, Blunk T, Muller RH, Verbabatz JM (1995) The controlled intravenous delivery of drugs using PEG coated sterically stabilized nanospheres. Adv Drug Deliv Rev 16:215–233

Heidelberger C, Chaudhuri NK, Danneberg P, Mooren D, Griesbach L (1957) Fuorinated pyrimidines, a new class of tumor-inhibitory compounds. Nature 179:663–666

Jain T, Marco AK, Sahoo SK, Leslie-Pelecky DL, Labhasetwar V (2005) Iron oxide nanoparticles for sustained delivery of anticancer agents. Mol Pharm 2:194–205

Jin S, Liu M, Chen S, Gao C (2010) A drug-loaded gel based on polyelectrolyte complexes of poly (acrylic acid) with poly(vinylpyrrolidone) and chitosan. Mater Chem Phys 123:463–470

Kaiwen Y, Sicheng W, Binbin C, Wenxia G et al (2016) Dual pH and temperature responsive Hydrogels based on β-cyclodextrin derivatives for atorvastatin delivery. Carbohyd Polym 136:300–306

Kinno R, Kii Y, Uchiyama M, Owan Y, Yamazaki T, Fukui T (2014) 5-fluorouracil-induced leukoencaphalopathy with acute stroke-like presentation fulfilling criteria for recombinant tissue plasmogen activator therapy. J Stroke Cerebrovas Dis 23:387–389

Korsmeyer RW, Gurny R, Doelker E, Buri P, Peppas NA (1983) Mechanisms of solute release from porous hydrophilic polymers. Int J Pharm 15:25–35

Lee JS, Chae GS, An TK, Khang G (2003) Prepartation of 5-fluorouracil-loaded poly (L-lactide-co-glycolide) wafer and evaluation of in vitro release behavior. Macromol Res 11:183–188

Li P, Wang Y, Peng Z, She F, Kong L (2011) Development of chitosan nanoparticlesas drug delivery systems for 5-fluorouracil and leucovorin blends. Carbohydr Polym 85:698–704

Logan RM, Stringer AM, Bowen JM, Gibson RI, Sonis ST, Keefe DM (2009) Is the pathobiology of chemotherapy-induced alimentary tract mucositis influenced by the type of mucotoxic drug administered. Cancer Chemother Pharmacol 63:239–251

Ma M, Zhang Y, Yu W, Shen H, Zhang H, Gu N (2003) Preparation and characterization of magnetite nanoparticles coated by aminosilane. Colloids Surf, A 212:219–226

Miculescu M, Thakur VK, Miculescu F, Voicu SI (2016) Graphene-based polymer nanocomposite membranes: a review. Polym Adv Technol 27(7):844–859

Mok H, Park JW, Park TG (2008) Enhanced intracellular delivery of quantum dot and adenovirus nanoparticles triggered by acidic pH via surface charge reversal. Bioconjug Chem 19:797–801

Nakahata M, Takashima Y, Yamaguchi H, Harada A (2011) Redox-responsive self healing materials formed from host-guest polymers. Nat Commun 2:511

Pan J, Yao H, Xu L, Ou H, Huo P, Li X, Yang Y (2011) Selective recognition of 2,4,6-trichlorophenol by molecularly imprinted polymers based on magnetic halloysitenanotubes composites. J Phys Chem 21:183–190

Rao KM, Mallikarjuna B, Rao KSVK, Siraj S, Rao KC, Subha MCS (2013) Novel thermo pH sensitive nanogels composed from poly(N-vinyl caprolactum) for controlled release of anticancer drug. Colloids Surf, B 102:891–897

Reis AV, Fajardo AR, Schuquel ITA, Guilherme MR, Vidotti GJ, Rubira AF, Muniz EC (2009) Reaction of glycidyl methacrylate at the hydroxyl and carboxylic groups of poly(vinyl alcohol) and poly(acrylic acid). Is this reaction mechanism still unclear? J Org Chem 74:3750–3757
Rejinold NS, Muthunarayanan M, Deepa N, Chennazhi SVN, Jayakumar R (2010) Development of novel fibrinogen nanoparticles by two-step co-acervation. Int J Biol Macromol 47:37–43
Rodriguez TC, Alvarez LA, Rodriguez PA, Concheiro JJ, Torres L (2007) Estradiol sustained release from high affinity cyclodextrin Hydrogels. Eur J Pharm Biopharm 66:55–62
Sakai S, Ueda K, Taya M (2015) Peritoneal adhesion prevention by a biodegradable hyaluronic acid-based hydrogel formed in situ through a cascade enzyme reaction initiated by contact with body fluid on tissue surfaces. Acta Biomater 24:152–158
Shukla S, Nidhi S, Pooja N, Charu A, Silvi M, Rizwana AB, Dubey GC, Tiwari A (2013) Preparation and characterization of cellulose derived from rice husk for drug delivery. Adv Mater Lett 4:714–719
Simoes S, Figueiras A, Francisco V (2012) Modular hydrogels for drug delivery. J Biomater Nanobiotechnol 3:185–199
Singh SA, Rana RK (2010) Enhancement of hydrophobic character of lignocellulosic fibers through graftcopolymerization. Adv Mater Lett 1:156
Soon SK, Bong JK, Soo NP (2015) Physicochemical properties of pH-sensitive hydrogels based on hydroxyethyl cellulose-hyaluronic acid for applications as transdermal delivery systems for skin lesions. Eur J Pharm Biopharm 92:146–154
Stefano S, Alessandra S, Sara B, Pietro M, Fabio R, Paolo C (2007) Cyclodextrin/PEG based hydrogels for multi-drug delivery. Int J Pharm 345:42–50
Sujan D, Pourali S, Dibakar D (2016) Temperature, pH and redox responsive cellulose based hydrogels for protein delivery. Int J Biol Macromol 87:92–100
Trache D, Hazwan Hussin M, Mohamad Haafiz MK, Kumar Thakur V (2017) Recent progress in cellulose nanocrystals: sources and production. Nanoscale 9(5):1763–1786
Van Kuilenberg AB (2004) Dihydropyrimidine dehydrogenase and the efficiency and toxicity of 5-fluorouracil. Eur J Cancer 40:939–950
Voicu SI, Condruz RM, Mitran V, Cimpean A, Miculescu F, Andronescu C, Thakur VK (2016) Sericin covalent immobilization onto cellulose acetate membrane for biomedical applications. ACS Sustain Chem Eng 4(3):1765–1774
Wilson AP (2000) Cytotoxicity and viability assay in animal cell culture: a practical approach, 3rd edn, vol 1. Oxford University Press, Oxford, pp 175–219
Ying E, Hwang H (2010) In vitro evaluation of the cytotoxicity of iron oxide nanoparticles with different coatings and different sizes in A3 human T lymphocytes. Sci Total Environ 408:4475–4481
Yuan Z, Ye Y, Gao F, Yuan H, Lan M, Lou K, Wang W (2013) Chitosan-graft-β-cyclodextrin nanoparticles as a carrier for controlled drug release. Int J Pharm 446:191–198

Chapter 8
Emerging Technology in Medical Applications of Hydrogel

G. Madhumitha, J. Fowsiya and Selvaraj Mohana Roopan

Abstract The term hydrogel used to describe the cross-linking long-chain polymer with high content water in the polymer matrix. The water-soluble and swollen property hydrogels were used in medical applications which include targeted delivery of drug, cell carriers, wound healing, and tissue engineering. Owing to different structures and cross-linking methods, the hydrogel has been utilized in biomedical and pharmacy industry. Hence, the preparation methods were important to prepare different types of hydrogels and it is useful in medical application based on their structural properties. The pursuit of this chapter is to concisely describe the recent development of hydrogel preparation, properties, and medical application. Furthermore, biomedical application and current clinical trial studies of hydrogel are summarized briefly.

Keywords Hydrogel · Preparation · Properties · Medical application

1 Introduction

Over the past 20 years, there is significant opportunity in examination of pharmaceutical drug delivery and this is not easy to release the drug at few characteristic rates that pharmaceutical researchers want. The recent development focused using water-swollen, cross-linked hydrogels as backbone of growth on new medicinal invention and drug delivery, peptides and proteins into targeted body site. The cross-linking structures and thermodynamic characteristics of materials play an important role in size of material, diffusion property, and stability of bioactive agents (Thakur et al. 2016). Scientists further developed more formulation in drug

G. Madhumitha · J. Fowsiya · S. M. Roopan (✉)
Chemistry of Heterocycles and Natural Product Research Laboratory,
Department of Chemistry, School of Advanced Science, Vellore Institute of Technology,
Vellore 632014, Tamil Nadu, India
e-mail: selvarajmohanaroopan@gmail.com; mohanaroopan.s@vit.ac.in

V. K. Thakur and M. K. Thakur (eds.), *Hydrogels*, Gels Horizons: From Science
to Smart Materials, https://doi.org/10.1007/978-981-10-6077-9_8

delivery which serves even better drug molecules and ended up with construction of biomaterials, hydrogel.

Hydrophilic gels are generally referred as hydrogels which consist of long-chain polymers that sometimes originate as colloidal gels where dispersion medium is water (Thakur and Thakur 2014a, b, 2015). Researchers defined hydrogel in different ways in which the most common hydrogel is defined as the polymer which is water distended and cross-linking polymeric composition formed by one or more monomers reactions (Ahmed et al. 2013). The hydrogel may also be defined as a polymeric material which has the capability of swelling and can maintain water in its structure. The water content of hydrogel arises due to the functional group, which are ionizable, present in the structure. These ionizable functional groups are responsible of characteristics of hydrogel like biocompatible nature, mechanical stability, and permeability (Kim et al. 2001). More recently hydrogels defined as two- or multi-constituent structure containing three-dimensional network linkages of long-chain polymers. The space between the hydrogel is covered with water or biological fluids. The network linkages consist of copolymers which are unsolvable due to chemical or physical cross-links namely entanglements or crystallites.

Over 50 years of research, hydrogel satisfied the particular need of biomedical field in the presence of high content of water. Due to this property of hydrogel, it is widely used as contact lenses, membrane for biosensors, lining for artificial hearts, and resources for artificial skin and drug supply devices (Ratner and Hoffman 1976). Hydrogel serves an improved therapeutic efficiency, drug safety controlled Hoare and Kohane (2008) by conventional methods and bringing genetically engineered pharmaceuticals, viz. protein and peptides. Nevertheless, these hydrogels also provide many applications which include tissue engineering due to biocompatibility, diagnostics, cellular immobilization, regenerative medicine, separation of biomolecules, and barrier materials to regulate biological adhesions (Bennett et al. 2003). Despite many advantages hydrogels have, few disadvantages like low tensile strength cause flow of loaded hydrogel from a targeted body site which is not important in many drug delivery applications such as subcutaneous injection. The amount and homogeneousness of hydrogel might be limited especially for hydrophobic drugs, and this could be resolved by formulating semi-interpenetrating networks to adjust the hydrophobicity to capture hydrophobic drugs. Most of the hydrogels are sufficiently deformable to inject and many do not need surgical implantation. Each of this importance of hydrogels leads to the restriction of drug delivery therapies in the clinic (Sutton 2005).

In the last two decades, naturally prepared hydrogels are completely substituted by synthetic hydrogel which has longer service life, water absorption capacity, and gel strength. Synthetic hydrogels are much well-defined structures that can be improved to yield tailorable degradability and functionality (Fig. 1). Synthetic hydrogel can be prepared by numerous chemical method including single-step procedures like polymerization and parallel cross-linking of monomers as well as multi-step procedures involving synthesis of polymer components containing reactive group (Burkert et al. 2007). A hydrogel/aqua gel consists of hydrophilic polymers that swell in water, in a three-dimensional (3D) network. Dried hydrogels

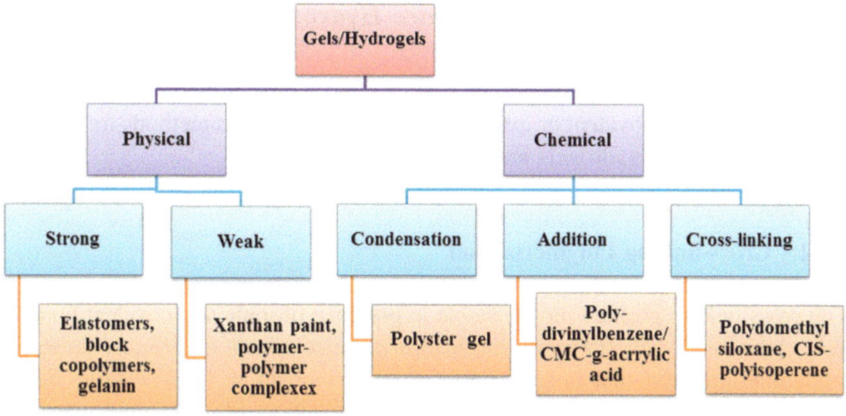

Fig. 1 A few examples for hydrogel

(or xerogels) absorb water to swell, and the water absorbance depends on the gel size. A hydrogel swells for the same reason that an analogous linear polymer dissolves in water to form an ordinary polymer solution. The swelling ratio of several hydrogels can easily reach greater than 100. In spite of having large quantity of water, solid forms are maintained by the highly swollen hydrogels. Hydrogels have played a major role in the anticancer drug discovery and development of controlled-release drug delivery system. This chapter is focused on the preparation, classification, properties, and anticancer drug delivery application.

2 Preparation of Hydrogel

Hydrogel is a large polymeric network, and this applies for the presence of cross-linkers to prevent suspension of the hydrophilic chain polymer in aqueous solution. The three-dimension of the hydrogel has to be degenerated into harmless products for their biocompatibility. Considering both of these reasons hydrogel prepared by chemical and physical methods such as cross-linking polymerization, graft polymerization, network formation of water-soluble polymer and radiation cross-linking, most of the hydrogels are slightly cross-linked copolymers of acrylate and acrylic acid and grafted starch–acrylic acid polymers formed by inverse suspension, emulsion polymerization, and solution polymerization (Pal et al. 2009).

2.1 Chemical Methods to Prepare Hydrogels

The chemically cross-linked hydrogels are formed by the addition of cross-linking agent which forms covalent or ionic bond in gels. The various methods for chemical methods to prepare hydrogels are as follows.

2.1.1 Cross-linking Polymerization

The cross-linking gels are mostly stable in presence of ionic and covalent bonds in gels. Additionally, the cross-linkers lead to adverse effects due to any poisonous compounds present that become more toxic to human body. The cross-linking process has been achieved by radical polymerization of low molecular mass monomers or homoco-polymers. Generally, few amount of cross-linkers are added during the formation of hydrogel. Higher the concentration of monomers generates higher rate of polymerization which liberates high heat energy during the process and can be avoided in low conversions (Achar and Peppas 1994).

2.1.2 Copolymerization

Poly(hydroxyalkyl methylacrylates) is one example which was prepared by copolymerization technique. In the typical preparation, azobisisobutyronitrile (AIBN) and benzoyl peroxide radicals have been used as an initiator (Fig. 2). The whole preparation method includes major four types such as initiation, propagation, cross-linking, and termination. The termination process occurs by combination or chain transfer to monomer (Achar and Peppas 1994).

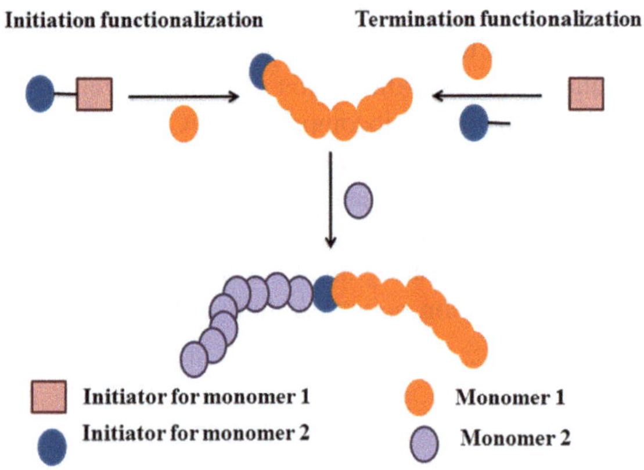

Fig. 2 Copolymerization mechanism

2.1.3 Cross-linking by High Energy Radiation

A beam of electron or gamma rays with high energy was achieved to polymerize the unsaturated compounds. The water solvable vinyl groups are transformed into hydrogels by using the high energy radiation. The preparation method by high energy radiation can be initiated by passing of gamma or beam of electron rays on aqueous solution and the polymer chain forms highly reactive free radicals. On the other hand, the water also forms hydroxyl radicals by radiolysis to form micro-radicals. The covalent and cross-linked polymers finally obtained by these combinations of micro-radicals. The formation of hydrogels by high energy radiation can be achieved in mild condition of water and also in the absence of any cross-linking agent. However, a few disadvantages were also faced by this method like damage to bioactive materials and biodegradation of hydrogels were not significantly observed.

2.1.4 Cross-linking Using Enzymes

The enzyme-catalyzed reaction (Fig. 3) of polyethylene glycol and lysine enzyme has been synthesized in the recent years. A tetrahydroxy polyethylene glycol with glutaminyl network was designed by adding of trans-glutaminase into a solution of

Fig. 3 Cross-linking using enzyme

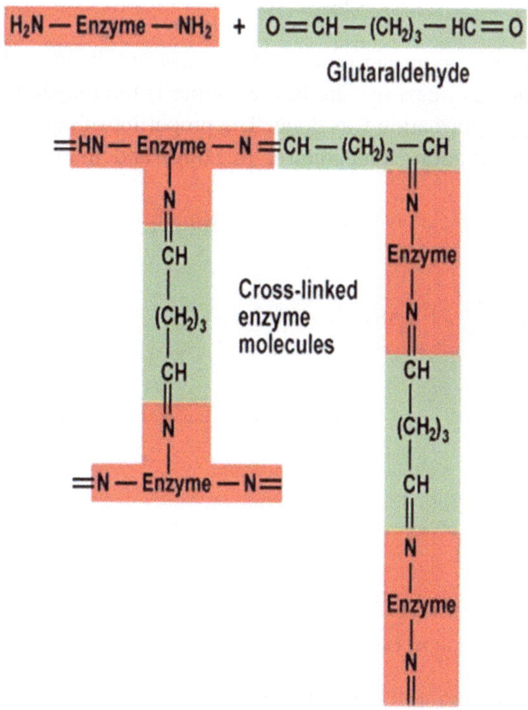

polyethylene glycol and poly lysine-co-phenylalanine (Chen 1995). The enzyme initiates the process to bonding between carboxamide group of polyethylene glycol and amine group of lysine.

2.2 *Physical Methods to Prepare Hydrogels*

The addition of cross-linking agent in chemical method is quite toxic, and it has to be removed from gels. Several methods have been reported on hydrogel preparation by physical methods. There are always pros and cons in each method due to the toxic nature of cross-linking, and physical process came into existence.

2.2.1 Cross-linking by Ionic Interaction

The ionic interaction occurs between two residues which have ions (Fig. 4). The alginate is one of the examples for the interaction of ions that consist of glucuronic and mannuronic residues linked by calcium ions at room temperature (Zhao et al. 2009).

2.2.2 Cross-linking by Crystallization

Another important protocol is to carry out the cross-linking in reduced temperature called as freezing. The best example is the preparation of polyvinyl alcohol where the gel formation was done. The properties of gel can be altered or can be varied by concentration, temperature, freezing, and thawing cycle times (Takamura et al. 1992).

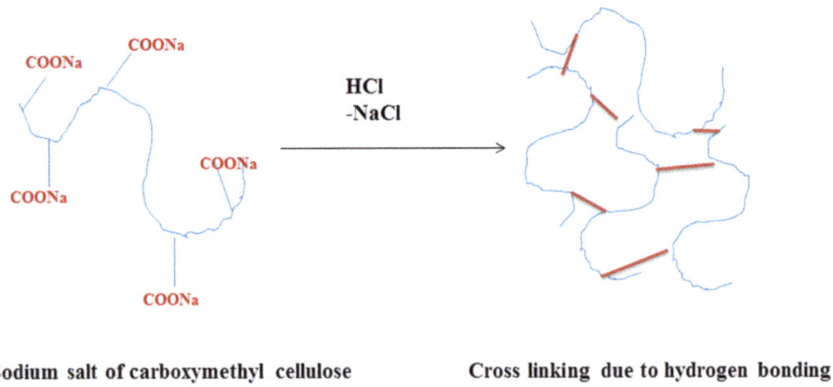

Fig. 4 Hydrogels formed by ionic interaction

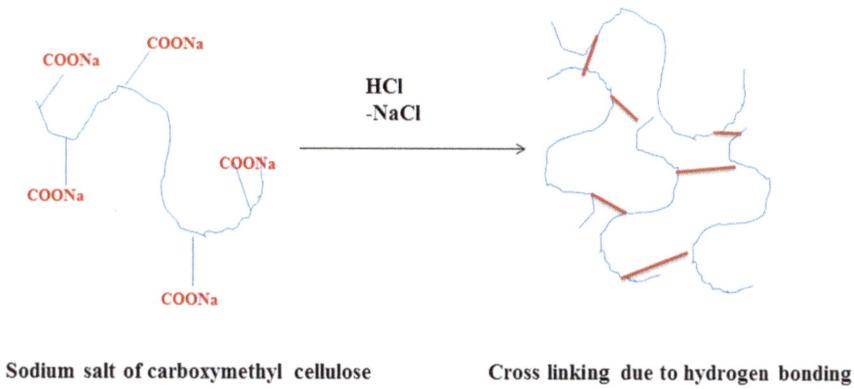

Fig. 5 Hydrogels formed due to hydrogen bonding

2.2.3 Hydrogen Bonding in the Cross-linking

Polyacrylic acid and poly(methacryclic acid) are a few examples that have been formed hydrogen bonding between oxygen of polyethylene glycol and carboxylic acid of poly(methacryclic acid). The hydrogen bonds were formed (Fig. 5) by protonation of carboxylic acid and also the swell properties depend on the pH (Eagland et al. 1994).

3 Types of Hydrogels

Depending on the properties of hydrogels, they are categorized into pH-responsive hydrogel, light-sensitive, ultrasonic, and temperature-sensitive hydrogels.

3.1 pH-Sensitive or Ion-Sensitive Hydrogels

The use of pH hydrogel has been widely studied in biomedical applications. The nano-sized polyethylenimine/sulfonamine micelles, poly(L-histidine), or sulfonamide with polysaccharides and many polymers or polyp lexes were used for pH-responsive hydrogel. Among all the hydrogel types pH-responsive hydrogel gained more attention since certain disease cause important revolution with respective pH of different parts in human body. Based on the nature of pH, hydrogels can be used in vivo and in vitro in drug delivery. The normal tissues extracellular fluid pH must be 7.4 constantly, whereas within the cell pH retained in 7.2, respectively. The results showed that changes in pH of many human parts such as skin, gastrointestinal tract, and intestine (Schmaljohann 2006) (Figs. 6 and 7).

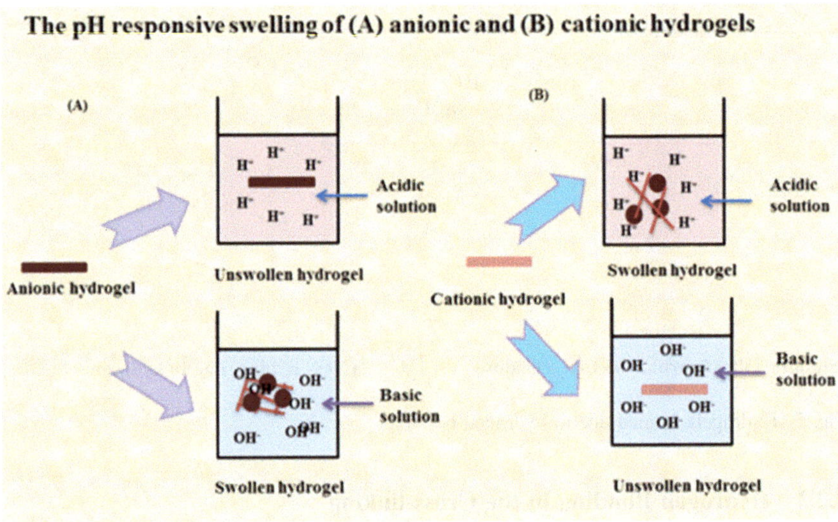

Fig. 6 pH-responsive hydrogel

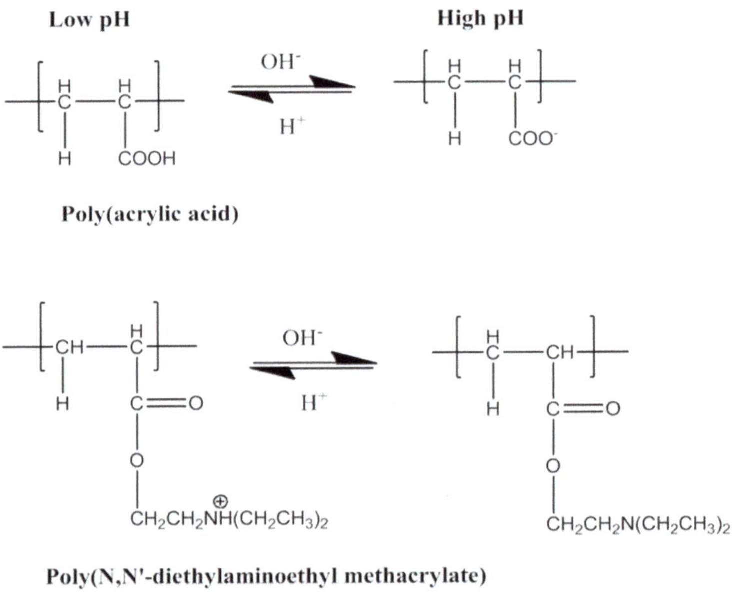

Fig. 7 Example of pH-responsive hydrogels

The chronic wounds show pH at 5.4–7.4, and tumors' pH lie below 7.2 by invasive or noninvasive methods. The acidic extracellular cancer tissues fluid of cancer tissues caused by hypoxia both in aerobic and anaerobic methods, and it was occurred by production of lactate and protons that make the metastasis, growth, and invasion of tumor. Therefore, the importance of pH hydrogel in biomedical application for drug delivery has been achieved more attention and it has been used for the controlled drug delivery including constant pH in blood for certain cardiovascular diseases (Lee et al. 2008).

Among all the potential application of pH hydrogel, design of drugs for treatment of cancer still a goal to develop a pH-responsive hydrogel than chemotherapy. Chemotherapy is one of the widely used medical technologies for several tumors, but it has many disadvantages like non-targeted toxicity, specific control of tumor cells, and resistance of drug. The first solution for these problems occurs from the available cancer drugs that cannot target cancer cells and affects healthy cells and leads to unexpected change in human body. Therefore, the main problem in an effective cancer treatment can be developed for distribution of drug to the tumor site. To address all these problems, pH hydrogels are considered as one of the potential approach (Li et al. 2011).

The pH-responsive hydrogels are very useful to sense the environment of cancer cells and controlled release of drug. Report on CdSe quantum dots in chitosan–poly (methacrylic acid) has showed an excellent colloidal and structural stability on B16F10 cells in pH range of 5–7.4 respectively (Wu et al. 2010). The pH sense in hydrogel has been reported on electron sharing bond interactions, e.g., repelling of water nature or opposite charge interactions found to be attractive than other drugs. The polyelectrolytes or ionomers are the major class of pH-responsive hydrogels. These kinds of gel contain amide or carboxylic acids in their structures that can sense the environmental pH in body and dissociate or ionize the functional group to extend the cross-linked structure of gels. The swelling of non-ionized state may be due to polymer-solvent mixing, polymer chain, and non-ionic polymer, whereas in the presence of water polyelectrolyte gels will ionize. The ionized gel generates electrostatic repulsive powers that are the reason for pH performance of hydrogel. The kinetics with twofold steps can be applied to proton spears into hydrogel. The first step includes a fast binding of charged particles on surface of gel, whereas the second is a slow diffusion of certain parts into the subsequent inner level of gel (Kazakov et al. 2006; Varga et al. 2006). For example, poly(acrylic acid) (PAA) ionized into acetate ions and protons at above the pKa value 4.25 and it generates acetate ions repulsion inside the polymer matrix. In the case of cationic hydrogels, it swells at lower pH in reversed manner. The polyacrylamide (PAAm), PAA, poly(methacrylic acid) (PMAA), poly(diethylaminoethyl methacrylate) (PDEAEMA), and poly(dimethylaminoethyl methacrylate) (PDMAEMA) are examples for widely used hydrogel with pH-responsive behavior (Oishi and Nagasaki 2007; Oishi et al. 2007).

The glycol chitosan with 3-diethylaminopropyl (DEAP) group was studied for self-assemble of pH as decrease with tumor cells. The destabilizing of tumor is due to the presence of DEAP protonation in nanogel matrix. The results have shown doxorubicin (DOX) release which is controlled with hydrogel, and it was found at low pH with increase of DOX uptake by cancer cells (Oh et al. 2010).

3.2 Thermo-Sensitive Hydrogel

The hydrogel shows changes in conformations, physical state, and properties with change in temperature. This kind of hydrogel was developed to increase the potential growth of bio-separation, tissue engineering, drug delivery, and targeted drug delivery to tumors. The following reasons are considered for the potential growth of hydrogel in drug delivery. Initially, temperature is important to many sites of living organism and it has been considered as safest external signal that can be tolerated by human at certain temperature (37 °C). For example, human body can be increased up to 45 °C without any side effect on the body. Another factor for the increased growth of hydrogel is available equipment for particular temperature in a specific tissue with predictable time. It was also believed that this medical equipment works depth from body superficial and particular time between every treatment (Taizo et al. 2010). The ultrasound diathermy devices, microwave, and radio frequency used clinically to produce heat on the targeted body site and used in hyperthermia therapy. From the report, on this type of techniques was found that the solid tumors were treated in high temperature to destroy cancer cells without affecting the non-cancer cells. Mostly the extracorporeal heating tools applied to generate temperature-sensitive polymer gels to get on–off switching tissues of drug without side effects. Therefore, the temperature controlled drug carrier and hyperthermia at targeted cells considered as a prospective tumor treatment (Fig. 8).

In addition, temperature controlled hydrogels can be categorized into two types depending on their solubility upon heating. The hydrogel becomes insoluble called as lower critical solution temperature (LCST), and it starts to dissolve with increase of temperature known as upper critical solution temperature (UCST). Still, a few gels showed both UCST and LCST. This kind of gels is not more applicable for potential effect in biomedical application. The presence of hydrophilic and hydrophobic moieties is responsible for the temperature sensitivity of hydrogel.

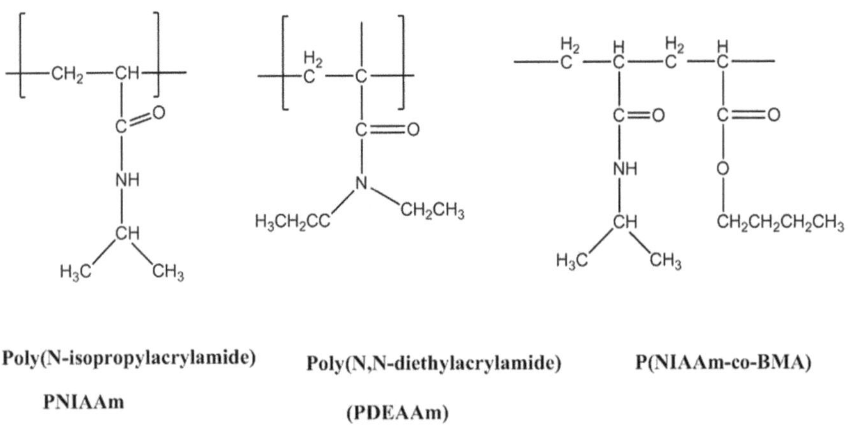

Fig. 8 Some temperature responsive hydrogels

The characteristics of hydrogels depend on the decrease of hydrogen bonding that lies between the water and polymer matrix. Most of polymers and structures show thermo-responsive efficiency such as polysaccharides, polymers based on *N*-isopropyl acrylamide, poloxamers, poly(ethylene oxide) (PEO)—poly(L-lactic acid)-PEO copolymers. (Medeiros et al. 2011). Generally, nanoparticle-loaded hydrogels have capacity to show both temperature-sensitive and change in properties due to increase and decrease in temperature. When the temperature below LCST, the characteristics such as hydrated, swollen, and hydrophilic changes, whereas high LCST hydrogels were formed by in situ and hydrophobic can be distorted and dehydrated. The biomedical application of temperature responsive hydrogel is due to its cross-linked polymers, lowest viscosities and fluidities in high temperature. The nanoparticle-loaded hydrogels are strengthen by the hydrophobic and hydrogen bonding between gel and particles. Among all the temperature-based hydrogels studied till date, poly(*N*-isopropylacrylamide) (PNIPAM) homopolymer and its copolymers are widely investigated. PNIPAM is one of the temperature responsive gels that show volume phase transition at 34 °C, and its temperature sensitivity has also been studied widely. The phase behavior of PNIPAM, dispersed in aqueous medium, has been studied with respective temperature in the presence of static and dynamic light scattering methods (Tata 2009). The PNIPAM efficiency can be improved by addition of nanoporous silica, and it induced drug into porous channel simultaneous opening of pore channel to outside media for slower diffusion from porous channel. The controlled drug delivery can be achieved by change in the composition of nanogel.

3.3 *Glucose-Sensitive Hydrogels*

The glucose-responsive hydrogels have been developed for diabetic patient especially for those patients who have lost blood glucose concentration control in body fluids. There are three types of glucose-responsive hydrogel including incorporation of glucose oxidase, combination of boronic acids derivatives, and incorporation of concanavalin A (Con A). The incorporation of glucose oxidase reacts along with pH-responsive hydrogel that regulates the insulin release (Zhu et al. 2005). Boronic acid and its derivatives bind reversibly to cis-diols such as glucose, carbohydrates, polyvinyl alcohol, and glycoproteins. The reaction between boronic acids and glucose stabilizes the ionic form of boronic acids which in turn makes difference in the swelling nature of gel particles (Siegel et al. 2010).

Another type is incorporation of concanavalin A (Con A) that competence free glucose and bonded glucose. Reduction in level of glucose tends to lower the extent of cross-linking which further affects the gel swelling (Miyata et al. 2004). Among all the glucose-responsive hydrogel Con A and glucose oxidase are very much targeted to glucose than boronic acid and its derivatives. The instability and inducing immune response have attracted more attention in glucose-responsive hydrogels. However, the use of such type of hydrogel is limited to pH and pKa values. To address this problem, few modifications toward aromatic ring and electron donor monomers have been suggested.

3.4 Photosensitive Hydrogels

The objective of photosensitive drug delivery system is the activated release of hydrophobically bonded drug on particular body site using light. Photo- or light-sensitive method can be reversibly switched in completely different rheological behaviors prior and post-light irradiation. Main benefit of photosensitive gel is unaltered composition that broadens the application in drug delivery. Generally, this type of gels contains photo-responsive segment that changes its physical and chemical properties when subjected into incident light (Peng et al. 2010). The azobenzene has attracted more attention that possesses two advantages. It has capability to change Nitrogen = Nitrogen transoid–cisoid isomerism after subjected to UV–vis photons. Trans and cis and thermally affected both photochemically without any side reactions. One of the isomers is furnished by photochemically, and another configuration was carried out by thermally. Azo-dextran is one the smart drug has been developed for Rhodamine B and Aspirin carrier. The controlled discharge of drug has been achieved by transoid–cisoid photo-isomerization of azobenzene group that presents in the cross-links. The isomer E and Z played an important role in release of biomolecules particularly E isomer showed slower release than azo in Z configuration (Patnaik et al. 2007).

3.5 Magnetically Responsive Hydrogels

Recently, magnetically sensitive hydrogels were used to target for a novel continuous development in chemotherapy. The aim of this gel is to lessen the side effects of cytotoxic drugs and decrease the distribution of drug dosage. The problem with this hydrogels is accounted of their fast clearance by macrophages of the reticuloendothelial system before arrived at the tumor cells. The superparamagnetic water-soluble nanoparticle-loaded hydrogels have been addressed this limitation due to the presence of specific functionalized groups on the surface. The polyacrylamide-coated Fe_3O_4 magnetic nanogel has been bonded with chymotrypsin (CT) to show better stability, storage stability, less side effects, and reusable. Another method used in magnetic responsive hydrogels is applying the magnetic field from outside (Hong et al. 2007b). As the magnetic field is switched off, a small quantity of the drug may disperse in the body but with no side effects. When magnetic field is increased, the linear blood flow rates in arteries (10 cm/s),

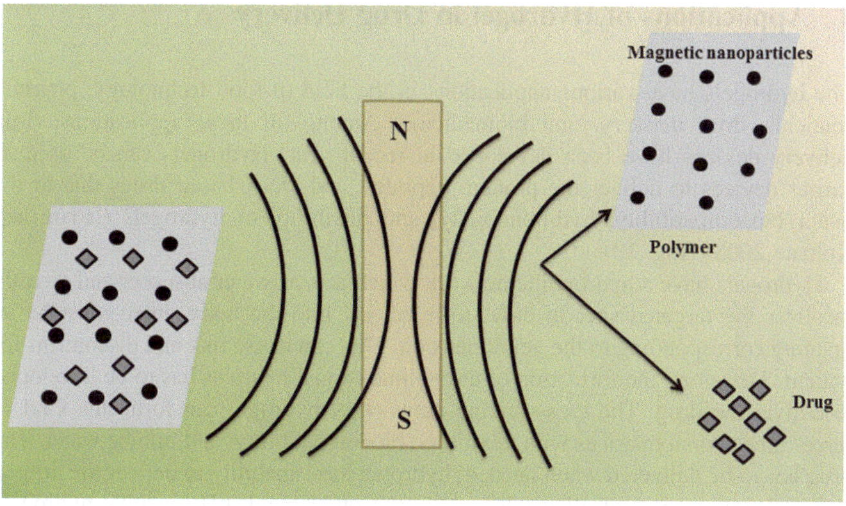

Fig. 9 Magnetically induced drug release from drug loaded hydrogel film

the particles may retain on the targeted cells to internalize the drug. However, the magnetic particles bounded drug can be released using enzyme or in the change of physiological conditions manly pH and temperature (Hong et al. 2007a; Bae et al. 2008) (Fig. 9).

3.6 *Ultrasound Responsive Hydrogels*

In this typical method, the drug may be induced by ultrasound for controlled release of drug. The pressure released from the ultrasound creates a mechanical disturbance at drug and leads to the increase of drug from carrier. The main advantage of this method is that there is no need for any distinct structural advantage of drug delivery system though the high rate of the release of dug is considered as one of the main limitations than non-ultrasound irradiated. Ultrasound includes ultrasonic energy usage for enhancing the transdermal delivery of solutes simultaneously or by means of pretreatment and is commonly referred to as sonophoresis or phonophoresis. The formation of gas cavity in the intercellular lipid on contact with ultrasound results in the disruption of stratum corneum (Borden et al. 2006).

4 Applications of Hydrogel in Drug Delivery

The hydrogels have various applications in the field of food technology, pharmaceuticals, drug delivery, and biomedicine. Among all these applications, drug delivery devices have been developed in recent year. Hydrogels can be used as carrier devices to deliver the protein, peptides, and DNA-based drugs due to the better biocompatibility, hydrophobicity, and flexibility of hydrogels (Hoare and Kohane 2008) (Fig. 10).

Hydrogels have a hydrophilic network which acts as water absorber and directly localizes the targeted site. In order to inject gel into the body, there must be an opening corresponding to the gel dimension. This can cause risk and discomfort for patient. Therefore, the formation of three-dimensional matrices has to be developed to deliver the drug. The cross-linking network of hydrogel can form this kind of three-dimensional matrices with flexibility, biocompatibility, and imbibe water. The drug has to be delivered when needed; hydrogel has capability to deliver the drug at required time with controlled release. On the other hand, pH-sensitive hydrogels have been considered as one of the oral drug delivery devices that can shield proteins in digestive system. The pH-responsiveness helps in lysosomal escape at the time of gene delivery. Secondly, hydrogels can also be manufactured to exhibit bioadhesiveness to assist drug targeting, especially through mucous membranes, for noninvasive drug administration (Hoare and Kohane 2008).

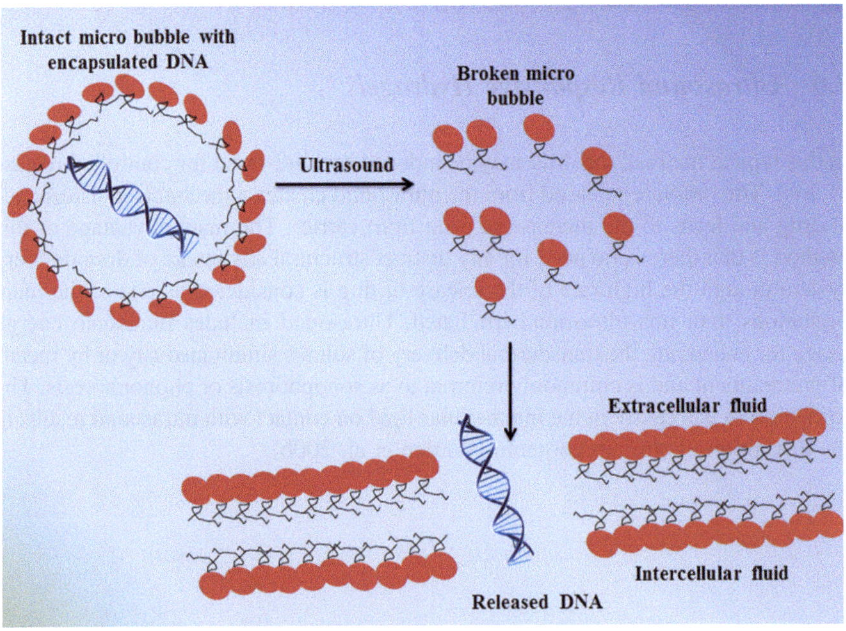

Fig. 10 Drug or gene delivery using ultrasound targeted therapy

(1) Easy to use, compliance, human factor
(2) Safety, efficacy, dosage frequency
(3) Stability, quality, material filling
(4) Synergy, manufacture preparation

The different types of hydrogels such as temperature, pH, magnetic, and ultrasonic hydrogels have biocompatibility and fast, reversible volume changes that give a potential application in drug delivery and biosensors. The pH hydrogels can be classified into two types, namely acidic and basic hydrogels. The acidic hydrogels get swollen, whereas basic hydrogels do not get swollen at lower pH condition. The acidic and basic groups like carboxylic acids, sulfonic acids, primary amines, and quaternary ammonium salts are responsible for pH sensitivity. At low pH carboxylic acid, groups remain uncharged, whereas at high pH, it gets charged; but for primary amine and quaternary ammonium salts, the reverse happens. The potential application of hydrogel in biosensors made to be very sensitive to biological agents that incorporated with hydrogel. The oxidoreductase is one of the enzymes used in biosensors for production of oxygen by the enzyme reaction. These enzyme incorporated hydrogels were particularly used to investigate glucose, cholinesterase, alcohol dehydrogenase, and penicillinase. The glucose biosensors are used to diffusional flux of oxygen on the surface of electrode by amperometric method where the current id produced by electrode. However, the glucose-sensitive hydrogel biosensors do not swell in responsive to pH changes (Traitel and Joseph 2005).

After the invention on the biosensors of hydrogels, there is another novelty increased by drug delivery. The potential application of hydrogel not only for drug delivery, it includes controlled and targeted drug delivery at specific site (Fig. 11).

Invention of conventional anticancer drug and its development has focused on cytotoxic agents with significant activity on tumor cell lines and can also cause tumor regression in murine tumor allografts or xenografts. Ability of anticancer

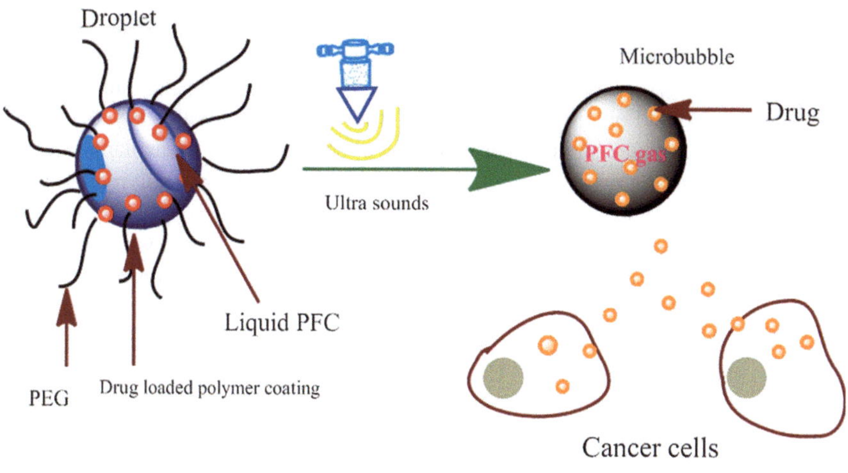

Fig. 11 Hydrogels act as anticancer drug on targeted cells

agents thereby inhibiting metabolic pathways crucial to cell division plays an important role. Anticancer drug has unique views and guidelines to accommodate and address those unique aspects. (Yanhua et al. 2013). Stimuli-sensitive hydrogels have been considered as better drug delivery material because of its capability to swell and de-swell according to the condition. This kind of hydrogel has three major functions in biomedical field namely, sensing an external signal, evaluation and action. Stimuli hydrogel can act as controlled drug delivery device with constant concentration of therapeutically active compounds in the blood with minimum fluctuations and predictable responsive release of drug in long time. It has no side effect, no waste of drug, use of frequent dosing, and better comfort to patience. Based on the stimuli, it has been classified into two types of hydrogel, namely temperature-sensitive hydrogel and pH-sensitive hydrogel (Swatantra et al. 2012).

5 Hydrogel in Tissue Engineering

In medical technologies, if certain organs or tissues are failed, there are many treatments to cure the problem such as replace of new organs or tissue or new tissue generation. But there are some limited techniques can solve such type of problems in which the surgical and replacement tissue have been gained good results. Nowadays, the implants have been showed a great success in tissue engineering using polymeric membrane exchange devices. These kinds of devices contain encapsulated cells from human or animal sources, and it is further termed as bioartificial or biohybrid organs. Hydrogels have been considered as one of the designed materials that contain more number of pores which is enough to provide living cells. For example, PLGA is very easy to bond with the cell membrane receptor peptide ligand to give stimulation, separation, and growth of cells in the hydrogel matrix (Lee and Mooney 2001). Still many limitations are there use of hydrogel in tissue engineering due to the low mechanical strength, difficult to handle and sterilization is very difficult in the case of hydrogels. Therefore it is clear that, the need of hydrogels may increase after overcoming of all problems (Hutmacher 2001). For instance, photo-polymerized poly(ethylene glycol) diacrylate, (PEG diacrylate)-based hydrogels were developed to transplant islets of Langerhans and cells were suspended in the photopolymerizable polymer solution. The PEG-based hydrogels showed suitable immunoisolation but the diffusion capacity of nutrients was limited. The photo-polymerized hydrogel was found to be feasible for continued time, and the particles are taken their immunoisolation function. The bioactive co-encapsulated system of metalloproteinase (MMP)-responsive PEG-based hydrogel has been studied for vascular cells, bioactive peptide, and thymosin beta4. The results showed favorable for human vein endothelial cell (HUVEC) where the HUVEC bonded with PEG hydrogels. This method is an inexpensive and favorable with encapsulation in the hydrogel (Nguyen and West 2002; Peppas 2006; Elisseeff et al. 1999) (Fig. 12).

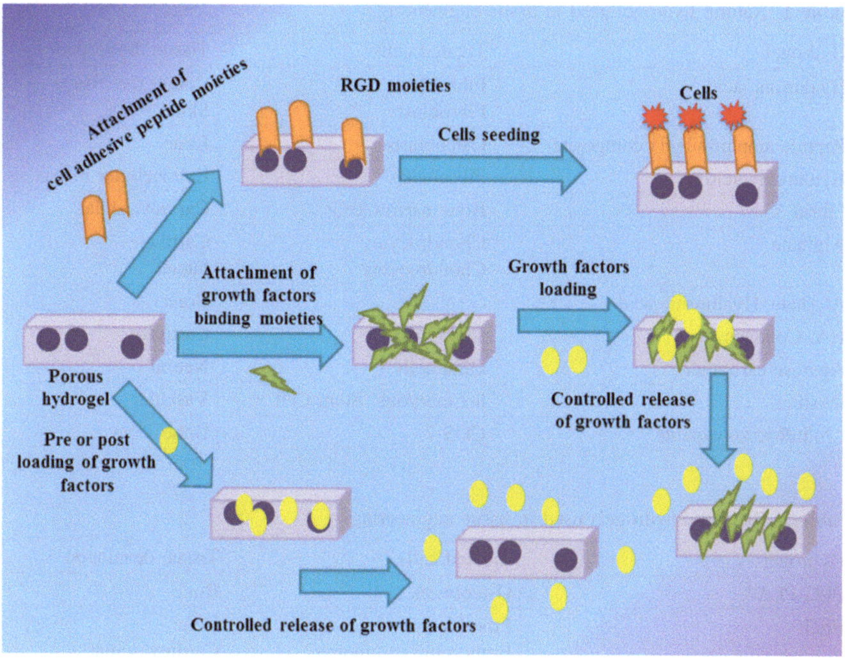

Fig. 12 Hydrogels in tissue engineering

Another type of hydrogel, based on self-assembly peptides (SAPs), namely hydrogel scaffolds is one of the major types in tissue engineering. SAPs consist of polypeptides which undergoes self-assembly at particular parameters including nanostructures and hydrophilic environment. The amphiphilic molecules can self-assemble long-chain alkyl tail and cell adhesion ligand. The polypeptide showed hydrophilic nature, whereas amphiphilic showed a hydrophobic nature with long-chain alkyl tail. The SAPs are useful with bioactive compounds for the controlled release. SAP-based hydrogels linked by conjugated moieties to make the signal to cell surface receptor and increase cellular adhesion (Adams et al. 2007; Williams et al. 2007; Hartgerink et al. 2001).

Conversely, many naturally available polymers were used to developed hydrogels for tissue engineering. A few naturally occurring hydrogels and their uses have been shown in Table 1. For example natural polynucleotides, polypeptides and polysaccharides were also used in tissue engineering. The collagen-based hydrogels scaffolds are achieved a great attention in tissue engineering. The collagen hydrogels were formed by cross-linking and self-aggregation in hydrated environment. Collagen molecules comprise of tropocollagen triple helixes, where each triple helix is formed as a result of self-arrangement of three polypeptide strands. The potential application of hydrogels from natural polymers is due to the intrinsic properties of biological system such as receptor, binding ligand, and cell-trigged proteolytic remodeling. However, the hydrogels from natural origin have shown

Table 1 Natural hydrogel used in tissue engineering

Hydrogel	Treated cells	Tissue developed
Hyaluronic acid	Fibroblasts Fibroblasts	Connective tissue Skin
Peptide amphiphile-Ti composite	Osteoblasts	Bone
Hyaluronic acid	Fibroblasts	Extracellular matrix
Fibrin	Bone narrow cells	Cardiovascular
Alginate	Chondrocytes Chondrocytes	Cartilage Facial
Alginate, Hyaluronic acid	–	Bone
HA, Collagen	Chondrocytes	Cartilage
Agarose carbomer	Glial cells	Neural
Dextran	h-Embryonic Stem Cell	Vascular
Chitin/hydroxyapatite	COS-7	Bone

Table 2 Synthetic hydrogels used in tissue engineering

Hydrogel	Treated cells	Tissue developed
PEG-PLA	Osteoblasts	Bone
PEG	Fibroblasts Embryonic carcinoma Chondrocytes Chondrocytes Chondrocytes	Bone Cardiovascular Cartilage Cartilage Cartilage
PEG-PLA	Chondrocytes	Cartilage
PHEMA	Myoblasts	Skeletal muscle
Dex-MA-LA and Gel MA	ECs and SMCs	Vascular
PEO semi-IPN	Chondrocytes	Cartilage
PVA	Chondrocytes	Cartilage

many limitations which involve purification, high complex structure, and pathogen transmission (Table 2).

6 Hydrogels in Wound Healing

The World Health Organization (WHO) has been declared that more than 300,000 people die each year due to induced burn and other forms of burns. The burned part causes many tissue damage, deep wound, abnormal skin function and also increases the local tissue damage. In such cases, particular clinical treatment is used to cure or reduce injury. In most of the cases, injured tissues are devastating, leading to plastic surgery on particular area on body. Many medical treatments have been developed for the wound healing property Ajji et al. (2005), but the hydrogel gained a great attention on wound healing treatment due to the hydrophilic and soft nature. Hydrogels are considered as one of the macromolecular networks, mostly stable with physical and chemical bonds with polymer matrix. This characteristic of hydrogel increases the water retain capacity in their mesh structure. The hybrid behavior of hybrid hydrogel with some mechanical property such as diffusive properties, it can absorb and release water in reversible manner and due to this property it promote fibroblast proliferation and keratinocyte migration, which are important in completion of epithelialization of the wound. Another size-dependent tight mesh size hydrogels are very active against bacteria and act as efficient transporter of bioactive molecules to the wound. In the case of burn, water is not available to wash and cool wound hydrogel can be used to cool and minimize the pain and damage. Hydrogels also considered as non-adhesive which cannot attack by hydrophilic surface. This property applies to dressing of wound where hydrogels do not stick on the wound and cause less pain to the patient. Still, there are several types of treatment using hydrogels that are available for wound and other skin burn in the form of amorphous gels, gel-impregnated gauzes, sheets, and plasters. Among all these, gels are mostly used for wounds; sheets were used to superficial burns. The hydrogel-based products are available in market to minimize wound, and still the need of hydrogels was increased for skin healing in medical applications. In particular, the regeneration of skin is taking into account to develop the hydrogel matrix that may instruct skin cells for regeneration rather than repair. The porous collagen and alginate scaffolds found to be good re-epithelialization and provide great inflammatory response. The hydrogel chemical composition, density, size, and cross-linking bond affect the cell behavior and wound healing efficiency.

7 Conclusion

The hydrogels are considered as one of the promising polymer matrices that can give a potential growth in biomedical applications. The different types of hydrogels have been classified and among all the types pH-responsive hydrogels achieved a great attraction in medical field due to its swell and non-swell behavior upon environment of cells at different pH. The drug delivery of hydrogel has been discussed, and the improved growth on the hydrogel is increased nowadays. Hydrogel-loaded nanoparticles and polymers or drug exhibit both hydrogel and nanoparticle properties for time controlled and targeted drug delivery. These particles are formed by physical and chemical cross-linking with polymer network and by utilizing the appropriate preparation techniques, the drug can be physically captured or bound chemically.

Acknowledgements The authors thank VIT University, Vellore, for their kind support and DST-FTYS (SB/FT/CS-113/2013) for the financial support.

Conflict of Interest

The authors declare that there is no conflict of interest.

References

Achar L, Peppas NA (1994) Preparation, characterization and mucoadhesive interactions of polymethacrylic acid copolymers with rat mucosa. J Controlled Release 31:271–276

Adams DJ, Holtzmann K, Schneider C, Butler MF (2007) Self-assembly of peptide surfactants. Langmuir 23:12729–12736

Ahmed M, Aggor FS, Awad A, Aref AT (2013) An innovative method for preparation of nanometal hydroxide superabsorbent hydrogel. Carbohydr Polym 91:693–698

Ajji Z, Othman I, Rosiak JM (2005) Production of hydrogel wound dressing using gamma radiation. Nucl Instrum Methods Phys Res Sect B 229:375–380

Bae KH, Mok H, Gwan Park T (2008) Synthesis, characterization, and intracellular delivery of reducible heparin nanogels for apoptotic cell death. Biomater 29:3376–3383

Bennett SL, Melanson DA, Torchiana DF, Wiseman DM, Sawhney AS (2003) Next-generation hydrogel films as tissue sealants and adhesion barriers. J Card Surg 18:494–499

Borden MA, Sarantos MR, Dayton PA (2006) Ultrasound radiation force modulates ligand availability on targeted contrast agents. Mol Imaging 437:139–147

Burkert S, Schmidt T, Gohs U, Dorschner H, Karl-Friedrich A (2007) Cross-linking of poly (N-vinyl pyrrolidone) films by electron beam irradiation. Radiat Phys Chem 76:1324–1328

Chen X (1995) Enzymatic and chemoenzymatic approaches to synthesis of sugar based polymer and hydrogels. Carbohydr Polym 28:15–21

Eagland D, Crowther NJ, Butler CJ (1994) Complexation between polyoxyethylene and polymethacrylic acid–the importance of the molar mass of polyoxyethylene. Eur Polym J 30:767–773

Elisseeff J, Anseth K, Sims D, McIntosh W, Randolph M, Langer R (1999) Transdermal photopolymerization for minimally invasive implantation. Proc Natl Acad Sci USA 96:3104–3107

Hartgerink JD, Beniash E, Stupp SI (2001) Self-assembly and mineralization of peptide-amphiphile nanofibers. Science 294:1684–1688

Hoare TR, Kohane DS (2008) Hydrogel in drug delivery: progress and challenges. Polymers 49 (8):1993–2007

Hong J, Gong P, Xu D, Dong L, Yao S (2007a) Stabilization of alpha-chymotrypsin by covalent immobilization on amine-functionalized superparamagnetic nanogel. J Biotechnol 128:597–605

Hong J, Xu D, Gong P, Sun H, Dong L, Yao S (2007b) Covalent binding of α-chymotrypsin on the magnetic nanogels covered by amino groups. J Mol Cat B Enzym 45:84–90

Hutmacher DW (2001) Scaffold design and fabrication technologies for engineering tissues—state of the art and future perspectives. J Biomater Sci Polymer Ed 12:107–124

Kazakov S, Kaholek M, Gazaryan I, Krasnikov B, Miller K, Levon K (2006) Ion concentration of external solution as a characteristic of micro- and nanogel ionic reservoirs. J Phys Chem B 110:15107–15116

Kim IS, Jeong YI, Kim DH, Lee YH, Kim SH (2001) Albumin release from biodegradable hydrogels composed of dextran and poly(ethylene glycol) macromere. Arch Pharm Res 24:69–73

Lee KY, Mooney DJ (2001) Hydrogels for tissue engineering. Chem Rev 101:1869–1879

Lee ES, Gao Z, Bae YH (2008) Recent progress in tumor pH targeting nanotechnology. J Control Release 132:164–170

Li N, Wang J, Yang X, Li L (2011) Novel nanogels as drug delivery systems for poorly soluble anticancer drugs. Coll Surf Biointerface 83:237–244

Medeiros SF, Santos AM, Fessi H, Elaissari A (2011) Stimuli-responsive magnetic particles for biomedical applications. Int J Pharm 403:139–161

Miyata T, Jikihara A, Nakamae K, Hoffman AS (2004) Preparation of reversibly glucose-responsive hydrogels by covalent immobilization of lectin in polymer networks having pendant glucose. J Biomater Sci Polym Ed 15:1085–1098

Nguyen KT, West JL (2002) Photopolymerizable hydrogels for tissue engineering applications. Biomaterials 23:4307–4314
Oh NM, Oh KT, Baik HJ, Lee BR, Lee AH, Youn YS, Lee ES (2010) A selforganized 3-diethylaminopropyl-bearing glycol chitosan nanogel for tumor acidic pH targeting: In vitro evaluation. Coll Sur B Biointerface 78:120–126
Oishi M, Nagasaki Y (2007) Synthesis, characterization, and biomedical applications of core–shell-type stimuli-responsive nanogels—Nanogel composed of poly[2-(N, N-diethylamino) ethyl methacrylate] core and PEG tethered chains. React Func Polym 67:1311–1329
Oishi M, Miyagawa N, Sakura T, Nagasaki Y (2007) pH-responsive PEGylated nanogel containing platinum nanoparticles: application to on–off regulation of catalytic activity for reactive oxygen species. React Funct Polym 67:662–668
Pal K, Banthia AK, Majumdar DK (2009) Polymeric hydrogels: characterization and biomedical applications—a mini review. Des monomer Polym 12:197–220
Patnaik S, Sharma AK, Garg BS, Gandhi RP, Gupta KC (2007) Photoregulation of drug release in azo-dextran nanogels. Int J Pharma 342:184–193
Peng K, Tomatsu I, Kros A (2010) Light controlled protein release from a supramolecular hydrogel. Chem Commun (Camb) 46:4094–4096
Peppas NA (2006) Intelligent biomaterials as pharmaceutical carriers in microfabricated and nanoscale devices. MRS Bull 31:888–893
Ratner BD, Hoffman AS (1976) Synthetic hydrogels for biomedical applications. In:Andrade JD (ed) Hydrogels for Medical and Related Applications. ACS Symposium Series, American Chem Soc 31: 1–36
Schmaljohann D (2006) Thermo- and pH-responsive polymers in drug delivery. Adv Drug Deliv Rev 58:1655–1670
Siegel RA, Gu Y, Lei M, Baldi A, Nuxoll EE, Ziaie B (2010) Hard and soft micro and nanofabrication: an integrated approach to hydrogel-based biosensing and drug delivery. J Control Release 141:303–313
Sutton C (2005) Obstet Gynaecol 7:168–176
Swatantra KS, Saxena P, Rai AK (2012) Stimuli hydrogel for ophthalmic drug delivery: a review. Int J Pharm Invest 2:54–60
Taizo H, Hisashi H, Yuehu P (2010) Charged particle beam irradiation apparatus. US Patent Application No. 20100288946
Thakur VK, Thakur MK (2014a) Recent trends in hydrogels based on psyllium polysaccharide: a review. J Clean Prod 82:1–15
Thakur VK, Thakur MK (2014b) Recent advances in graft copolymerization and applications of chitosan: a review. ACS Sustain Chem Eng 2(12):2637–2652
Thakur VK, Thakur MK (2015) Recent advances in green hydrogels from lignin: a review. Int J Biol Macromol 72:834–847
Thakur MK, Thakur VK, Gupta RK, Pappu A (2016) Synthesis and applications of biodegradable soy based graft copolymers: a review. ACS Sustain Chem Eng 4(1):1–17
Takamura A, Ishii F, Hidaka H (1992) Drug release from poly(vinyl alcohol) gel prepared by freeze-thaw procedure. J Controlled Release 20:21–27
Tata BV (2009) Phase behavior of poly(N-isopropylacrylamide) nanogel dispersions: temperature dependent particle size and interactions. J Nanosci Nanotechnol 9:5323–5328
Traitel T, Joseph K (2005) pH-responsive hydrogels: swelling model. Biomaterials 553:29–43
Varga I, Szalai I, Mészaros R, Gilanyi T (2006) Pulsating pH-responsive nanogels. J Phys Chem B 110:20297–20301
Williams BAR, Lund K, Liu Y, Yan H, Chaput JC (2007) Self-assembled peptide nanoarrays: an approach to studying protein-protein interactions. Angew Chem Int Ed 46:3051–3054
Wu W, Shen J, Banerjee P, Zhou S (2010) Chitosan-based responsive hybrid nanogels for integration of optical pH-sensing, tumor cell imaging and controlled drug delivery. Biomater 31:8371–8381

Yanhua L, Wenping W, Jianhong Y, Chengming Z, Jin S (2013) pH-sensitive polymeric micelles triggered drug release for extracellular and intracellular drug targeting delivery. Asian J Pharm Sci 8:159–167

Zhao Q, Ji QX, Xing K, Li XY, Liu CS, Chen XG (2009) Preparation and characteristics of novel porous hydrogel films based on chitosan and glycerophosphate. Carbohydr Polym 76:410–416

Zhu H, Srivastava R, Brown JQ, McShane MJ (2005) Combined physical and chemical immobilization of glucose oxidase in alginate microspheres improves stability of encapsulation and activity. Bioconjug Chem 16:1451–1458

Chapter 9
Electrospinning of Hydrogels for Biomedical Applications

Gabriel Goetten de Lima, Sean Lyons, Declan M. Devine and Michael J. D. Nugent

Abstract The field of biomedical applications for hydrogels requires the development of nanostructures with specific controlled diameter and mechanical properties. Nanofibers are ideal candidates for these advanced requirements, and one of the easiest techniques that can produce one-dimensional nanostructured materials in fibrous form is the electrospinning process. This technique provides extremely thin fibers with controlled diameter and highly porous microstructure with interconnected pores. Electrospinning demonstrates extreme versatility allowing the use of different polymers for tailoring properties and applications. It is a simple cost-effective method for the preparation of scaffolds. In this section, we will discuss recent and specific applications with a focus on their mechanisms. As such, we conclude this section with a discussion on perspectives and future possibilities on this field.

Keywords Electrospinning · Hydrogel · Tissue engineering · Drug delivery
Wound healing

1 Introduction

Electrospinning is an easy technique for the production of nanoscale polymer fibers from a variety of materials in one-dimensional, two-dimensional, and three-dimensional configurations (Li and Xia 2004; Greiner and Wendorff 2007). This technique provides extremely thin fibers with controlled diameters and highly porous microstructure with interconnected pores; versatility allows the use of various polymers for tailoring various applications requirements. The ability to

G. G. de Lima · S. Lyons · D. M. Devine · M. J. D. Nugent (✉)
Materials Research Institute, Athlone Institute of Technology, Athlone, County Westmeath, Ireland
e-mail: mnugent@ait.ie

D. M. Devine
Rehabilitation Medicine Center, The Mayo Clinic, Rochester, MN, USA

V. K. Thakur and M. K. Thakur (eds.), *Hydrogels*, Gels Horizons: From Science to Smart Materials, https://doi.org/10.1007/978-981-10-6077-9_9

produce nanofibers composites of blend polymers, composites with metals, and/or ceramics with high surface area than regular fibers, it provides various areas of applications such as nanocatalysis, tissue engineering, biomedical, pharmaceutical, and environmental engineering (Bhardwaj and Kundu 2010). One of the limitations of electrospinning is their low production rate which several groups are trying now to overcome this by producing multiple spinnerets arranged in an ordered position such as circle or lines (Hou et al. 2009; Persano et al. 2013).

Fields of applications on electrospinning largely studied are drug delivery, tissue engineering, and wound healing. One of the main reasons on these applications is that polymeric scaffolds produced by electrospinning have the advantage of mimicking the natural extracellular matrix (ECM). However, nanofibers produced by electrospinning on tissue engineering have irregular distribution of cells and poor migration at the interior of the scaffold on normal behaviors (Bhardwaj and Kundu 2010). Furthermore, the increase in packing density of nanofibers produced by electrospinning limits cells to enter on the inner part of the fibers. Several techniques have been performed to modify the fiber parameters to overcome these issues (Li et al. 2014) which will be discussed in this chapter.

Although not yet deeply discussed, hydrogel nanofibers can overcome the limitation of simple polymeric nanofibers. Hydrogels are hydrophilic cross-linked polymers structured in three dimensions that can swell under aqueous conditions but not dissolve instantly. Hydrogels swell to a greater or lesser extent in water due to hydrophilic power of the group chains. On lower hydrophilicity, the polymer will swell in water, but with a further increase in hydrophilicity the polymer becomes water soluble (de Lima et al. 2015a, b). Hydrogel nanofibers have the advantage of combining both techniques, and it is being widely investigated in wound healing (Fogaça and Catalani 2013; Choi et al. 2015) since they can maintain a healing environment and adequate humidity on the region of the wound and absorb the exudate with exchange of minimal pain. In addition, their inherent abilities help on diffusion of extended release of drugs, in particular on the case of hydrophobic drugs (McKenzie et al. 2015).

The focus on this chapter will be treated, along with the fundamentals on electrospinning, on the polymers employed for the application intended, modifications of parameters on fibers for tissue engineering. Finally, a focus on incorporation of drugs into nanofibers and the diverse applications with novel techniques have been investigated with a perspective for future prospects.

2 Overview of Electrospinning

Electrospinning process is a simple and very controllable technique which produces fine fibers using electrostatic forces. These fibers are either produced by polymer melt or solution with fibers' diameters in the nanometer scale with a large surface area. A typical electrospinning apparatus usually involves syringe (attached with the emitter) with feed pump, a grounded collector, and power supply with high

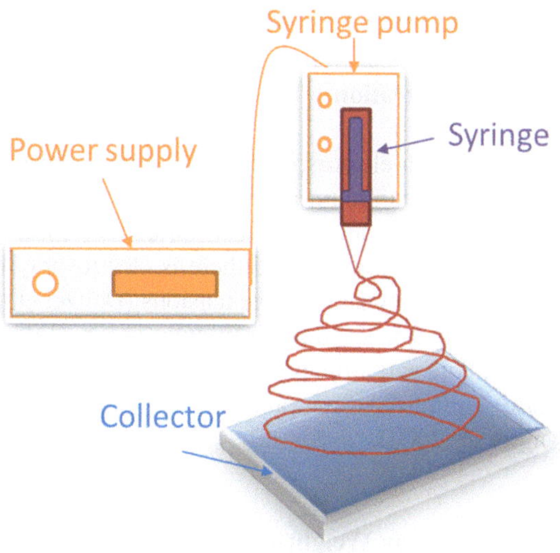

Fig. 1 Schematic drawing of basic setup for electrospinning

voltage as described in Fig. 1 (Bhardwaj and Kundu 2010). The process of electrospinning involves applying high voltage in the emitter and collector and charging the polymeric solution. When this polymer reaches the electric emitter, the droplet that is being held by its surface tension deforms once the electric field reaches an important value to a cone formation labeled as "Taylor cone"; the polymeric solution is then expelled in a jet shape from the emitter and accelerated toward the collector which is of opposite polarity. During the traveling toward the collector, the solvent is evaporated in the air and dried fibers form at the collector. However, it is essential to adjust the electrospinning parameters due to instability of polymeric jet to obtain uniform nanofibers (Greiner and Wendorff 2007).

2.1 *Influence of Process Parameters on Electrospinning*

The process parameters are important when aiming to obtain continuous non-beaded fibers, and manipulating improves it when designing nanofibers; some of these processes can be governed by the solution such as surface tension, concentration, viscosity, molecular weight, and conductivity. Alternatively, processing parameters of the electrospinning apparatus also plays important roles, which are flow rate, voltage, type of collector, and collector–emitter (tip) distance. Finally, the ambient when producing nanofibers affects the morphology and structure.

2.2 *Characteristics of the Solution*

2.2.1 Concentration

Concentration in electrospinning is crucial for fiber formation because it cannot be formed until a minimum is required. According to Li and Wang (2013), there are four critical concentrations that affect the morphology of fibers: very low—nanoparticles are formed, and there is electrospray rather than electrospinning due to limited viscosity and surface tensions (Deitzel et al. 2001). At slightly increasing concentration, fibers and beads mixture occurs (Eda and Shivkumar 2007), and with a suitable concentration evenly nanofibers are obtained (Fong et al. 1999). Finally, in case of high concentrations, "helix-shaped" microribbons can be observed (Yang et al. 2004).

2.2.2 Molecular Weight (MW)

The polymeric solution entangled chains are indicative of molecular weight and along with concentration contribute to the solution viscosity and affect the morphology of fibers. Electrospinning is essential to aim for higher molecular weight since there are enough chain entanglements to stabilize the jet and form nanofibers (Bhardwaj and Kundu 2010). An equation was obtained by (Gupta et al. 2005), for the exact transition from beads to fibers as it will be discussed in Sect. 2.2.

2.2.3 Viscosity

As another crucial parameter, viscosity can also determine the fiber morphology (Larrondo and St. John Manley 1981). The balance of viscosity is crucial since low viscosity excludes the formation of continuous fibers and just beads are normally obtained, whereas high viscosity makes difficult the jet formation of polymers on the emitter. For melt electrospinning, the variation of polymer viscosity varies depending on the spinning (Bhardwaj and Kundu 2010). The viscosity is also connected to the concentration and molecular weight parameters. Various works investigated the effect of viscosity onto electrospinning polymeric solutions (Geng et al. 2005; Kim et al. 2010; Binulal et al. 2014).

2.2.4 Surface Tension

Surface tension values might be different for different solvents. Higher values of surface tension might inhibit the electrospinning generating unsteady droplets (Moses et al. 2001). A lower surface tension values on the other hand strongly depend on the other parameters; for example, electrospinning can be formed on low electric field with low values of surface tension (Rogina 2014). According to Bhardwaj et al., the surface tension determines the boundary of nanofibers' condition on electrospinning if all other variables are held constant.

2.2.5 Conductivity/Surface Charge Density

With a few exceptions, polymers are polyelectrolyte and this ability highly influences the jet formation. Usually, higher values of conductivity of the solution decrease the nanofiber diameter and lower conductivity results in non-uniform fibers with beads (Chuachamsai et al. 2008). Increasing conductivity can be achieved by adding ionic salts on the polymeric solution like NaCl or KH_2PO_4 (Li and Xia 2004) resulting in fibers with decreasing diameters and beadless fibers (Huang et al. 2006).

2.3 Processing Parameters

2.3.1 Voltage

Voltage is one of the most important parameters since electric jets released from the emitter only occur after a voltage threshold. Higher voltage on polymers for electrospinning influences fiber diameter, but it depends on the concentration and the emitter-to-collector distance (Deitzel et al. 2001; Li and Wang 2013).

2.3.2 Flow Rate

Flow rate affects the velocity of the jet and the amount of polymer transferred. Lower flow rate is generally required since it leads to more time for polarization and evaporation of the polymeric solution (Yuan et al. 2004). Higher values of flow rate may induce the formation of bead fibers with thick diameter due to short drying period before reaching the collector (Zhang et al. 2005).

2.3.3 Type of Collector

Collectors are conductive substrates that collect the charged fibers during the electrospinning process. The simplest collector is aluminum foil, but it is difficult to remove the fibers on this type of collector, so different arrangements have been performed such as conductive paper, gridded bar, rotating rod, and wheel. Such collectors can also contribute to the alignment of fibers which are useful in tissue engineering (Vaquette and Cooper-White 2011).

2.3.4 Distance Between the Collector and Emitter (Tip)

The distance between the emitter and collector also affects the fiber diameter and structure. The required distance is usually minimal so that fibers dry before it reaches the collector, otherwise beaded fibers will be formed (Ki et al. 2005).

2.4 Ambient Parameters

As explained previously, the ambient conditions affect the morphology of the nanofibers. Humidity seems to have a great impact on some polymers in which after a critical value it can dissolve in the cross-points of the structure and fuse together forming a dispersal layer (Yao et al. 2013). On the other hand, it has been shown that lower temperatures led to a slower evaporation of the solvent and fibers could not be formed. The average diameter of the nanofibers has a complex relationship in terms of temperatures (De Vrieze et al. 2009) since the formation of nanofibers on electrospinning depends on the solvent and rigidity of the polymeric chains.

2.5 Theoretical Foundation for Electrospinning Polymers Capable of Physical Gelation

The process formation of fibers via electrospinning is affected by various parameters such as flow rate (Uchko et al. 1999), voltage (Yuan et al. 2004; Zhang et al. 2005), polymer molecular weight (MW) (Casper and Stephens 2004; Tao and Shivkumar 2007), and concentration (Fong et al. 1999; Deitzel et al. 2001; Eda and Shivkumar 2007; Felice et al. 2015). To tailor these parameters to obtain a window for range of fiber formations is not straightforward, since these processes are interdependent on each other. For these reasons, an integrated approach must be used to identify the range of fiber formation. For gel polymers, the nanofiber formation is dependent on the entanglements of the structure of the gel and the solution (Shenoy et al. 2005a). For this reason, a semiempirical method, described by Shenoy et al. (2005a), can presume the shift over electrospraying to spinning based on the amount of chain entanglements—physical link of polymeric chains which are similarly to chemical cross-links (Shenoy et al. 2005b).

Equation (1) correlates the polymer MW to its entanglement solution MW. Due to the preparation method of polymers, normally the average MW is considered.

$$(n_e)_{So\ln} = \frac{M_w}{(M_e)_{so\ln}} = \frac{\phi M_w}{M_e} \tag{1}$$

For polymers solutions, its entanglement value is dependent on the solution concentration and the polymer volume fraction ϕ, described in Eq. (2)

$$\phi_p = \frac{\rho_s W_p}{\rho_s W_p + (1 - W_p)\rho_p} \tag{2}$$

where ρ_p and ρ_s denote the densities of polymer and solvent and W_p is polymer weight concentration of the solution.

With these equations, it is possible to obtain a single value (n_e) which corresponds to viability of formation of fibers in electrospinning based on their polymeric chain entanglements (Shenoy et al. 2005a; Husain et al. 2016).

The number of chain entanglements, due to variations of polymer solution concentration (C) and molecular weight, causes morphology transitions on most electrospinnable polymers (Felice et al. 2015). This variation of chain entanglement leads to alteration of viscosity (Bock et al. 2012) and can be used as a determinant for solution regimes of polymers (Gupta et al. 2005). The different solution regimes, particles to fiber, can be defined by the Berry in Eq. (3) (Hager and Berry 1982) which correlates the polymer concentration with the intrinsic viscosity:

$$B_e = [\eta]C \tag{3}$$

where B_e is the berry number, ($[\eta] = \lim\limits_{c \to 0} \frac{\eta_{sp}}{C}$) the intrinsic viscosity; η_{sp} is the specific viscosity, and C is the polymer concentration.

The solution regimes are classified according to the critical overlap concentration (C^*) and the chain entanglement concentration (C_e) described as diluted, semi-diluted unentangled, and semi-diluted entangled also represented in Fig. 2 (Gupta et al. 2005). Chain overlap is absent in dilute system, and thus, chain entanglement does not exist, limiting the system with significantly weak entanglements (Shenoy et al. 2005a). When the concentration of the solution is equal to the concentration inside a single macromolecular chain, the critical concentration overlap occurs, A $\rightarrow$ B process in Fig. 2.

As the concentration increases, the chain reaction overlap occurs, but it is not enough to produce a degree of entanglement. Although some entanglements are observed, the chain entanglements within the drop formed on electrospray are not enough to stabilize the particle structure formation and these are not optimum as it leads to inferior and non-reproducible morphology since they lose their shape when impacting over the collector showing a beaded and non-continuous fiber formation (Husain et al. 2016). For entangled systems, $B_e > 4$ or C_e is generally accepted as the formation of pure fibers and the effect is due to the increase in entanglement chains so the jet formation can be stabilized through jet breakup inhibition because of the increased surface tension (Felice et al. 2015; Husain et al. 2016).

In practice, electrospun beaded fibers occur unexpectedly, but it was reported in the literature (Taepaiboon et al. 2006; Husain et al. 2016). Formation of beads, as explained before, usually is dependent on the parameters of the polymeric solution, which include the polymer structure, concentration, and salt content (Li and Wang 2013).

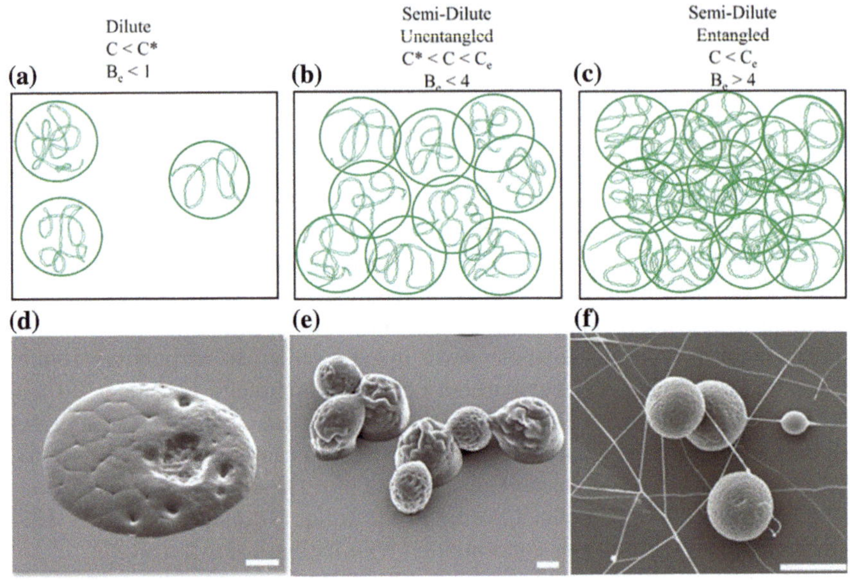

Fig. 2 Physical representation of the three solutions regimes with corresponding SEM images. **a**, **d** dilute; **b**, **e** semi-dilute unentangled; **c**, **f** semi-dilute entangled. $C^* = 1/[\eta]$. Adapted with permission from Bock et al. (2012). Copyright (2016), Elsevier

3 Applications

3.1 Tissue Engineering Applications

Tissue engineering creates artificial materials for regeneration of tissues based on the approach from materials engineering and life science (Ma 2004; Lanza et al. 2011; Okamoto and John 2013). Mostly, the repair or replacement of failing organs and tissues is facilitated by tissue engineering creating biological substitutes, like the growing of cells on scaffolds to support the regeneration of the desired tissue (Okamoto and John 2013). Cell interactions are extremely important for cell growth since it can change the cell functions via complex pathways (Kai et al. 2013; Temenoff and Mikos 2014).

Seeding of cells on scaffolds normally progresses with attachment followed by proliferation and differentiation. The cell agility to proliferate is one of the most important factors for tissue regeneration. With faster proliferation, the development of the injured tissue is improved and reduces scar tissue formation in vivo (Prakash et al. 2010; Kai et al. 2013).

Electrospinning for tissue engineering is one of the most used techniques since its fibrous structure is comparable to the tissue extracellular matrix (ECM) (Kim et al. 2005; Li et al. 2014; Guarino et al. 2015; Weng and Xie 2015). Ito et al. (2005) have shown that the attachment of fibroblast-like cell line (COS-7) was higher to nanofiber mats of poly(3-hydroxybutyrate-co-3-hydroxyvalerate) than on regular scaffolds of the same material after a short incubation time. The increased surface area of nanofibers helped the improvement on cell attachment due to its 3D features (Ito et al. 2005). Moreover, Chua et al. (2006) also demonstrated that nanofiber scaffolds containing amine groups had higher cell proliferation and adhesion, in comparison with films containing amine groups.

3.1.1 Choice of Polymer Hydrogel for Tissue Engineering

The behavior of cells can be adjusted by focal adhesion and signaling complexes (Wozniak et al. 2004), and such behaviors are intimately related to the properties of the scaffold (Kai et al. 2013). In addition, high-surface area nanofibrous scaffolds are preferably used on tissue engineering and different cells react different to the surface where it is attached (Leung and Ko 2011; Kai et al. 2015). Moreover, materials have to be biocompatible—the body must not reject the implant or scaffolds and must interact with it, so the spreading of cells is achieved faster (Yang et al. 2010; Van Vlierberghe et al. 2011; Rogina 2014). Furthermore, it must not elicit inflammatory response and cytotoxicity with a three-dimensional environment for cell proliferation and adhesion (Langer and Peppas 2003; Leung and Ko 2011; Parratt and Yao 2013). For these reasons, the choices of polymer and/or hydrogel designed for tissue engineering are important. Figure 3 shows the different polymers employed in electrospinning along with its properties. Currently, biodegradable scaffolds are the most investigated materials for production of nanofibers via electrospinning tissue engineering (Rogina 2014). Biodegradable nanofibers can have tailored degradation rate in the body so that it will be metabolized or excreted. The degradation mechanism normally consists of chemical degradation, and the most important parameters are the polymer structure and the environment which can also influence the degradation rate such as the pH and chemical structure (Simionescu and Ivanov 2016).

3.1.2 Natural Polymer Hydrogel Nanofibers for Tissue Engineering

Natural-based hydrogels/polymers are normally obtained from living plants and animals and are hydrophilic in nature; they can be extracted on polysaccharides and cellulose from plants (Simionescu and Ivanov 2016). Although most are obtained from vegetal sources, microorganisms can also synthesize biodegradable polymers. Examples are: fish—fish sin serves as a collagen; crustaceans—shells have large amounts of chitin; corals-algae—rich source of polysaccharides. These polymers are biocompatible, biodegradable, and non-toxic. In addition, natural polymers have an organized structure and it helps on cell viability and tissue ingrowth (Wolf et al. 2015; Simionescu and Ivanov 2016).

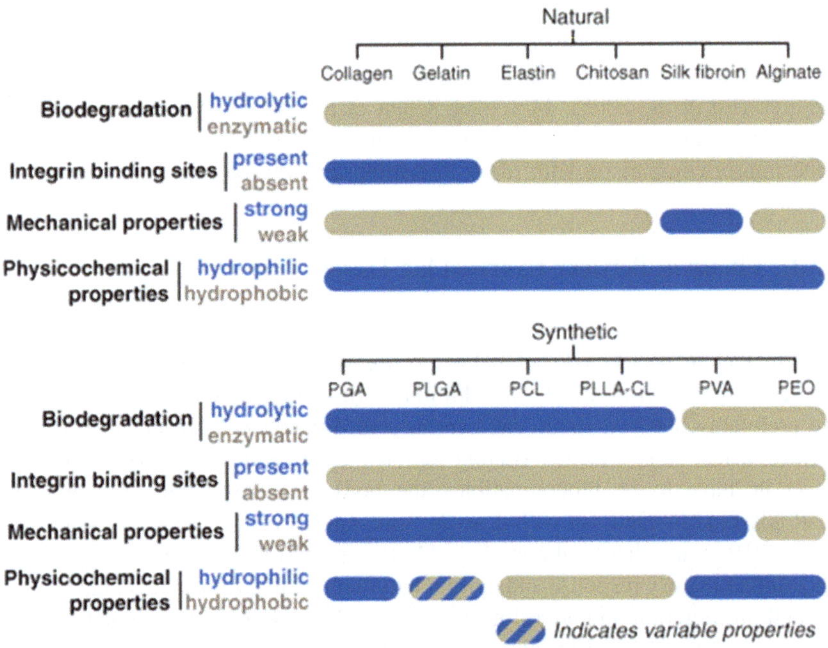

Fig. 3 Most common used natural and synthetic polymers on electrospinning along with its biological, mechanical, and physiochemical properties. Reprinted with permission from Gunn and Zhang (2010). Copyright (2016), Elsevier

The main advantages of natural polymers (collagen, chitosan, silk fibroin, gelatin, etc.) are their inherent cellular interaction and similar chemical versatility—greatly mimicking the extracellular matrix structure (ECM) for bone cell support matrix, which encourages protein adsorption and cellular adhesion, surface migration, and proliferation (Lai et al. 2014). However, ECM of different tissues has unique characteristics (Li et al. 2014; Khorshidi et al. 2015).

Collagen nanofiber scaffold can be used as a matrix for osteogenic progenitor cells to adhere, proliferate, and differentiate into osteoblasts (Chen and Lv 2015). Some studies performed with collagen nanofibers exhibited an adhesion of over 45% for mesenchymal stem cell population in a quickly response time of 30 min at room temperature (Chan et al. 2009).

In addition, recent studies suggest that stiffness plays an important part in bone tissue engineering for nanofiber scaffolds. Collagen nanofibers when tested under MG63 osteoblast-like cells produce bone-ECM proteins such as osteocalcin, responsible for ECM deposition and mineralization; leading to superior values of cell maturation and their cellular adhesion as well as response is increased based on the value of stiffness of the scaffold (Torres-Giner et al. 2009; Tsai et al. 2012).

Chitosan, which is biocompatible and biodegradable, presents low toxicity and is widely investigated on electrospinning as nanofibers (Rogina 2014). However, pure electrospun chitosan fibers are difficult to obtain (Rogina 2014) and chemical modifications into derivatives that are soluble in common organic solvents are used (Neamnark et al. 2006). The morphology of chitosan fibers is altered by the acetic acid concentration, whereas an increase in the solvent changes the chain entanglements and is easy to form uniform fibers (Geng et al. 2005). The properties of chitosan nanofibers for bone regeneration have been studied (Shin et al. 2005), and cells MG63 proliferated on the nanofiber with collagen and osteocalcin induction after two weeks in vitro. In addition, chitosan can also maintain its integrity for as long of six weeks and enhance bone regeneration without inflammatory reaction due to its antifungal and antibacterial properties, which can be improved by UV irradiation (Shin et al. 2005). Charged reactions can occur on chitosan, leading to a network between polymeric chains (Berger et al. 2004). This network acts as a hydrogel. However, one of the drawbacks is its poor mechanical strength and, therefore, it is normally used with other material or cross-linked with a copolymer.

Alginate, a naturally biodegradable polysaccharide material, is obtained from brown seaweed and is biocompatible, non-toxic, and non-immunogenic. However, it is hard to obtain continuous and uniform nanofiber scaffolds by electrospinning due to the rigid and extended chain conformation in aqueous solution and lack of chain entanglement (Chen and Lv 2015). Therefore, the addition of another copolymer is usually performed on alginate nanofibers. The ability of alginates to absorb and retain water is superior to that of natural gums (Berger et al. 1953).

However, nanofibers composed of natural polymers are compared poorly in terms of their mechanical properties with low resistance to aqueous solutions that limits their use as tissue engineering. (Frenot and Chronakis 2003; Khadka and Haynie 2012). Many researchers reported a number of cross-linking procedures to stabilize the nanofibers synthesized using a wide range of natural polymers and their blends (Delmar and Bianco-Peled 2016; Jalaja et al. 2016). The effect of cross-linking on nanofibers creates a network of 3D polymer chains which can possibly enhance fiber stability and other physical and mechanical properties (Miraftab et al. 2015; Laha et al. 2016). Cross-linking on different natural polymers have been performed (Torres-Giner et al. 2009; Gualandi et al. 2016; Jalaja et al. 2016), such as gelatin in order to decrease its solubility for being able to use in long term (Zhang et al. 2006a). Laha et al. (2016) developed a gelatin with saturated vapor of glutaraldehyde as cross-link. The effect of cross-link induces the nanofibers to a more hydrogellic state (Fig. 4).

3.1.3 Synthetic Hydrogels/Polymer for Tissue Engineering

One of the drawbacks of using natural polymers is their inherent brittleness, in addition to their restricted flexibility (Chen and Lv 2015). Additionally, synthetic polymers can be easily modified in contrast to natural polymers which have

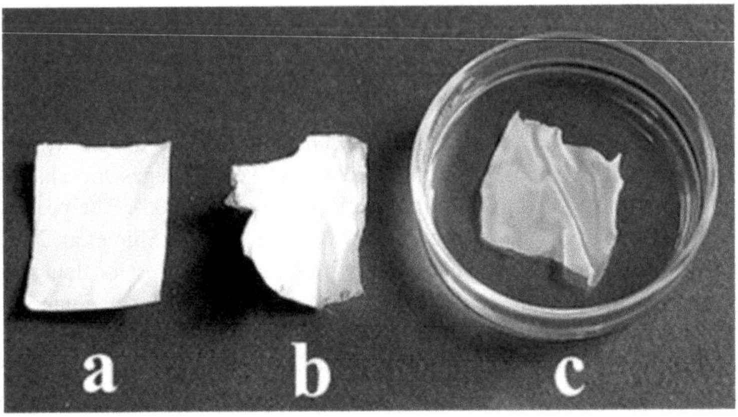

Fig. 4 Digital images representing **a** non-cross-linked gelatin, **b** gelatin after cross-link, and **c** gelatin with cross-link immersed in water. Reprinted with permission from Laha et al. (2016). Copyright (2016), Elsevier

sensitivity to processing conditions such as pH or UV radiations, water. (Simionescu and Ivanov 2016).

Synthetic polymers are, as the name suggests, synthetized in the laboratory. Between the synthetic nanofibers, the most attractive ones used in tissue engineering include polyvinyl alcohol (PVA), poly(ethylene glycol) (PEG) and poly (ethylene oxide) (PEO), polycaprolactone (PCL), poly(Lactic Acid) (PLA), and poly-(N-vinyl-2-pyrrolidone) (PVP) (Wolf et al. 2015). With most of the biodegradable synthetic materials approved by FDA, the interest in using these materials has drawn great attention.

PVA polyvinyl alcohol is created as a result of free radical polymerization of vinyl acetate with subsequent hydrolysis of acetate groups to hydroxyl moieties resulting in a wide molecular weight distribution (Hassan and Peppas 2000). PVA is biocompatible material and non-toxic with useful mechanical properties for tissue engineering and has the ability to swell to a large extent in solutions similar to those of human tissues. Nanofibers of PVA can be used as single scaffold (Felice et al. 2015) or combined (Yang et al. 2008; Vashisth and Pruthi 2016), especially for drug delivery systems and tissue engineering (Felice et al. 2015). PVA can be easily dissolved in aqueous solutions and produced as nanofibers via electrospinning. Furthermore, cross-linking approaches have been performed to produce PVA hydrogels, such as chemical cross-linking, freeze/thawed, and UV irradiation (Franco et al. 2012; Canillas et al. 2015). The cross-linking via methanol and chemical with glutaraldehyde seems to be the mostly used method for PVA nanofiber hydrogel. However, glutaraldehyde and chemical cross-links seem to have toxicity problems, which could impact and damage cells. Therefore, authors tried to elaborate different physical and chemical cross-linking methods (Torres-Giner et al. 2009).

PEG and PEG-containing block copolymers are used in many synthetic forms, due to its non-degradability by simple swelling and its limited metabolism in the body. In addition, PEG with molecular weight inferior of 50 kDa is used in tissue engineering applications (Yamaoka et al. 1994) so that they can be totally degraded on the kidneys (<30 kDa) or the liver (>30 kDa) (Veronese and Pasut 2005).

PCL nanofibers have been extensively used (Torres-Giner et al. 2009) due to their bioresorb ability. Aliphatic polyesters are associated with low cost and slow degradation. Mesenchymal stem cells (MSCs), which show potential to treat large bone defects (Quarto et al. 2001), proliferate and grow well on PCL scaffold (Yoshimoto et al. 2003), showing bone-like appearance when implanted on rat models (Shin et al. 2004). In addition, in comparison with normal PCL substrates, PCL nanofibers support higher MSC adhesion and viability (Ruckh et al. 2010). In addition, PCL nanofiber demonstrates the deposition of HAp on simulated body fluid (SBF) (Araujo et al. 2008).

Poly-(N-vinyl-2-pyrrolidone) (PVP) has excellent biocompatibility with high ability to absorb water. PVP nanofibers are excellent candidates for fabrication of nanofibers with non-spinnable materials due to various inorganic ions that can be produced and its ability to disperse particles acting as covering agent, which makes it a polymer of choice for the fabrication of electrospun fibers with non-spinnable materials. However, PVP nanofibers are very soluble in water showing poor properties as hydrogel (Lubasova et al. 2015). Consequently, PVP hydrogel nanofibers with further cross-linking are required for improvements and acting as hydrogels. Recent methods have been performing through Fenton reaction and UV irradiation (Fogaça and Catalani 2013), also with polyacrylic acid, since this reaction forms a strong hydrogen bond interaction (Lubasova et al. 2015).

3.1.4 Blend of Natural and Synthetic Polymer Nanofibers and Integration of Nanofibers with Hydrogels for Tissue Engineering

The polyblend of natural biodegradable polymers (chitosan and silk fibroin) with synthetic polymers offers a major advancement in tissue engineering by the simple and economical approach of the favorable biological properties on natural polymers and the excellent mechanical properties of the synthetic polymers that favor cell growth and proliferation (Gunn and Zhang 2010; Abdal-Hay et al. 2016). Various polyblend polymers have been researched in the past years (Cosme et al. 2016; Hu et al. 2016a; Mahoney et al. 2016; Zhijiang et al. 2016; Ziaee et al. 2016).

A common polyblend scaffold is made from PCL/gelatin mixture and demonstrates increased hydrophilicity and promotes bone osteogenesis and mineralization of MSCs in vitro (Alvarez Perez et al. 2012). In addition, with the support of PCL the scaffold provided improved mechanical and biochemical properties to guide bone regeneration. Currently, various types of polyblends have been researched for the application intended, such as polyester urethane urea (PEUU) with gelatin for application in myocardial tissue engineering. Since cardiac tissue challenges are the development of scaffold with suitable Young's modulus, the polyblend PEUU/ gelatin provided relatively desirable values for the application intended. In addition, cardiomyocytes were able to proliferate and grow in this nanofiber scaffold (Jamadi et al. 2016).

The combination of nanofibers and pure hydrogels may offer the advantages of incorporating both structures and minimize these problems. Although this concept is in early stages, some researchers developed different designs for tissue engineering applications (Bosworth et al. 2013; Xu et al. 2015).

Laminated composites The easiest way to incorporate both structures is to fabricate individually and mix them by the layer-by-layer method, and this method is adjustable by the layers, class of fibers, and hydrogels which can be tuned to the morphology and mechanical properties required. Another technique is to cross-link the hydrogels directly onto nanofiber films, dropping the solution onto the nanofiber (Quinn et al. 2007; Manna and Patil 2009; Shi et al. 2015). In this application, the voids that separate each nanofiber are filled with hydrogels. Yang et al. (2011) developed a layer by layer of oriented nanofibers of PLA with collagen type 1 hydrogels. These were separated by filter paper with cells deposited onto the fibers and incubated before adding another layer. The authors investigated the different fiber directions on the arrangement of the layers. The results found that there was a difference in direction of cells which were dependent on the orientation of the deposited fiber. These suggested further studies were possible into developing complicated structures to simulate ECM found in some tissues, such as skin and cartilages.

Encapsulating fibers in hydrogel Controlled placement of fibers with an ordered structure can be achieved by encapsulating the fibers in hydrogel. McMahon et al. (2011) achieved a nanotubular scaffold with nanofibers of PEUU (poly(ester urethane) urea and PEG-fibrin hydrogel (Mcmahon et al. 2011) for the application of coronary artery bypass grafts. Basically, rectangular segments of aligned electrospun fibers were revolved around latex mandrels. Smooth muscular progenitor cells were embedded onto the fibers and incubated for two days. After incubation, the fibers were revolved again through a latex tube with the cells facing outwards. The tube was added in a hollow Teflon cylinder, and a thrombin/fibrinogen cell was added to encapsulate the fibrous layers before polymerization and support enough mechanical strength so that after the removal of the latex, the composite hydrogel would be intact. After incubation, the latex and Teflon tubes were removed and the final composite was immersed in diacrylate-derivatized polyethylene glycol (PEG) solution for the final construct of PEG on fibrin gel. Finally, polymerization was achieved by UV. This structure was hypothesized to mimic the coronary artery vessels, showing a biphasic layer of hydrogel fibers and collagen. Suture strength results showed that this construct has similar values to those of human artery. In addition, smooth muscle cells were able to proliferate and migrate on this hydrogel nanofiber construct.

Injectable Composites A recently technique has been developed by Brown et al. (2011) named as melt electrospinning writing which can produce complex porous fiber structure using an automated stage, and this technique can create scaffolds by stacking melt electrospun fibers on top of each other which is similar to melt extrusion based on direct writing but in a sub-micrometer magnitude. (Visser et al. 2015) use the innovation of this technique to produce scaffolds for cartilage tissue and via an injection mold prepared a reinforced hydrogel scaffold nanofiber by adding the polymer GelMA with an etched PCL nanofiber scaffold produced by melt electrospinning. These constructs have increased stiffness compared with hydrogels or nanofiber scaffolds alone approaching that of articular cartilage tissue. The scaffolds were also embedded with human chondrocytes and show viable responses, retaining their morphology, and can respond to biological regime in terms of matrix production and gene expression, making it a feasible material to culture cells in various environments mechanically diverse.

3.2 Control of Fiber Parameters for Tissue Engineering

With recent progress of cells interactions and tissue microenvironment, important investigations on the structural nanofiber scaffolds on different cells environment have been performed (Liu et al. 2012; Li et al. 2014). For the application intended, a control of the fiber parameters is important. Between these parameters, four majors, as pointed by a great recent article Xu et al. (2013), have been deeply investigated: fiber diameter, packing, orientation, and 3D shape.

Researchers have been trying to control the nanofiber diameter (Du et al. 2008) since it can regulate cell behavior influence like adhesion, proliferation, migration, differentiation, and protein adsorption (Christopherson et al. 2009). Currently, controllable diameter size of the electrospun fibers produced has a range of 150 nm–5 μm (Ishii et al. 2008; Nasouri et al. 2012). These ranges of fiber are in much smaller size than those produced by conventional electrospun methods including some cells which help facilitate contact guidance of cells (Wang et al. 2010a). Normally, the size controls are dependent on the polymer solution parameters (Zhang et al. 2005) and processing conditions (Deitzel et al. 2001).

The conventional electrospun method forms 3D tightly packed structures, and studies indicate that cells can only grow and migrate on the superficial surface, resulting in a 2D membrane rather than a 3D structure. Additionally, these scaffolds restrict cell infiltration and limit nutrient exchange. This effect results in loss of cells and the unsuccessful or partial regeneration of tissues (Li et al. 2014). To overcome this issue, the control of fiber packing is important when designing the structure for tissue engineering culture. One strategy is to increase the pore size and porous structure of these scaffolds (Baker et al. 2008; Kim et al. 2008). Various techniques to increase the porosity and pore sizes have been developed such as increasing the diameter of nanofibers (Rnjak-Kovacina and Weiss 2011), evaporating constituents of a mixed polymer after electrospun process (Wu et al. 2014). These allow a low packing density of nanofibers which helps the cells to infiltrate the scaffold.

Researchers are trying to obtain oriented fibers since the extracellular matrix in tissues of human body has anisotropic architectures, so the scaffold needs to display the same anisotropic behavior for the tissue application intended (Li et al. 2014). In addition, cell migration and extension are improved when nanofibers are oriented to a single-axis direction, such as neural cells, in which oriented nanofibers help direct the growth of axons in healing process on neural tissue engineering. However, random orientation is typical in electrospun fibers which are attributed to the bending instability related to the spinning jet. To overcome this, fiber orientation can be produced on electrospun fibers by using two grounded rods (Li et al. 2003) or a rotating drum (Kim and Reneker 1999). A rotating drum at high speed can be used to obtain aligned fibers, but their orientation is not perfect, and disk collectors have also been used (Xu et al. 2004). Usually, on this configuration the electric field is higher at the disk edge and the fibers are well aligned along the edge of the collector. Another technique is the use of two conductive strips separated by a void gap, and due to this void gap, which acts as an insulated region, it results in electrostatic interactions on the nanofibers, and they are stretched to form a parallel array across the gap.

Since then, different designs have been developed to improve this system such as conductive coil for the collector (Lee et al. 2016) and addition of finite-length hollow cylindrical electrode along the jet trajectory, to suppress the coil formation. Parallel-plate electrodes in between the collector and a cylindrical electrode to orientate the nanofiber scaffold at the collector have also been used (Karatay et al. 2014).

Furthermore, research is ongoing to produce fibers with similar morphologies and characteristics of the native tissue ECM. As an example, a nanotubular scaffold—that simulates the muscle layer of blood vessel while mimicking elasticity, mechanical strength, and high surface area—is possible to be obtained through electrospinning. Tubular electrospinning structures give aligned fibrous scaffolds with a high surface area that can induce the proliferation and adhesion of loads of cells for a faster and complete healing of tissues (Wang et al. 2014). In addition, recent studies have shown that patterned nanofibers in 3D tubular scaffolds can be produced easily, using designed collector templates. The collector template can be designed based on varying nanopatterns with different shapes, producing various nanofiber structures that can be tailored by the collector template (Daming and Jiang 2008). These specific controlling parameters tend to be important for specific tissue engineering.

3.2.1 Control of Fibers for Neural Tissue Engineering

Neural cells are affected by the 3D shape of electrospun membranes (Christopherson et al. 2009; He et al. 2010; Wang et al. 2010a), and its fiber diameter affects adhesion, proliferation, migration of rat hippocampus-derived adult neural stem cells (rNSC) (Christopherson et al. 2009; He et al. 2010; Li et al. 2014). Studies performed by (Meng et al. 2015) tried to simulate the ectoplasm of nerve (polyanionic nanofibrous cortical layer) by producing a tubular-shape PAA nanofiber hydrogels. The results from Meng et al. concludes that PAA fibrilar structure could be a potential candidate for aligned fibers that can mimic the cortical layer structure.

The effect of alignment was reported by several authors (He et al. 2010; Wang et al. 2010a; Hu et al. 2016b), and aligned fibers appear to enhance Schwann cell maturation more than randomly oriented fibers (Wang et al. 2010a).

Finally, hydrogels seem to affect the neural tissue. In the innovative work of Hodde et al. (2015), layers of oriented PCL nanofibers were embedded with fibrin hydrogel. Fibrin was selected because it could act as targeting tissue healing, depositing fibrin at the site of the injury and acting as extracellular matrix for migration and proliferation of perineural fibroblasts, Schwann cells, and regenerating axons during nerve healing (Weis et al. 1994). The hydrogel fibrin nanofiber structure was similar to the ECM of peripheral nerves. In addition, the fiber hydrogel on 2D structures retained the deficient simple flattened, unipolar morphologies of Schwann cells, where the 3D construct had a complex, highly branched morphology similar to neuron-like morphology, and the maximum outgrowth was observed after one day of incubation. In addition, Schwann cells were oriented to the direction of PCL fibers (Hodde et al. 2015).

3.2.2 Control of Fibers for Vascular Tissue Engineering

Alignment of fibers seems to improve the proliferation of vascular cells (Del Gaudio et al. 2009; Ma et al. 2012; Li et al. 2014; Shalumon et al. 2015; Ercolani et al. 2015). Shalumon et al. (2015) studied PLLA/gelatin electrospun aligned fibers in smooth muscle cells (SMCs), and its results indicate that aligned fibers support SMCs cells and improve the proliferation (Shalumon et al. 2015).

3D shape of electrospun fibers also seems to strongly affect vascular tissue cells (Ma et al. 2012; Merkle et al. 2015) with a preferable tubular shapes for the reason that it mimics the natural blood vessels (Stitzel et al. 2006).

Electrospun fibers produced in a core–shell design promote NIH 3T3 fibroblasts (FBs) and rat smooth muscle cells (rSMC) viability (Merkle et al. 2014, 2015). In addition, the packing of fibers impacts on the overall effect, and according to Zilla et al. (2007), tubular scaffolds must possess a porous structure with pore diameter of 10 μm in average and 20–80 μm^2 minimum area to allow penetration of cells so as to obtain a fast regeneration of vascular tissues (Zilla et al. 2007). Vascular tissue grafts also need to possess proper mechanical properties and a confluent endothelialized lumen to resist thrombosis (Lin 2011). Hydrogel-based systems such as hydrogel nanofiber scaffolds are generally more biocompatible in the peritoneum than hydrophobic polymeric devices (Yeo et al. 2007). Stefani et al. produced a novel hydrogel nanofiber with PCL with copolymer of acrylated poly ((l-lactide-co-trimethylene carbonate), aPLA-co-TMC, the mixing of the two incompatible polymers as melts with UV-cross-link produced core–shell tubular fibers and the scaffolds supported the migration, alignment, and proliferation of human mesenchymal stem cells (hMSCs). In addition, cells followed the fiber alignment; random fibers give raise to diverse migration and adhesion, and more spread and disorganized cell orientation, while aligned fibers promoted cell elongation with high organized and oriented pattern (Stefani and Cooper-White 2016).

3.2.3 Control of Fibers for Bone Tissue Engineering

Diameter of fibers influence bone forming cell behavior, Badami et al., investigated the use of electrospun mats on MC3T3-E1 cells and found that the optimum range of fibers diameters for cell proliferation is usually between 0.14 and 2.1 μm and they are likely to reproduce on larger diameter fibers; optimization of the fiber dimension helped in the regeneration of bone tissues (Badami et al. 2006). The fiber packing also affects bone tissue, and enhanced porosities significantly improve the cell penetration and distribution (Baker et al. 2008; Vaquette and Cooper-White 2011). The alignment of fibers takes an important place in bone tissue because the bone ECM is oriented in the same direction of the collagen nanofibers in basic structures of the bone (Cai et al. 2012). The orientation and pattern of the fibrous scaffold have to resemble the fibrous structure of the natural extracellular matrix (ECM) because it can improve osteoblast cell migration, proliferation and simulate the ECM structure (Wang et al. 2009a, 2010b).

Hydrogel nanofibers with PVA/gelatin followed by cross-linking based on methanol, for the improvement in the mechanical structure, show a well-spaced 3D structure with growth and proliferation of MG-63 (human osteosarcoma cells) (Linh and Lee 2012). Recently, Pangon et al. (2016) developed a chitosan/chitin whisker with HAp hydrogel nanofiber for the objective of promote bone cell response, and the results indicate that after the simulated body fluid (SBF) the samples were randomly mineralized with Ca-deficient HAp. In addition, the hydrogel presented non-toxicity to osteoblast cells with proliferation and viability enhanced by increasing the chitin whisker (Pangon et al. 2016).

3.2.4 Control of Fibers for Ligament and Tendon

Alignment of electrospun nanofibers structure is important on ligaments and tendons because it mimics the dimensionality of collagen fibrils that comprise native tendons and ligaments (Li et al. 2007; Choi et al. 2008). In this case, it is extremely important that the orientation of the nanofibers is optimized with the mechanical properties, since it has to mimic the native tissue structure. Deepthi et al. aligned PCL nanofibers coated with a hydrogel chitosan-hyaluronic acid in a cross-link of N,N-(3-dimethylaminopropyl)-N-ethyl carbodiimide (EDC) and lyophized by stacking multiple nanofiber mats. The cell results with rabbit ligament fibroblasts exhibited better migration, attachment, and proliferation along the direction of alignment of fibers with elongation in comparison with PCL random ones where cells just spread onto the fibers (Deepthi et al. 2015). However, the results from

Deepthi et al. (2015) showed that PCL fibers with random orientation infiltrated more onto the scaffolds owing to the increased porosity. However, ligament tissues require strong scaffolds and the random PCL fibers exhibit poor mechanical properties. In addition, cell arrangement along the direction of the acting force is a prerequisite for ligament tissue engineering and the mechanical strengths of most of the scaffolds presented in the literature are yet inferior compared to the ligament tissue (ACL) (Liu et al. 2012; Deepthi et al. 2015).

3.2.5 Control of Fibers for Skin Tissue Engineering

Skin tissue consists of multilayers categorized in terms of the epidermis, dermis, and hypodermis. These diverse tissue layers comprised of diverse cells and diverse functions, so complex structured scaffolds are needed for this tissue engineering. The effect of electrospun nanofibers has been extensively studied (Cui et al. 2008; Huang and Fu 2010). Studies have been showed that fibroblasts proliferated best at 350–1100 nm. Few results have been reported on the parameters for fiber orientation and fiber packing on skin regeneration (Li et al. 2014). Hydrogel nanofibers containing gellan/PVA and its effects on human dermal fibroblast (3T3L1) cells were investigated showing positive results (Vashisth and Pruthi 2016). Various similar results have been reported (Cui et al. 2008; Loh et al. 2010).

3.2.6 Control of Fibers for Cartilage Tissue Engineering

Cartilage wounds can advance to osteoarthritis and are acute challenges for regenerative medicine (McCullen et al. 2012; Steele et al. 2014). One of the unique features of cartilage tissues is that they have anisotropic mechanical properties due to differences in density and structural arrangements (Becerra et al. 2010). This leads to three distinct zones with collagen fibers varying their orientation, developing from aligned in the surface regions, to random and orientating perpendicular to deep zones (Steele et al. 2014). These profiles for cartilage zones contribute in terms of mechanical properties (Klein et al. 2009; Becerra et al. 2010).

Authors tried to mimic this cartilage unique zone behavior. Steele et al. developed trilayered electrospun fibers (Fig. 5) with scaffold nanofibers regions by using the same polymer with varying the alignment, packing, and diameter of fibers. The combination of the two distinct zones is designed to yield an anisotropic scaffold with a smooth articulating surface and a more porous region for ECM deposition. The morphological changes were able to provide desired functionality, in terms of mechanical stability, while impacting both chondrocyte gene expression and ECM accumulation. Hydrogels with PLA and chitosan have also been successfully reported for the regeneration of cartilage tissue (Mallick et al. 2016). The water adsorption of hydrogel nanofiber is a significant parameter during the chondrocyte culture and cartilage regeneration (Mallick et al. 2016). Chitosan has a similar structure than of glycosaminoglycan (GAG)—a natural biopolymer found in tissues

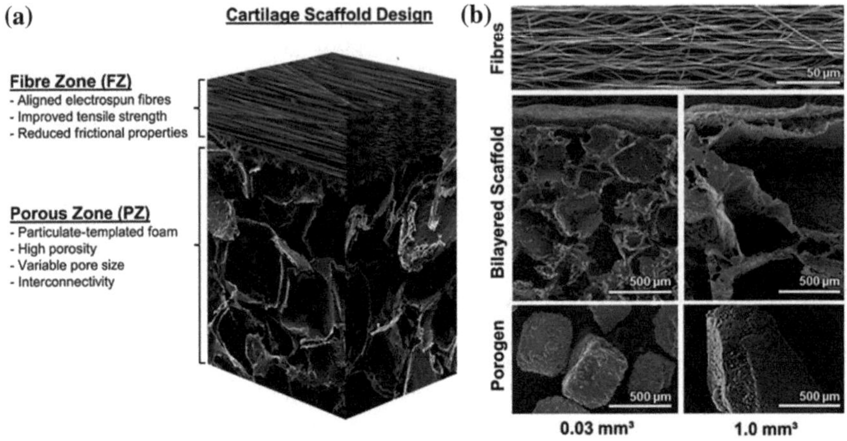

Fig. 5 Bilayered cartilage scaffold schematic. **a** A diagram illustrating the electrospun fiber zone (FZ) and a porous zone (PZ). **b** SEM images of (top) the aligned fiber zone that is shared between both scaffold varieties, (middle) the complete bilayered scaffolds with 0.03 mm^3 (left) and 1.0 mm^3 (right) pores and (bottom) the sodium chloride porogens used to produce higher porous zones. Reprinted from Steele et al. (2014)

and ECM. Results from Mallick et al. (2016) using nanofiber scaffolds for cartilage engineering show that rabbit chondrocytes can attach well and proliferate throughout the nanofibers. In addition, the glycosaminoglycan values resulted from chondrocyte culture from a specific design scaffold exhibited a desired enhancement in the GAG release, which the authors suggest it might be due to suitable modulation and maturation of the chondrocytes (Mallick et al. 2016).

3.3 *Dressings for Wound Healing*

Wounds, if exposed and untreated, are susceptible to water loss, toxins, and bacterial infections and could promote unnecessary and continued inflammatory response that limits the regeneration process. In this way, wound dressing materials are used for regeneration and repair of the dermal tissues, protecting it mainly against microorganism (Mogoşanu and Grumezescu 2014).

Ideal dressings must contain certain characteristics, including absorption ability of wound exudates, bacterial barrier, functional adhesion—which can adhere to healthy tissue but not adhere to wound tissue, ease of removal, and low cost (Thomas 1990; Bhardwaj and Kundu 2010).

Hydrogel nanofibers for wound dressing applications meet most of the requirements as dressing material due to their microfibrous and/or nanofibrous structure related to the electrospinning, producing a moist environment and helping the regeneration of skin with no scar (Alvarez et al. 1983). In addition, hydrogels have great ventilation ability and it can be suitable and efficient on absorption of contaminated exudates (Dumville et al. 2013). Furthermore, electrospun nanofibers can prevent and control microbial biofilms, cleanse the injured tissue, and eliminate/minimalize pain (Lee et al. 2003; Mogoşanu and Grumezescu 2014).

Xu et al. (2016) electrospun a mixture of chitosan/PLA and PEG, and the formed nanofibers were evaporated to remove the solvent at 60 °C. The cross-link of nanofibers was done with Glutaraldehyde vapor at room temperature followed by 0.1 M glycine aqueous solution to block unreacted aldehyde groups. These fibers demonstrated high wicking rates and equilibrium water absorptions. These fibers can hold excess exudates and create a wet wound healing environment for the wound. The swollen fibers reduced the size of the pores and permeability of air, but the cross-linked reaction allowed the ingress of oxygen and carbon dioxide. Finally, Xu et al. demonstrated that these hydrogel nanofiber mats showed good antibacterial activities (Xu et al. 2016).

3.3.1 Choice of Polymer for Wound Healing

Natural polymers are widely used for wound and burn dressing due to their biocompatible properties and similarity to ECM. Natural polymers can stimulate the healing process and repair the damaged tissues and skin regeneration (Huang and Fu 2010). Between the natural polymers, the most common used are cellulose—used in chronic wound dressing (Hunt et al. 1984; Montesano and Orci 1988), chitin and chitosan—anti-inflammatory and wound healing properties (Anitha et al. 2014; Mogoşanu and Grumezescu 2014), alginates—hemostatic properties in exudation/bleeding wounds and burns (Wang et al. 2002; Paul and Sharma 2004). Numerous hydrophilic polymers such as polyethylene glycol (PEG) have the potential hydrogel properties. Due to the 3D cross-linked networks, polymeric hydrogels are extensively used in pharmaceutical and biomedical area, tissue engineering, drug delivery (Peppas and Sahlin 1996; Peppas 2000; Samchenko et al. 2011). Therefore, hydrogels cross-linked natural polymers can be used for wound and burn dressings (Mogoşanu and Grumezescu 2014; Wolf et al. 2015). Synthetic polymers on the other hand are usually selected as carriers for drugs (Peppas 2000; Chaterji et al. 2007; Nguyen and Alsberg 2014; de Lima et al. 2015a, b) and can be designated for the choice as hydrogel blend.

3.4 Drug Delivery Applications

3.4.1 Methods of Drug Loading unto Electrospun fibers

Several methods can be used in order to incorporate antibacterial drugs into nanofibers.

1. One of the most common methods is to mix the drug with the polymer solution directly following the electrospun fibers. This method does not require any

additional steps and is one of the simplest systems to obtain effective nanofibers as long as the drug and the polymer lipophilic or hydrophilic interactions are considered (Zeng et al. 2005).

2. The drug is mixed with the polymer solution, as per method 1, and another polymer electrospun layer is added to encapsulate the drug and act as a shell. This method is particularly useful for sensitive drugs. In addition, shell layers improve the sustainable release of the drug and overcome the initial rush initial release typical on drug delivery systems of nanofibers produced with electrospinning. This method is effective for the delivery of active pharmaceutical ingredients and permits a biphasic drug release (Miao and Liu 2015). The effective prolonged release of core–shell structure is demonstrated by Zhang et al. (2006b), and core–shell nanofibers BSA loaded onto PEG slow the release of BSA for one month. In addition, this technique shows improved adhesion to fibers allowing uniform fiber structure and the protection of the drug in the early activity with the biological ambient (Miao and Liu 2015).
3. Another method is where nanoparticles of drug are loaded onto the polymer and finally electrospun forming the fiber matrix. Due to the specific properties that are obtained by some materials at nanoscale, such as silver nanoparticles (Zhao et al. 2012) and bioactive europium-doped hydroxyapatite (Hap:Eu3 +) (Hou et al. 2009), it is necessary to process these materials to obtain nanoparticles before processing the electrospinning.
4. Adsorption of the drug by immersing the fiber mats into a required amount of drug. This method can be achieved by using nanofiber mats that have a high surface, resulting in greater drug loading amount. This method allows immediate release of drugs from the surface of the scaffold and allows facile dosage control (Yoo et al. 2009). Chen et al. (2007) developed a PLA nanofiber immersed in TiO_2 nanoparticles. Subsequently, the authors immersed these nanofiber mats in daunorubicin drug, for the treatment of cancer. The AFM results show that after immersed in daunorubicin, the TiO_2 nanoparticles and drug packed together and formed spherical particles on PLA nanofibers. This surface incorporated with anticancer drug can adhere to the surface of the targeted cancer cells and approach the damaged cell surface slightly improving the metabolic system.
5. Immobilization of the drug using surface activation on the polymer mats and following immersion in a required amount of drug (Yoo et al. 2009). Immobilization of bioactive molecules on the exterior of electrospinning nanofibers is also performed to produce reactive functional groups. Owing to benefits of polymers process, greater selection of molecules with distinct biological features can be immobilized onto the nanofiber mat without compromising the bulk properties. Chemical immobilization reduces release rate of the biomolecule, but also allows a precise control (Goonoo et al. 2014). The immobilization can be performed either by physical and chemical methods (Yoo et al. 2009); however, it is usually done with chemical method since the drugs are covalently attached to nanofibers and they are not easily removed from this modified nanofiber mat when incubated over long periods of time. Zomer Volpato et al. (2012)

tested chitosan nanofiber mats where it was surface activated and coated with heparin-containing polyelectrolyte complex nanoparticles (PCN) which presents basic fibroblast growth factor (FGF-2). The PCN helps protect FGF-S for over 30 days, and it could be modulated.

6. Drug nanoparticles and the polymer solutions are electrospun in a side-by-side method to form a biphasic layer of fiber mats. In this method, two liquids (polymer and drug solutions) are loaded in two parallel metal capillaries. These form an interesting morphology behavior (Liu et al. 2007). Such fibers are also called Janus fibers (Yu et al. 2016). However, with different nozzles, the fiber is formed with the junction of the drug and the polymer. In addition, the drug is trapped between each fiber connection.

Yu et al. (2016) tested a new design of using a Teflon barrier between the nozzles in the side-by-side method, and it was possible to observe for the same fiber two different morphologies with a poorly soluble drug amorphously distributed, with biphasic controlled release and achieving an initial burst and a slower sustained release phase.

All these methods are different approaches that can be used to incorporate various active pharmaceutical ingredients (API) in electrospun fibers that can control the release profile through changes in the fibers morphology, porosity, and composition.

3.4.2 Choice of Polymer/Hydrogels for Nanofibers in Drug Delivery Systems

Target-drug delivery nanofibers can be produced by using matrices with either biodegradable or non-degradable polymers. The drug release mechanism may vary depending of the type of polymer, diffusion for non-degradable and matrix erosion for biodegradable (Pillay et al. 2013). The attainable delivery of the drug can be achieved depending on the polymer used. However, the parameters polymer type, solvent, and drug compatibility are important process variables when designing stable nanofibers (Zeng et al. 2005; Pillay et al. 2013). However, recent studies have shown that it is possible to design hydrophobic drug in electrospun carriers (Laha et al. 2016) without losing the properties of the polymer matrix.

The release of drug in nanofibers produced by electrospinning is controlled by diffusion of drug and/or degradation of the polymer matrix (Cui et al. 2006; Loh et al. 2010; Laha et al. 2016). However, Laha et al. (2016) tested the effect of cross-link on hydrophobic drug-loaded gelatin nanofibers. Laha et al. (2016) found that nanofibers in the polymer matrix, without cross-linking, usually led to deficient exchanges with the poorly water-soluble drug molecules and as such, the hydrophilic nanofiber mat result in rapid release of drug within few hours. When cross-link was achieved on the nanofiber mats, their swelling nature and osmotic behavior provided the principal mechanism for delivery of the hydrophobic drug in the medium and consequently, even at longer periods of time, and there was sustained release of drug as it diffused to the release medium through the carrier gradually (Laha et al. 2016).

3.4.3 Electrospun Nanofibers that Contain Natural Products for Drug Delivery Systems and Tissue Engineering Applications

Recently, blended nanofibers with plant-derived natural biomaterials have gathered great interest since it is possible to incorporate the curative, regenerative, antimicrobial, and anti-inflammatory properties of plants with the incorporation of good mechanical properties and slow biodegradation of synthetic polymers. One of the advantages of blending plant-derived materials into polymers is that plants have usually large molecular sizes and most compounds are unable to cross the lipid membranes of the cells which results in poor adsorption and loss of bioavailability and efficacy (Venugopal et al. 2014). Electrospun fibers deliver the active ingredients of the plants at sufficient concentration during the entire treatment period to the host site (Venugopal et al. 2014).

Many plants have been tested in electrospun nanofibers for drug delivery, and as example there is *Aloe vera*—antioxidant having good medicinal properties for tissue engineering. Nanofibers with *Aloe vera* show that human dermal fibroblasts can have better and faster attachment, proliferation, and guided growth (Tam et al. 2014). Moreover, it can accelerate the healing of open wounds in type 2 diabetic radiation-exposed rats (Venugopal et al. 2014).

Asian ginseng (Panax ginseng root) has numerous applications for central nervous systems, cardiovascular and human skin applications (Lee et al. 2007). Panax ginseng extracts can also promote collagen in human dermal fibroblast cells (Lee et al. 2007; Pajoumshariati et al. 2015) and positive effect on osteogenesis and cell proliferation. On nanofiber scaffolds, (Pajoumshariati et al. 2015) results indicate that ginseng extracts show an improvement in cell attachment and proliferation, and it also enhanced the MSCs osteogenic differentiation with high level of calcium content deposited on the surface of fibers which shows a potential candidate for bone tissue engineering.

The range of nanofibers medicinal plant is vast, and this field has enormous potential.

3.4.4 Electrospun Nanofibers with Drugs for Tissue Engineering Applications

In addition to bone regeneration that nanofibers can induce, drug loading can be added. This is of interest since the number of patients with infections is rising owing to the risk of bacterial contamination on the implant (Zimmerli et al. 2004). Infections associated with guided tissue regeneration (GTR) and guided bone regeneration (GBR) implants are mainly caused by anaerobic bacterial infections (Ulubayram et al. 2015). In this way, nanofiber scaffolds have been currently used as an anti-infection barrier with loadings of various different drugs such as metronidazole (MNA) and silver nanoparticles, and since the barrier had a sustainable release of those drugs, the membranes maintained its antibacterial effect for a long term (Xue et al. 2014). In addition, recent studies have shown that it is also possible to obtain a biphasic release kinetics with good antibiotic encapsulation ($\sim$75–100%) on PLGA electrospun fibers; the researchers used 10% (w;w) fusidic acid and 5% (w;w) rifampicin which help reduce the number of adherent bacteria by 99.9% in an in vivo rodent model of implant associated infection (Gilchrist et al. 2013; Ulubayram et al. 2015).

3.4.5 Drug-Encapsulated Nanofibers for Wound Recovery

Nanofiber scaffolds and hydrogel nanofibers have been successfully employed as vehicles for antibacterial agents for wound recovery (Lee and Yoo 2008; Choi et al. 2015; Vashisth et al. 2016). Antimicrobial agents are preferred to tailor the role in the wound healing process, and preferably such products usually must protect against gram-negative, gram-positive, and antibiotic-resistant bacteria. One natural product that has protection from gram-negative and gram-positive is propolis (de Lima et al. 2015b), and from our knowledge, it has not been overlooked as nanofiber scaffolds. On the other hand, strong antibiotics have been encapsulated into the nanofiber scaffolds, such as neomycin (Nitanan et al. 2013), ampicillin (Sabitha and Rajiv 2015), ciprofloxacin (Canillas et al. 2015). Cefixime is a very effective antibactericidal antibiotic for *Escherichia coli* and *Staphylococcus aureus* bacterial strains, which are very common in wound infections (Bergeron and Turcotte 1986; Arshad et al. 2012). Shahzad et al. produced nanofiber mats of chitosan, PVA, and HAp with Cefixime for wound healing, and the nanofibers were cross-linked via freeze-thawed method, freezing the mats at −80 °C for 24 h; freeze dried to form porous scaffolds and finally heat treated at 80 °C for 10 min. Shahzad et al. demonstrated that heat treatment affects the structure of the mats, has excellent interconnected porous structure, high swelling capabilities with sustained release of Cefixime, inhibition against *S. aureus* and *E. coli* and cytocompatibility with VERO cell line (Shahzad et al. 2015). Ciprofloxacin has also been studied with PVA and PVA/PAA hydrogel structures for the objective of reducing infection against osteomyelitis (Canillas et al. 2015); however, its investigation still needs to be analyzed for nanofibers.

4 Recent Strategy Developments in Electrospinning for Drug Delivery

Although many works have been described in terms of drug delivery on wound healing and tissue engineering, it still needs more investigation in terms of poorly water-soluble drugs and localized targeting delivery. The effect of initial burst release also needs to be improved, so the next session describes innovations on recent fronts of targeting and poorly water-soluble drugs.

4.1 Multilayered Nanofibers

The ability to engineer drug release produced by nanofibers can also be improved or modified by the polymers matrix. If the drug is incorporated in separate nanofiber matrix, the diffusion is different and can promote a controlled release of the drug (Wang et al. 2010d; Huang et al. 2012). In this manner, complex nanofiber mats can be produced with various polymers to attain multiphased drug delivery. In this configuration, after the first layer of electrospun is deposited with the first polymer, another polymer is sequentially deposited on the same collector. This process can be repeated multiple times to produce meshes of multilayered fibers with ordered structure and different kinds of polymer. Recently, controlled release systems for oral delivery of poorly water-soluble drug have been studied with multilayered nanofibers, such as ketoprofen (KET) incorporated in a trilayered electrospun with two different polymers (Huang et al. 2012).

In addition, a recently new triaxial electrospinning strategy has been adopted with a gradual layered structure mesh that is dependent on the various working fluids containing varied concentrations of the drug which was designed to incrementally increase the content of drug moving from the exterior of the fibers inwards resulting in a gradient distribution of the drug—linear release of KET. By incorporating this formulation into an enteric-coated capsule, a linear release colon-targeted oral drug delivery system can be produced as shown in Fig. 6.

4.2 Hollow and Core–Shell Electrospun Fibers

In contrast to solid nanofibers, the hollow cores have interesting advantages in terms of higher specific surface area, lower density, and multiphase interfaces (Wu et al. 2011). Hollow spheres as compared to normal nanoscaffolds have gathered interest as drug delivery systems owing to this hollow core can encapsulate large amounts of drugs. In addition, it also controls the release rate, to prevent the initial burst and short-term release that occurs in solid nanofibers (Wang et al. 2010c). Wu et al. created a hollow hydroxyapatite fiber for protein delivery systems, the hydroxyapatite nanofibers produced an ultrafine fiber diameter with interconnected pores providing large specific surface area and demonstrated great protein adsorption ability with long-term sustained release. However, hollow fibers have poor mechanical properties (Kang et al. 2015). To overcome the deficiencies of this technique, core–shell nanofibers have been formed on the exterior of the hollow

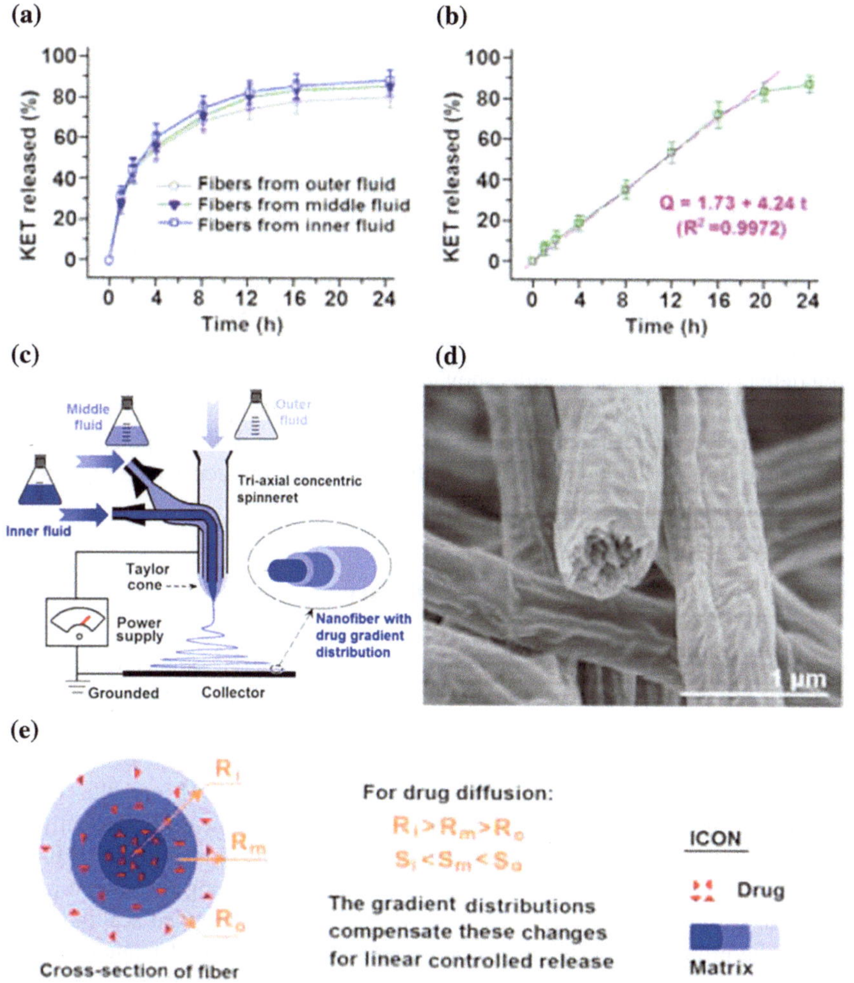

Fig. 6 **a** In vitro dissolution test results for KET incorporated in each monolithic nanofibers and **b** the tri-layer nanofibers, **c** a schematic of the triaxial electrospinning process **d** FESEM images of the tri-layer nanofibers after release of all the KET loading, and **e** a diagram explaining how the gradient drug distribution can yield a linear release profile. Adapted with permission from Yu et al. (2015). Copyright (2016) American Chemical Society

fibers (Kang et al. 2015). The principle behind using core–shell fibers as a layer on hollow fibers was that since large quantities of drugs can be encapsulated onto the hollow fibers, it might be useful to protect them, in which the core–shell technique preserves the biological activity of the drug (Su et al. 2012; Zhang et al. 2013).

4.3 *Patterned and Alignment Structures of Nanofibers*

There are several methods to produce nanofibers with controllable patterns (Zhang and Chang 2007), and as a simple method, it is based on the theory that the orientation of nanofibers is affected by the topography of the collector, so a patterned nanofibrous scaffold can be generated by using a patterned conductive collector (Zhang and Chang 2007; Li et al. 2014). Alternating the patterned collector and distribution, nanofibers can be assembled into well-ordered nanofiber meshes with both types of topographies, random and parallel alignment existing in the same mesh (Daming and Jiang 2008; Wang et al. 2009b).

The orientation effect and pattern on nanofiber have also been investigating for drug delivery applications. Results show that the drug delivery comportment can be adjusted by the surface wettability of the drug carrier. Meng et al. (2011) showed that aligned scaffolds of nanofibers had a lower release rate compared to radon orientation, suggesting that alignment of fibers could also influence the drug release.

5 Conclusion and Future Perspective

Nanofibers produced by electrospinning are versatile materials that can have many applications in the biomedical field. Nevertheless, polymer jets are relatively unstable, so controlling the solution and electrospinning parameters is vitally important to obtain uniform nanofibers.

The electrospun nanofiber scaffolds mimic the natural ECM, and this feature suggests that the materials are ideal candidates for tissue engineering. Presently, the focus and specific knowledge on different tissues of the body is helping for further progression in this field immensely. Such knowledge is helping to achieve regeneration of tissues at a relatively fast rate. Nonetheless, the designed scaffold needs appropriate properties for the application intended, so a deep understanding of the nanofibrous scaffold material is crucial to application of the technologies. Currently, the blend of natural with synthetic polymers offers the appropriate properties for biomedical applications. Furthermore, the incorporation of nanofibers with hydrogels seems to be a solution for the current problems and challenges in this particular field.

The control of the nanofiber diameter, packing, orientation, and 3D shape impacts on different tissues environment. This will direct effect cell behaviors such as attachment, orientation, proliferation, and migration. For example, cartilage has three different layers and requires a complex nanofibrous scaffold with varying porosities and alignment. However, this field still needs further investigation for better understanding on how tissue regeneration is impacted by physiological signals.

Nanofibers can also incorporate drugs by different methods, and it can not only help in terms of tissue engineering by creating antimicrobial barriers (Huang et al. 2012; Xue et al. 2014) but also help on wound healing. Hydrogel nanofibers are suitable candidates for wound healing applications, and their swelling ability coupled with permeability for gaseous transfers can promote a perfect condition for healing environment. In this manner, the ability to encapsulate drugs and control the release rate can reduce infections. As technology advances and studies progress, novel techniques for incorporation of drugs in specific target delivery will be at the forefront. The field of drug delivery on nanofibers impacts on the main current problems on biomedical field such as poorly water-soluble drugs and biofilms of bacteria (Ahire et al. 2015; Paaver et al. 2015).

References

Abdal-Hay A, Hussein KH, Casettari L et al (2016) Fabrication of novel high performance ductile poly(lactic acid) nanofiber scaffold coated with poly(vinyl alcohol) for tissue engineering applications. Mater Sci Eng, C 60:143–150. https://doi.org/10.1016/j.msec.2015.11.024

Ahire JJ, Neveling DP, Hattingh M, Dicks LMT (2015) Ciprofloxacin-Eluting nanofibers inhibits biofilm formation by Pseudomonas aeruginosa and a methicillin-resistant Staphylococcus aureus. PLoS ONE 10:1–13. https://doi.org/10.1371/journal.pone.0123648

Alvarez Perez MA, Guarino V, Cirillo V, Ambrosio L (2012) In vitro mineralization and bone osteogenesis in poly(ε-caprolactone)/gelatin nanofibers. J Biomed Mater Res A 100:3008–3019. https://doi.org/10.1002/jbm.a.34233

Alvarez OM, Mertz PM, Eaglstein WH (1983) The effect of occlusive dressings on collagen synthesis and re-epithelialization in superficial wounds. J Surg Res 35:142–148

Anitha A, Sowmya S, Kumar PTS et al (2014) Chitin and chitosan in selected biomedical applications. Prog Polym Sci. https://doi.org/10.1016/j.progpolymsci.2014.02.008

Araujo JV, Martins A, Leonor IB, et al (2008) Surface controlled biomimetic coating of polycaprolactone nanofiber meshes to be used as bone extracellular matrix analogues. J Biomater Sci Polym Ed 19:1261–78. https://doi.org/10.1163/156856208786052335

Arshad HM, Mohiuddin OA, Bilal M (2012) Comparative in vitro antibacterial analysis of different brands of cefixime against clinical isolates of Staphylococcus aureus and Escherichia coli. 02:109–113

Badami AS, Kreke MR, Thompson MS et al (2006) Effect of fiber diameter on spreading, proliferation, and differentiation of osteoblastic cells on electrospun poly(lactic acid) substrates. Biomaterials 27:596–606. https://doi.org/10.1016/j.biomaterials.2005.05.084

Baker BM, Gee AO, Metter RB et al (2008) The potential to improve cell infiltration in composite fiber-aligned electrospun scaffolds by the selective removal of sacrificial fibers. Biomaterials 29:2348–2358. https://doi.org/10.1016/j.biomaterials.2008.01.032

Becerra J, Andrades JA, Guerado E et al (2010) Articular cartilage: structure and regeneration. Tissue Eng Part B Rev 16:617–627

Berger FM, Ludwig BJ, Wielich KH (1953) The hydrophilic and acid binding properties of alginates. Am J Dig Dis 20:39–42

Berger J, Reist M, Mayer JM et al (2004) Structure and interactions in covalently and ionically crosslinked chitosan hydrogels for biomedical applications. Eur J Pharm Biopharm 57:19–34. https://doi.org/10.1016/S0939-6411(03)00161-9

Bergeron MG, Turcotte A (1986) penetration of cefixime into fibrin clots and in vivo efficacy against *Escherichia coli*, *Klebsiella pneumoniae*, and *Staphylococcus aureus*. Antimicrob Agents Chemother 30:913–916

Bhardwaj N, Kundu SC (2010) Electrospinning: a fascinating fiber fabrication technique. Biotechnol Adv 28:325–347. https://doi.org/10.1016/j.biotechadv.2010.01.004

Binulal NS, Natarajan A, Menon D et al (2014) PCL-gelatin composite nanofibers electrospun using diluted acetic acid-ethyl acetate solvent system for stem cell-based bone tissue engineering. J Biomater Sci Polym Ed 25:325–340. https://doi.org/10.1080/09205063.2013.859872

Bock N, Dargaville TR, Woodruff MA (2012) Electrospraying of polymers with therapeutic molecules: state of the art. Prog Polym Sci 37:1510–1551. https://doi.org/10.1016/j.progpolymsci.2012.03.002

Bosworth LA, Turner LA, Cartmell SH (2013) State of the art composites comprising electrospun fibres coupled with hydrogels: a review. Nanomed Nanotechnol Biol Med 9:322–335. https://doi.org/10.1016/j.nano.2012.10.008

Brown TD, Dalton PD, Hutmacher DW (2011) Direct writing by way of melt electrospinning. Adv Mater 23:5651–5657. https://doi.org/10.1002/adma.201103482

Cai YZ, Zhang GR, Wang LL, et al (2012) Novel biodegradable three-dimensional macroporous scaffold using aligned electrospun nanofibrous yarns for bone tissue engineering. J Biomed Mater Res A 100:1187–1194. https://doi.org/10.1002/jbm.a.34063

Canillas M, de Lima GG, Rodríguez MA, et al (2015) Bioactive composites fabricated by freezing-thawing method for bone regeneration applications. J Polym Sci Part B Polym Phys. https://doi.org/10.1002/polb.23974

Casper CL, Stephens JS (2004) Controlling surface morphology of electrospun polysterene fibers: effect of humidity and molecular weight in electrospinning process. Macromolecules 37:573–578

Chan CK, Liao S, Li B et al (2009) Early adhesive behavior of bone-marrow-derived mesenchymal stem cells on collagen electrospun fibers. Biomed Mater 4:035006. https://doi.org/10.1088/1748-6041/4/3/035006

Chaterji S, Kwon IK, Park K (2007) Smart polymeric gels: redefining the limits of biomedical devices. Prog Polym Sci 32:1083–1122. https://doi.org/10.1016/j.progpolymsci.2007.05.018

Chen C, Lv G, Pan C et al (2007) Poly(lactic acid) (PLA) based nanocomposites—a novel way of drug-releasing. Biomed Mater 2:L1–L4. https://doi.org/10.1088/1748-6041/2/4/L01

Chen G, Lv Y (2015) Immobilization and application of electrospun nanofiber scaffold-based growth factor in bone tissue engineering. Curr Pharm Des 21:1967–1978

Choi JS, Kim HS, Yoo HS (2015) Electrospinning strategies of drug-incorporated nanofibrous mats for wound recovery. Drug Deliv Transl Res 5:137–145. https://doi.org/10.1007/s13346-013-0148-9

San Choi J, Lee SJ, Christ GJ, et al (2008) The influence of electrospun aligned poly (ε-caprolactone)/collagen nanofiber meshes on the formation of self-aligned skeletal muscle myotubes. Biomaterials 29:2899–2906. https://doi.org/10.1016/j.biomaterials.2008.03.031

Christopherson GT, Song H, Mao HQ (2009) The influence of fiber diameter of electrospun substrates on neural stem cell differentiation and proliferation. Biomaterials 30:556–564. https://doi.org/10.1016/j.biomaterials.2008.10.004

Chua K-N, Chai C, Lee P-C et al (2006) Surface-aminated electrospun nanofibers enhance adhesion and expansion of human umbilical cord blood hematopoietic stem/progenitor cells. Biomaterials 27:6043–6051. https://doi.org/10.1016/j.biomaterials.2006.06.017

Chuachamsai A, Lertviriyasawat S, Danwanichakul P (2008) Spinnability and defect formation of chitosan/ poly vinyl alcohol electrospun nanofibers. Water 13:24–29

Cosme JGL, Silva VM, Nunes RRC, Picciani PHS (2016) Development of biobased poly (lactic acid)/epoxidized natural rubber blends processed by electrospinning: morphological, Structural and thermal properties. 210–219

Cui W, Jin Y, Zhu X, Li X (2008) Electrospun fibrous mats with high porosity as potential scaffolds for skin tissue engineering. Biomacromol 9:1795–1801. https://doi.org/10.1021/bm800476u

Cui W, Li X, Zhu X et al (2006) Investigation of drug release and matrix degradation of electrospun poly (DL-lactide) fibers with paracetanol inoculation. Biomacromol 7:1623–1629. https://doi.org/10.1021/bm060057z

Daming Z, Jiang C (2008) Electrospinning of three-dimensional nanofibrous tubes with controllable architectures. Nano Lett 8:3283–3287. https://doi.org/10.1021/nl801667s

de Lima GG, De Souza RO, Bozzi AD et al (2015a) Extraction method plays critical role in antibacterial activity of propolis-loaded hydrogels. J Pharm Sci 105:1248–1257. https://doi.org/10.1016/j.xphs.2015.12.027

de Lima GG, Kanwar D, Macken D, et al (2015) Smart hydrogels: therapeutic advancements in hydrogel technology for smart drug delivery applications. In: Thakur VK, Thakur MK (eds) Handbook of polymers for pharmaceutical technologies, 1st edn. Wiley, New York, pp 1–16

De Vrieze S, Van Camp T, Nelvig A et al (2009) The effect of temperature and humidity on electrospinning. J Mater Sci 44:1357–1362. https://doi.org/10.1007/s10853-008-3010-6

Deepthi S, Jeevitha K, Nivedhitha Sundaram M et al (2015) Chitosan-hyaluronic acid hydrogel coated poly(caprolactone) multiscale bilayer scaffold for ligament regeneration. Chem Eng J 260:478–485. https://doi.org/10.1016/j.cej.2014.08.106

Deitzel J, Kleinmeyer J, Harris D, Beck Tan N (2001) The effect of processing variables on the morphology of electrospun nanofibers and textiles. Polymer (Guildf) 42:261–272. https://doi.org/10.1016/S0032-3861(00)00250-0

Del Gaudio C, Bianco A, Folin M et al (2009) Structural characterization and cell response evaluation of electrospun PCL membranes: micrometric versus submicrometric fibers. J Biomed Mater Res A 89A:1028–1039. https://doi.org/10.1002/jbm.a.32048

Delmar K, Bianco-Peled H (2016) Composite chitosan hydrogels for extended release of hydrophobic drugs. Carbohydr Polym 136:570–580. https://doi.org/10.1016/j.carbpol.2015.09.072

Du J, Shintay S, Zhang X (2008) Diameter control of electrospun polyacrylonitrile/iron acetylacetonate ultrafine nanofibers. J Polym Sci, Part B: Polym Phys 46:1611–1618

Dumville JC, O'Meara S, Deshpande S, Speak K (2013) Hydrogel dressings for healing diabetic foot ulcers

Eda G, Shivkumar S (2007) Bead-to-fiber transition in electrospun polystyrene. J Appl Polym Sci 106:475–487

Ercolani E, Del Gaudio C, Bianco A (2015) Vascular tissue engineering of small-diameter blood vessels: reviewing the electrospinning approach. J Tissue Eng Regen Med 9:861–888. https://doi.org/10.1002/term.1697

Felice B, Prabhakaran MP, Zamani M et al (2015) Electrosprayed poly(vinyl alcohol) particles: preparation and evaluation of their drug release profile. Polym Int 64:1722–1732. https://doi.org/10.1002/pi.4972

Fogaça R, Catalani LH (2013) PVP hydrogel membranes produced by electrospinning for protein release devices. Soft Mater 11:61–68. https://doi.org/10.1080/1539445X.2011.580411

Fong H, Fong H, Chun I et al (1999) Beaded nano bers formed during electrospinning. Polymer (Guildf) 40:4585–4592

Franco RA, Min YK, Yang HM, Lee BT (2012) On stabilization of PVPA/PVA electrospun nanofiber membrane and its effect on material properties and biocompatibility. J Nanomater. https://doi.org/10.1155/2012/393042

Frenot A, Chronakis IS (2003) Polymer nanofibers assembled by electrospinning. Curr Opin Colloid Interface Sci 8:64–75

Geng X, Kwon OH, Jang J (2005) Electrospinning of chitosan dissolved in concentrated acetic acid solution. Biomaterials 26:5427–5432. https://doi.org/10.1016/j.biomaterials.2005.01.066

Gilchrist SE, Lange D, Letchford K et al (2013) Fusidic acid and rifampicin co-loaded PLGA nanofibers for the prevention of orthopedic implant associated infections. J Control Release 170:64–73. https://doi.org/10.1016/j.jconrel.2013.04.012

Goonoo N, Bhaw-Luximon A, Jhurry D (2014) Drug loading and release from electrospun biodegradable nanofibers. J Biomed Nanotechnol 10:2173–2199. https://doi.org/10.1166/jbn.2014.1885

Greiner A, Wendorff JH (2007) Electrospinning: a fascinating method for the preparation of ultrathin fibers. Angew Chemie—Int Ed 46:5670–5703. https://doi.org/10.1002/anie.200604646

Gualandi C, Torricelli P, Panzavolta S et al (2016) An innovative co-axial system to electrospin in situ crosslinked gelatin nanofibers. Biomed Mater 11:025007. https://doi.org/10.1088/1748-6041/11/2/025007

Guarino V, Cirillo V, Altobelli R, Ambrosio L (2015) Polymer-based platforms by electric field-assisted techniques for tissue engineering and cancer therapy. Expert Rev Med Devices 12:113–129. https://doi.org/10.1586/17434440.2014.953058

Gunn J, Zhang M (2010) Polyblend nanofibers for biomedical applications: perspectives and challenges. Trends Biotechnol 28:189–197. https://doi.org/10.1016/j.tibtech.2009.12.006

Gupta P, Elkins C, Long TE, Wilkes GL (2005) Electrospinning of linear homopolymers of poly (methyl methacrylate): exploring relationships between fiber formation, viscosity, molecular weight and concentration in a good solvent. Polymer (Guildf) 46:4799–4810. https://doi.org/10.1016/j.polymer.2005.04.021

Hager BL, Berry GC (1982) Moderately concentrated solutions of polystyrene. I. Viscosity as a function of concentration, temperature, and molecular weight. J Polym Sci Polym Phys Ed 20:911–928

Hassan CM, Peppas N (2000) Structure and morphology of freeze/thawed PVA hydrogels. Macromolecules 33:2472–2479. https://doi.org/10.1021/ma9907587

He L, Liao S, Quan D et al (2010) Synergistic effects of electrospun PLLA fiber dimension and pattern on neonatal mouse cerebellum C17.2 stem cells. Acta Biomater 6:2960–2969. https://doi.org/10.1016/j.actbio.2010.02.039

Hodde D, Gerardo-Nava J, Wöhlk V et al (2015) Characterisation of cell-substrate interactions between Schwann cells and three-dimensional fibrin hydrogels containing orientated nanofibre topographical cues. Eur J Neurosci 43:376–387. https://doi.org/10.1111/ejn.13026

Hou Z, Yang P, Lian H et al (2009) Multifunctional hydroxyapatite nanofibers and microbelts as drug carriers. Chem—A Eur J 15:6973–6982. https://doi.org/10.1002/chem.200900269

Hu J, Li H-Y, Williams GR et al (2016a) Electrospun poly(N-isopropylacrylamide)/ethyl cellulose nanofibers as thermoresponsive drug delivery systems. J Pharm Sci 105:1104–1112. https://doi.org/10.1016/S0022-3549(15)00191-4

Hu J, Tian L, Prabhakaran MP et al (2016b) Fabrication of nerve growth factor encapsulated aligned poly (ε-caprolactone) nanofibers and their assessment as a potential neural tissue engineering scaffold. Polymers (Basel) 8:54

Huang C, Chen S, Lai C et al (2006) Electrospun polymer nanofibres with small diameters. Nanotechnology 17:1558

Huang LY, Branford-White C, Shen XX et al (2012) Time-engineeringed biphasic drug release by electrospun nanofiber meshes. Int J Pharm 436:88–96. https://doi.org/10.1016/j.ijpharm.2012.06.058

Huang S, Fu X (2010) Naturally derived materials-based cell and drug delivery systems in skin regeneration. J Control Release 142:149–159. https://doi.org/10.1016/j.jconrel.2009.10.018

Hunt TK, Knighton DR, Thakral KK et al (1984) Studies on inflammation and wound healing: angiogenesis and collagen synthesis stimulated in vivo by resident and activated wound macrophages. Surgery 96:48–54

Husain O, Lau W, Edirisinghe M, Parhizkar M (2016) Investigating the particle to fibre transition threshold during electrohydrodynamic atomization of a polymer solution. Mater Sci Eng, C 65:240–250. https://doi.org/10.1016/j.msec.2016.03.076

Ishii Y, Sakai H, Murata H (2008) A new electrospinning method to control the number and a diameter of uniaxially aligned polymer fibers. Mater Lett 62:3370–3372. https://doi.org/10.1016/j.matlet.2008.03.038

Ito Y, Hasuda H, Kamitakahara M et al (2005) A composite of hydroxyapatite with electrospun biodegradable nanofibers as a tissue engineering material. J Biosci Bioeng 100:43–49. https://doi.org/10.1263/jbb.100.43

Jalaja K, Naskar D, Kundu SC, James NR (2016) Potential of electrospun core-shell structured gelatin-chitosan nanofibers for biomedical applications. Carbohydr Polym 136:1098–1107. https://doi.org/10.1016/j.carbpol.2015.10.014

Jamadi ES, Ghasemi-Mobarakeh L, Morshed M et al (2016) Synthesis of polyester urethane urea and fabrication of elastomeric nanofibrous scaffolds for myocardial regeneration. Mater Sci Eng, C 63:106–116. https://doi.org/10.1016/j.msec.2016.02.051

Kai D, Jin G, Prabhakaran MP, Ramakrishna S (2013) Electrospun synthetic and natural nanofibers for regenerative medicine and stem cells. Biotechnol J 8:59–72. https://doi.org/10.1002/biot.201200249

Kai D, Liow SS, Loh XJ (2015) Biodegradable polymers for electrospinning: towards biomedical applications. Mater Sci Eng, C 45:659–670. https://doi.org/10.1016/j.msec.2014.04.051

Kang MS, Kim JH, Singh RK et al (2015) Therapeutic-designed electrospun bone scaffolds: mesoporous bioactive nanocarriers in hollow fiber composites to sequentially deliver dual growth factors. Acta Biomater 16:103–116. https://doi.org/10.1016/j.actbio.2014.12.028

Karatay O, Dogan M, Uyar T et al (2014) An alternative electrospinning approach with varying electric field for 2-D-aligned nanofibers. IEEE Trans Nanotechnol 13:101–108. https://doi.org/10.1109/TNANO.2013.2293704

Khadka DB, Haynie DT (2012) Protein- and peptide-based electrospun nanofibers in medical biomaterials. Nanomed Nanotechnol Biol Med 8:1242–1262. https://doi.org/10.1016/j.nano.2012.02.013

Khorshidi S, Solouk A, Mirzadeh H, et al (2015) A review of key challenges of electrospun scaffolds for tissue-engineering applications

Ki CS, Baek DH, Gang KD et al (2005) Characterization of gelatin nanofiber prepared from gelatin-formic acid solution. Polymer (Guildf) 46:5094–5102. https://doi.org/10.1016/j.polymer.2005.04.040

Kim ES, Kim SH, Lee CH (2010) Electrospinning of polylactide fibers containing silver nanoparticles. Macromol Res 18:215–221. https://doi.org/10.1007/s13233-010-0316-4

Kim J, Reneker DH (1999) Polybenzimidazole nanofiber produced by electrospinning. Polym Eng Sci 39:849–854. https://doi.org/10.1002/pen.11473

Kim KH, Jeong L, Park HN et al (2005) Biological efficacy of silk fibroin nanofiber membranes for guided bone regeneration. J Biotechnol 120:327–339. https://doi.org/10.1016/j.jbiotec.2005.06.033

Kim TG, Chung HJ, Park TG (2008) Macroporous and nanofibrous hyaluronic acid/collagen hybrid scaffold fabricated by concurrent electrospinning and deposition/leaching of salt particles. Acta Biomater 4:1611–1619. https://doi.org/10.1016/j.actbio.2008.06.008

Klein TJ, Malda J, Sah RL, Hutmacher DW (2009) Tissue engineering of articular cartilage with biomimetic zones. Tissue Eng Part B Rev 15:143–157

Laha A, Yadav S, Majumdar S, Sharma CS (2016) In-vitro release study of hydrophobic drug using electrospun cross-linked gelatin nanofibers 105:481–488

Lai G-J, Shalumon KT, Chen S-H, Chen J-P (2014) Composite chitosan/silk fibroin nanofibers for modulation of osteogenic differentiation and proliferation of human mesenchymal stem cells. Carbohydr Polym 111:288–297. https://doi.org/10.1016/j.carbpol.2014.04.094

Langer R, Peppas NA (2003) Advances in biomaterials, drug delivery, and bionanotechnology. AIChE J 49:2990–3006

Lanza R, Langer R, Vacanti JP (2011) Principles of tissue engineering. Academic press

Larrondo L, St. John Manley R (1981) Electrostatic fiber spinning from polymer melts. I. Experimental observations on fiber formation and properties. J Polym Sci Polym Phys Ed 19:909–920

Lee J, Jung E, Lee J et al (2007) Panax ginseng induces human Type I collagen synthesis through activation of Smad signaling. J Ethnopharmacol 109:29–34. https://doi.org/10.1016/j.jep.2006.06.008

Lee JH, Shin DW, Nam KB et al (2016) Continuous bundles of aligned electrospun PAN nano-fiber using electrostatic spiral collector and converging coil. Polymer (Guildf) 84:52–58. https://doi.org/10.1016/j.polymer.2015.11.046

Lee JI, Yoo HS (2008) Biodegradable microspheres containing poly (epsilon-caprolactone)-Pluronic block copolymers for temperature-responsive release of proteins. Colloids Surf B Biointerfaces 61:81–87

Lee P-Y, Li Z, Huang L (2003) Thermosensitive hydrogel as a Tgf-β1 gene delivery vehicle enhances diabetic wound healing. Pharm Res 20:1995–2000

Leung V, Ko F (2011) Biomedical applications of nanofibers. Polym Adv Technol 22:350–365

Li D, Wang Y, Xia Y (2003) Electrospinning of polymeric and ceramic nanofibers as uniaxially aligned arrays. Nano Lett 3:1167–1171. https://doi.org/10.1021/nl0344256

Li D, Xia Y (2004) Electrospinning of nanofibers: reinventing the wheel? Adv Mater 16:1151–1170. https://doi.org/10.1002/adma.200400719

Li HY, Xu YC, Xu H, Chang J (2014) Electrospun membranes: control of the structure and structure related applications in tissue regeneration and drug delivery. J Mater Chem B 2:5492–5510. https://doi.org/10.1039/c4tb00913d

Li WJ, Mauck RL, Cooper JA et al (2007) Engineering controllable anisotropy in electrospun biodegradable nanofibrous scaffolds for musculoskeletal tissue engineering. J Biomech 40:1686–1693. https://doi.org/10.1016/j.jbiomech.2006.09.004

Li Z, Wang C (2013) Effects of working parameters on electrospinning. In: One-dimensional nanostructures. Springer, pp 15–28

Lin T (2011) Nanofibers-production, properties and functional applications. InTech

Linh NTB, Lee B-T (2012) Electrospinning of polyvinyl alcohol/gelatin nanofiber composites and cross-linking for bone tissue engineering application. J Biomater Appl 27:255–266. https://doi.org/10.1177/0885328211401932

Liu W, Thomopoulos S, Xia Y (2012) Electrospun nanofibers for regenerative medicine. Adv Healthc Mater 1:10–25. https://doi.org/10.1002/adhm.201100021

Liu Z, Sun DD, Guo P, Leckie JO (2007) An efficient bicomponent TiO_2/SnO_2 nanofiber photocatalyst fabricated by electrospinning with a side-by-side dual spinneret method. Nano Lett 7:1081–1085. https://doi.org/10.1021/nl061898e

Loh XJ, Peh P, Liao S et al (2010) Controlled drug release from biodegradable thermoresponsive physical hydrogel nanofibers. J Control Release 143:175–182. https://doi.org/10.1016/j.jconrel.2009.12.030

Lubasova D, Niu H, Zhao X, Lin T (2015) Hydrogel properties of electrospun polyvinylpyrrolidone and polyvinylpyrrolidone/poly(acrylic acid) blend nanofibers. RSC Adv 5:54481–54487. https://doi.org/10.1039/C5RA07514A

Ma G, Fang D, Liu Y et al (2012) Electrospun sodium alginate/poly(ethylene oxide) core-shell nanofibers scaffolds potential for tissue engineering applications. Carbohydr Polym 87:737–743. https://doi.org/10.1016/j.carbpol.2011.08.055

Ma PX (2004) Scaffolds for tissue fabrication. Mater Today 7:30–40. https://doi.org/10.1016/S1369-7021(04)00233-0

Mahoney C, Conklin D, Waterman J, et al (2016) Electrospun Nanofibers of Poly (ε-caprolactone)/Depolymerized Chitosan for Respiratory Tissue Engineering Applications. J Biomater Sci Polym Ed 1–21

Mallick SP, Pal K, Rastogi A, Srivastava P (2016) Evaluation of poly(L-lactide) and chitosan composite scaffolds for cartilage tissue regeneration. Des Monomers Polym 19:271–282. https://doi.org/10.1080/15685551.2015.1136535

Manna U, Patil S (2009) Borax mediated layer-by-layer self-assembly of neutral poly(vinyl alcohol) and chitosan. J Phys Chem B 113:9137–9142. https://doi.org/10.1021/jp9025333

McCullen SD, Autefage H, Callanan A et al (2012) Anisotropic fibrous scaffolds for articular cartilage regeneration. Tissue Eng Part A 18:2073–2083. https://doi.org/10.1089/ten.TEA.2011.0606

McKenzie M, Betts D, Suh A et al (2015) Hydrogel-based drug delivery systems for poorly water-soluble drugs. Molecules 20:20397–20408. https://doi.org/10.3390/molecules201119705

McMahon RE, Qu X, Jimenez-Vergara AC, et al (2011) Hydrogel—electrospun mesh composites for coronary artery bypass grafts. https://doi.org/10.1089/ten.tec.2010.0427

Meng L, Klinkajon W, K-hasuwan PR et al (2015) Electrospun crosslinked poly(acrylic acid) fiber constructs: towards a synthetic model of the cortical layer of nerve. Polym Int 64:42–48. https://doi.org/10.1002/pi.4793

Meng ZX, Zheng W, Li L, Zheng YF (2011) Fabrication, characterization and in vitro drug release behavior of electrospun PLGA/chitosan nanofibrous scaffold. Mater Chem Phys 125:606–611. https://doi.org/10.1016/j.matchemphys.2010.10.010

Merkle VM, Tran PL, Hutchinson M et al (2015) Core-shell PVA/gelatin electrospun nanofibers promote human umbilical vein endothelial cell and smooth muscle cell proliferation and migration. Acta Biomater 27:77–87. https://doi.org/10.1016/j.actbio.2015.08.044

Merkle VM, Zeng L, Slepian MJ, Wu X (2014) Core-shell nanofibers: integrating the bioactivity of gelatin and the mechanical property of polyvinyl alcohol. Biopolymers 101:336–346. https://doi.org/10.1002/bip.22367

Miao Y-E, Liu T (2015) Electrospun biopolymer nanofibers and their composites for drug delivery applications. Biodegrad Polyesters 275–298. https://doi.org/10.1002/9783527656950.ch11

Shin Michael, Yoshimoto H, Vacanti JP (2004) In vivo bone tissue engineering using mesenchymal stem cells on a novel electrospun nanofibrous scaffold. Tissue Eng 10:33–41. https://doi.org/10.1089/107632704322791673

Miraftab M, Saifullah AN, Çay A (2015) Physical stabilisation of electrospun poly(vinyl alcohol) nanofibres: comparative study on methanol and heat-based crosslinking. J Mater Sci 50:1943–1957. https://doi.org/10.1007/s10853-014-8759-1

Mogoşanu GD, Grumezescu AM (2014) Natural and synthetic polymers for wounds and burns dressing. Int J Pharm 463:127–136. https://doi.org/10.1016/j.ijpharm.2013.12.015

Montesano R, Orci L (1988) Transforming growth factor beta stimulates collagen-matrix contraction by fibroblasts: implications for wound healing. Proc Natl Acad Sci 85:4894–4897

Moses H, Gregory R, Michael P (2001) Electrospinning and electrically forced jets. II, Applications

Nasouri K, Bahrambeygi H, Rabbi A et al (2012) Modeling and optimization of electrospun PAN nanofiber diameter using response surface methodology and artificial neural networks. J Appl Polym Sci 126:127–135

Neamnark A, Rujiravanit R, Supaphol P (2006) Electrospinning of hexanoyl chitosan. Carbohydr Polym 66:298–305. https://doi.org/10.1016/j.carbpol.2006.03.015

Nguyen MK, Alsberg E (2014) Bioactive factor delivery strategies from engineered polymer hydrogels for therapeutic medicine. Prog Polym Sci. https://doi.org/10.1016/j.progpolymsci.2013.12.001

Nitanan T, Akkaramongkolporn P, Rojanarata T et al (2013) Neomycin-loaded poly(styrene sulfonic acid-co-maleic acid) (PSSA-MA)/polyvinyl alcohol (PVA) ion exchange nanofibers for wound dressing materials. Int J Pharm 448:71–78. https://doi.org/10.1016/j.ijpharm.2013.03.011

Okamoto M, John B (2013) Synthetic biopolymer nanocomposites for tissue engineering scaffolds. Prog Polym Sci 38:1487–1503. https://doi.org/10.1016/j.progpolymsci.2013.06.001

Paaver U, Heinämäki J, Laidmäe I, et al (2015) Electrospun nanofibers as a potential controlled-release solid dispersion system for poorly water-soluble drugs. Int J Pharm 479:252–260. https://doi.org/10.1016/j.ijpharm.2014.12.024

Pajoumshariati S, Yavari SK, Shokrgozar MA (2015) Physical and biological modification of polycaprolactone electrospun nanofiber by panax ginseng extract for bone tissue engineering application. Ann Biomed Eng 44:1808–1820. https://doi.org/10.1007/s10439-015-1478-1

Pangon A, Saesoo S, Saengkrit N et al (2016) Hydroxyapatite-hybridized chitosan/chitin whisker bionanocomposite fibers for bone tissue engineering applications. Carbohydr Polym 144:419–427. https://doi.org/10.1016/j.carbpol.2016.02.053

Parratt K, Yao N (2013) Nanostructured biomaterials and their applications. nanomaterials 3:242–271. https://doi.org/10.3390/nano3020242

Paul W, Sharma CP (2004) Chitosan and alginate wound dressings: a short review. Trends Biomater Artif Organs 18:18–23

Peppas N (2000) Hydrogels in pharmaceutical formulations. Eur J Pharm Biopharm 50:27–46. https://doi.org/10.1016/S0939-6411(00)00090-4

Peppas N a., Sahlin JJ (1996) Hydrogels as mucoadhesive and bioadhesive materials: a review. Biomaterials 17:1553–1561. https://doi.org/10.1016/0142-9612(95)00307-x

Persano L, Camposeo A, Tekmen C, Pisignano D (2013) Industrial upscaling of electrospinning and applications of polymer nanofibers: a review. Macromol Mater Eng 298:504–520. https://doi.org/10.1002/mame.201200290

Pillay V, Dott C, Choonara YE, et al (2013) A review of the effect of processing variables on the fabrication of electrospun nanofibers for drug delivery applications 22. https://doi.org/10.1155/2013/789289

Prakash S, Khan A, Paul A (2010) Nanoscaffold based stem cell regeneration therapy: recent advancement and future potential. Expert Opin Biol Ther 10:1649–1661. https://doi.org/10.1517/14712598.2010.528387

Quarto R, Mastrogiacomo M, Cancedda R et al (2001) Repair of large bone defects with the use of autologous bone marrow stromal cells. N Engl J Med 344:385–386

Quinn JF, Johnston APR, Such GK et al (2007) Next generation, sequentially assembled ultrathin films: beyond electrostatics. Chem Soc Rev 36:707–718. https://doi.org/10.1039/b610778h

Rnjak-Kovacina J, Weiss AS (2011) Increasing the pore size of electrospun scaffolds. Tissue Eng B, Rev 17:365–372. https://doi.org/10.1089/ten.teb.2011.0235

Rogina A (2014) Electrospinning process: versatile preparation method for biodegradable and natural polymers and biocomposite systems applied in tissue engineering and drug delivery. Appl Surf Sci 296:221–230. https://doi.org/10.1016/j.apsusc.2014.01.098

Ruckh TT, Kumar K, Kipper MJ, Popat KC (2010) Osteogenic differentiation of bone marrow stromal cells on poly(ε-caprolactone) nanofiber scaffolds. Acta Biomater 6:2949–2959. https://doi.org/10.1016/j.actbio.2010.02.006

Sabitha M, Rajiv S (2015) Preparation and characterization of ampicillin-incorporated electrospun polyurethane scaffolds for wound healing and infection control. Polym Eng Sci 55:541–548

Samchenko Y, Ulberg Z, Korotych O (2011) Multipurpose smart hydrogel systems. Adv Colloid Interface Sci 168:247–262. https://doi.org/10.1016/j.cis.2011.06.005

Shahzad S, Yar M, Siddiqi SA et al (2015) Chitosan-based electrospun nanofibrous mats, hydrogels and cast films: novel anti-bacterial wound dressing matrices. J Mater Sci Mater Med 26:1–12. https://doi.org/10.1007/s10856-015-5462-y

Shalumon KT, Deepthi S, Anupama MS, et al (2015) Fabrication of poly (l-lactic acid)/gelatin composite tubular scaffolds for vascular tissue engineering. Int J Biol Macromol 72:1048–1055. https://doi.org/10.1016/j.ijbiomac.2014.09.058

Shenoy SL, Bates WD, Frisch HL, Wnek GE (2005a) Role of chain entanglements on fiber formation during electrospinning of polymer solutions: good solvent, non-specific polymer-polymer interaction limit. Polymer (Guildf) 46:3372–3384. https://doi.org/10.1016/j.polymer.2005.03.011

Shenoy SL, Bates WD, Wnek G (2005b) Correlations between electrospinnability and physical gelation. Polymer (Guildf) 46:8990–9004. https://doi.org/10.1016/j.polymer.2005.06.053

Shi Q, Fan Q, Ye W et al (2015) Binary release of ascorbic acid and lecithin from core-shell nanofibers on blood-contacting surface for reducing long-term hemolysis of erythrocyte. Colloids Surfaces B Biointerfaces 125:28–33. https://doi.org/10.1016/j.colsurfb.2014.11.013

Shin S-Y, Park H-N, Kim K-H et al (2005) Biological evaluation of chitosan nanofiber membrane for guided bone regeneration. J Periodontol 76:1778–1784

Simionescu BC, Ivanov D (2016) Natural and synthetic polymers for designing composite materials. Handb Bioceram Biocomposites 233–286

Steele JAM, McCullen SD, Callanan A et al (2014) Combinatorial scaffold morphologies for zonal articular cartilage engineering. Acta Biomater 10:2065–2075. https://doi.org/10.1016/j.actbio.2013.12.030

Stefani I, Cooper-White JJ (2016) Development of an in-process UV-crosslinked, electrospun PCL/aPLA-co-TMC composite polymer for tubular tissue engineering applications. Acta Biomater 36:231–240. https://doi.org/10.1016/j.actbio.2016.03.013

Stitzel J, Liu J, Lee SJ et al (2006) Controlled fabrication of a biological vascular substitute. Biomaterials 27:1088–1094. https://doi.org/10.1016/j.biomaterials.2005.07.048

Su Y, Su Q, Liu W et al (2012) Controlled release of bone morphogenetic protein 2 and dexamethasone loaded in core-shell PLLACL-collagen fibers for use in bone tissue engineering. Acta Biomater 8:763–771. https://doi.org/10.1016/j.actbio.2011.11.002

Taepaiboon P, Rungsardthong U, Supaphol P (2006) Drug-loaded electrospun mats of poly(vinyl alcohol) fibres and their release characteristics of four model drugs. Nanotechnology 17:2317–2329. https://doi.org/10.1088/0957-4484/17/9/041

Tam K, Cheyyatraviendran S, Venugopal J et al (2014) A nanoscaffold impregnated with human wharton's jelly stem cells or its secretions improves healing of wounds. J Cell Biochem 115:794–803

Tao J, Shivkumar S (2007) Molecular weight dependent structural regimes during the electrospinning of PVA. Mater Lett 61:2325–2328. https://doi.org/10.1016/j.matlet.2006.09.004

Temenoff JS, Mikos AG (2014) Biomaterials: the intersection of biology and materials science

Thomas S (1990) Wound management and dressings. Pharmaceutical Press

Torres-Giner S, Gimeno-Alcañiz JV, Ocio MJ, Lagaron JM (2009) Comparative performance of electrospun collagen nanofibers cross-linked by means of different methods. ACS Appl Mater Interfaces 1:218–223. https://doi.org/10.1021/am800063x

Tsai SW, Liou HM, Lin CJ et al (2012) MG63 osteoblast-like cells exhibit different behavior when grown on electrospun collagen matrix versus electrospun gelatin matrix. PLoS ONE 7:1–11. https://doi.org/10.1371/journal.pone.0031200

Uchko CJ, Chen LC, Shen Y, Martina DC (1999) Processing and microstructural characterization of porous biocompatible\rprotein polymer thin films. Polymer (Guildf) 40:7397–7407. https://doi.org/10.1016/S0032-3861(98)00866-0

Ulubayram K, Calamak S, Shahbazi R, Eroglu I (2015) Nanofibers based antibacterial drug design, delivery and applications. Curr Pharm Des 21:1930–1943

Van Vlierberghe S, Dubruel P, Schacht E (2011) Biopolymer-based hydrogels as scaffolds for tissue engineering applications: a review. Biomacromol 12:1387–1408. https://doi.org/10.1021/bm200083n

Vaquette C, Cooper-White JJ (2011) Increasing electrospun scaffold pore size with tailored collectors for improved cell penetration. Acta Biomater 7:2544–2557. https://doi.org/10.1016/j.actbio.2011.02.036

Vashisth P, Pruthi V (2016) Synthesis and characterization of crosslinked gellan/PVA nanofibers for tissue engineering application. Mater Sci Eng, C 67:304–312. https://doi.org/10.1016/j.msec.2016.05.049

Vashisth P, Srivastava AK, Nagar H et al (2016) Drug functionalized microbial polysaccharide based nanofibers as transdermal substitute. Nanomed Nanotechnol Biol Med 12:1375–1385. https://doi.org/10.1016/j.nano.2016.01.019

Venugopal JR, Sridhar S, Ramakrishna S (2014) Electrospun plant-derived natural biomaterials for tissue engineering. Plant Sci Today 1:151–154. https://doi.org/10.14719/pst.2014.1.3.65

Veronese FM, Pasut G (2005) PEGylation, successful approach to drug delivery. Drug Discov Today 10:1451–1458. https://doi.org/10.1016/S1359-6446(05)03575-0

Visser J, Melchels FPW, Jeon JE et al (2015) Reinforcement of hydrogels using three-dimensionally printed microfibres. Nat Commun 6:6933. https://doi.org/10.1038/ncomms7933

Wang HB, Mullins ME, Cregg JM et al (2010a) Varying the diameter of aligned electrospun fibers alters neurite outgrowth and Schwann cell migration. Acta Biomater 6:2970–2978. https://doi.org/10.1016/j.actbio.2010.02.020

Wang J, Valmikinathan CM, Liu W et al (2010b) Spiral-structured, nanofibrous, 3D scaffolds for bone tissue engineering. J Biomed Mater Res A 93:753–762. https://doi.org/10.1002/jbm.a.32591

Wang KW, Zhu YJ, Chen XY et al (2010c) Flower-like hierarchically nanostructured hydroxyapatite hollow spheres: facile preparation and application in anticancer drug cellular delivery. Chem—Asian J 5:2477–2482. https://doi.org/10.1002/asia.201000463

Wang L, Khor E, Wee A, Lim LY (2002) Chitosan-alginate PEC membrane as a wound dressing: assessment of incisional wound healing. J Biomed Mater Res 63:610–618

Wang Q, Ogawa K, Toma K, Tamiaki H (2009a) Smart pH sensitive luminescent hydrogel based on Eu(III) β-diketonate complex and its enhanced photostability. J Photochem Photobiol A Chem 201:87–90. https://doi.org/10.1016/j.jphotochem.2008.10.005

Wang Y, Shi H, Qiao J, et al (2014) Electrospun tubular scaffold with circumferentially aligned nano fibers for regulating smooth muscle cell growth. doi:http://pubs.acs.org/doi/abs/10.1021/am405556x

Wang Y, Wang B, Qiao W, Yin T (2010d) A novel controlled release drug delivery system for multiple drugs based on electrospun nanofibers containing nanoparticles. J Pharm Sci 99:4805–4811. https://doi.org/10.1002/jps.22189

Wang Y, Wang G, Chen L et al (2009b) Electrospun nanofiber meshes with tailored architectures and patterns as potential tissue-engineering scaffolds. Biofabrication 1:15001–15009. https://doi.org/10.1088/1758-5082/1/1/015001

Weis J, May R, Schröder JM (1994) Fine structural and immunohistochemical identification of perineurial cells connecting proximal and distal stumps of transected peripheral nerves at early stages of regeneration in silicone tubes. Acta Neuropathol 88:159–165

Weng L, Xie J (2015) Smart electrospun nanofibers for controlled drug release : recent advances and new perspectives. 1944–1959. https://doi.org/10.2174/13816128216661503021 51959

Wolf MT, Dearth CL, Sonnenberg SB et al (2015) Naturally derived and synthetic scaffolds for skeletal muscle reconstruction. Adv Drug Deliv Rev 84:208–221. https://doi.org/10.1016/j.addr.2014.08.011

Wozniak MA, Modzelewska K, Kwong L, Keely PJ (2004) Focal adhesion regulation of cell behavior. Biochim Biophys Acta (BBA)-Mol Cell Res 1692:103–119

Wu J, Huang C, Liu W et al (2014) Cell infiltration and vascularization in porous nanoyarn scaffolds prepared by dynamic liquid electrospinning. J Biomed Nanotechnol 10:603–614

Wu L, Dou Y, Lin K et al (2011) Hierarchically structured nanocrystalline hydroxyapatite assembled hollow fibers as a promising protein delivery system. Chem Commun (Camb) 47:11674–11676. https://doi.org/10.1039/c1cc14709a

Xu CY, Inai R, Kotaki M, Ramakrishna S (2004) Aligned biodegradable nanofibrous structure: a potential scaffold for blood vessel engineering. Biomaterials 25:877–886. https://doi.org/10.1016/S0142-9612(03)00593-3

Xu H, Li H, Chang J (2013) Controlled drug release from a polymer matrix by patterned electrospun nanofibers with controllable hydrophobicity. J Mater Chem B 1:4182. https://doi.org/10.1039/c3tb20404a

Xu S, Deng L, Zhang J, et al (2015) Composites of electrospun-fibers and hydrogels: a potential solution to current challenges in biological and biomedical field. J Biomed Mater Res B Appl Biomater 1–17. https://doi.org/10.1002/jbm.b.33420

Xu X, Zhou G, Li X et al (2016) Solution blowing of chitosan/PLA/PEG hydrogel nanofibers for wound dressing. Fibers Polym 17:205–211. https://doi.org/10.1007/s12221-016-5800-9

Xue J, He M, Liu H et al (2014) Drug loaded homogeneous electrospun PCL/gelatin hybrid nanofiber structures for anti-infective tissue regeneration membranes. Biomaterials 35:9395–9405. https://doi.org/10.1016/j.biomaterials.2014.07.060

Yamaoka T, Tabata Y, Ikada Y (1994) Distribution and tissue uptake of poly(ethylene glycol) with different molecular weights after intravenous administration to mice. J Pharm Sci 83:601–606. https://doi.org/10.1002/jps.2600830432

Yang C, Xu L, Zhou Y et al (2010) A green fabrication approach of gelatin/CM-chitosan hybrid hydrogel for wound healing. Carbohydr Polym 82:1297–1305. https://doi.org/10.1016/j.carbpol.2010.07.013

Yang Q, Zhenyu LI, Hong Y et al (2004) Influence of solvents on the formation of ultrathin uniform poly(vinyl pyrrolidone) nanofibers with electrospinning. J Polym Sci B: Polym Phys 42:3721–3726. https://doi.org/10.1002/polb.20222

Yang X, Liu Q, Chen X et al (2008) Investigation of PVA/ws-chitosan hydrogels prepared by combined γ-irradiation and freeze-thawing. Carbohydr Polym 73:401–408. https://doi.org/10.1016/j.carbpol.2007.12.008

Yang Y, Wimpenny I, Ahearne M (2011) Portable nanofiber meshes dictate cell orientation throughout three-dimensional hydrogels. Nanomed Nanotechnol Biol Med 7:131–136. https://doi.org/10.1016/j.nano.2010.12.011

Yao S, Wang X, Liu X et al (2013) Effects of ambient relative humidity and solvent properties on the electrospinning of pure hyaluronic acid nanofibers. J Nanosci Nanotechnol 13:4752–4758. https://doi.org/10.1166/jnn.2013.7197

Yeo Y, Adil M, Bellas E et al (2007) Prevention of peritoneal adhesions with an in situ cross-linkable hyaluronan hydrogel delivering budesonide. J Control Release 120:178–185. https://doi.org/10.1016/j.jconrel.2007.04.016

Yoo HS, Kim TG, Park TG (2009) Surface-functionalized electrospun nanofibers for tissue engineering and drug delivery. Adv Drug Deliv Rev 61:1033–1042. https://doi.org/10.1016/j.addr.2009.07.007

Yoshimoto H, Shin YM, Terai H, Vacanti JP (2003) A biodegradable nanofiber scaffold by electrospinning and its potential for bone tissue engineering. Biomaterials 24:2077–2082. https://doi.org/10.1016/S0142-9612(02)00635-X

Yu DG, Li XY, Wang X et al (2015) Nanofibers fabricated using triaxial electrospinning as zero order drug delivery systems. ACS Appl Mater Interfaces 7:18891–18897. https://doi.org/10.1021/acsami.5b06007

Yu D-G, Yang C, Jin M et al (2016) Medicated Janus fibers fabricated using a Teflon-coated side-by-side spinneret. Colloids Surfaces B Biointerfaces 138:110–116

Yuan XY, Zhang YY, Dong C, Sheng J (2004) Morphology of ultrafine polysulfone fibers prepared by electrospinning. Polym Int 53:1704–1710. https://doi.org/10.1002/pi.1538

Zeng J, Yang L, Liang Q et al (2005) Influence of the drug compatibility with polymer solution on the release kinetics of electrospun fiber formulation. J Control Release 105:43–51. https://doi.org/10.1016/j.jconrel.2005.02.024

Zhang C, Yuan X, Wu L et al (2005) Study on morphology of electrospun poly(vinyl alcohol) mats. Eur Polym J 41:423–432. https://doi.org/10.1016/j.eurpolymj.2004.10.027

Zhang D, Chang J (2007) Patterning of electrospun fibers using electroconductive templates. Adv Mater 19:3662–3667. https://doi.org/10.1002/adma.200700896

Zhang H, Jia X, Han F et al (2013) Dual-delivery of VEGF and PDGF by double-layered electrospun membranes for blood vessel regeneration. Biomaterials 34:2202–2212. https://doi.org/10.1016/j.biomaterials.2012.12.005

Zhang YZ, Venugopal J, Huang ZM et al (2006a) Crosslinking of the electrospun gelatin nanofibers. Polymer (Guildf) 47:2911–2917. https://doi.org/10.1016/j.polymer.2006.02.046

Zhang YZ, Wang X, Feng Y et al (2006b) Coaxial electrospinning of (fluorescein isothiocyanate-conjugated bovine serum albumin)-encapsulated poly (ε-caprolactone) nanofibers for sustained release. Biomacromol 7:1049–1057

Zhao Y, Zhou Y, Wu X et al (2012) A facile method for electrospinning of Ag nanoparticles/poly (vinyl alcohol)/carboxymethyl-chitosan nanofibers. Appl Surf Sci 258:8867–8873. https://doi.org/10.1016/j.apsusc.2012.05.106

Zhijiang C, Yi X, Haizheng Y et al (2016) Poly (hydroxybutyrate)/cellulose acetate blend nanofiber scaffolds: preparation, characterization and cytocompatibility. Mater Sci Eng, C 58:757–767

Zhou F, Gong R, Porat I (2009) Mass production of nanofibre assemblies by electrostatic spinning. Polym Int 58:331–342

Ziaee M, Zahedi P, Abdouss M et al (2016) Electrospun poly (N-isopropylacrylamide-co-acrylic acid)/cellulose laurate blend nanofibers containing adapalene: morphology, drug release, and cell culture studies. Int J Polym Mater Polym Biomater 65:477–486

Zilla P, Bezuidenhout D, Human P (2007) Prosthetic vascular grafts: wrong models, wrong questions and no healing. Biomaterials 28:5009–5027. https://doi.org/10.1016/j.biomaterials.2007.07.017

Zimmerli W, Trampuz A, Ochsner PE (2004) Prosthetic-joint infections. N Engl J Med 351:1645–1654

Zomer Volpato F, Almodóvar J, Erickson K et al (2012) Preservation of FGF-2 bioactivity using heparin-based nanoparticles, and their delivery from electrospun chitosan fibers. Acta Biomater 8:1551–1559. https://doi.org/10.1016/j.actbio.2011.12.023

Chapter 10
Self-assembling Hydrogels from pH-Responsive Ionic Block Copolymers

Constantinos Tsitsilianis

Abstract Hydrogels are three-dimensional (3D) soft materials that consist of a solid matrix (usually a three-dimensional network) entrapping high content of water (more than 90 wt%). This remarkable feature makes them suitable for many applications especially in medicine as drug carriers and tissue engineering scaffolds. As far as polymeric matrices are concerned, two main strategies for achieving 3D network structures can be distinguished. The first one relies on the covalent bonding of hydrophilic polymer chains, leading to hydrogels referred as chemical networks. The second approach deals with the self-assembly of tailor-made segmented macromolecules via reversible weak interactions, namely hydrophobic, ionic, π–π staking, and so on, that leads to the so-called self-assembling hydrogels. The use of reversible (physical) cross-links allows the design of "smart" soft materials that can response to their environment (e.g., pH, ionic strength, temperature, shear). This chapter is devoted to the self-assembling hydrogels arising from associative block copolymers bearing ionic or ionogenic blocks, namely polyelectrolytes or polyampholytes. This specific feature endows the hydrogels with responsiveness to pH and ionic strength which make them attractive soft materials for potential biomedical applications.

Keywords Hydrogel · Block copolymers · Self-assembly · pH responsive
3D network · Polyelectrolyte · Polyampholyte

1 Introduction

Hydrogels represent a very attractive class of soft matter which has found great development during the last decades, thanks to their wide range of (potential) applications in several fields of modern technologies, including biomedical applications such as cell transplantation, tissue engineering, and controlled drug delivery

C. Tsitsilianis (✉)
Department of Chemical Engineering, University of Patras, 26504 Patras, Greece
e-mail: ct@chemeng.upatras.gr

V. K. Thakur and M. K. Thakur (eds.), *Hydrogels*, Gels Horizons: From Science to Smart Materials, https://doi.org/10.1007/978-981-10-6077-9_10

(Thakur and Thakur 2014a, b, 2015). The hydrogel is a soft and wet bi-continuous material, constituted of a solid substrate highly swollen by water which cannot flow, due to the three-dimensional (3D) cross-linked network afforded by the solid constituent. The latter may be an associative small molecule that forms supramolecular structures (usually of high aspect ratio, e.g., fibers) or a macromolecular chain (polymer). Depending on the nature of crosslinking, the hydrogels can be divided into two classes, namely chemical (irreversible) and physical (reversible) hydrogels. In the first class, the crosslinking of the macromolecular chains is achieved by a chemical reaction that leads to permanent covalent bonds, forming an infinite molecular weight compound. Thus, the chemical hydrogels need a template (mold) to get its shape prior crosslinking, which can be swelled/deswelled, depending on the nature of the polymer used. On the other hand, the physical hydrogels are formed through non-covalent bonds arising from secondary molecular interactions, namely hydrophobic, ionic (Coulomb), H-bonding, π–π staking, host/guest which drives well-designed tailor-made macromolecules to self-assemble in water, forming a 3D structure. In this case, the hydrogel can selfadapt to its environment, filling the available space in which it is applied. More importantly, the physical crosslinking can be directed by a trigger (e.g., pH, temperature, light) in the course of suitable design of the functionalities, carried by the macromolecule, as to be able to form the physical crosslinks. The so-resulting hydrogels are referred as "smart" or "intelligent" or "stimulus responsive" hydrogels which are very promising materials for biomedical applications (Tsitsilianis 2010; Koetting et al. 2015). For instance, the "in situ" gelling systems (injectable hydrogels) hold in a non-associative situation (sol) prior injection while they jellify after injection into a living tissue undergoing a sol–gel transition triggered by a stimulus (Van Tomme et al. 2008).

As far the physical hydrogel are concerned, the 3D network is created through the self-organization of associative macromolecules of special architecture and functionality. The involved macromolecules (also named gelators) are in most of the cases block copolymers comprising a long hydrophilic block carrying a number of short hydrophobic blocks (also named stickers) which are responsible for the intermolecular hydrophobic association of the macromolecules into an infinite network. The so-formed hydrogels are referred also as self-assembling hydrogels (Xu and Kopeček 2007). One of the important features of the self-assembling hydrogels, that govern their rheological properties, is the exchange dynamics of the stickers correlated with the lifetime of the bridges. For strictly rheological point of view, hydrogels exhibit very long relaxations times (very slow exchange dynamics) much higher than the experimental time, exhibiting hence, immeasurable zero-shear viscosity and yield behavior. However, a broader definition of hydrogels, comprising systems that exhibit viscoelasticity with measurable relaxation times and viscosities, has been established in the literature. Along this chapter, the broader definition will be followed.

Depending on the functionality of the hydrophilic block, we could discriminate two main categories of gelators: the non-ionic and the ionic (or ionogenic) macromolecules (polyelectrolytes) which bear along their chain ionic functions, that affect strongly the chain conformation and in turn the gelation process and functionality. Moreover, and in the case of weak polyelectrolytes, the gelators and the properties they confer are sensitive to pH which makes them very promising candidates for biomedical applications.

This chapter is devoted to the recent developments of the self-assembling hydrogels and especially to those arising from the ionic gelators, bearing charged groups either of the same charge (polyelectrolytes) or of opposite charges (polyampholytes). It comprises a variety of segmented (blocky) macromolecules differing in their topology, e.g., linear block copolymers or terpolymers, as well as branched macromolecules such as graft and star-shaped copolymers, which are associated through hydrophobic and/or ionic attractive interactions. Gels arising from the jamming effect of the micelles (amorphous structure) or the micelle crystallization (absence of bridges in both cases) that requires high polymer volume fraction will not be considered.

2 Gelation Procedure and Association Mechanism

Self-assembling hydrogels are created through "bottom-up" strategies according to which the rational design of the gelator is the first critical step. The gelator design is dictated from the targeting associative and/or responsive properties of the macromolecule which is not other than the hierarchical self-organization process of the macromolecular entities to spontaneously form a three-dimensional network through non-covalent bonding. Moreover, additional secondary network functionalities have to be taken into account like for instance the chain conformation of the hydrophilic parts, as well as their ability to incorporate therapeutic agents through dynamic covalent bonding (labile bonds easily hydrolyzed reversibly).

Thus, attention has to be paid to the architecture/topology/functionality of the involved macromolecules and the nature of interactions, among the various monomer repeating units and the solvent, which are going to be developed in the aqueous medium. Block copolymers and terpolymers (comprise three different kinds of monomer units) are suitable macromolecular species for such "bottom-up" strategies since they can be designed at the molecular level through macromolecular engineering (Matyjaszewski et al. 2007). The enormous development of synthetic polymerization techniques named controlled/living polymerization methods together with click chemistry offer plentiful possibilities in designing of tailor-made block copolymers (Kahveci et al. 2016) enabling to self-assemble in nanostructured networks of nearly predictable properties, e.g., mesh size, mechanical strength, dynamics, responsiveness.

Water-soluble macromolecules bearing attractive groups that can develop physical bonds through intermolecular associations, when they are dissolved in aqueous media, belong to a broad class of polymeric systems named "associative" polymers, APs, (Winnik and Yekta 1997) which are the best-suited gelators for the temporary network formation. In Fig. 1, a representative variety of block copolymer APs differing in the macromolecular topology is demonstrated.

In principle, the AP gelators are constituted of a long hydrophilic chain (A), bearing a number of short hydrophobic (or potentially hydrophobic) chains (stickers, B) either as end-blocks (B-*b*-A-*b*-B-type block copolymers) or as pendant grafting chains (A-*g*-B-type graft copolymers). More complicated polymer topologies have been also used as gelators, as will be presented below. This kind of amphiphiles self-associates in water which acts as a selective solvent, i.e., good solvent for the central block and bad solvent for the stickers, above a certain concentration, named critical micelle concentration (*cmc*) or critical aggregation

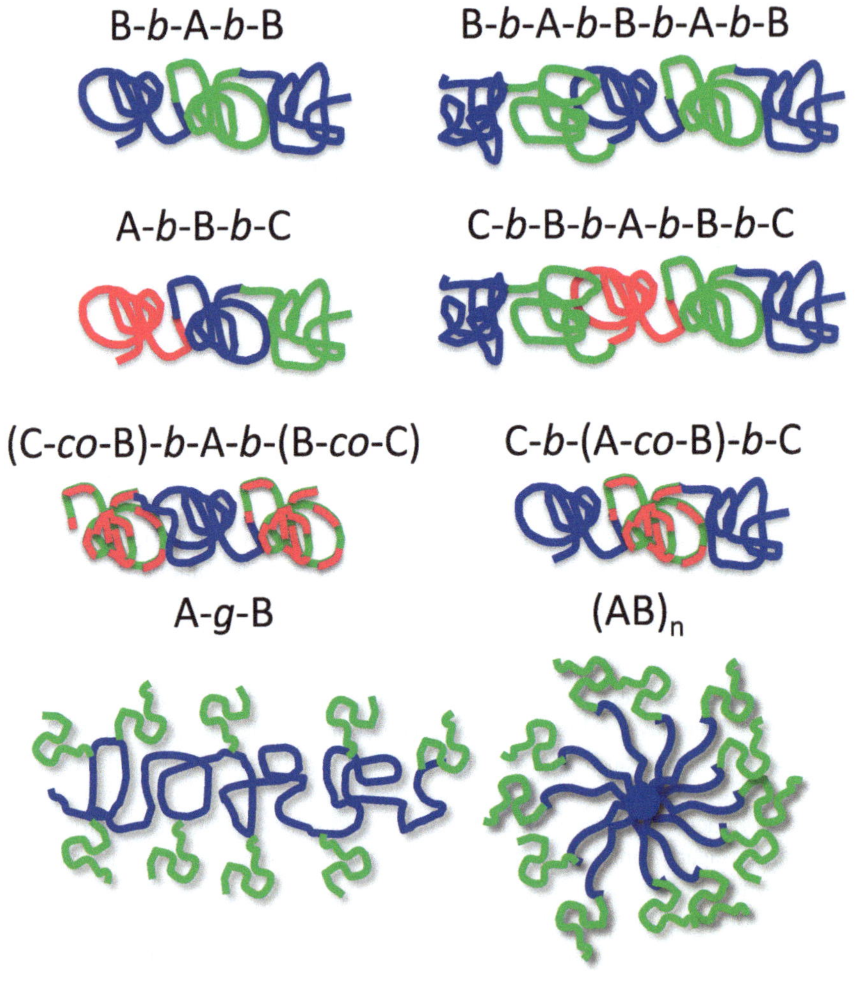

Fig. 1 Schematic representation of associative polymer topologies capable to create a three-dimensional network

concentration (*cac*). The *cmc* is the concentration at which the entropic loss due to localization of the chains in the self-assembly is equal to the energetic gain arising from this association. The driving force is the minimization of the unfavorable contacts of the hydrophobic sticky ends with water (Rubinstein and Dobrynin 1997). The resulted flower-like micelle morphology comprises a hydrophobic core and hydrophilic swollen looping chains. The aggregation number (number of stickers per core) is determined by the balance between the steric repulsion of the looping chains in the corona and the surface energy gain of the stickers in the core. By increasing concentration, micellar bridging (the stickers of the same chain lie in two adjacent micellar cores) occurs leading to clusters of micelles. The bridge formation increases the configurational entropy of the middle hydrophilic chains while decreases the translational entropy of the micelles. Therefore, the number of bridges increases with concentration, which leads to growing clusters. Above a percolation concentration, namely critical gelation concentration (*cgc*), a three-dimensional (3D) transient network of micelles is formed (Fig. 2).

The association process and the gel properties are influenced and eventually determined by the molecular characteristics of the blocks of the gelator that is the length (degree of polymerization) the chain conformation (end to end distance) and the interactions with water (interaction parameter). Concerning the chain conformation of the hydrophilic block, we can distinguish two main categories; the non-ionic water-soluble polymers and the ionic ones that bear a number of charged groups along the chain (e.g., polyelectrolytes or polyampholytes). In the latter case, the chain conformation can be varied from coil (neutral) to extended conformation

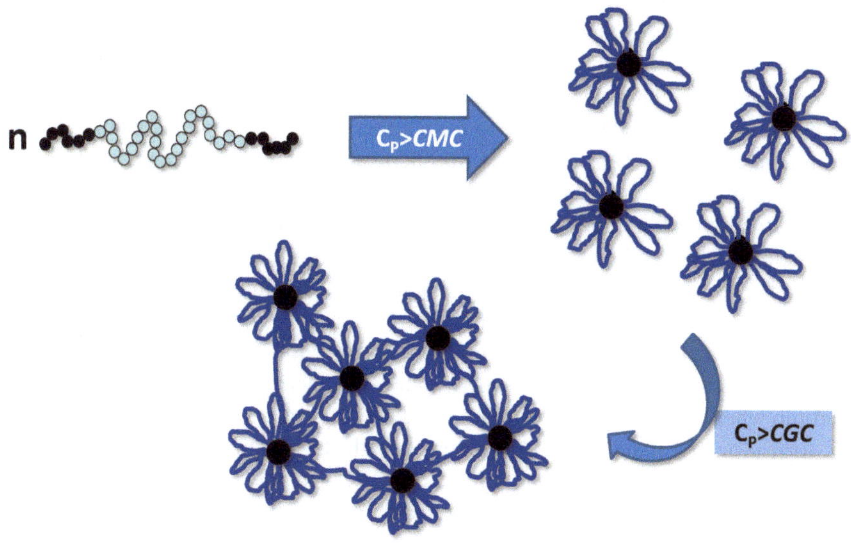

Fig. 2 Schematic representation of the association procedure of a model BAB amphiphilic triblock copolymer with B hydrophobic sticky blocks

(charged) due to the repulsive coulomb interactions along the chain. For the weak polyelectrolytes, the degree of ionization (fraction of charged monomer units) is determined by the *pK* of the polyelectrolyte and the environmental conditions such as the pH and the ionic strength (presence of salts) of the medium. Thus, the ionic polymer gelators are richer and the same time more complicated systems exhibiting unique properties that have attracted considerable attention the recent years.

3 Theoretical and Simulation Approaches

The simplest macromolecular architecture of the block copolymer ionic gelators is the asymmetric BAB triblock topology, comprising a longer weak polyelectrolyte as the A block, end-capped by the hydrophobic (insoluble) B blocks (named telechelic polyelectrolytes, TP). Mean-field theory predicts that the phase behavior of the system is governed by the interplay between the attraction of the stickers and the repulsion originated by the charged moieties along the hydrophilic block and their mobile counterions (Potemkin et al. 1999). Thus, the free energy of the system was written as a sum of three terms.

$$F_m = F_a + F_{\mathrm{el}} + F_{\mathrm{ch}} \quad \text{where} \quad F_{\mathrm{ch}} = F_{\mathrm{coulomb}} + F_{\mathrm{trans}}$$

F_a is the energy gain due to the sticker association, F_{el} is the elastic free energy describing the entropy losses due to the extension of the central chain and F_{ch} accounts for the contribution of the charges of the chain (F_{coulomb}, energy of Coulombic interactions) and their counterions (F_{trans}, energy of counterion translational motion). It should be noticed that the main effect leading to chain swelling are the osmotic pressure of the counterions and the repulsive forces among the charges located along the chain. If the attraction between the stickers is strong enough, the telechelic polyelectrolytes are able to form either thermodynamically stable clusters of finite size (microgel) or a 3D network, depending on polymer concentration and the degree of ionization of the polyelectrolyte block. In the case where the attraction energy is much higher than the thermal energy ($\varepsilon \gg kT$) (i.e., strong hydrophobic interactions), the resulted 3D structure is similar to a chemically cross-linked gel resembling to an elastic soft solid (Tsitsilianis and Iliopoulos 2002).

The degree of ionization (α), formulated as $\alpha = 1/\sigma$, where σ is the number of neutral monomer units among two charged groups along the hydrophilic chain, is an important factor and can be easily tuned by varying the pH of the aqueous medium when weak polyelectrolytes are used. Two scenarios were predicted for the gel formation. At fixed association energy, with decreasing α, both the Coulomb energy and the entropy of counterions decrease and the gel formation becomes possible when these contributions become smaller than the attraction energy. By increasing α, the Coulomb energy ($F_{\mathrm{coulomb}} \sim 1/\sigma^{4/3}$) increases faster than the energy of the counterions ($F_{\mathrm{trans}} \sim 1/\sigma$). When the corresponding free energy cost becomes too high, intermolecular association occurs, leading to gel formation. Simultaneously, part of the outer counteranions was absorbed by the chains stimulating macroscopic gelation (Potemkin et al. 1999).

The association process toward a 3D network of TP was also explored by Monte Carlo (MC) simulations (Zhang et al. 2010) using percolation theory. The percolation probability $P(x)$ (x is a variable) was used to determine whether a gel is formed or not. For $P(x) = 0$, the system is in the sol phase, whereas for $P(x) = 1$, the system is in the gel phase. Note that the percolation model is valid for the criteria of gelation only when the attraction energy, $\varepsilon_{\text{attr}}$, is much larger than kT (Kumar and Panagiotopoulos 1999).

In order to look at the role of Coulomb interactions involved in TPs relative to the neutral telechelic systems, the percolation probability was investigated as a function of the charging fraction f at high enough $\varepsilon_{\text{attr}}/kT = 8$ to form a gel. The percolation $P(f)$ drops rapidly to zero for $f < 0.6$. In Fig. 3, snapshots of simulation under fully charged and noncharged conditions are demonstrated (Zhang et al. 2010). Obviously, the 3D network is favored in the charged system due to stretched conformation adopted by the ionized chains. Thus, the TPs are able to form hydrogels at lower concentrations than the non-ionic counterparts. Figure 4 demonstrates the percolation probability $P(\varphi, \varepsilon)$ as a function of concentration φ at different dimensionless hydrophobic attraction energies, $\varepsilon = \varepsilon_{\text{attr}}/kT$. Upon increasing φ, a sol–gel transition was observed which is shifted to lower concentrations by increasing the hydrophobicity of the stickers for a fully charged polyelectrolyte central block (Zhang et al. 2011).

More importantly, Monte Carlo simulations can give insights dealing with the telechelic chain conformations existing in the solution as a function of TP concentration. Four types of chain conformations can be distinguished: free, dangling

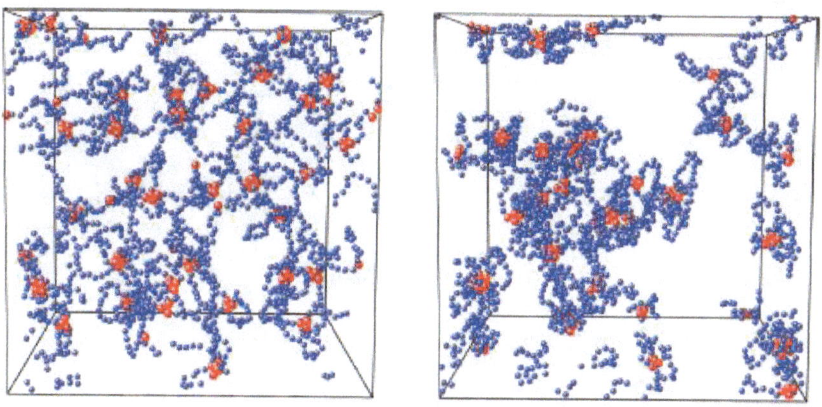

Fig. 3 Snapshots of simulation at $\varepsilon = 8$, $\xi = 0.5$: 100% charged (left) neutral telechelic-associative polymers (right). The stickers are shown in red, and the middle block is shown in blue. From Zhang et al. (2010) by permission of ACS

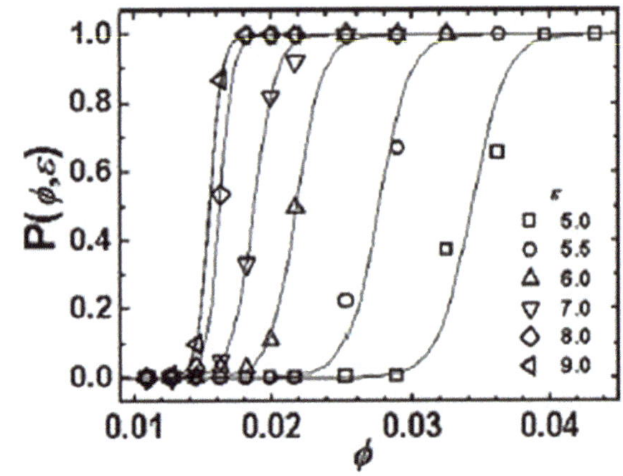

Fig. 4 Percolation probability, $P(\varphi, \varepsilon)$, plotted versus concentration φ at different hydrophobic interaction energy ε. The solid line is the Sigmoidal-Boltzmann fit of the percolation data. From Zhang et al. (2011) by permission of AIP

ends, loops, and bridges (Fig. 5a). As can be seen in Fig. 5, for sufficient large hydrophobic interactions ($\varepsilon = 8$), the fraction of free chains decreases abruptly upon increasing φ, transformed to loop and bridges. During the sol–gel transition, loop to bridge transitions occur leading to a network dominated by bridges at higher concentrations where dangling ends and free chains have been disappeared. The physical picture inferred from Fig. 5 depends on the hydrophobic interactions, ε, and the charge density of the hydrophilic part of the telechelic chains, i.e., the higher ε the larger fraction of bridges. However, for complete uncharged chains, the picture is substantially different dominated by looping chains (flower-like micelles) as reported by Li et al. (2004).

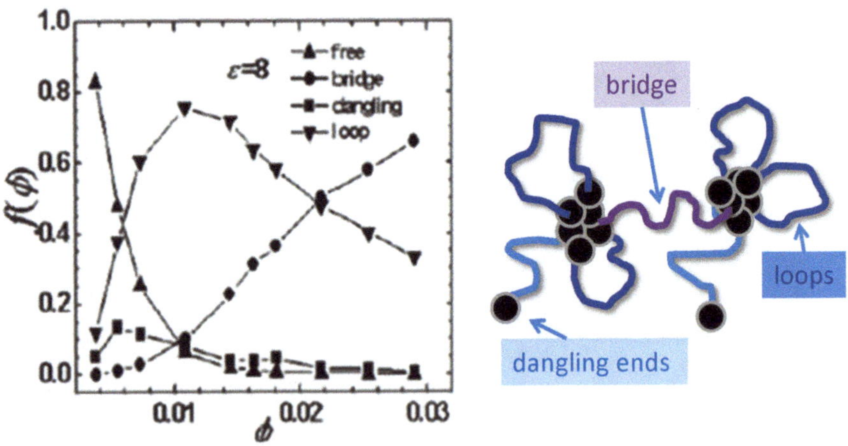

Fig. 5 Schematic representation of TP chain conformations (bridge, loop, dangling end) (right) and the fraction of chain conformations as a function of concentration for high hydrophobic interaction energy $\varepsilon = 8$ (left). From Zhang et al. (2011) by permission of AIP

Just recently, Langevin dynamics simulation of coarse-grained BAB copolymer chains (MC technique) were involved to study the effect of various parameters, namely midblock charge fraction, valency of neutralizing counterions, and stiffness and length of hydrophobic end-blocks, on the self-assembly behavior of hydrophobic-capped polyelectrolytes in aqueous solution, at fixed polymer concentration and temperature (Ghelichi and Qazvini 2016). Again, the increase of the charge fraction in the polyelectrolyte transforms the solution of flower-like micelles to a 3D network of hydrophobic nanodomains, formed by the sticky end-blocks, interconnected by stretched polyelectrolyte mid-chains (Fig. 6a). Moreover, the model was able to monitor the conformational change of the midblock chain, in terms of their radius of gyration and the persistence length, upon increasing the degree of ionization α. Figure 6b illustrates that both increase with α as the result of the increasing chain rigidity, due to the repulsive Coulomb interactions along the increasing charged chains. This effect is accompanied by the enhancement of the fraction of bridges in the expense of loops. For fully charged chains (α = 1), the fraction of bridges approaches 70% of the involved macromolecules, in excellent agreement with the MC-percolation theoretical predictions depicted in Fig. 5. Note that the number density of bridges of the network and their degree of rigidity are

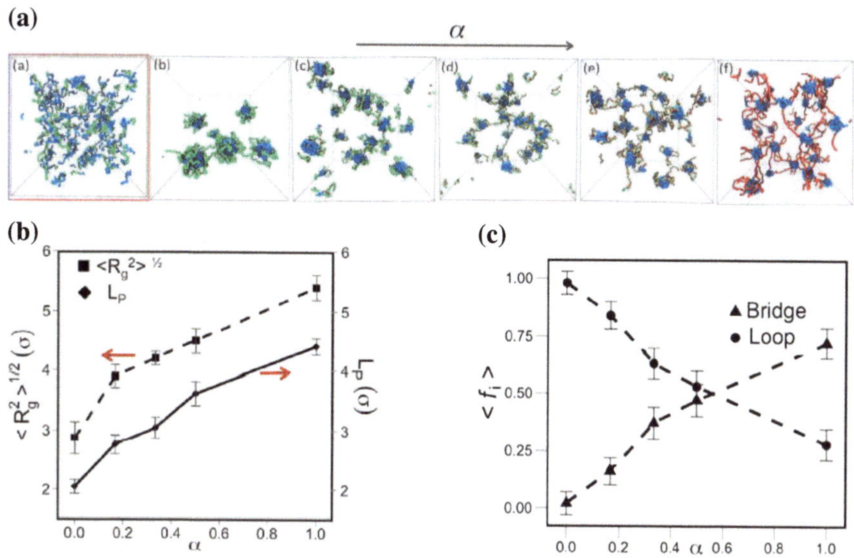

Fig. 6 **a** Simulation snapshots demonstrated the effect of midblock charge content, α on the micellar organization of ABA copolymer solutions: (*a*) beginning random structure, (*b*) α = 0, (*c*) α = 1/6, (*d*) α = 1/3, (*e*) α = 1/2, (*f*) α = 1. Color codes: A beads in blue. B beads either hydrophilic in green or ionized in red. Counterions are not shown for clarity. **b** α dependence of the radius of gyration and persistence length of the hydrophilic B block. **c** Fraction of looped and bridged midblock chains as a function of α (error bars represent the standard deviations over the equilibrium trajectories). From Ghelichi and Qazvini (2016) with permission of RSC

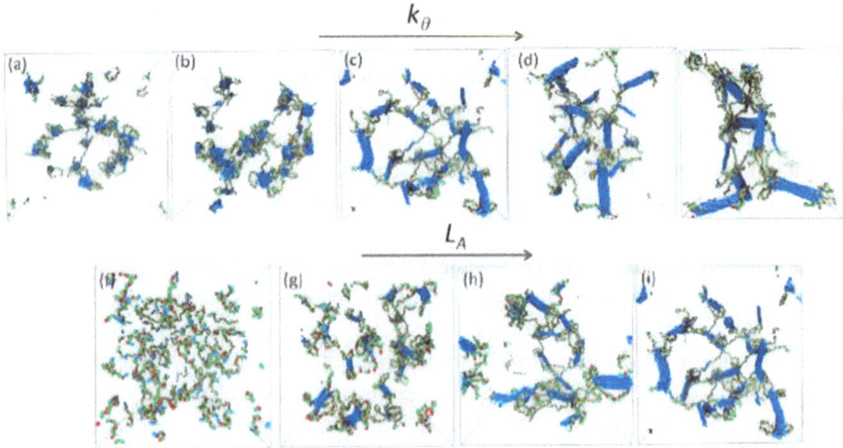

Fig. 7 From **a** to **e**, simulation snapshots for different k_Θ (rigidity), **f** to **i** show the simulated copolymer systems at different L_A (length). Baseline charge fraction (α = 1/3) is considered in **a** to **i**. From Ghelichi and Qazvini (2016) with permission of RSC

significant factors affecting the modulus of elasticity of the hydrogel (Tsitsilianis and Iliopoulos 2002).

Furthermore, by increasing the valency of the counterions resulted in a disruption of the interconnections of hydrophobic domains (3D network). The multivalent counterions screen the opposite charges and introduce a bridging effect between the ionized groups of the chain that eventually lead to the formation of bulky spherical aggregates.

Finally, it was shown that the rigidity of the stickers also exhibits a remarkable effect on the network nanostructure. Stiffer hydrophobic blocks yield bundle-like hydrophobic structures while the increase of the end-block length is accompanied by transformation of the network structure from a solution of non-associating chains into disk-like isolated clusters with looped hydrophilic chains and eventually into a spanning 3D network of cylindrical clusters. Figure 7 shows a number of snapshots revealing the micellar morphology transformation upon increasing the rigidity (k_Θ) and the end-block length, for fixed rigidity and charge fraction of the middle block ($\alpha = 1/3$) (Ghelichi and Qazvini 2016).

4 Polyelectrolyte-Based Hydrogels

In this section, self-assembling hydrogels arising from the association of gelators, bearing ionic and/or ionizable moieties of the same kind of charge (anionic or cationic) along the macromolecular chain, in various topologies, are presented and evaluated.

4.1 BAB-Type Hydrophobically Associated Polyelectrolytes

Polyelectrolytes, end-capped by hydrophobic blocks (TP), have attracted the attention of experimentalists, due to their well-defined (simplest) macromolecular topology that aid insights into the molecular origin of the phase behavior of their aqueous solutions and the rheological properties of the resulted hydrogels. The molecular futures of the two components of the BAB gelators govern their properties in the sol and gel state. The use of weak polyelectrolytes renders the TP pH responsive. In this case, the chain conformation of the central polyelectrolyte and its water solubility depend on the degree of ionization and can be tuned by pH. In addition, the strength of the hydrophobicity of the end-blocks affects strongly the structural and mechanical properties. The hydrophobicity depends on the length (degree of polymerization, N_B) of the end-block and the interaction parameter with water. The latter was formulated in a MC study as degree of incompatibility ($2N_B\beta\varepsilon$) for the non-ionic telechelic APs, where ε is the interaction energy and $\beta = 1/kT$ (Nguyen-Misra and Mattice 1995). As was shown, the *cmc* and the critical percolation concentration (also referred as *cgc*) decrease as the incompatibility factor increases following scaling laws, while the middle block size N_A effect on the critical concentrations was found insignificant. However, in the ionic TPs, the degree of ionization of the central block which affects its water solubility and its effective length (degree of stretching) affects the *cgc*.

Symmetric polystyrene-poly(acrylic acid)-polystyrene in its fully neutralized form (PS_{23}-$PANa_{1134}$-PS_{23}) (close to physiological pH) was the first TP explored in salt-free aqueous solutions (Tsitsilianis et al. 2000a). As it was observed, the particular molecular characteristics of this copolymer, (long, fully charged polyectrolyte block, highly hydrophobic stickers) drives the aqueous solutions to form a transparent free-standing gel from very low concentrations (*cgc* < 1 wt%), considerably lower than the non-ionic counterpart, in accordance with the theoretical and simulation predictions described above. The physical picture of the underpinning 3D spanning network is spherical hydrophobic PS nanodomains, interconnected by stretched polyelectrolyte bridging chains. For the latter, clear evidence has been provided by direct visualization of the network bridges using atomic force microscopy (Tsitsilianis et al. 2008a, b; Gotzamanis et al. 2006).

The rheological properties of the so-formed hydrogels are controlled by the lifetime, the number and the conformation of the elastically active (bridging) chains. The lifetime of the bridges depends on the exchange rate of the stickers which is related to the activation energy, *E*, needed a sticker to escape from the hydrophobic nanodomains. *E* is proportional to $\gamma N_B^{2/3}$ where γ is the interfacial tension between the sticker and the medium (Halperin and Alexander 1989) which is a combination of the length and the interaction parameter with water of the *B* end-blocks and can be tuned by macromolecular engineering. Thus, the PS_{23}-$PANa_{1134}$-PS_{23} self-assembling hydrogel exhibits long relaxation times (very low exchange dynamics) and nearly frequency-independent shear storage (G′) and loss (G″) modulus, rendering it in practice a kinetically "frozen" network (Nicolai et al. 2010) resembling in mechanical response, at the linear viscoelastic regime, to the chemically crosslinking hydrogels (Tsitsilianis and Iliopoulos 2002). Similar results have been reported for cationic-based TPs, bearing weak tertiary amine repeating units that can be ionized by protonation at low pH (Bossard et al. 2006; Gotzamanis et al. 2006).

The TP hydrogels exhibit several exceptional properties concerning their linear and nonlinear rheological response to shear. When the sticky end-blocks are highly hydrophobic (e.g., PS, $N_B > 15$), the shear viscosity profile is characterized by an apparent yield stress and complex shear thinning. The viscosity drops several orders of magnitude upon shearing, while the recovery of the structure upon decreasing shear exhibits remarkable hysteresis due to long relaxation time of the system. The zero-shear viscosity (or low-shear viscosity) of the TP solutions increases steeply with concentration (e.g., eight orders of magnitude for C_p between 0.1 and 2 wt%) (Katsampas et al. 2005) and the *cgc* (percolation concentration) appears, in most of the cases, lower than 1 wt%, for sufficient long polyelectrolyte midblocks.

Another interesting and unique property of these hydrogels is that their viscosity remains stable (Tsitsilianis et al. 2000b) or even slightly increases (Stavrouli et al. 2008a) with temperature up to 60 °C. This implies stability of the 3D structure and that the slow exchange dynamics of the stickers hold over the temperature range examined.

More importantly, since the conformation of the midblock weak polyelectrolyte of the TP gelators depends on their degree of ionization, the gel properties can be controlled by pH, rendering these gels stimulus responsive. This behavior has been demonstrated by studying the $PMMA_{32}$-b-$PDMAEMA_{224}$-b-$PMMA_{32}$ aqueous solutions as a function of pH (Bossard et al. 2006). The pH dependence of zero-shear viscosity of 1 wt% polymer aqueous solutions is shown in Fig. 8. The viscosity passes through a maximum at pH 4 in where the PDMAEMA central block is close to its fully ionized state. Under these conditions, a free-standing transparent gel was formed (inset of Fig. 8). By increasing pH above 4, a progressive deprotonation of PDAEMA occurs resulted in a rigid to flexible transition of the chain conformation,

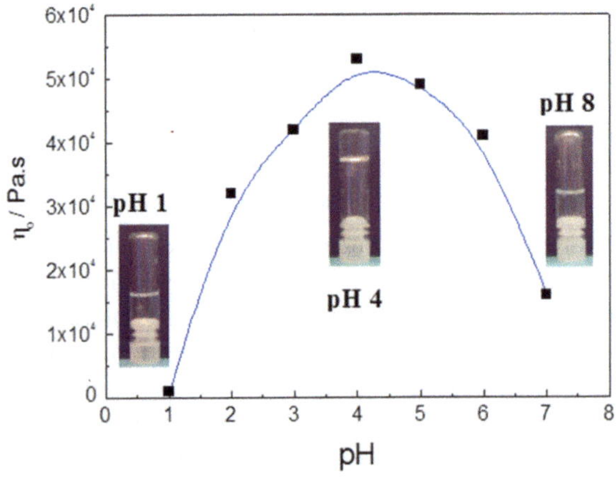

Fig. 8 pH dependence of the zero-shear viscosity of a 1 wt % $PMMA_{32}$-$PDMAEMA_{224}$-$PMMA_{32}$ polymer solution. In the inset, digital photographs of the solutions showing the formation of a stiff gel at pH 4. From Bossard et al. (2006) with permission of RSC

which in turn imposes a decrease of η_0, leading eventually to a gel to sol transition at pH 8. By decreasing pH below 4, the viscosity drops again, owing to the ionic strength augmentation of the medium, which screens the electrostatic repulsive interactions along the chain and thus decreasing the rigidity of the PDAEMA. This effect again leads to a gel to sol transition at pH 1.

In order to control the exchange dynamics of the stickers of the associative TP gelators and in turn the network structure, along with the rheological properties of the formed hydrogel, an interesting strategy was developed the recent years. Instead of using as stickers pure hydrophobic blocks (e.g., PS, PnBMA), random copolymers constituted of hydrophobic and hydrophilic repeating units, the latter being ionic and thus pH sensitive, were incorporated as end-blocks to TP gelators [in fact (B-*co*-C)-*b*-A-*b*-(B-*co*-C) terpolymer topology]. Two types of gelators were explored, comprising both PAA as the central block end-capped either by statistical P(nBA-*stat*-AA) (Charbonneau et al. 2011) or gradient P(S-*grad*-AA) copolymers (Borisova et al. 2011).

The rheological properties of the hydrogels made by a $P(nBA_{50\%}\text{-}stat\text{-}AA_{50\%})_{101}\text{-b-}PAA_{204}\text{-b-}P(nBA_{50\%}\text{-}stat\text{-}AA_{50\%})_{101}$ were thoroughly investigated as a function of pH which controls the degree of ionization, α, and thus the strength of the hydrophobicity of the stickers (Charbonneau et al. 2011). The study of the concentration dependence of the steady-state shear viscosity of the aqueous solutions (Fig. 9a) revealed a remarkable effect of α on the percolation concentration (equivalent to *cgc*). As can be seen in Fig. 9b, C_p increased dramatically with α about 25 times in the range of $0.3 < \alpha < 0.9$. The increase of α results to lowering of the hydrophobicity of the stickers, and the same time to increasing of the hydrophilicity (ionization too) of the central PAA block, which both shift the hydrophobic/hydrophilic balance toward lower values. This should affect the functionality of the network (number of bridging chains coming out from the reticulation nods). Thus, as this balance decreases with α, the association ability of the copolymer weakens and more polymers are needed to form a 3D spanning network.

Even more dramatic changes were observed in the relaxation times, τ, versus α (Fig. 9c). For the TP with 50% AA incorporated in the end-blocks (TH50 sample), τ decreased exponentially over 9 decades from 10^6 to 10^{-3} s in the range $0.3 < \alpha < 0.7$. This effect is fully reversible implying that the hydrogel can become frozen (long τ) or dynamic (short τ) by suitably varying α, controlled by pH. The effect of increasing α on τ was ascribed to the decreasing hydrophobicity of the P (nBMA-co-AA) stickers, due to repulsive interactions exerted by the created charged AA units, which facilitates their escape from the hydrophobic nanodomains (physical crosslinks). It was also shown that the hydrophilic content (AA percentage in the end-blocks) of the stickers affect significantly their exchange dynamics (Shedge et al. 2014). For a given degree of ionization, τ increases remarkably with decreasing the AA content (Fig. 9c). For instance, at $a = 0.5$ (pH 5), the relaxation time increases more than 11 orders of magnitude by decreasing the percentage of AA from 60 (TH60) to 40% (TH40). Simultaneously, the percolation concentration decreases remarkably with decreasing AA percentage.

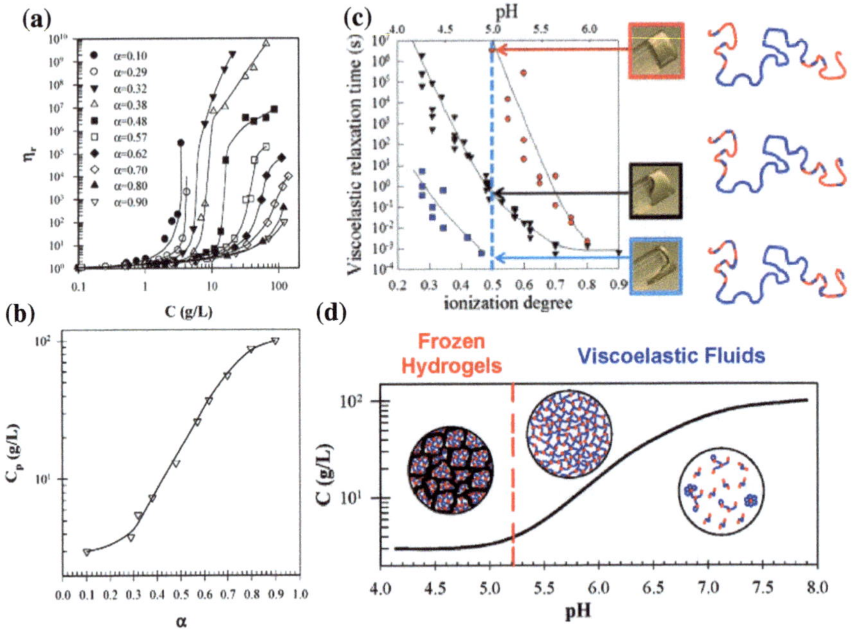

Fig. 9 **a** Concentration dependence of the steady-state shear viscosity of the P(nBA$_{50\ \%}$-*stat*-AA$_{50\ \%}$)$_{101}$-b-PAA$_{204}$-b-P(nBA$_{50\ \%}$-*stat*-AA$_{50\ \%}$)$_{101}$ aqueous solutions at different degrees of ionization (α) of AA units. **b** Percolation concentration C_p with. **c** relaxation time, τ, versus α at different hydrophilic contents (AA percentage) of the stickers (schematics). **d** State diagram of the percolation concentration as a function of pH: for $C > C_p$ and pH > 5.2 the system forms a transient network behaving as a viscoelastic liquid; for $C < C_p$ and pH > 5.2, free polymers, micelles, and clusters coexist; for $C > C_p$ and pH < 5.2, the system is dynamically frozen behaving as an elastic soft solid. From Charbonneau et al. (2011), Shedge et al. (2014) with the permission of ACS

The whole set of data suggested a transition between kinetically frozen hydrogels, exhibiting elastic response, and dynamic ones exhibiting viscoelastic response. A state diagram of the percolation concentration as a function of pH was suggested for TH50 as it is demonstrated in Fig. 9D. Several regions could be distinguished. For $C > C_p$ and pH > 5.2, the system forms a transient network behaving as a viscoelastic liquid. For $C < C_p$ and pH > 5.2, free polymers, micelles, and clusters coexist. Finally, for $C > C_p$ and pH < 5.2, the system is dynamically frozen behaving as an elastic soft solid. In this state, the structure depends on the preparation method and probably consists of densely packed microgels (Charbonneau et al. 2011). Evidently, the state diagram will be differentiated by varying the AA content of the end-blocks as discussed above.

Another strategy to tune the exchange dynamics and in turn the rheological properties of TP hydrogels is to decrease γ by using solvent mixtures rather than pure water. This idea was exemplified studying the rheological properties of salt-free aqueous solutions of PS$_{23}$-PANa$_{1134}$-PS$_{23}$ in water/DMF binary solvent mixture. DMF is a good solvent of PS, and thus by increasing its content in the solvent mixture, γ decreases facilitating the exchange dynamics of the stickers (Tsitsilianis et al. 2010).

4.2 Associative Polyelectrolytes of Complex Macromolecular Topologies

Macromolecular engineering offers numerous possibilities to design associative polyelectrolytes capable to form 3D transient networks with macromolecular architecture other than the simplest BAB triblock copolymer topology. Thus, multiblock copolymers (BABABAB), ABC block terpolymers, and branched topologies (star-shaped or graft copolymers) have been designed, studied, and evaluated as polyelectrolyte-based hydrogelators.

In order to explore the potential influence of the macromolecular topology on the self-association and gelation ability of the associative polyelectrolytes and to improve the design of efficient ionic polymeric hydrogelators, a series of well-defined linear amphiphilic polyelectrolyte multiblock copolymers were considered. Particularly, model ABA triblock, BABAB pentablock, ABABABA heptablock, and BABABABAB nonablock copolymers, with uniform length of their building blocks, were synthesized by "living" group transfer polymerization (GTP), where the B block is the ionisable poly[2-(dimethylamino)ethyl methacrylate] (PDMA) and the A block is the hydrophobic poly(*n*-butyl methacrylate) (PBuMA). Two series of such copolymers were studied: copolymers with constant length of the A, building blocks varying in overall molecular weight and composition, and copolymers with constant overall molecular weight and hydrophobic content. In the latter case, the size of the A, B blocks was varied (Popescu et al. 2010).

The ability of the multiblock copolymer to form a 3D network was governed by the interplay of the hydrophobic attractive interactions of the hydrophobic A blocks and the electrostatic repulsive interactions of the polyelectrolyte B blocks. Although they form small aggregates at low concentrations, the polymers with polyelectrolyte segments at both ends (e.g., D_{25}-*b*-B_5-*b*-D_{25}-*b*-B_5-D_{25}, B_5-*b*-D_{25}-*b*-B_5-*b*-D_{25}-*b*-B_5-D_{25}-*b*-B5) cannot thicken the solution even at a 10 wt% polymer concentration which could be attributed to the repulsive electrostatic interactions of the outer charged segments that prevented further growth of the aggregates. It should be noticed here that this effect should not be taken as generic because the relative block lengths play important role (see below). On the other hand, for the same architecture (e.g., heptablocks: B_5-*b*-D_{25}-*b*-B_5-*b*-D_{25}-*b*-B_5-*b*-D_{25}-*b*-B_5 and B_3-*b*-D_{20}-*b*-B_3-*b*-D_{20}-*b*-B_3-*b*-D_{20}-*b*-B_3) of nearly the same hydrophobic content, the heptablock with longer B_5 stickers exhibits remarkable higher gelation efficiency with improved rheological properties in agreement with theoretical and experimental reports.

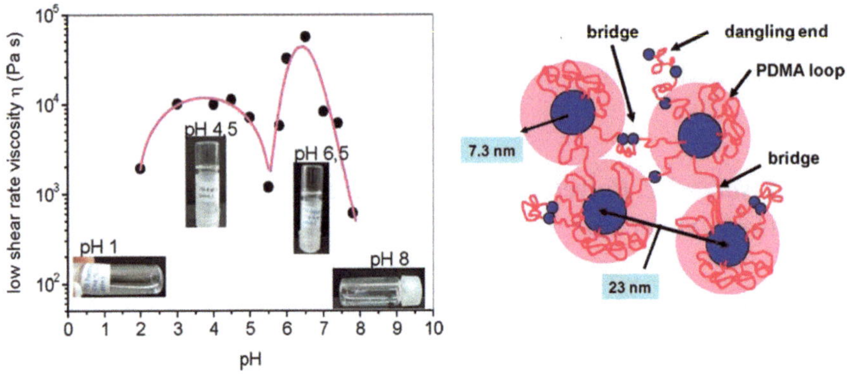

Fig. 10 Variation of the low-shear viscosity, η (together with digital photographs of the samples) as a function of pH of a 4 wt% heptablock copolymer aqueous solution at 25 °C and schematic representation of the structure of the micellar transient network as deduced from the analysis of the sans data. From (Popescu et al. 2012) with the permission of ACS

The pH dependence of the heptablock copolymer was accordingly investigated by rheology and SANS (Popescu et al. 2012). A systematic investigation of the pH dependence of the low shear viscosity of 4 wt% heptablock copolymer aqueous solutions revealed a complex behavior which did not appear in the triblock TPs. As shown in Fig. 10, the viscosity passes through two maxima, at pH 4.5 and 6.5, where free supporting gels were formed. The highest viscosity (and elastic modulus) was observed at pH 6.5 followed by dramatic drop, down to pH 8, where the PDMA blocks have been almost deionized ($\alpha \sim 0.1$). Yet, the viscosity drops in low pH probably are due to electrostatic screening arising from elevated ionic strength. The different pH-responses of the heptablock versus the triblock copolymer (Fig. 10) should be account for the different macromolecular topologies of the polymer which affects the internal structure of the 3D network (Popescu et al. 2012).

Multiblock-associative polyelectrolytes of very high molecular weights, prepared with a micellar radical polymerization method, exhibit strong thickening ability in aqueous media (Kujawa et al. 2004). These APs are in fact random terpolymers bearing blocky hydrophobic sequence (stickers) of *N,N*-dihexyl acrylamide monomers (2 mol%) randomly distributed along a hydrophilic random copolymer constituted of acrylamide (86 mol%) and the negatively charged sodium 2-acrylamido-2-methylpropane sulfonate (12 mol%). The synthetic method allows control of the length (denoted as the number of repeating units, N_H) and thus of the hydrophobic strength of the stickers, along with the very high molecular weight achievement. The latter is advantageous, since it is hard to be accomplished by controlled polymerization techniques. The length of the stickers, N_H, affects significantly the rheological properties. As the length of the hydrophobic sequence increases from N_H 3 to 7, the percolation threshold for the network formation shifts to lower concentrations. The viscosity augmentation above this critical concentration is more pronounced for N7, characterized by a steeper (almost jump) viscosity enhancement, which attains eight orders of magnitude higher than the medium at concentration as low as 1 wt%, showing strong thickening capability. The study of the dynamics of the polymers in the semidilute unentangled regime revealed that the relaxation process of the polymer solutions strongly slows down as N_H increases. For N7, long relaxation times, of the order of thousands of seconds classify this polymer to strong gelators. In contrast, much shorter relaxation times (40–50 s) were observed in the same concentration range for N3 with shorter hydrophobic blocks.

Due to the presence of strong ionized groups along the chain, these APs are sensitive to ionic strength. Comparing the viscosity versus concentration profiles between the salt-free solutions and those in 0.1 N NaCl (Fig. 11), the following remarks can be concluded. The effect of the ionic strength on the rheological properties depends on the polymer concentration range and the N_H. For N7, the percolation threshold shifts to higher concentration in 0.1 N NaCl and almost coincides with that of N3 in salt-free solutions, which is not affected by the presence of salt. Thus, under ionic strength of 0.1 N NaCl, it seems that the critical concentration is independent on N_H, which is in strong contrast to the behavior in salt-free solutions. The effect of the ionic strength on the rheological properties depends on the polymer concentration range, and it is not appeared in the entangled

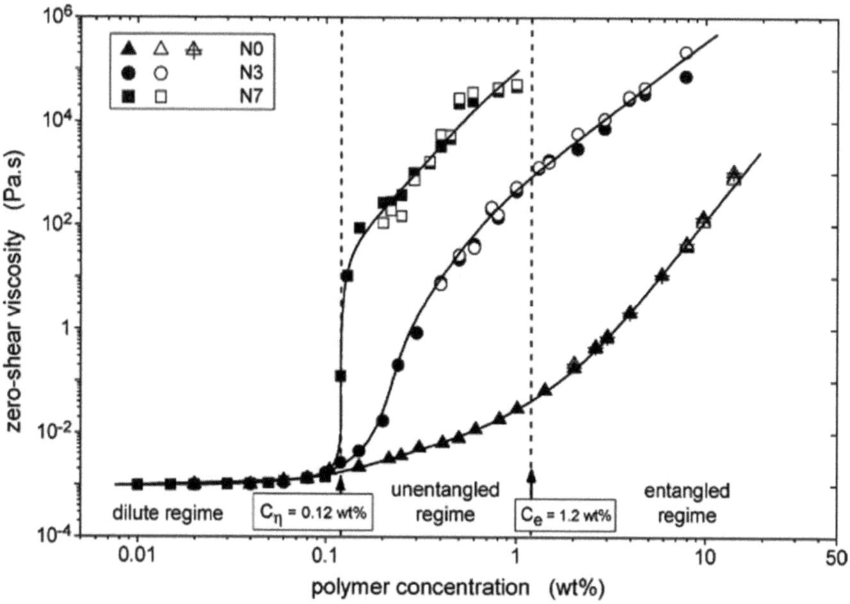

Fig. 11 Concentration dependence of the zero-shear viscosity of multiblock associative polyelectrolytes in 0.1 M NaCl aqueous solutions and the hydrophobe-free homologue. Data from steady state (solid symbols), creep (open symbols), and oscillatory (cross-centered symbols). From (Kujawa et al. 2006) with the permission of ACS

regime. Above the critical C_η, in the unentangled regime, the zero-shear viscosity and the plateau modulus decreased upon the addition of salt but the terminal relaxation time (correlated to the lifetime of the junctions) seems to be slightly affected. Thus, the effect on the modulus could be attributed to the screening electrostatic effect which influences the effective size (rigidity) of the macromolecular chain and in turn the number of elastically active chains (bridges) of the network, below the entanglement concentration. Above this concentration, the dynamics is no longer controlled by the coil dimensions but is mainly controlled by the entanglements and the mesh size of the temporary network (Kujawa et al. 2006).

Amphiphilic ionic star-block copolymers, of the type $(AB)_4$, with Poly(acrylic acid) as internal A blocks and PS outer sticky blocks were explored aiming to investigate the effect of star polymer topology and the length of the stickers (strength of hydrophobicity) (Hietala et al. 2007). The $(PAA_{54}\text{-}PS_6)_4$ stars self-associate forming 3D networks, as in the case of TP, with percolation concentrations of the order of 2 wt% at 20 °C. Comparing samples with similar molar mass and composition, star-block copolymers exhibited more elastic behavior than the linear triblock counterpart. Moreover, the length of the PS end-blocks of the arms affected the rheological properties of the formed hydrogels. Stars with shorter stickers resulted in gels with longer linear response followed by strain hardening prior shear thinning. The plateau modulus increased with the length of stickers. However, the longer the stickers, the more "fragile" the gel, as characterized by short linear viscoelastic regime. In the light of the recent findings (Charbonneau et al. 2011), the stars with the higher length of PS blocks leads to "frozen" gels, (very low exchange dynamics of stickers) which implies that the network structure and the properties depend on the preparation conditions. Thus, for the stars with the shorter PS blocks, a homogeneous network is expected to form which in turn exhibits linear regime expansion. This effect is similar to that observed when DMF (good solvent for PS) was added to the aqueous media of PS-PAA-PS solutions (Tsitsilianis et al. 2010). In both cases, the underpinning effect is the augmentation of the sticker exchange dynamics, leading to more homogeneous 3D network structure.

Triblock terpolymers constituted of three different monomeric units (ABC type) have been considered to design associative polymeric hydrogelators. The ABC architecture offers additional possibilities since the three blocks integrate three different functionalities on the macromolecules. The simplest ionic associative terpolymer comprises a weak polyelectrolyte long middle block, end-capped by two, different in chemistry, hydrophobic stickers, also named heterotelechelic polyelectrolytes (Tsitsilianis et al. 2000b; Katsampas and Tsitsilianis 2005). These APs self-assemble through hydrophobic interactions forming percolated 3D networks with rheological response similar to those of TPs. Comparing with the ABA TP, the heterotelechelic TP affords some advantages as it will be presented below.

4.3 Ionogenic pH-Responsive Blocks as Stickers in Associative Gelators

Weak polyelectrolyte blocks that undergo a hydrophilic–hydrophobic phase transition, triggered by switching pH, have been used as building blocks to design "smart" hydrogels in various topologies. For instance, poly(*tertiary*-amine methacrylates) like poly(diisopropylo amino)ethyl methacrylate) PDPA or PDEA becomes hydrophobic above their pK_as inducing gelation when these blocks end-cap a hydrophilic chain (BAB type) (Medsen and Armes 2012). Thus, biomimetic Poly (methacryloyl)ethyl phosphoryl choline)-(PMPC)-based block copolymers of the type PDPA-PMPC-PDPA, PDEA-PMPC-PDEA, and $(PMPC\text{-}PDPA)_3$ 3-armed star-block copolymers have shown fine gelation ability upon increasing pH. The more efficient gelator was evaluated to be the star-block counterpart, since it exhibited the lower critical gelation concentration (i.e., 5 wt%). In addition, the gel was formed at physiological pH (Li et al. 2005). This kind of hydrogelators seems to be promising candidates for biomedical applications thanks to their good biocompatibility (Lomas et al. 2008). For instance, PDPA-PMPC-PDPA has been used as drug delivery system to release the cardiovascular hydrophobic drug dipyridamole in a very well-controlled manner which depends on the polymer concentration (Ma et al. 2003).

Analogous pH-inducing hydrogelator has been prepared with tertiary amine methacrylates in both blocks (PDPA-PDMA-PDPA) (Taktak and Bütün 2010). The central PDMA exhibits higher pK_a, and hence, its hydrophilicity is preserved around the physiological pH while the end-blocks was turned to hydrophobic. This hydrogelator was tested as drug delivery system. Preliminary studies indicated that sustained release of a model hydrophobic drug can be achieved by tuning the solution pH, polymer molecular weight, and polymer concentration. In addition, the thermoresponsive nature of the middle PDMA block (exhibits LCST) affects the drug release kinetics.

P2VP exhibiting lower pK_a (c.a. pH 5) has also been used as the pH-triggered block in a P2VP-b-PAA-b-poly(n-butyl methacrylate) ($P2VP_{25}$-PAA_{576}-$PnBMA_{36}$) ABC triblock terpolymer topology, resulted in a highly efficient pH-responsive gelator (Katsampas et al. 2005). Particularly, at physiological pH, P2VP becomes neutral, and thus, hydrophobic PAA is nearly fully ionized. Thus, the triblock terpolymer is transformed to a hydrophobically end-capped heterotelechelic polyelectrolyte which self-assembles forming flower-like aggregates at low concentrations with compartmentalized cores consisted of PnBMA internal and P2VP external nanodomains, stabilized by ionized PAA looping chains, that bridge adjacent cores at higher concentrations forming a 3D network (Tsitsilianis et al. 2008a). Indeed at a polymer concentration as low as 1 wt%, a sol–gel transition was observed at pH 5 rendering this formulation an injectable hydrogel. Popescu et al. have taken the advantage of this pH-triggered hydrogelator to design a controlled drug delivery system (Popescu et al. 2011). A solution of the terpolymer at low pH was mixed with liposomes loaded with calcein as model hydrophilic drug.

Upon raising pH to physiological, a hydrogel, hosting the liposomes, was formed (Fig. 12). The resulted nanocomposite gel exhibited a nearly elastic soft solid behavior, with improved viscoelastic features with respect of the plain hydrogel while the liposomes remained intact within the gel matrix. Sustained drug release was achieved which can be controlled by regulating the gelator concentration; i.e., from 1 to 1.5 wt%, the drug release period was significantly prolonged from 14 to 32 days (Fig. 12). Cell viability test showed minimal cytotoxicity of the terpolymer, which makes this liposome/hydrogel nanocomposite formulation promising as tunable sustained drug release system.

4.4 Dual pH/Thermoresponsive Gelators

Thermo-sensitivity can be induced to block copolymers and/or terpolymers by incorporating a hydrophilic block exhibiting lower critical solution temperature (LCST) in aqueous solutions. Thus, by replacing the hydrophobic end-blocks of a telechelic polyelectrolyte with such LCST-exhibiting blocks a dual pH/thermoresponsive gelator can be designed. The first attempt toward this strategy

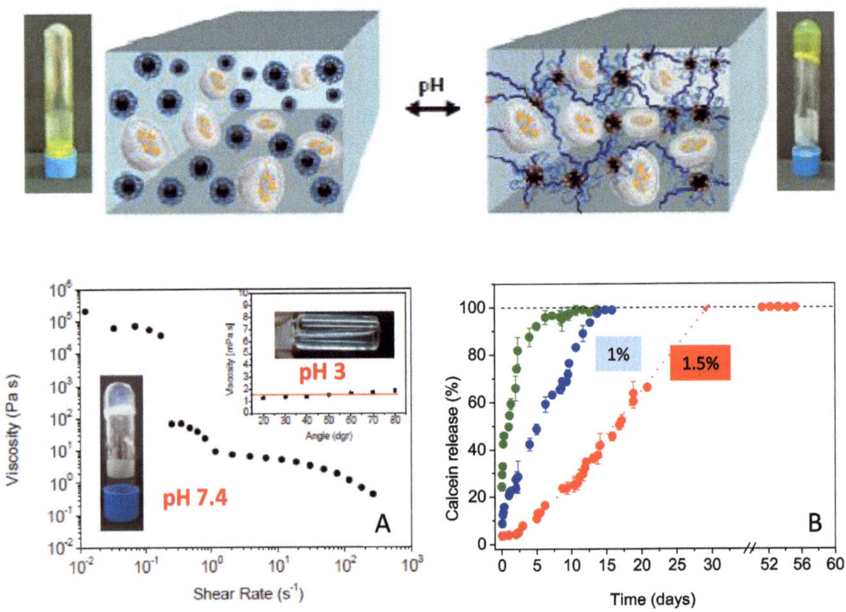

Fig. 12 Schematic representation of the hydrogel (encapsulating liposomes) formation triggered by switching pH. **a** Viscosity profiles of P2VP-PAA-PnBMA (gelator) aqueous solutions at pH 3 and 7.4. **b** Sustained calcein release from the liposome/hydrogel formulation at physiological pH at different gelator concentrations. From Popescu et al. (2011) with the permission of ACS

was accomplished by synthesizing a poly(*N*,*N*-diethyl acrylamide)-b-poly(acrylic acid)-b-poly(*N*,*N*-diethyl acrylamide) (PDEAm-PAA-PDEAm) triblock copolymer by anionic polymerization (Angelopoulos and Tsitsilianis 2006). This copolymer is in fact a double hydrophilic TP, exhibiting a sol–gel transition in physiological pH upon increasing temperature above the LCST of PDEAm blocks, transformed to hydrophobically end-capped polyelectrolyte. At concentration as low as 3 wt%, a transient 3D network was formed comprising PDEAm hydrophobic domains interconnected by PAA negatively charged elastic chains in a similar manner with the TPs described above. The system is also sensitive to ionic strength which affects both the LCST of PDEAm and the conformation of PAA. More importantly, the presence of salt decreases the sol–gel transition temperature as it lowers the LCST (salting-out effect). Although this particular system exhibits sol–gel transition, out of the physiological temperature and ionic strength, hence not exhibiting potentials for biomedical applications, it provides proof of concept, for the design of thermoresponsive TPs efficient gelators with tunable LCST end-blocks.

A good example of the above concept was reported recently involving an O-phosphoethanolamine(PEA)-grafted poly(acrylic acid) (PAA(PEA)) flanked at both ends by poly(N-isopropylacrylamide) (PNIPAM) blocks. The partial phosphorilation of the PAA central block was dictated from two reasons: first to avoid strong interpolymer complexation (H-bonding) between PNIPAM and PAA which leads to precipitation second to provide higher affinity for calcium ions which will benefit the biomineralization process, targeting potential application as scaffold for bone tissue engineering (Lin et al. 2013). At a relatively low concentration (2 w/v% in phosphate buffered saline), the triblock copolymers can form freestanding gels at physiological temperature 37 °C. The gel properties are highly dependent on both the length of the PNIPAM thermo-stickers and the substitution degree of phosphate group.

ABC linear triblock terpolymer topologies offer also strong potential to design pH-thermoresponsive gelators. Hence stimuli-responsive P2VP-b-PEO-b-P(GME-co-EGE) triblock terpolymers, constituted of the pH-responsive P2VP, non-ionic hydrophilic PEO and the thermo-sensitive third block, consisting of a glycidyl methyl ether (GME)/ethyl glycidyl ether (EGE) random copolymer, were evaluated as hydrogelators (Reinicke et al. 2009). At the physiological pH regime, P2VP is hydrophobic and the P(GME-co-EGE) turns hydrophobic upon heating thus transforming the terpolymer to hydrophobically end-capped, capable to form a 3D network. Yet, the gelation temperature depends on the LCST of the P(GME-co-EGE) which can be adjusted by the comonomer ratio. At remarkably elevated concentrations (18 wt%) and pH 7, an unexpected gel–sol–gel transition was observed upon increasing temperature. Although the high sol–gel transition is due to the P(GME-co-EGE) coil-globule transition, the low gel phase was attributed to a cubic arrangement of spherical core–shell–corona micelles with P2VP cores, as revealed by SANS experiments. Moreover, at pH 3.5, a thermo-induced gelation was observed due to dense packing of the core-shell-corona micelles with P(GME-co-EGE) cores and ionized (extended) P2VP corona chains.

As we have shown above, star-block copolymers are suitable topologies to form 3D networks and thus can be used to design pH/thermoresponsive gelators. This was accomplished by combining pH-sensitive weak polyelectrolyte (PDMA) as inner blocks and thermo-sensitive (PDEGMA) outer blocks in stars with 4 and 6 arms (Schmalz et al. 2012). Therefore, in concentrated aqueous solutions, 3D networks were formed upon increasing temperature above the LCST of the outer PDEGMA blocks, which are transformed to hydrophobic stickers. Interestingly, the inner PDMA exhibits LCST, in the non- or very low ionized form, which is strongly dependent on pH in the range of 7–9. Provided that at pH 7 the LCST of PDMA is high (about 80 °C) the thermo-sensitivity of the copolymer is manifested only from the outer (PDEGMA) blocks. When the gels are prepared at pH 9 and the temperature is increased further above the transition temperature of the PDMA, they exhibit significantly reduced gel strength due to the shrinkage of the PDMA inner block chains of the arms.

By incorporating ionic groups in segmented macromolecules, exhibiting thermal triggered association toward 3D networks, enrich their properties rendering them pH responsive as well. Due to the fact that these hydrogelators are constituted mainly of non-ionic hydrophilic segments and the ionic moieties aid to adjust the LCST properties, they fall outside of the scope of this chapter and will not be further considered. The readers can find some interesting thermo/pH-responsive systems in a recent review (Chassenieux and Tsitsilianis 2016).

4.5 Effect of Secondary Ionic Interactions on Telechelic Polyelectrolytes

The polyelectrolyte nature of the middle block of the associative TPs, enables complexation with multivalent cations, increasing further the crosslinking density of the network through an additional mechanism other than the hydrophobic association of the hydrophobic end-blocks. This strategy has been used for enhancing mechanical performance in an attempt to achieve properties similar to those reported for double network hydrogels with permanent crosslinks. The above idea was exemplified with a PMMA-PMAA-PMMA physical gel, further cross-linked by divalent acetates of different cations (Zn, Ca, Ni, Co, Cu) in controlled pH environments (Henderson et al. 2010). High strength hydrogels were achieved, the properties of which depend on a number of factors (i.e., concentration and type of cation, pH) affecting the magnitude of crosslinks and the bond stability and in turns the hydrogel stiffness. Concerning cation identity, the bond strength increased according to the sequence: Ca < Co(I) < Ni(II) < Zn(II) < Cu(II). It should be underlined here, that the observed differences in bond strengths between Ca and Zn cross-linked systems can be attributed to the different interactions exerted by the cations of group II metals, (e.g., Ca) and the transition metals (e.g., Zn). In the former case, only electrostatic interactions are developed with the ionized polyelectrolytes, while transition metals adopt more complicated interactions according to the electronic state of the cations' d-orbital configuration. Finally, pH shifts the required concentration for cross-linking but does not affect the final cross-linked state. In fact, pH influences the associative ability of midblock carboxylates through the protonation-deprotonation equilibrium.

5 Hydrogels from Macromolecules Bearing Oppositely Charged Groups

Hydrogels formed by polymers bearing positively and/or negatively charged moieties along the macromolecular chain have been developed the recent years. These systems can be classified to several categories depending on the macromolecular architecture, i.e., block or random polyampholytes, as well as on the number of polymer constituents involved to the gelling system, i.e., associative polyampholytes or oppositely charged polyelectrolyte mixtures.

5.1 *Block Polyampholyte-Based Hydrogels*

Double hydrophilic PAA-P2VP-PAA block polyampholytes exhibited an interesting associative behavior by varying pH from the acidic to basic aqueous solutions (Sfika and Tsitsilianis 2003). Due to the ionic interactions of the oppositely charged segments and the relative low *pK* of P2VP (becomes hydrophobic at pH > 5), this copolymer precipitates in the pH region between 4 and 7, centered by the isoelectric point (*iep*: equal number of opposite charges at pH 5.5) of the polyampholyte. By investigating the association capability of PAA_{134}-$P2VP_{628}$-PAA_{134} in salt-free solutions in the soluble pH regimes, a gel phase was observed in the acidic regime (pH < 4 where the triblock behaves as polyampholyte) in the vicinity of *iep*, despite the lack of hydrophobic stickers. Particularly at pH 3.4 the aqueous solutions exhibited a critical concentration of 2.5 wt% polymer, above which the viscosity raised steeply. Freestanding gels form at concentrations above 4 wt% where the relaxation times are higher (τ > 300 s) than the observation time (Bossard et al. 2004). Evidently, a 3D reversible network was formed which was attributed to intermolecular association through attractive interactions of oppositely charged moieties between the ionized protonated P2VP and a small amount of deprotonated negatively charged acrylic acid moieties of the outer blocks. The suggested physical picture of such a network comprises the physical crosslinks, constituted of the so-called interpoly electrolyte complex (IPEC) which are electro-neutral and thus precipitate out of the aqueous solution, interconnected by the P2VP positively charged uncomplexed long segments.

In the quaternized P2VP version of this triblock, where the P2VP was transformed to strong polyelectrolyte and pH variation controls only the charge density of the PAA outer blocks, again a freestanding gel was formed at the same pH region corroborating the proposed association mechanism for this kind of ABA gelators (Dyakonova et al. 2014). More importantly, the as-form charge-driven network structure, appeared at a minimum charge molar ratio, $[AA^-]/[2VPR^+]$, of ca. 6%. Upon increasing pH, the charge asymmetry decreases and extensive ionic interactions between the opposite charges resulted to uncorrelated dense microgels at pH 7, as revealed by SANS. This is the reason why the viscosity of the solution drops abruptly about 3 orders of magnitude even at pH 5, where core-shell particles aggregate into weakly connected fractal clusters.

Although the PAA-P2VP-PAA polyampholyte hydrogels form in acidic media fall outside of potential biomedical applications, they can inspire designing other ABA triblocks since this macromolecular topology is highly versatile and synthetically easy to be implemented by controlled polymerization methods. For instance, a negatively charged long polyelectrolyte, end-capped by positively charged weak poly(*tertiary* amine methacrylates) blocks could give analogous gels in the vicinity of physiological pH.

Provided that the PAA-P2VP-PAA associative polyampholyte arises from its PtBA-P2VP-PtBA precursor of the same degree of polymerization, which in fact is an associative telechelic polyelectrolyte, it was possible to compare the properties of these kinds of gelators in terms of gelation efficiency, mechanical response, stimuli responsiveness (Stavrouli et al. 2008a) and network nanostructure (Dyakonova et al. 2014). Thus comparing the hydrogel properties made by the self-assembly of TP ($PtBA_{109}$-b-$P2VP_{695}$-b-$PtBA_{109}$) and block polyampholytes (BP) (PAA_{109}-b-$P2VP_{695}$-b-PAA_{109}) and (PAA_{109}-b-$qP2VP_{695}$-b-PAA_{109}) at low pH conditions, the following results were observed. The TP exhibits lower C_{gel} higher elastic modulus and longer relaxation times than those of BPs, implying higher gelation efficiency and stronger gels for TP. These differences should be account for the different network nanostructures formed through different interactions, i.e., hydrophobic for TP versus ionic for BP. Thus the resulted networks differ remarkably as revealed by SANS experiments. Particularly, the size of the hydrophobic domains, their mean distance (hard sphere radius) and thus the mesh size are higher in TP (Dyakonova et al. 2014).

On the other hand, weak BP is much more sensitive to pH, exhibiting also an unexpected thermo-thickening effect, whereas TP remained stable upon heating (Bossard et al. 2005). The rheological data revealed a sol/gel-like transition, manifested by more than 2 orders of magnitude viscosity and relaxation time enhancement upon heating from 10 to 35 °C, while beyond this temperature the hydrogel exhibited Arrhenius behavior. The fact that none of the polymeric components of the copolymer exhibit LCST, which was the only reason known so far to induce a thermo-thickening effect in APs, render this phenomenon worthy of attention, while its interpretation should deal with the extent of complexation of the oppositely charged components of the polyampholyte. This peculiar thermal response was attributed mainly to the UCST behavior of the PAA end-blocks which is about 15 °C. Thus coil expansion, occurring upon heating, allowed the intermolecular interactions among the oppositely charged moieties to develop, favoring intermolecular bridging enhancement as corroborated by the elastic modulus augmentation. Yet, shift of pK_a of P2VP to lower pH with temperature will increase the degree of ionization, favoring thus complexation with PAA.

The influence of ionic strength on the network and rheological properties of weak (PAA-P2VP-PAA) and strong (PAA-qP2VP-PAA) BPs were investigated recently (Dyakonova et al. 2015). The PAA-P2VP-PAA hydrogel preserves its elastic response even at 0.5 M NaCl, since G′ remained higher than G″, and both were independent of frequency. Upon increasing salt concentration, the elastic modulus initially increased passing through a maximum at 0.05 M NaCl, while it becomes lower than that in salt-free solution above 0.2 M. Simultaneously, the zero-shear viscosity and the relaxation time was continuously and remarkably decreased with ionic strength. This behavior implies that the expected screening of the electrostatic interactions (either repulsive along the P2VP segments or attractive between oppositely charged segments), imposes structural rearrangements, namely number density of the elastically active chains of the network, and junction size and exchange dynamics of the end-blocks along with the increasing of the hydrophobic interactions exerted by the non-ionized 2VP moieties, resulted eventually to a softer hydrogel. Computer simulation and SANS investigations revealed rearrangements in the network nanostructure induced by the presence of salt and pointed out differences between weak (P2VP) and strong (qP2VP) polyampholytes. Overall, due to the lower charge density and higher hydrophobicity, the weak BP exhibited less sensitivity to the ionic strength than the strong counterpart.

Integrating TP and BP into a single macromolecule, a novel associative terpolymer of pentablock topology, composed of the triblock polyampholyte (PAA-P2VP-PAA) end-capped by PMMA hydrophobic outer blocks (stickers), was designed and its ability to form hydrogels at low and high pH was explored (Tsitsilianis et al. 2008b). Particularly, a $PMMA_{119}$-PAA_{355}-$P2VP_{296}$-PAA_{355}-$PMMA_{119}$ was investigated in aqueous solutions in the presence of MeOH (30 wt%) which is a good solvent of the inner blocks, facilitating solubility which is not easy in pure water. This terpolymer self-assembles, mainly through hydrophobic association of the PMMA stickers, creating 3D networks both in acidic and basic conditions. Thanks to pH-responsiveness of the central triblock polyampholyte and the secondary ionic interactions developed by varying pH, the resulted networks exhibited distinct internal structures and functionalities. At low pH, the hydrogel is positively charged (protonated P2VP predominates) while it is transformed to negatively charged hydrogel (deprotonated PAA) at high pH. Moreover, the central P2VP block is functioning as additional sticker since it becomes hydrophobic above pH 5. Thus, the different internal network structures at different pH regions induced significant differences on the rheological properties of the hydrogels. In basic conditions the gel exhibited significant lower C_{gel} (0.7 wt% vs. 1.5%), much higher elastic modulus G′ (1 Hz) (175 Pa vs. 25 Pa), apparent yield stress (200 Pa vs. 5 Pa) and shear viscosity (two orders of magnitude higher) regarding those in acidic conditions.

In the high pH regime the hydrophobic P2VP central block is swellable in MeOH motivating thus investigation as a function of MeOH content in MeOH/H_2O binary solvent mixtures. In 1 wt% polymer concentration, a sol–gel–sol transition was observed, induced by varying the MeOH content which affects the solvent selectivity and the dielectric permittivity of the medium. Thus the increase of MeOH in the binary solvent mixture, resulted to two opposite effects dealing with the conformation of the central blocks, i.e., P2VP swelling (due to solubility improvement) and PAA unstreching (due to dielectric permittivity decrease). Thus, the gel appearance in 20–40% v.v. MeOH was attributed to the conformational changes of the PAA-P2VP-PAA central part (potential bridging chains) of the pentablock terpolymer which influence remarkably the network connectivity.

Interestingly, a thermo-thickening effect was also observed in the low pH regime. More precisely, the zero-shear viscosity increased about one order of magnitude and the apparent yield stress about four times upon heating from 25 to 35 °C, suggesting a substantially strengthening of the network.

Another pentablock terpolymer ($PMMA_{43}$-PAA_{24}-$P2VP_{650}$-PAA_{24}-$PMMA_{43}$) with different relative lengths of the various blocks was also explored in acidic aqueous dilute solutions (Stavrouli et al. 2008b). Again the gelling efficiency of the terpolymer was very high since a freestanding gel was achieved at concentration as low as 0.8 wt%. More importantly, a remarkable thermo-sensitivity (3 orders of magnitude viscosity increase upon heating) was observed at pH 3.6 whereas this effect was not observed at pH 2. This phenomenon resembles to that observed in the bare PAA-P2VP-PAA polyampholyte at the same pH, as described above, which suggests being of the same origin. Thus, the pentablock seems to integrate the thermoresponsive properties of the triblock polyampholyte, combining the intermolecular hydrophobic and ionic interactions (affected by temperature) among the different blocks of the terpolymer.

In conclusion, the suggested pentablock topology with the particular block sequence exhibits multi-functionality that leads to multi-responsive self-assembling hydrogels, the properties of which can be tuned by adjusting the monomer composition (block size) and the various external conditions like pH, temperature and dielectric permittivity of the medium.

5.2 Random Polyampholyte-Based Hydrogels

Instead of block type, statistical copolymers bearing oppositely charged moieties, randomly distributed along the chain, have been used to design ABA hydrogelators either as the hydrophilic central block or as pH-triggered end-block stickers.

A Poly(2-(diethylamino)ethyl methacrylate-*co*-methacrylic acid) end-capped by hydrophobic shorter PMMA blocks, PPMMA-*b*-(DEA-*co*-MAA)-*b*-PMMA, named telechelic polyampholyte, in analogy to telechelic polyelectrolytes, was synthesized and investigated in aqueous media of controlled pH (Gotzamanis et al. 2016; Dyakonova et al. 2017). This associative polymer self-assembles in the bulk-forming cylindrical hydrophobic PMMA nanodomains within the P(DEA-*co*-MAA) matrix, which exhibits an *iep* at pH 8.2 thus able to absorb water out of the vicinity of *iep*. Indeed, thin films can be swollen by aqueous media achieving highest swelling ratio (e.g., 72) in pH 3, where the degree of ionization of the predominant DEA component (DEA/MAA molar ratio 2) reaches its maximum value. Rheological investigation of swollen films (with swelling ratio 33, corresponding to 3 wt% aqueous solutions) at pH 3 and 11.5 (far below and above *iep*) showed the formation of 3D networks, in which the cylindrical nanodomains were highly bridged by the soluble P(DEA-*co*-MAA) central blocks.

By solubilizing this telechelic polyelectrolyte directly to aqueous media at different pH, hydrogels were formed in the acidic regime (pH < *iep*) with rheological properties strongly dependent on pH, especially in approaching *iep* where intensive ionic interactions between the opposite charged groups take place. On the other hand, hydrogel could not be formed in the basic conditions (pH > *iep*), attributed to inability of bridging the flower-like micelles. This is due to the additional hydrophobicity of DEA units in this pH range and consequently the low degree of ionization of the hydrophilic P(DEA-*co*-MAA) blocks. Thus the properties of these telechelic polyampholyte hydrogelators depend on their *iep* that is tunable by the anionic to cationic molar ratio and the corresponding *pK*s. Thanks to the available controlled polymerization methods like ATRP, RAFT etc., a large variety of monomer combination and compositions can be used, aiding to tailor the properties and potential applications of these "smart" hydrogelators.

P(DEA-*co*-MAA) blocks were designed as the end-blocks as well, capping a hydrophilic p(ethylene glycol methyl ether methacrylate) block, [(P(DEA-*co*-MAA)-*b*-PEGMA-*b*-P(DEA-*co*-MAA)] (Iatridi et al. 2011). These end-blocks can be neutralized at a certain pH window, in the vicinity of their *iep*, transformed to hydrophobic stickers, driving the polymer to self-assemble forming flower-like micelles and a 3D micellar network at higher concentrations. Thus, this system exhibits gelation efficiency in a certain pH window which can be tuned at will by the monomer composition and the choice of the oppositely charged electrolyte monomers (of a given *pK*) forming the polyampholyte blocks.

Moreover, thanks to the thermo-sensitivity of PEGMA central block, the formed hydrogel reveal a complex thermoresponse on a heating–cooling cycle. As the temperature increased, the gel weakened as the elastic and storage modulii decreased parallel up to 53 °C. Above this temperature, steep increase of both modulii was observed strengthening the gel. This accounts for the additional attractions exerted by the PEGMA chains as the LCST effect is evident. Upon decreasing temperature, the hydrogel preserves higher elasticity than that before the thermal treatment, showing an increase of the bridging elastic chains. Finally, this polymer was evaluated as drug nanocarrier for potential biomaterial applications. Micelles formed at low concentrations were successfully loaded with an anticancer drug, doxorubicin and preliminary studies showed that the drug release can be controlled by tuning the solution pH.

5.3 *Charge-Driven Co-assembling Hydrogels*

Another strategy to create 3D networks in aqueous media, developed the recent years, is the co-assembly of two different polyelectrolyte-based macromolecules bearing opposite charges. The physically cross-linked nodes of the network arise from the intermolecular IPECs between the oppositely charged segments, located at the different macromolecules. Two different systems were investigated so far (Fig. 13). In the first system, a double hydrophilic ABA triblock copolymer, with non-ionic water-soluble central B block end capped by polyelectrolyte blocks, and an oppositely charged homopolyelectrolyte, C, were employed (ABA/C) (Lemmers et al. 2010). The second one consisted of ABA and CBC triblock copolymers with the same neutral central block and oppositely charged end-blocks (ABA/CBC) of the same degree of polymerization (Hunt et al. 2011).

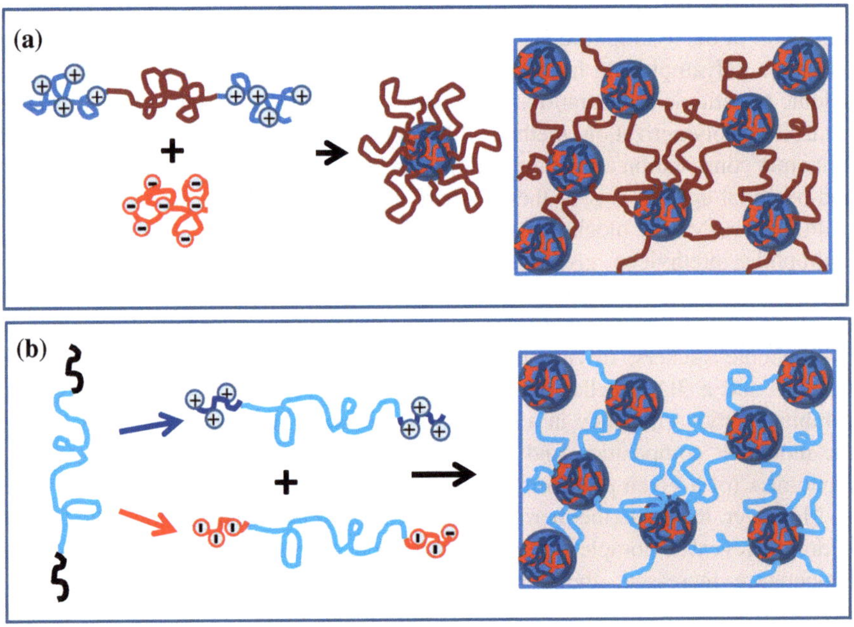

Fig. 13 Schematic representation of **a** co-assembly between triblock copolymers bearing charged end-blocks and oppositely charged homopolymers (ABA/C system) leading to flower-like micellar 3D network [inspired from Lemmers et al. (2010)] and **b** coacervate-driven 3D network formation from equimolar dilute aqueous solutions of oppositely charged ionic ABA/CBC triblock copolyelectrolytes derived from a common triblock copolymer precursor, from (Hunt et al. 2011) by permission of ACS

In the ABA/C system, a $PSPMA_{28}$-PEO_{239}-$PSPMA_{28}$ triblock copolymer with negatively charged end-blocks and the poly(allylamine hydrochloride) (PAH_{160}) cationic weak polyelectrolyte were mixed in aqueous environment forming a transparent hydrogel. Prerequisites for gelling was elevated concentrations (C_{gel} ca 4 wt%) and pH < 8 providing ionization of PEI. The as-formed hydrogel resulted from the oppositely charged (PSMA/PAH) complexation assembled into flower-like micelles, with an IPEC core surrounded by the PEO looped chains of the triblock. At polymer concentrations above C_{gel}, extensive bridging of micelles led to a reversible 3D network (Fig. 13a) (Lemmers et al. 2010).

This charge-driven co-assembling system is very sensitive to the charge ratio of the opposite charges, as well as, to the ionic strength and pH of the medium, since all these factors affect remarkably the interaction strength between the associating polyelectrolytes and in turn the formation of the physical crosslinks (micellar cores) of the network. As it was reported, maximum mechanical performance of the gel was achieved at charge stoichiometry, i.e., $f^{+} = [+]/[+] + [-] = 0.5$, where [+], [−] denote the concentration of positively and negatively charged units, respectively. In addition, due to electrostatic screening which disintegrate the IPECs, the viscosity of the hydrogel decreased exponentially with increasing salt concentration. Finally, a gel–sol transition was observed above pH 9 where PAH was deionized (Lemmers et al. 2012).

A novel charge-driven ABA/C assembling system, constituted of poly[4-(2,2,6,6-tetramethyl piperidine-N-oxyl) aminomethyl styrene]-b-poly(ethylene glycol)-bpoly[4-(2,2,6,6-tetramethyl piperidine-N-oxyl)aminomethyl styrene] (PMNT−PEG−PMNT) triblock copolymer and poly-(acrylic acid) (PAA), were designed and evaluated as injectable hydrogel recently (Ishii et al. 2015). The PMNT/PAA complexation led to flower-like micelles which were bridged irreversibly upon increasing temperature, at 0.15 M NaCl, forming a 3D network (Fig. 14). Rheological investigation of complex viscosity as a function of temperature, varying micelle concentration, ionic strength, and oppositely charged molar ratio showed that the gelation critical temperature shifted to lower values upon increasing ionic strength and concentration while the molar ratio seems to have small effect on gelation temperature. Hence, the above factors can be used to tune the gelation properties as to design an injectable hydrogel, i.e., sol at room temperature and gel under physiological conditions. This formulation was evaluated as drug delivery system, showing sustained release of fluorescein-labeled poly (acrylic acid) (FL-PAAc) for more than 4 weeks (Fig. 14). Provided that the cationic PMNT−PEG–PMNT developed ionic interactions with the anionic FL-PAAc throughout the gel, it was suggested that gradually gel degradation from the surface induced the drug release (Ishii et al. 2015).

Following the second strategy (Fig. 13b), hydrogels were obtained by mixing two triblock copolymers with an identical PEO central block, end-capped by cationic (ammonium or guanidinium) or anionic (sulfonate or carboxylate) blocks of the same degree of polymerization, in equal amount as to provide exact charge stoichiometry (Fig. 13b). The preparation of hydrogels appears to be convenient, very fast, and robust (Hunt et al. 2011). The synthetic strategy followed, i.e.,

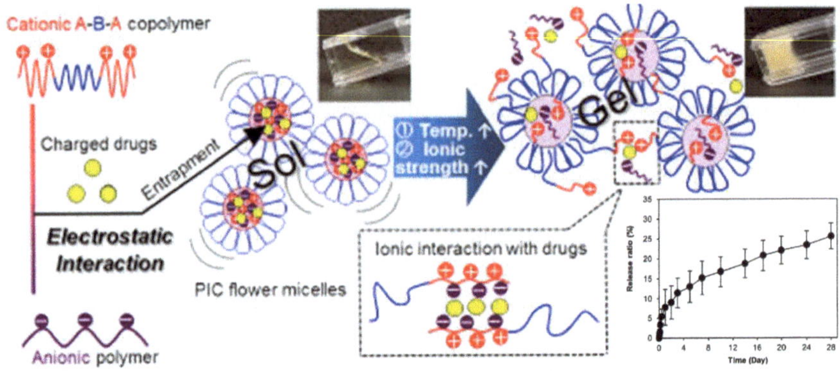

Fig. 14 Schematic illustration of temperature- and ionic strength-responsive redox-active injectable gel (RIG) system. Polyion complex (PIC) flower micelles are formed by self-assembly via electrostatic interaction between cationic (PMNT–PEG–PMNT) and anionic PAAc. PIC flower micelle solution shows irreversible sol–gel transition with increasing temperature and ionic strength. The RIG was able to provide sustained release of an anionic model drug for more than 4 weeks (bottom/right). From Ishii et al. (2015) by permission of ACS

common triblock copolymer precursor with outer blocks transformed to the oppositely charged partners by post-polymerization "click" reaction, was chosen as to eliminate any uncertainty that charge imbalance might have on the resulting properties of the hydrogel.

The structure of the formed network comprises complex coacervate domains (fluid phase), arising from the ionic interactions between the two oppositely charged end-blocks of the involved triblocks, bridged by the central hydrophilic blocks. The mechanical properties of the obtained hydrogels ranged from viscoelastic to fully elastic by carefully choosing the nature of the cation/anion pair. Stiffer hydrogels were formed when the strongest acidic and basic units were used, i.e., sulfonate/guanidinium ion pair. By matching the stoichiometry of positive and negative groups, colorless and transparent hydrogels were formed within 30 s.

Two important factors, namely concentration and ionic strength, govern the network structure and the mechanical response. By keeping the polymer length and end-block fraction fixed and increasing polymer concentration, the underpinning structure of the hydrogel changed from disordered spheres to body-centered cubic (BCC) spheres and finally to hexagonally packed cylinders. These structure rearrangements affected the mechanical response as well. The elastic modulus enhanced in BCC spheres, while in the hexagonal packing cylinders, the gel weakened (Krogstad et al. 2013).

The presence of salt affects remarkably both the phase behavior and the rheological properties. These coacervate gels remained robust to salt concentrations up to 0.25 M, whereas their mechanical properties were weakened at higher concentrations, and above 1.0 M NaCl the gel formation prohibited. The combined effect of polymer and salt concentration was demonstrated in details in a phase diagram depicted in Fig. 15 (Krogstad et al. 2013).

The structure of these systems was also simulated using an embedded fluctuation (EF) model in conjunction with self-consistent field theory. Qualitative agreement was observed between simulating and SAXS results in terms of phase behavior and domain spacing. Using both the EF model and SAXS, phase diagrams were generated as a function of end-block fraction and polymer concentration. Finally, the EF model was used to explore how parameters that directly relate to polymer chemistry can be tuned to modify the resulting phase diagram (Audu et al. 2015).

Recently, dynamical mechanical analysis and SAXS experiments in triblock and diblock copolymers as well showed slow kinetics and difficulties in reaching equilibrium phase structures, especially in triblocks. Detailed analysis of the SAXS data revealed that the triblock copolymer formed ordered phases via a nucleation and growth pathway. Moreover, when a small amount of corresponding diblock

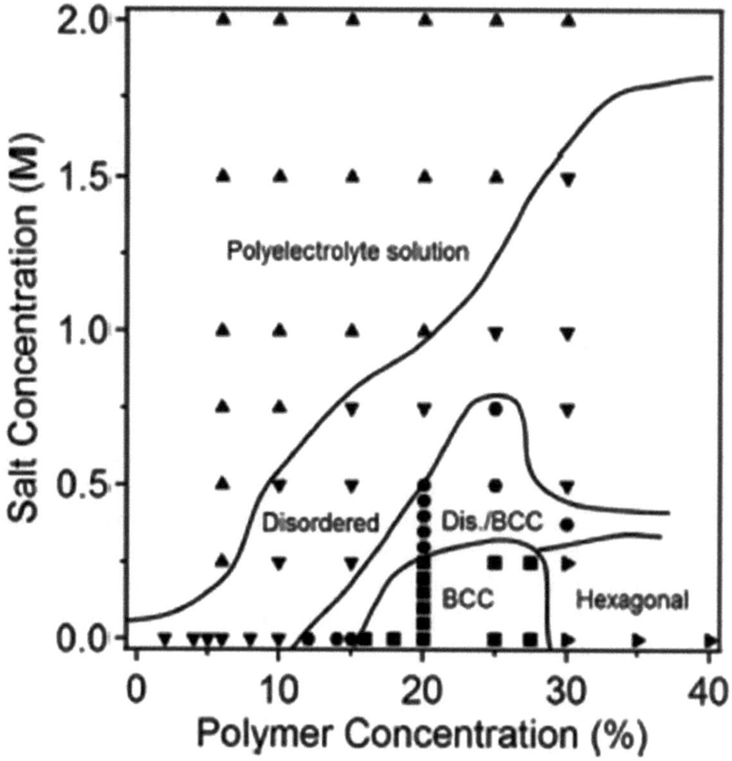

Fig. 15 Phase diagram showing the structure of equimolar dilute aqueous solutions of guanidinium-functionalized and sulfonate-functionalized PAGE-b-PEO-b-PAGE triblock copolymers by varying both polymer and salt concentration. The phase diagram is a compilation of the results from the SAXS and dynamic mechanical data. The lines have been drawn to guide the eye. From Krogstad et al. (2013) by permission of ACS

copolymer were mixed in with the triblock copolymer, the rate of structure formation increased and the mechanical properties of a 80% triblock copolymer mixture were improved compared to those of the pure triblock copolymer (Krogstad et al. 2014).

It comes out that a number of factors (e.g., overall polymer concentration, ionic strength, end-block length, and functionality) can be adjusted as to develop coacervate hydrogels of tunable structures and mechanical properties, suitable for specific potential applications.

Biodegradable charge-driven coacervate hydrogels were recently designed following the strategy of mixing oppositely charged ABA/CBC triblocks. The involved triblocks were poly(L-glutamic acid)-PEG-poly(L-glutamic acid) (PGA-PEG-PGA) and poly(L-lysine)-PEG-poly(L-lysine) (PLL-PEG-PLL). Due to the strong and highly efficient ionic interactions of the PGA with PLL, robust and high modulus hydrogels were obtained within a few seconds at concentrations of 3–5 wt% polymer after mixing the cationic and anionic components. The ionic polypeptide blocks bear pH-responsive ionic functions, and thus, the gel formation and the rheological properties are pH-dependent. In the physiological range of pH, the gel appears stable and the modulus of elasticity exhibits maximum at pH 7. These peptide-based hydrogels encompass excellent cytocompatibility and can also encapsulate and release a model protein, BSA, in a controllable manner. In vivo evaluation, performed in rats, indicated that the hydrogels were formed subcutaneously after injection and degraded completely within four weeks. It seems that these kinds of peptide-based formulations could be used as injectable hydrogels for potential biomedical applications (Cui et al. 2015).

6 Concluding Remarks

Self-assembling hydrogels constitute a specific category of the broad field of hydrogels, resulted from a hierarchical association of macromolecules, mainly of block copolymer topology, that creates a 3D network, through non-covalent reversible physical bonds. This chapter focused to an even narrower class of hydrogels in which the involved associative macromolecules bear ionic repeating units. This particular family of gelators belongs to the class of "smart" soft materials as well, displaying pH-responsive properties, thus exhibiting a number of benefits regarding the non-ionic gelators.

Telechelic polyelectrolytes, comprising a long weak polyelectrolyte flanked by two hydrophobic end-blocks, are the simplest ionic gelators and are widely explored the recent years. Macromolecular engineering afforded a variety of such responsive gelators of tunable molecular futures, while rheological investigations combined with theoretical and computational considerations have enriched our knowledge about these gelling systems. Recent computer simulation developments showed details of the network structure and the adopted conformations of the ionic sequences that could not be afforded by the experiment. On the other hand, experimental efforts demonstrated that we are able to precisely tune the hydrophobic strength of the TP stickers by replacing the homopolymer hydrophobic end-blocks with random hydrophobic/ionic copolymers. This way, the exchange dynamics of the sticky ends, which highly affect the network structure, can be adjusted simply by varying the medium pH.

The research on the ionic hydrogelators was extended to other macromolecular topologies of more complex architectures, including ABC terpolymers and star-shaped block copolymers. The macromolecular architecture seems to play an important role in the 3D network formation affecting therefore the rheological properties of the resulted hydrogels. Moreover, the use of polyelectrolyte end-blocks, exhibiting a hydrophilic/hydrophobic transition, endows the hydrogels with a sharp sol–gel transition in a certain pH window, controlled by the *pK* of the chosen monomer electrolyte. For instance, P2VP becomes hydrophobic above pH 5, and thus, it was used as end-block in ABC topology to design an injectable hydrogel for sustained drug delivery potential applications.

The presence of polyelectrolytes of oppositely charged repeating units, either as random copolymers or as homopolyelectrolytes incorporated on the same macromolecule, constitutes another family of ionic gelators developed in the recent years. The ionic interactions between the oppositely charged moieties endow these gelators with novel additional properties. For instance, it was possible to integrate the properties of anionic and cationic TPs into a single macromolecule. Depending on pH, three regimes were recognized in which the associative polymer behaves like anionic TP (basic conditions) or cationic TP (acidic conditions) or associative polyampholyte in the vicinity of their *iep* (intermediate pH). These regimes can be tuned at will, by macromolecular engineering through rational selection of the monomer nature (dictated by its *pK*) and the monomer composition (determining *iep*).

Finally, gelling systems comprising two partner macromolecules (co-assembly), based on association through electrostatic interactions, have emerged the recent years. Two different systems were explored so far. The first one concerns the mixture of a double hydrophilic ABA triblock copolymer with A polyelectrolyte and a homopolyelectrolyte bearing opposite charges. The second one comprises two ABA and CBC triblocks arising from a common precursor, followed by suitable modification of the end-blocks into oppositely charged polyelectrolytes. The latter strategy provides identical degree of polymerization of all blocks. In both cases, charge stoichiometry affords maximum mechanical performance of the hydrogels. Several factors including the nature of the cation/anion pair, relative length of blocks, overall concentration, stoichiometry as well as the environmental conditions, pH, and ionic strength control the network structure and the rheological properties.

Research efforts have been undertaken the recent years toward potential applications, as drug delivery systems and/or tissue engineering scaffolds, using the above-described ionic gelators, especially those systems that display injectability, suitable for “in situ” gelling. Although some systems are not really biocompatible and/or biodegradable or exhibit the desired properties out of the appropriate conditions (physiological pH, temperature, ionic strength) for biomedical use, they provide proof of concept, aiding the design of polymeric materials, meeting all the requirements for this purpose. A promising strategy toward the above direction is the suggested system, where the oppositely charged polyelectrolytes of the ABA/CBC two-component system was replaced with charged polypeptides, providing biocompatibility and biodegradability. Therefore, it seems that the perspectives for the design of hydrogels with specific tailor-made structural and mechanical properties toward specific applications are very promising, thanks to the macromolecular engineering, and the development of the required interdisciplinary.

References

Angelopoulos SA, Tsitsilianis C (2006) Thermo-reversible hydrogels based on poly(*N*,*N*-diethylacrylamide)-block-poly(acrylicacid)-block-poly(*N*,N-diethyl acrylamide) double hydrophilic triblock copolymer. Macromol Chem Phys 207:2188–2194

Audu DJ, Gopez JD, Krogstad DV, Lynd NA, Kramer EJ, Hawker CJ, Fredickson GH (2015) Phase behavior of electrostatically complexed polyelectrolyte gels using an embedded fluctuation model. Soft Matter 2015:1214–1225

Borisova O, Billon L, Zaremski M, Grassl B, Bakaeva Z, Lapp A, Stepanek P, Borisov O (2011) pH-triggered reversible sol–gel transition in aqueous solutions of amphiphilic gradient copolymers. Soft Matter 7:10824–10833

Bossard F, Sfika V, Tsitsilianis C (2004) Rheological properties of physical gel formed by triblock polyampholyte in salt-free aqueous solutions. Macromolecules 37:3899–3904

Bossard F, Tsitsilianis C, Yannopoulos SN, Petekidis G, Sfika V (2005) A novel thermo-thickening phenomenon exhibited by a triblock polyampholyte in aqueous salt-free solutions. Macromolecules 38:2883–2888

Bossard F, Aubry T, Gotzamanis GT, Tsitsilianis C (2006) pH-Tunable rheological properties of a telechelic cationic polyelectrolyte reversible hydrogel. Soft Matter 2:510–516

Charbonneau C, Chassenieux C, Colombani O, Nicolai T (2011) Controlling the dynamics of self-assembled triblock copolymer networks via the pH. Macromolecules 44:4487–4495

Chassenieux C, Tsitsilianis C (2016) Recent trends on pH/thermo-responsive self-assembling hydrogels: from polyions to peptide-based polymeric gelators. Soft Matter 12:1344–1359

Cui H, Zhuang X, He C, Wei Y, Chen X (2015) High performance and reversible ionic polypeptide hydrogel based on charge-driven assembly for biomedical applications. Acta Biomater 11:183–190

Dyakonova MA, Stavrouli N, Popescu M-T, Kyriakos K, Grillo I, Philipp M, Jaksch S, Tsitsilianis C, Papadakis CM (2014) Physical Hydrogels via charge driven self-organization of a triblock polyampholyte: rheological and structural investigations. Macromolecules 47:7561–7572

Dyakonova M, Berezkin AV, Kyriakos K, Gkermpoura S, Popescu M-T, Filippov SK, Stepanek P, Di Z, Tsitsilianis C, Papadakis CM (2015) Salt-induced changes in triblock polyampholyte hydrogels: computer simulations and rheological, structural, and dynamic characterization. Macromolecules 48:8177–8189

Dyakonova MA, Gotzamanis G, Niebuur B-J, Vishnevetskaya NS, Raftopoulos KN, Di Z, Filippov SK, Tsitsilianis C, Papadakis CM (2017) pH Responsiveness of hydrogels formed by telechelic polyampholytes. Soft Matter 13:3568–3579

Ghelichi M, Qazvini NT (2016) Self-organization of hydrophobic-capped triblock copolymers with polyelectrolyte midblock: a coarse-grained molecular dynamics simulation study. Soft Matter 12:4611–4620

Gotzamanis GT, Tsitsilianis C, Hadjiyannakou SC, Patrickios CS, Lupitskyy R, Minko S (2006) Cationic telechelic polyelectrolytes: synthesis by group transfer polymerization and self-organization in aqueous media. Macromolecules 39:678–683

Gotzamanis GT, Papadimitriou K, Tsitsilianis C (2016) Design of a C-*b*-(A-*co*-B)-*b*-C telechelic polyampholyte pH-responsive gelator. Polym Chem 7:2121–2129

Halperin A, Alexander S (1989) Polymeric micelles: their relaxation kinetics. Macromolecules 22:2403–2412

Henderson KJ, Tian TC, Otim KJ, Shull KR (2010) Ionically cross-linked triblock copolymer hydrogels with high strength. Macromolecules 43:6193–6201

Hietala S, Monomen P, Strandman S, Järvi P, Torkkeli M, Jankova K, Hvilsted S, Tenhu H (2007) Synthesis and rheological properties of an associative star polymers in aqueous solutions. Polymer 48:4087–4096

Hietala S, Strandman S, Järvi P, Torkkeli M, Jankova K, Hvilsted S, Tenhu H (2009) Rheological properties of associative star polymers in aqueous solutions: effect of hydrophobe length and polymer topology. Macromolecules 42:1726–1732

Hunt JN, Feldman KE, Lynd NA, Deek J, Campos LM, Spruell JM, Hernandez BM, Kramer EJ, Hawker CJ (2011) Tunable, high modulus hydrogels driven by ionic coacervation. Adv Mater 23:2327–2331

Iatridi Z, Mattheolabakis G, Avgoustakis K, Tsitsilianis C (2011) Self-assembly and drug delivery studies of pH/thermo-sensitive polyampholytic (A-co-B)-*b*-C-*b*-(A-co-B) segmented terpolymers. Soft Matter 7:11160–11168

Ishii S, Kaneko J, Nagasaki Y (2015) Dual stimuli-responsive redox-active injectable gel by polyion complex based flower micelles for biomedical applications. Macromolecules 48:3088–3094

Kahveci MU, Yagci Y, Avgeropoulos A, Tsitsilianis C (2016) Polymeric materials—well defined block copolymers. In: Reference module in materials science and materials engineering. Elsevier, Amsterdam

Katsampas I, Tsitsilianis C (2005) Hierarchical self-organization of ABC terpolymer constituted of a long polyelectrolyte end-capped by different hydrophobic blocks. Macromolecules 38:1307–1314

Katsampas I, Roiter Y, Minko S, Tsitsilianis C (2005) Multifunctional stimuli responsive ABC terpolymers: from 3-compartment micelles to 3-dimentional network. Macromol Rapid Commun 26:1371–1376

Koetting MC, Peters JP, Steichen SD, Peppas NA (2015) Stimulus-responsive hydrogels: theory, modern advances, and applications. Mater. Sci. Eng R 93:1–49

Krogstad DV, Lynd NA, Choi S-H, Spruell JM, Hawker CJ, Kramer EJ, Tirrell MV (2013) Effects of polymer and salt concentration on the structure and properties of triblock copolymer coacervate hydrogels. Macromolecules 46:1512–1518

Krogstad DV, Lynd NA, Miyajim D, Gopez J, Hawker CJ, Kramer EJ, Tirrell MV (2014) Structural evolution of polyelectrolyte complex core micelles and ordered-phase bulk materials. Macromolecules 47:8026–8032

Kujawa P, Audibert-Hayet A, Selb J, Candau F (2004) Rheological properties of multisticker associative polyelectrolytes in semidilute aqueous solutions. J Polym Sci Part B Polym Phys 42:1640–1655

Kujawa P, Audibert-Hayet A, Selb J, Candau F (2006) Effect of ionic strength on the rheological properties of multisticker associative polyelectrolytes. Macromolecules 39:384–392

Kumar SK, Panagiotopoulos AZ (1999) Thermodynamics of reversibly associating polymer solutions. Phys Rev Lett 82:5060–5064

Lemmers M, Sprakel J, Voets IK, van der Gucht J, Cohen Stuart MA (2010) Multiresponsive reversible gels based on charge-driven assembly. Angew Chem Int Ed 49:708–711

Lemmers M, Spruijt E, Beun L, Fokkink R, Leermakers F, Portale G, Cohen Stuart MA, van der Gucht J (2012) The influence of charge ratio on transient networks of polyelectrolyte complex micelles. Soft Matter 8:104–117

Li Y, Sun Z, Shi T, An L (2004) Conformation studies on sol-gel transition in triblock copolymer solutions. J Chem Phys 121:1133–1140
Li Y, Tang Y, Narain R, Lewis AL, Armes SP (2005) Biomimetic stimulus-responsive star diblock gelators. Langmuir 21:9946–9954
Lin Z, Cao S, Chen X, Wu W, Li J (2013) Thermoresponsive hydrogels from phosphorylated ABA triblock copolymers: a potential scaffold for bone tissue engineering. Biomacromolecules 14:2206–2214
Lomas H, Massignani M, Abdullah KA, Canton I, Lo Presti C, MacNeil S, Du J, Blanazs A, Madsen J, Armes SP, Lewis AL, Battaglia G (2008) Non-cytotoxic polymer vesicles for rapid and efficient intracellular delivery. Faraday Discuss 139:143–159
Ma Y, Tang Y, Billingham NC, Armes SP (2003) Synthesis of biocompatible, stimuli-responsive, physical gels based on ABA triblock copolymers. Biomacromolecules 4:864–868
Matyjaszewski K, Gnanou Y, Leibler L (2007) Macromolecular engineering: precise synthesis, materials properties, applications. Wiley-vch, Weinheim
Medsen J, Armes SP (2012) (Meth)acrylic stimulus-responsive block copolymer hyrogels. Soft Matter 8:592–605
Nguyen-Misra M, Mattice WL (1995) Micellization and gelation of symmetric triblock copolymers with insoluble blocks. Macromolecules 28:1444–1457
Nicolai T, Colombani O, Chassenieux Ch (2010) Dynamic polymeric micelles versus frozen nanoparticles formed by block copolymers. Soft Matter 6:3111–3118
Popescu M-T, Athanasoulias I, Tsitsilianis C, Hadjiantoniou NA, Patrickios CS (2010) Reversible hydrogels from amphiphilic polyelectrolyte model multiblock copolymers: the importance of macromolecular topology. Soft Matter 6:5417–5424
Popescu M-T, Mourtas S, Pampalakis G, Antimisiaris SG, Tsitsilianis C (2011) pH-Responsive hydrogel/liposome soft nanocomposites for tuning drug release. Biomacromolecules 12:3023–3030
Popescu M-T, Tsitsilianis C, Papadakis CM, Adelsberger J, Balog S, Busch P, Hadjiantoniou NA, Patrickios CS (2012) Hydrogels: an unusual pH-response. Macromolecules 45:3523–3530
Potemkin II, Vasilevskaya VV, Khokhlov AR (1999) Associating polyelectrolytes: finite size cluster stabilization versus gel formation. J Chem Phys 11:2809–2817
Reinicke S, Schmelz J, Lapp A, Karg M, Hellweg T, Schmalz H (2009) Smart hydrogels based on double responsive triblock terpolymers. Soft Matter 5:2648–2657
Rubinstein M, Dobrynin AV (1997) Solutions of associative polymers. Trends Polym Sci 5:181–186
Schmalz A, Schmalz H, Müller AHE (2012) Smart hydrogels based on responsive star-block copolymers. Soft Matter 8:9436–9445
Sfika V, Tsitsilianis C (2003) Association phenomena of poly(acrylic acid)-*b*-poly (2-vinylpyridine)-*b*-poly(acrylic acid) triblock polyampholyte in aqueous solutions: from transient network to compact micelles. Macromolecules 36:4983–4988
Shedge A, Colombani O, Nicolai T, Chassenieux C (2014) Charge dependent dynamics of transient networks and hydrogels formed by self-assembled pH-sensitive triblock copolyelectrolytes. Macromolecules 47:2439–2444
Stavrouli N, Aubry T, Tsitsilianis C (2008a) Polymer rheological properties of ABA telechelic polyelectrolyte and ABA polyampholyte reversible hydrogels: a comparative study. Polymer 49:1249–1256
Stavrouli N, Katsampas I, Angelopoulos S, Tsitsilianis C (2008b) pH/Thermo-sensitive hydrogels formed at low pH by a PMMA-PAA-P2VP-PAA-PMMA pentablock terpolymer. Macromol Rapid Commun 29:130–135
Taktak FF, Bütün V (2010) Synthesis and physical gels of pH- and thermo responsive tertiary amine methacrylate based ABA triblock copolymers and drug release studies. Polymer 51:3618–3626
Thakur VK, Thakur MK (2014a) Recent trends in hydrogels based on psyllium polysaccharide: a review. J Clean Prod 82:1–15

Thakur VK, Thakur MK (2014b) Recent advances in graft copolymerization and applications of chitosan: a review. ACS Sustain Chem Eng 2(12):2637–2652
Thakur VK, Thakur MK (2015) Recent advances in green hydrogels from lignin: a review. Int J Biol Macromol 72:834–847
Tsitsilianis C (2010) Responsive reversible hydrogels from associative “smart” macromolecules. Soft Matter 6:2372–2388
Tsitsilianis C, Iliopoulos I (2002) Viscoelastic properties of physical gels formed by associative telechelic polyelectrolytes in aqueous media. Macromolecules 35:3662–3667
Tsitsilianis C, Iliopoulos I, Ducouret G (2000a) An associative polyelectrolyte end-capped with short polystyrene chains. Synthesis and rheological behavior. Macromolecules 33:2936–2943
Tsitsilianis C, Katsampas I, Sfika V (2000b) ABC heterotelechelic associative polyelectrolytes. Rheological behavior in aqueous media. Macromolecules 33:9054–9059
Tsitsilianis C, Roiter Y, Katsampas I, Minko M (2008a) Diversity of nanostructured self-assemblies from a pH-responsive ABC terpolymer in aqueous media. Macromolecules 41:925–934
Tsitsilianis C, Stavrouli N, Bocharova V, Angelopoulo S, Kiriy A, Katsampas I, Stamm M (2008b) Stimuli responsive associative polyampholytes based on ABCBA pentablock terpolymer architecture. Polymer 49:2996–3006
Tsitsilianis C, Aubry T, Iliopoulos I, Norvez S (2010) Effect of DMF on the rheological properties of telechelic polyelectrolyte hydrogels. Macromolecules 43:7779–7784
Van Tomme SR, Storm G, Hennink WE (2008) In situ gelling hydrogels for pharmaceutical and biomedical applications. Intern. J. of Pharmaceutics 355:1–18
Winnik MA, Yekta A (1997) Associative polymers in aqueous solution. Curr Opin Coll Interface Sci 2:424–436
Xu C, Kopeček J (2007) Self-assembling hydrogels. J Polym Bull 58:53–63
Zhang R, Shi T, An L, Sun Z, Tong T (2010) Conformational study on sol-gel transition in telechelic polyelectrolytes. J Phys Chem B 114:3449–3456
Zhang R, Shi T, Li H, An L (2011) Effect of the concentration on sol-gel transition in telechelic polyelectrolytes. J Chem Phys 134(034903):1–7

Chapter 11
Cellulose Hydrogels; Fabrication, Properties, and Their Application to Biocompatible and Tissue Engineering

Takaomi Kobayashi

Abstract Cellulose hydrogels made of agro-industrial bagasses of sugarcane and other are introduced in this chapter for the fabrication, properties, and their the biocompatible materials with cytocompatibility for tissue engineering. To obtain the cellulose hydrogels, firstly cellulose was regenerated from bagasse wastes by chemical pretreatments and bleaching. The renewable cellulose was converted to hydrogels by phase inversion process under ethanol vapor. To evaluate the biocompatibility, the hydrogel was implanted in the intraperitoneal of mice. The results were shown as small influence of the implanted hydrogel on the growth of mice. The implanted hydrogel was somewhat decreased in the molecular weight in 3–4 weeks, meaning biodegradable materials. However, the hydrogels kept enough mechanical strength in the living body. This indicated that the cellulose hydrogel regenerated waste bagasse showed acceptable biocompatibility and durability in the body. In addition, hydrogels are excellent in regeneration of cytocompatible property for tissue regeneration.

Keywords Cellulose hydrogel · Biomass waste · Cytocompatibility
Biocompatible material · Tissue regeneration

1 Introduction

1.1 Hydrogels Made of Hydrophilic Polymers with Crosslinking

In late years, development and application of the bio-based polymer hydrogels are pushed forward in various kinds of fields because of deep consideration of an influence to environment society and especially for medical uses. Hydrogels

T. Kobayashi (✉)
Department of Materials Science and Technology, Nagaoka University of Technology,
1603-1 Kamitomioka, Nagaoka, Niigata, Japan
e-mail: takaomi@vos.nagaokaut.ac.jp; takaomi@nagaokaut.ac.jp

V. K. Thakur and M. K. Thakur (eds.), *Hydrogels*, Gels Horizons: From Science to Smart Materials, https://doi.org/10.1007/978-981-10-6077-9_11

Table 1 Comparison of natural polymer and synthetic polymer for hydrogel materials

Classification	Sources	Advantages disadvantages	Polymer species
Natural polymer	Vegetable	Good biocompatibility	Agarose, carrageenan, coconut
	Animal	Weak mechanical property	Collagen, chitosan, chitin
	Microorganisms	Weak product stability	Bio-cellulose, Nata de coco
		Weak processability	
Synthetic polymer	Monomers polymerization	Weak biocompatibility	Acrylamide-based gel, PHEMA, PEG, PVA
	Crosslinking reaction	Good mechanical property	
	Interpenetrating polymer network structure	Good product stability	
		Good processability	

composed of three-dimensional networks of hydrophilic polymer have high affinity for water, because of the hydrophilic segments of the hydrogel. A large amount of water is retained in the polymer chains but hydrogels are insoluble in water. This is because that swelling polymer segments are crosslinked each other. Therefore, such hydrogel is a favorable candidate for biological application and biocompatible materials, which are similar to the environment of natural tissues (Van Vlierberghe et al. 2011; Chang et al. 2013; Giri et al. 2012). For this reason, hydrogels have attracted for biological and biomedical engineering including regenerative medicine (Tchemtchoua et al. 2011), drug release (Qiu and Park 2012; Li et al. 2012), and nanomaterial (Lee and Mooney 2012). Table 1 compares characteristics of natural-sourced polymers and synthetic polymer for hydrogel materials. As seen, mechanical strength in synthetic polymers is higher than bio-based polymers, and their structure control is much easier in synthetic polymers. Also, an advantage is seen in superior plasticity of synthetic polymers, but biocompatibility is inferior relative to natural polymer. So, this is reason for increasing environmental concerns into uses of natural polymers, as related to their properties of sustainability and bioactivity.

1.2 Hydrogels Composed of Agro-Waste Cellulose

In agro-industrial wastes including sugarcane bagasse, cassava bagasse, rice husk, wheat straw, and coconut fiber, the regeneration strongly contributes to the progress of sustainable renewability (Canilha et al. 2013; Fernandes et al. 2013). Because of foreseeable depletion of fossil fuels and the environmental pollution, these wastes have been used as alternative energy source with bioethanol and biofuels by product especially after the extraction of sweet juice (Garcia-Soto et al. 2011; Velmurugan and Muthukumar 2011; Nigam and Singh 2011; Murillo-Alvarado et al. 2014). As well known for the agro-industrial waste, for example, sugarcane bagasse has characteristics composed of fibrous residues including valuable chemical components like cellulose, hemicellulose, and lignin (Saucedo-Luna et al. 2010; Chandel et al. 2012). The sugarcane waste was produced about 1.69×10^3 million tons in the world in 2010, and generally 280 kg of humid bagasse was generated from 1 ton of sugarcane (Hofsetz and Silva 2012). So, the regeneration of such chemical material is useful for the bagasse with the largest possibility in cellulosic agro-industrial by-products. Similar situation was seen in *Agave tequilana* bagasse for the production of tequila in Mexico (Kestur et al. 2013; Linan-Montes et al. 2014). As well as sugarcane bagasse waste, the fibrous cellulose tissue of the blue agave has been discharged after extraction of the sap (Iñuguez-Covarrubias et al. 2001; Kobayashi et al. 2015). In Mexico, more than 800,000 tons of agave bagasse was wasted per year. As an important problem, people have considered attention to development of regeneration materials from their industrial wastes. At present, one of parts is provided and reused for manure, feedstock, and fuel of the boiler in the sugar industry. However, the remaining residual substance is left and unused, causing environmental and social problem. Therefore, for approaching sustainable future, the reuse is the important problem that is essential in conjunction with biomass residual utilization.

1.3 Biomass Hydrogels for Medical Application

As indicated in Table 1, such agro-waste cellulose is expected to be advantageous in biocompatible materials for tissue engineering (Kobayashi 2015). Thus, the medical application used for advanced materials is interesting, if such biomass can converted to hydrogels, which come into safety contact with the living body for improving biocompatibility. In hydrogels sourced from biomass, attention has been increasingly devoted to bio-based polymers that are environmentally friendly materials. In addition, biomass hydrogel component strongly relates to therapy (Nguyen and Alsberg 2014) and diagnosis for drug delivery systems (Lee et al. 2012; Qiu and Park 2012; Kost and Langer 2012; Jiang et al. 2016) and supportive scaffolds for guided tissue growth (Rao et al. 2013; Chen and Liu 2016). In recent years, considerable interesting in regeneration of biomass polymers as seen in Scheme 1 is paid for biocompatible hydrogel materials, which can substitute damaged tissues and organs as tissue regeneration, biomaterial hydrogels play a key role in most approaches for natural polymers such as collagen (Macaya et al. 2011; Lowe et al. 2016), hyaluronic acid (Burdick and Prestwich 2011; Rodell et al. 2015; Sideris et al. 2016), alginate (Lee and Mooney 2012; Zhao et al. 2012), agarose

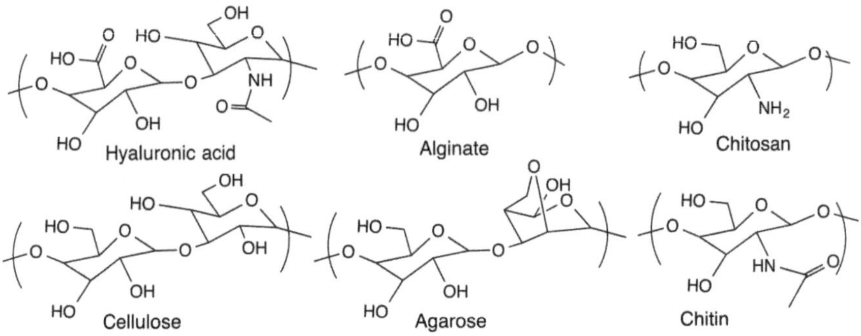

Scheme 1 Chemical structure of biopolymers

(Kometani et al. 2015; Forget et al. 2016), chitin (Madhumathi et al. 2010; Benhabiles et al. 2012), and chitosan (Liu et al. 2011; Elsabee and Abdou 2013; Ding et al. 2016) for the development of hydrogels.

Among them, cellulose is also one of a naturally occurring polymer consisting of β-glucose and the most abundant material in nature (Chang and Zhang 2011; Jorfi and Foster 2015). An improvement study such as raising biocompatible affinity having high enough mechanical and viscoelastic biomass hydrogels would be attractive for the future materials in medical application.

1.4 Cellulosic Hydrogels Used as Strategic Biocompatible Materials

In biomedical application, it is seemed that cellulose and its derivatives should be employed more and more in practical uses. However, still very limited works are performed by using native cellulose for the preparation of hydrogels. This is due to the difficulty of dissoluble property of cellulose in several solvents for the regeneration process (Rojas 2016). The present review describes such cellulose hydrogel in fabrication, properties, and application to biocompatible and less cytotoxic materials for a new type of environmentally friendly "green hydrogel sheets" containing large amounts of water (Kobayashi et al. 2014a; Kobayashi 2015; Torres-Rendon et al. 2016). The renewable cellulose purified from agro-industrial waste was used to prepare cellulose hydrogel films and applied for biocompatible (Nakasone et al. 2016) and cytocompatible materials (Tovar-Carrillo et al. 2013a, 2014; Kobayashi and Tovar-Carrillo 2015).

2 Cellulose Hydrogels in Fabrication and Properties

2.1 Conversion Processes of Agro-Industrial Bagasses to Celluloses and then Hydrogels

In sustainable source of plants, cellulose-based materials have attractively studied for bioplastics in advanced materials including cellulose composite (Peng et al. 2016; Anyaporn et al. 2017) and cellulose hybrid hydrogels have some possibilities to apply for several fields (Chang and Zhang 2011; Kobayashi et al. 2014b; Shen et al. 2016; Tummala et al. 2016). The reuses of the agro-industry bagasses were reported in fabrication and properties of cellulosic materials (Kobayashi et al. 2015). In the regeneration process of the wastes, chemical treatments of the bagasse were shown by sodium hydroxide (NaOH) and sodium hypochlorite (NaOCl), and these were important for good cellulose hydrogel films (Tovar-Carrillo et al. 2014; Kobayashi et al. 2015; Rojas 2016; Nakasone and Kobayashi 2016a, b). The pretreatment processes seem to influence strongly the molecular weight, degree of crystallinity, and amount of trace lignin, which influenced resultant films (Nakasone

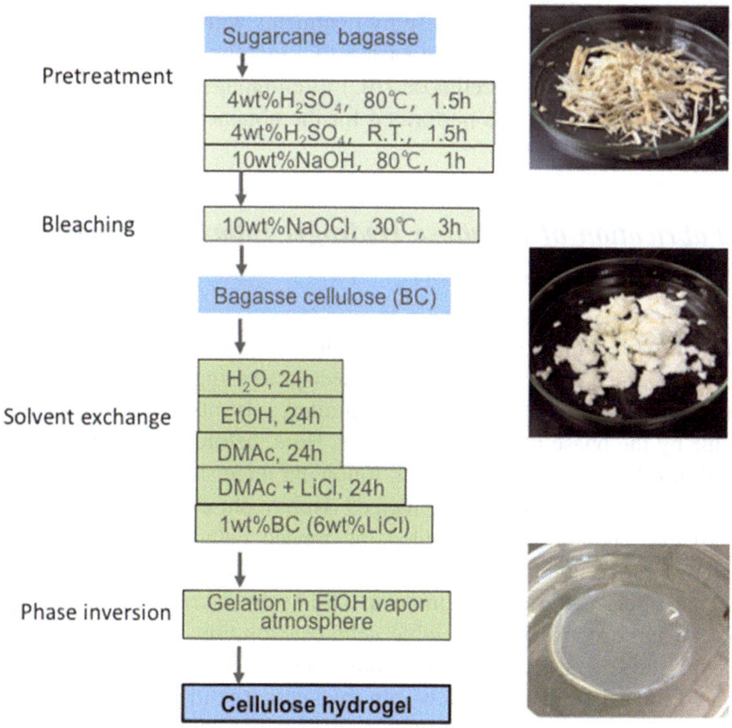

Fig. 1 Flow process of purification sugarcane bagasse and preparation to hydrogels. In the pictures, sugarcane bagasse fiber and that treated with pretreatment and bleaching with NaOCl. Bottom picture is sugarcane cellulosic hydrogel

and Kobayashi 2016a). As seen in Fig. 1, regeneration of agro-industry bagasses was taken in advantageous conversion of cellulosic fibers. In order to obtain the renewable fibers from the agro-waste sugarcane bagasse, pretreatment of the bagasse (inserted picture in Fig. 1) was performed by following procedure (Nakasone and Kobayashi 2016b).

For example, hot water was used to wash the bagasse firstly at 80 °C, in order to remove remaining sugar components. Then the resident fibers were dried in oven at 50 °C. The bagasse (10 g) was stirred in 300 mL of 4 vol.% H_2SO_4 aqueous solution at 90 °C for 1.5 h, and then the acidic aqueous solution was replaced with 300 mL of 10 wt% NaOH aqueous solution at 90 °C for 12 h sequentially. The fiber residue was washed with excess distilled water until neutral pH. After that, the resulted bagasse was filtrated and dried in vacuum oven at room temperature overnight. Additionally, bleaching in 10 g of structure and strong hydrogen bonds between the cellulose segments (Cao et al. 2010). It was reported that cellulose-based films were already prepared (Huq et al. 2012; Takegawa et al. 2010), when dissolving cellulose was used in DMAc (*N,N*-dimethyl acetamide)/ LiCl (Sjoholm et al. 2000; Henniges et al 2010), NaOH–water with or without additives (Moigne and Navard 2010), and ionic liquids (Mahadeva and Kim 2011; Xiong et al. 2012). As seen in Figure 1, cellulose used for the hydrogel films was prepared from agro-industrial bagasse by treatments with 4 vol.% H_2SO_4, 10 wt% NaOH and then 10 vol.% NaOCl at each 80°C. Finally, excess water was used to wash the remaining bagasse cellulose (BC) to eliminate soluble residential chemicals, and vacuum drying was performed for the obtained BC.

2.2 Fabrication of Cellulose Hydrogel Films by Using Phase Inversion Process

Difficulty in fabricating cellulose to create films and other products is well known (Rojas 2016), since generally, such film sheet processing can be conducted using such a cellulose solution, which is very limited solvents. This is because of the crystalline by the phase inversion procedure. Here, before the cellulose fibers were dissolved in DMAc solution with LiCl, following three steps were important in the processes with stepwise solvent exchanges (Kobayashi et al. 2015; Tovar-Carrillo et al. 2013b). The first step included the treated cellulose fibers suspended in distilled water overnight and then immersion in ethanol till swelling for 24 h. Finally, the fibers were swelled more in DMAc by stirring overnight. After the solvent exchange steps, the dissoluble cellulose solution, in DMAc/LiCl the cellulose hydrogels, was prepared with homogenously 1 wt% concentration. When this poured into glass dish with few mm thickness, the dish was stored in ethanol vapor atmosphere at room temperature and kept for 24 h in a plastic container, which was filled with 40 mL of ethanol. In the phase inversion step, the cellulose coagulation was found gradually as progressed precipitation under the vapor

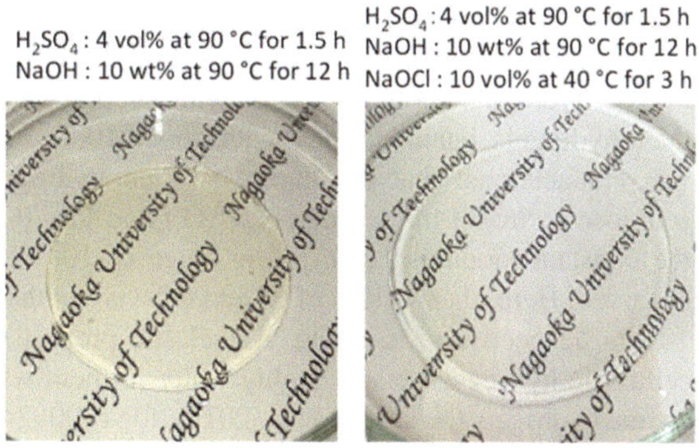

Fig. 2 Pictures of cellulosic hydrogels prepared from different chemical treatments. Less bleaching remains lignin in the hydrogel (left)

condition. Finally, the swelled cellulose hydrogel film was obtained as a result of the liquid phase change to solid gel. Then excess water was used to remove the DMAc and ethanol from the geleous films (Fig. 2).

2.3 *Properties of Cellulosic Hydrogels*

Several fibrous celluloses were purified from waste bagasses of agave (Tovar-Carrillo et al. 2013b), sugarcane (Nakasone and Kobayashi 2016a), cottonseed hulls (Anyaporn et al. 2017), and bamboo (Tovar-Carrillo et al. 2013a) and could be used to fabrication to similar hydrogels as seen in Fig. 2. Here, the following chemical pretreatments and bleaching depended upon the hydrogel properties. Actually, difference in the transparent and strength of the resulted cellulose hydrogels was obtained in the coagulated hydrogels with transparent films. As shown in the pictures for sugarcane cellulose hydrogels (Nakasone and Kobayashi 2016b), each hydrogel had flexible and highly transparent nature. Only NaOH-treated fibers provided slightly yellowish color due to residence of the trace lignin (Nakasone and Kobayashi 2016b), while the bleached fibers noted transparent colorless in the hydrogels. This meant that the chemical influenced the residence of the trace lignin and the absence of the crystalline cellulose fibers in the hydrogels. The XRD measurement of the bagasse and the purified fibers had the patterns that exhibited typical crystalline lattice of cellulose I at 22.6° and 16.1°. The crystallinity of bagasse and cellulose fibers decreased the degree of the chemical treatments by the NaOH and NaOCl for the fibers. In the cases of the regenerated cellulose hydrogels prepared from each cellulose fiber, XRD pattern showed significant change in the broad peaks at around 20° attributed to the crystalline of cellulose II, meaning that transformation from cellulose I to cellulose II was occurred during the phase inversion process in the liquid cellulose. As a result, the regeneration process decreased the crystalline structure of the cellulose in the hydrogel in the range of 50–56%. So the hydrogel was composed of amorphous celluloses which were easily interacted with water molecule for high water loading in the cellulose network environment (Fig. 3) (Kobayashi et al. 2014a).

The hydrogels showed flexible with enough mechanical strength even though their hydrogels retained water with about 1000–3000 wt% (Kobayashi 2015). It was characterized that the highly swelled hydrogels had viscoelastic nature of cellulose hydrogel films. In storage (G′) and loss (G″) modulus versus strain (%), the constant modulus range in the range of 10^{-1}–2×10^{0}% strain typically was observed. Here, their values of G′ and G″ were in the range of 10^4–10^5 Pa and 10^3–10^4 Pa, respectively, meaning viscoelastic properties of the soft materials. Their cellulosic hydrogels retains highly water in them with about 100–2000% for sugarcane hydrogels (Nakasone et al. 2016), for 1000–2400% for defatted cotton (Jiang et al. 2016), 250–300% for agave (Tovar-Carrillo et al. 2013a), 200–320% for pulp (Kobayashi and Tovar-Carrillo 2015), and 140% for bamboo (Tovar-Carrillo et al. 2013a). It was clear that the water retention depends on the source of the plants. To analyze the presence of water in the hydrogel, FT-IR analysis of cellulose hydrogels was carried out for OH stretching band at 3000–3700 cm^{-1}. This was evaluated water–hydrogel interactions by appearance of the OH group and water as broaden band thin the region (Venegas-Sanchez et al. 2013a; Jiang et al. 2016; Li et al. 2016). In the spectral data of hydrated cellulose (Fig. 3), water molecules strongly bonded with free OH group of the polysaccharide chain with intermolecular and intramolecular bondings (as described in the figure caption) (Venegas-Sanchez et al. 2013b). For the FT-IR spectroscopy, the spectra of the OH group of cellulose

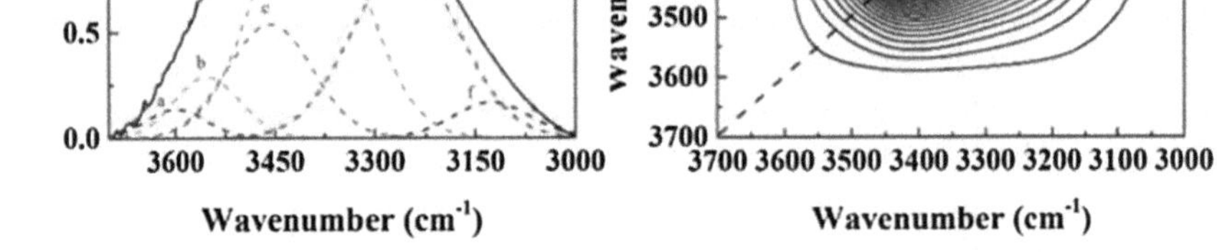

Fig. 3 FT-IR spectra of the hydrogel films after the cultivation (right) for 14 days (left) and 2D spectrum of the wet hydrogel (right). The LiCl concentration was 6 wt% in the phase inversion process of the DMAc solution. The peak f (2900 cm^{-1}), peak e (3045 cm^{-1}), peak d (3300 cm^{-1}), peak c (3440 cm^{-1}), peak b (3550 cm^{-1}), and peak a (3600 cm^{-1}) were attributed for CH stretching, hydrogen bonds of interpolymer and intrapolymer interaction in the cellulose, free OH group of the cellulose and water-cellulose hydrogen bond and water hydrogen bond, respectively (Jiang et al. 2016)

hydrogel were deconvoluted into six Gaussian components, according to our previous reports (Venegas-Sanchez et al. 2014; Li and Kobayashi 2016a) and the results of 2D correlation (Guo and Wu 2008; Li and Kobayashi 2016b). The spectral analysis resulted the peak separation of the 3599 and 3551 cm^{-1} range belonged to free water in hydrogel and the OH groups in cellulose (Venegas-Sanchez et al. 2013b). These indicated that water was well retained in the hydrogel by hydrogel bonding as drawn in Fig. 3.

3 Cellulose Hydrogels in Their Bio and Cytocompatible Applications

3.1 Evaluation of the Biocompatibility by In Vivo Test

We have reported that cellulose hydrogel films composed of natural renewable products have excellent biocompatibility (Nakasone et al. 2016) as well as that of other biopolymers actually exhibited for collagen and others (Macaya et al. 2011; Burdick and Prestwich 2011). However, there has been very limited report on biocompatibility of the cellulosic hydrogels (Kobayashi et al. 2014b). Our group has been investigating cellulose hydrogels prepared from various biomasses including agave bagasse and bamboo fibers (Tovar-Carrillo et al. 2013a), aiming for a scaffold for a fibroblast culture and biocompatibility. It was reported that these cellulose hydrogels showed excellent cytocompatibility and facilitated cell propagation (Tovar-Carrillo et al. 2013b, 2014). In their examples, cellulose hydrogels prepared from the regenerated sugarcane bagasse expressed excellent in vivo biocompatibility. The biocompatibility of the hydrogel film was tested by implantation of the hydrogel film in female ICR mice (Nakasone et al. 2016). For the in vivo test (Fig. 4), the film was implanted during 4 weeks. In the experiments, the films (10×10 mm^2) having 0.75–1.3 mm thickness were washed by 70% alcohol. Then PBS solution was used for a complete solvent exchange by immersing it well. Ten mice were randomly chosen for four in vivo testing groups, and the implantation of the hydrogel films was carried out. The influences of the implanted cellulose hydrogel film on living body were determined by monitoring intakes of food and water and body weight during 4 weeks after the implantation. Actually, there were no any deaths in treatment mice during implantation test for 4 weeks. The mice implanted showed increased body weight after implantation as well as the reference mice. As seen in Fig. 4, the implanted cellulose hydrogel films did not show significant differences in signs of internal inflammation and injury in comparison with the control group. In these pictures, the existence of inflammation in the intraperitoneal was not seen in the postmortem examination, meaning quietly low rejection reaction (Tovar-Carrillo et al. 2015) at the places where the hydrogel film was contacted. It was known that such inflammatory reaction was occurred for one of the biological responses in the body when the low biocompatible material implanted (Onuki et al. 2008).

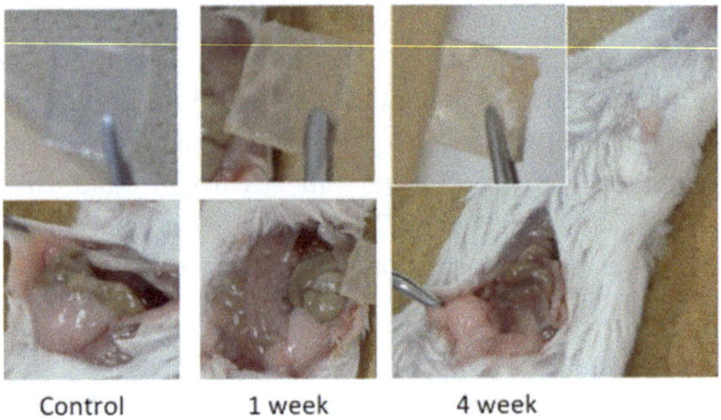

Fig. 4 Postmortem views of the mice at different week post implantation: control, 1 week, and 4 weeks for implantation of the hydrogels. All procedures involving use of animals were approved by Local Animal Ethics Committee at Okinawa National College of Technology

Furthermore, the implanted cellulose hydrogel films were examined in viscoelastic behaviors (Fig. 5). The viscoelastic values of storage modulus and loss elastic modulus of the cellulose hydrogel films were compared at various strains. The implanted hydrogel film had G′ value of 7.2×10^4 Pa for the 4-week implantation sample. As seen, the control hydrogel film showed G′ value of the 5.8×10^4 Pa. So there was insignificant change in the viscoelasticity of the films before and after implantation. But, G″ in the hydrogel film implanted was shifted to lower modulus, suggesting that the hydrogel flexibility decreased over 3-week implantation. The molecular weight determination showed that a certain level of degradation was observed in the intraperitoneal of mouse. As noted in Fig. 4, the

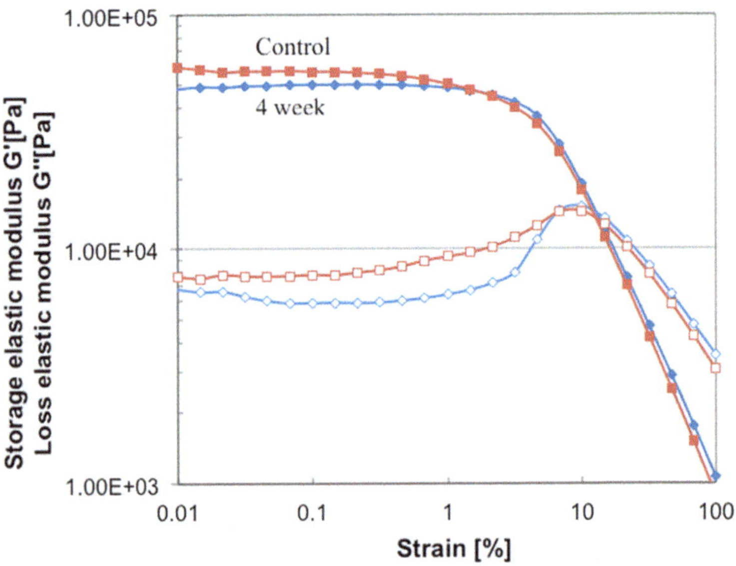

Fig. 5 Viscoelastic behaviors of sugarcane cellulose hydrogel films before and after implantation for 4 weeks at 37 °C. Storage elastic (G′) and loss elastic (G″) moduli against strain varied from 10^{-2} to 10^2 were measured at fixed frequency of 1 Hz

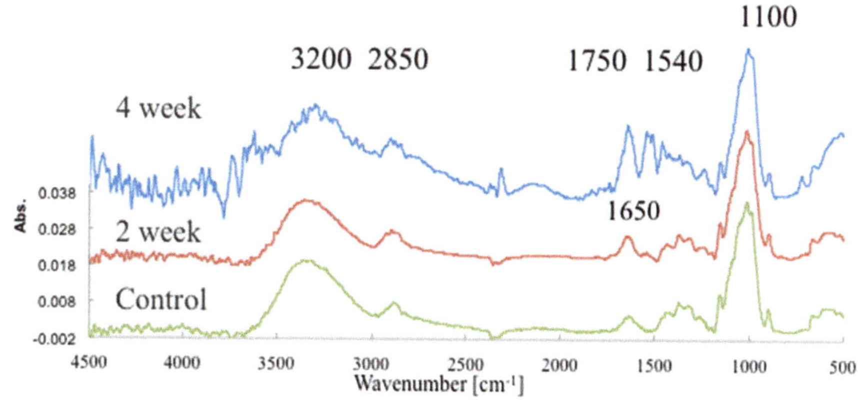

Fig. 6 ATR FT-IR spectra of sugarcane cellulose hydrogels before (control) and after implantation for 2 week and 4 weeks

removed cellulose hydrogel films became clouded in contrast with the transparent control. This could be due to the adsorption of proteins to the film body. FT-IR spectroscopy applied for their samples (Fig. 6) strongly suggested the protein adhesion in the film surface as followed.

In the IR absorption spectra of implanted cellulose hydrogel films, several extra peaks appeared at 3200, 2850, 1730, 1540, and 1100 cm^{-1}. The absorption peaks at 2850 and 1730 cm^{-1} corresponded to symmetric $-CH_2-$ stretching vibration of long alkyl chain and ester C–O stretching band mainly observed in lipid. The peak at around 3200 and 1540 cm^{-1} was assigned to N–H stretching and amide II (mainly N–H bending) of the protein, while the absorption of amide I (mainly C–O stretching) overlapped with that of adsorbed water molecules at 1650 cm^{-1} in the cellulose (Cestelli Guidi et al. 2012).

3.2 *Cytocompatibility for Tissue Regeneration in Cellulosic Hydrogel Scaffold*

It is known that the fibroblast cell is the most common type of cell found in connective tissue, because it can synthesizes collagen, which maintains extracellular matrix in animal tissue. Thus, it is important to evaluate fibroblast compatibility of the hydrogel film to test the cytocompatibility to fibroblast cells (Kobayashi 2015; Tovar-Carrillo et al. 2013b; Nakasone and Kobayashi 2016a, b). After the fibroblast cell seeding on the hydrogel films at 37 °C, the picture images were taken at 4 and 72 h (Fig. 7). The comparison suggested that fibroblast cells' spreading was influenced by the segmental microstructure of the cellulose segments in the hydrogels (Fig. 8). As shown in Fig. 9, atomic force microscopy (AFM) images showed that the hydrogel films with lower LiCl apparently contained ordered and aggregated

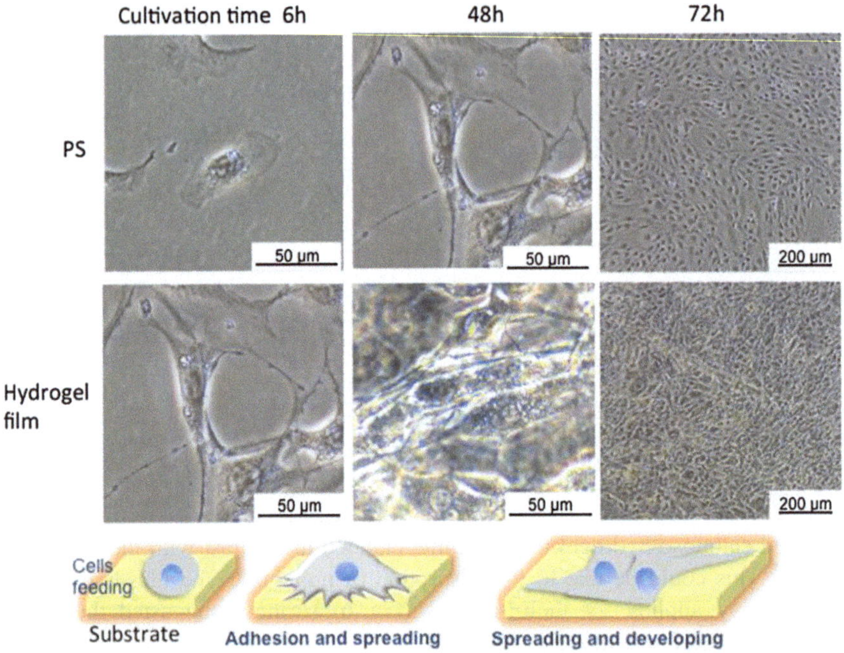

Fig. 7 Phase-contrast light images of hydrogel film and commercial polystyrene dish (PS) as control for fibroblast cell cultivation. The hydrogel films were prepared with 6 and 12 wt% LiCl. The cell culture times were 46, 48 and 72 h at 37 °C (Tovar-Carrillo et al. 2013a)

fiber orientation. As seen in Fig. 8, the fibers aggregation in the hydrogel depends on the LiCl concentration used in the DMAc-cellulose solution significantly for the phase inversion gelation. The roughness areas showed a highly ordered fiber pattern with the periodic size of 150–300 nm, when lower LiCl was used. But, the ordered morphology was not seen in the hydrogels for the higher LiCl concentration. There was tendency for enhancement of the culture glowing in the aggregated ordered surface of the hydrogels.

As shown in Fig. 9, the relationship of pore size distribution calculated by BJH method indicated that mesopores having nm order size were present in the hydrogel films having different contents of lignin (Nakasone and Kobayashi 2016a, b). The comparison between the hydrogel A and C was clear that higher pore volume was observed in the hydrogel A. Here, it should be noted that pore volume in 2–4 nm range was not observed in the hydrogel C with lower lignin content. This might be due to that the hydrogel films containing trace lignin were easily swollen allowed by easy access of water interaction with the cellulose segments. But high lignin content enhanced the aggregation of fibrous celluloses as drawn in Fig. 8. The water condition in the hydrogel behavior influenced to the 3D network of celluloses and increase average pore size of the hydrogel films. For example, the decrease of the lignin content from 1.62 to 0.68% in the hydrogel influence to the water retention and average pore size was increased from 7.5 to 11.7 nm, respectively. In the inserted microscope images for adhering cells on the hydrogel at the initial 6 h culture time, it was noted that the cultivated cells had the footprints which were well

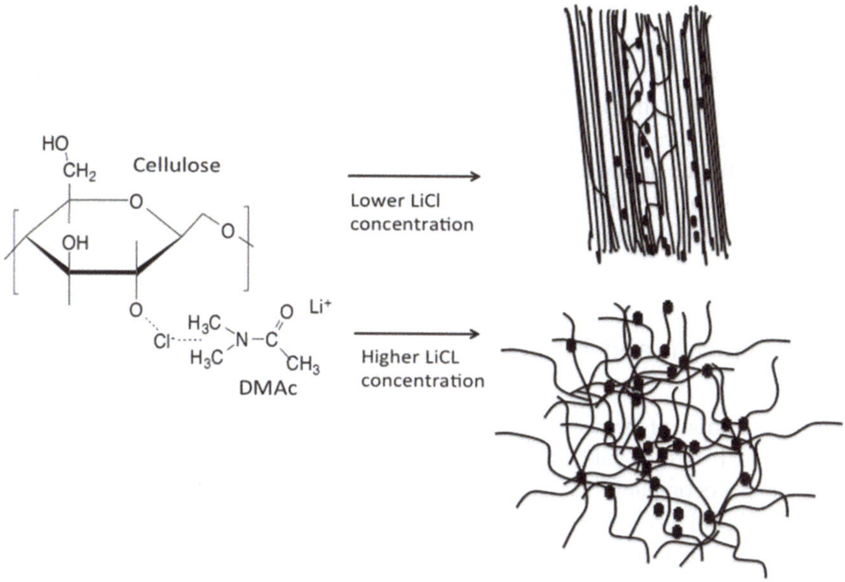

Fig. 8 Drawing of cellulose segmental aggregation of the hydrogels prepared from DMAc-cellulose solutions having different LiCl concentration

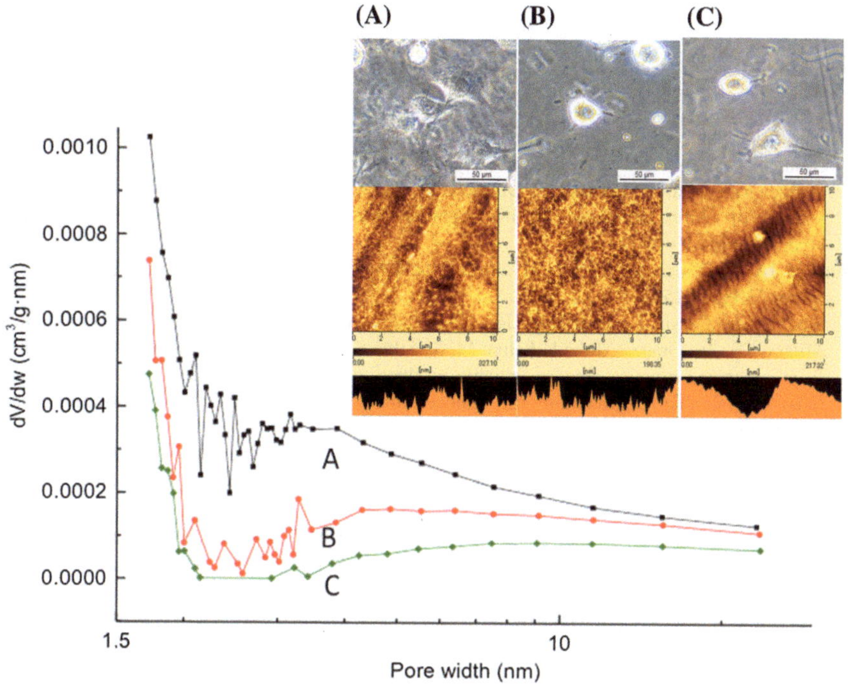

Fig. 9 Pore size distribution of cellulose hydrogel films and AFM morphology and pictures of adhered cells on (**A**) 12 h NaOH treatment, (**B**) 1 h NaOH treatment and NaOCl bleaching, and (**C**) 12 h NaOH treatment and NaOCl bleaching at the culture time of 6 h. The swollen hydrogel films were freeze-dried for the measurement

extended in the case of the hydrogel as seen in A (Fig. 9). In the initial adhesion of the cells, the round-shaped cell was found in the case of the hydrogel C for less-expanded pseudopod. This suggested that the hydrophobic lignin in the gel medium formed such rough surface ranging in the few nm. This led that difference in porous structure of the hydrogel films and such gel morphology influenced strongly both the initial cell adhesion and then proliferated the eventual cells. Similar phenomena were shown when different LiCl concentration in the range of 4–12 wt% was used in the gelation process (Tovar-Carrillo et al. 2013a).

4 Conclusion and Future Perspective

For regeneration of bagasse cellulosic wastes, chemically purified cellulose was obtained and used to fabricate cellulosic hydrogels. The hydrogel film exhibited acceptable biocompatibility and durability in living body in addition with better mechanical and viscoelastic properties. This also demonstrated an utilization of waste biomass like sugarcane bagasse as a biomaterial and raised the possibility of further applications of this material. The factors in the gelation condition caused change in the hydrogel morphology and the nontoxic properties of the hydrogels to the fibroblast cell. Consequently, the cellulose hydrogels are acceptable in biocompatible regenerative medicine.

References

Anyaporn B, Nakajima L, Kobayashi T (2017) Cellulose hydrogels nanocomposited silica and cellulose of cottonseed hulls. J Appl Polym Sci. https://doi.org/10.1002/app.44557

Benhabiles MS, Salah R, Lounici H, Drouiche N, Goosen MFA, Mameri N (2012) Antibacterial activity of chitin, chitosan and its oligomers prepared from shrimp shell waste. Food Hydrocolloids 29:48–56

Burdick JA, Prestwich GD (2011) Hyaluronic acid hydrogels for biomedical applications. Adv Mater 23:H41–H56

Canilha L, de Cássia L, Brambilla RR, Fernandes Antunes FA, Chandel AK, dos Santos Milessi TS, das Graças Almeida Felipe M, da Silva SS (2013) Bioconversion of hemicellulose from sugarcane biomass into sustainable products (Chap. 2). In: Chandel AK, da Silva SS (eds) Sustainable degradation of lignocellulosic biomass-techniques, applications and commercialization

Cao Y, Li H, Zhang Y, Zhang J, He J (2010) Structure and properties of novel regenerated cellulose films prepared from cornhusk cellulose in room temperature ionic liquids. J Appl Polym Sci 116:547–554

Cestelli Guidi M, Mirri C, Fratini E, Licursi V, Negri R, Marcelli A, Amendola R (2012) In vivo skin leptin modulation after 14 MeV neutron irradiation: a molecular and FT-IR spectroscopic study. Anal Bioanal Chem 404:1317–1326

Chandel AK, da Silva SS, Carvalho W, Singh Om V (2012) Sugarcane bagasse and leaves: foreseeable biomass of biofuel and bio-products. J Chem Technol Biotechnol 87:11–20

Chang C, Zhang L (2011) Cellulose-based hydrogels: present status and application prospects. Carbohydr Polym 84:40–53

Chang C, Peng N, He M, Teramoto Y, Nishio Y, Zhang L (2013) Fabrication and properties of chitin/hydroxyapatite hybrid hydrogels as scaffold nanomaterials. Carbohydr Polym 91:7–13

Chen FM, Liu X (2016) Advancing biomaterials of human origin for tissue engineering. Prog. Polym Sci 53:86–168

Ding B, Gao H, Song J, Li Y, Zhang L, Cao X, Xu M, Cai J (2016) Tough and cell-compatible chitosan physical hydrogels for mouse bone mesenchymal stem cell in vitro. ACS Appl Mater Interfaces 8:19739–19746

Elsabee MZ, Abdou ES (2013) Chitosan based edible film and coating: a review. Mater Sci Eng C 33:1819–1841

Fernandes EM, Pires RA, Mano JF, Reis RL (2013) Bionanocomposites from lignocellulosic resources: properties, applications and future trends for their use in the biomedical field. Prog Polym Sci 38:1415–1441

Forget A, Arya N, Randriantsilefisoa R, Miessmer F, Buck M, Ahmadi V, Jonas D, Blencowe A, Shastri VP (2016) Nonwoven carboxylated agarose-based fiber meshes with antimicrobial properties. Biomacromolecules 17:4021–4026

Garcia-Soto MJ, Jimenez-Isas H, Navarrete-Bolanos JL, Rico-Martinez R, Mirinda-Lopez R, Botello-Alvarez JE (2011) Kinetic study of the thermal hydrolysis of *Agave salmiana* for mescal production. J Agric Food Chem 59:7333–7340

Giri TK, Thakur A, Alexander A, Badwaik H, Tripathi DK (2012) Modified chitosan hydrogels as drug delivery and tissue engineering systems: present status and applications. Acta Pharm Sin B 2:439–449

Guo Y, Wu P (2008) Investigation of the hydrogen-bond structure of cellulose diacetate by two-dimensional infrared correlation spectroscopy. Carbohydr Polym 74:509–513

Henniges U, Schiehser S, Rosenau T, Potthast A (2010) Cellulose solubility and analysis of problematic cellulose pulp in the solvent system DMAc/LiCl. ASC Symp Ser 1033:165–177

Hofsetz K, Silva MA (2012) Brazilian sugarcane bagasse: energy and non-energy consumption. Biomass Bioenerg 46:564–573

Huq T, Salmieri S, Khan A, Khan RA, Le Tien C, Riedl B, Fraschini C, Bouchard J, Uribe-Calderon J, Kamal MR, Lacroix M (2012) Nanocrystalline cellulose (NCC) reinforced alginate based biodegradable nanocomposite film. Carbohyd Polym 90:1757–1763

Iñuguez-Covarrubias G, Lang SE, Rowel RM (2001) Utilization of by-products from the tequila industry. Part 1: agave bagasse as a raw material for animal feeding and fiberboard production. Bioresour Technol 77:25–32

Jiang H, Tovar-Carrillo K, Kobayashi T (2016) Ultrasound stimulated release of mimosa medicine from cellulose hydrogel matrix. Ultrason Sonochem 32:398–406

Jorfi M, Foster EJ (2015) Recent advances in nanocellulose for biomedical applications. J Appl Polym Sci 132:41719

Kestur SG, Flores-Sahgun THS, Dos Santos LP, Dos Santos J, Mazzaro I, Mikowski A (2013) Characterization of blue agave bagasse fibers of Mexico. Compos A 45:153–161

Kobayashi T (2015) Fabrication of cellulose hydrogels and characterization of their biocompatible films. In: Rahman A (ed) Studies in natural products chemistry, vol 45. Elsevier B.V., Amsterdam, pp 1–15

Kobayashi T, Tovar-Carrillo K (2015) Fibroblast cell cultivation on wooden pulp cellulose hydrogels for cytocompatibility scaffold method. Pharm Anal Acta 6:423

Kobayashi T, Tovar-Carrillo KL, Nakasone K, Tagaya M (2014a) Biopolymer hydrogels regenerated from *Agave tequilana* waste for cytocompatable materials. IMRC Meet MRS Proc 1613:75–82

Kobayashi T, Tovar-Carrillo K, Tagaya M (2014b) Biohydrogels interpenetrated with hydroxyethyl cellulose and wooden pulp for biocompatible materials. Ind Eng Chem Res 53:4650–4659

Kobayashi T, Tovar-Carrillo K, Tagaya M (2015) Bagasse sustainable polymers for cellulose hydrogel sheets showing tissue regeneration. In: Thakur VK, Thakur MK (eds) Handbook of sustainable polymers structure and chemistry. Pan Stanford Publishing Pte Ltd.

Kometani N, Tanabe M, Su L, Yang K, Nishinari K (2015) Insitu observations of thermoreversible gelation and phase separation of agarose and methylcellulose solutions under high pressure. J Phys Chem B 119:6878–6883

Kost J, Langer R (2012) Responsive polymeric delivery systems. Adv Drug Deliv Rev 64:327–341

Le Moigne N, Navard P (2010) Dissolution mechanisms of wood cellulose fibres in NaOH–water. Cellulose 17:31–45

Lee KY, Mooney DY (2012) Alginate: properties and biomedical applications. Prog Polym Sci 37 (1):106–126

Lee H, Ahn S, Kim GH (2012) Three-dimensional collagen/alginate hydrid scaffolds functionalized with a drug delivery system (DDS) for bone tissue regeneration. Chem Mater 24:881–891

Li K, Kobayashi T (2016a) Ultrasound response of aqueous poly(ionic liquid) solution. Ultrason Sonochem 30:52–60

Li K, Kobayashi T (2016b) FT-IR spectroscopy of ultrasound effect on aqueous imidazole ionic liquids having different counter ions. Ultrason Sonochem 28:39–46

Li Y, Xiao W, Xiao K, Berti L, Luo J, Tseng HP, Fung G, Lam KS (2012) Well defined, reversible boronate crosslinked nanocarriers for targeted drug delivery in response to acidic pH values and cis-Diols. Angew Chem Int Ed 124:2918–2923

Li K, Noguchi S, Kobayashi T (2016) Ultrasound-responsive behavior of gelatinous ionic liquid/ poly(vinyl alcohol) composites. Ind Eng Chem Res 55:9915–9924

Linan-Montes A, de la Parra-Arciniega SM, Garza-Gonzalez MT, Garcia-Reyes RB, Soto-Regalado E, Cerino-Cordova FJ (2014) Characterization and thermal analysis of agave bagasse and malt spent grain. J Therm Anal Calorim 115:751–758

Liu X, Ma L, Mao Z, Gao C (2011) Chitosan-based biomaterials for tissue repair and regeneration. Adv Polym Sci 244:81–127

Lowe CJ, Reucroft IM, Grota MC, Shreiber DI (2016) Production of highly aligned collagen scaffolds by freeze-drying of self-assembled, fibrillar collagen gels. ACS Biomater Sci Eng 2:645–651

Macaya D, Ng KK, Spector M (2011) Injectable collagen-genipin gel for the treatment of spinal cord injury: in vitro studies. Adv Funct Mater 21:4788–4797

Madhumathi M, Sudheesh-Kumar PT, Abhilash S, Sreeja V, Tamura H, Manzoor K, Nair SV, Jayakumar R (2010) Development of novel chitin/nanosilver composite scaffolds for wound dressing applications. J Mater Sci Mater Med 21:807–813

Mahadeva SK, Kim J (2011) Addition of 1-butyl-3-methylimidazolium bis (trifluoromethylsulfonyl) imide to improve the thermal stability of regenerated cellulose. J Appl Polym Sci 121:750–755

Murillo-Alvarado PE, Santibanez-Aguilar JE, Ponce-Ortega JM, Castro-Montoya AJ, Serna-Gonzalez M, El-Haiwagi MM (2014) Optimization of the supply chain associated to the production of bioethanol from residues of agave from the tequila process in Mexico. Ind Eng Chem Res 53:5524–5538

Nakasone K, Kobayashi T (2016a) Effect of pre-treatment of sugarcane bagasse on the cellulose solution and application for the cellulose hydrogel films. Polym Adv Technol 27:973–980

Nakasone K, Kobayashi T (2016b) Cytocompatible cellulose hydrogels containing trace lignin. Mater Sci Eng C 64:269–277

Nakasone K, Ikematsu S, Kobayashi T (2016) Biocompatibility evaluation of cellulose hydrogel film regenerated from sugarcane bagasse waste and its in vivo behavior in mice. Ind Eng Chem Res 55:30–37

Nguyen MK, Alsberg E (2014) Bioactive factor delivery strategies from engineered polymer hydrogels for therapeutic medicine. Prog Polym Sci 39:1235–1265

Nigam PS, Singh A (2011) Production of liquid biofuels from renewable resources. Prog Energy Combust Sci 37:52–68

Onuki Y, Bhardwaj U, Papadimitrakopoulos F, Burgess DJ (2008) A review of the biocompatibility of implantable devices: current challenges to overcome foreign body response. J Diabetes Sci Technol 2:1003–1015

Peng N, Hu D, Zeng J, Li Y, Liang L, Chang C (2016) Superabsorbent cellulose-clay nanocomposite hydrogels for highly efficient removal of dye in water. ACS Sustain Chem Eng 4:7217–7224

Qiu Y, Park K (2012) Environment-sensitive hydrogels for drug delivery. Adv Drug Deliv Rev 64:49–60

Rao SS, Dejesus J, Short AR, Otero JJ, Sakar A, Winter JO (2013) Glioblastoma behavior in three-dimensional collagen-hyaluronan composite hydrogels. ACS Appl Mater Interfaces 5:9276–9284

Rodell CB, Wade RJ, Purcell BP, Dusai NN, Burdick JA (2015) Selective proteolytic degradation of guest-host assembled, injectable hyaluronic acid hydrogels. ACS Biomater Sci Eng 1:277–286

Rojas OJ (2016) Cellulose chemistry and properties: fibers. Nanocellulose and advanced materials. Springer, Swizerland, pp 22–26

Saucedo-Luna J, Castro-Montoya AJ, Campos-Garcia J, Rico JL (2010) Optimization of acid hydrolysis bagasse from *Agave tequilana* Weber. Rev Mex Ing Quim 9:91–97

Shen X, Shamshina JL, Berton P, Bandomir J, Wang H, Gurau G, Rogers RD (2016) Comparison of hydrogels prepared with ionic liquid-isolated vs commercial chitin and cellulose. ACS Sustain Chem Eng 4:471–480

Sideris E, Griffin DR, Ding Y, Li S, Weaver WM, Carlo DD, Hsiai T, Segura T (2016) Particle hydrogels based on hyaluronic acid building blocks. ACS Biomater Sci Eng 2:2034–2041

Sjoholm E, Gustafsson K, Eriksson B, Brown W, Colmsjo A (2000) Aggregation of cellulose in lithium chloride/N,N-dimethylacetamide. Carbohydr Polym 41:153–159

Takegawa A, Murakami M, Kaneko Y, Kadokawa J (2010) Preparation of chitin/cellulose composite gels and films with ionic liquids. Carbohydr Polym 79:85–90

Tchemtchoua VT, Atanasova G, Aqil A, Filee P, Garbacki N, Vanhooteghem O, Deroanne C, Noel A, Jerome C, Nushens B, Poumay Y, Colige A (2011) Development of a chitosan nanofibrillar schaford for skin repair and regeneration. Biomacromolecules 12:3194–3204

Torres-Rendon JG, Kopf M, Gehlen D, Blaeser A, Fischer H, Laporte LD, Walter A (2016) Cellulose nanofibril hydrogel tubes as sacrificial templates for freestanding tubular cell constructs. Biomacromolecules 17:905–913

Tovar-Carrillo KL, Tagaya M, Kobayashi T (2013a) Bamboo fibers elaborating cellulose hydrogel films for medical applications. J Mater Sci Chem Eng 1:7–12

Tovar-Carrillo K, Sugita SS, Tagaya M, Kobayashi T (2013b) Fibroblast compatibility on scaffold hydrogels prepared from *Agave tequilana* Webber bagasse for tissue regeneration. Ind Eng Chem Res 52:11607–11613

Tovar-Carrillo KL, Tagaya M, Kobayashi T (2014) Effects of sodium hypochlorite on *A. tequilana* Weber bagasse fibers used to elaborate cyto and biocompatible hydrogel films. Mater Sci Eng C 42:808–815

Tovar-Carrillo KL, Tamayo G, Donohue A, Kobayashi T, Saucedo RA (2015) Obtaining of hydrogels using PVA and HEC for adipose tissue regeneration. J Tissue Sci Eng 6:152

Tummala GK, Joffre T, Lopes VR, Liszka A, Buznyk O, Ferraz N, Persson C, Griffith M, Mihranyan A (2016) ACS Biomater Sci Eng 2:2072–2079

Van Vlierberghe S, Dubruel P, Schacht E (2011) Biopolymer-based hydrogels as scaffolds for tissue engineering applications: a review. Biomacromolecules 12:1387–1408

Velmurugan R, Muthukumar K (2011) Utilization of sugarcane bagasse for bioethanol production: sono-assisted acid hydrolysis approach. Biores Technol 102:7119–7123

Venegas-Sanchez JA, Tagaya M, Kobayashi T (2013a) Ultrasound effect used as external stimulus for viscosity change of aqueous carrageenans. Ultrason Sonochem 20:1081–1091

Venegas-Sanchez JA, Tagaya M, Kobayashi T (2013b) Effect of ultrasound on the aqueous viscosity of several water-soluble polymers. Polym J 45:1–9

Venegas-Sanchez JA, Tagaya M, Kobayashi T (2014) Ultrasound stimulus inducing change in hydrogen bonded crosslinking of aqueous polyvinyl alcohols. Ultrason Sonochem 21:295–309

Xiong R, Hameed N, Guo Q (2012) Cellulose/polycaprolactone blends regenerated from ionic liquid 1-butyl-3-methylimidazolium chloride. Carbohydr Polym 90:575–582

Zhao D, Liu C, Zhuo R, Cheng S (2012) Alginate/Caco3 hybrid nanoparticles for efficient codelivery of antitumor gene and drug. Mol. Pharmaceutics 9:2887–2893

Chapter 12
Injectable Hydrogels for Cartilage Regeneration

Cenk Celik, Vishal T. Mogal, James Hoi Po Hui, Xian Jun Loh and Wei Seong Toh

Abstract Articular cartilage injuries have a limited potential to heal, which over time, may lead to osteoarthritis, an inflammatory and degenerative joint disease associated with activity-related pain, swelling, and impaired mobility. Regeneration and restoration of joint tissue and function remain unmet challenges. Intra-articular injections of therapeutic agents are effective to some extent, but often require multiple injections. In the past decade, injectable hydrogels have emerged as promising biomaterials, due largely to their biocompatibility, tissue extracellular matrix (ECM) mimicry, excellent permeability, and easy adaptation for minimal-invasive procedures. Moreover, hydrogels can be designed as carriers for sustained release of therapeutic agents and protective matrices for cell delivery.

C. Celik
Department of Orthopaedic Surgery, Yong Loo Lin School of Medicine, National University of Singapore, 10 Kent Ridge Crescent, Singapore 119260, Singapore

V. T. Mogal · W. S. Toh (✉)
Faculty of Dentistry, National University of Singapore, 11 Lower Kent Ridge Road, Singapore 119083, Singapore
e-mail: dentohws@nus.edu.sg

J. H. P. Hui
Cartilage Repair Program, Therapeutic Tissue Engineering Laboratory, National University Health System, 1E Kent Ridge Road, Singapore 119288, Singapore

J. H. P. Hui · W. S. Toh
Tissue Engineering Program, Life Sciences Institute, National University of Singapore, 27 Medical Drive, Singapore 117510, Singapore

X. J. Loh
Institute of Materials Research and Engineering (IMRE), 3 Research Link, Singapore 117602, Singapore

X. J. Loh
Department of Materials Science and Engineering, National University of Singapore, 9 Engineering Drive 1, Singapore 117576, Singapore

X. J. Loh
Singapore Eye Research Institute, 11 Third Hospital Avenue, Singapore 168751, Singapore

V. K. Thakur and M. K. Thakur (eds.), *Hydrogels*, Gels Horizons: From Science to Smart Materials, https://doi.org/10.1007/978-981-10-6077-9_12

This chapter provides an overview of the injectable hydrogel systems currently being applied together with therapeutic drug delivery and/or cell therapy for treatment of cartilage lesions and osteoarthritis.

Keywords Hydrogels · Stem cells · Drug delivery · Tissue regeneration

1 Introduction

Articular cartilage is a unique hypocellular, aneural, alymphatic, and avascular load-bearing tissue that composed of chondrocytes embedded within a dense extracellular matrix (ECM) of collagens, proteoglycans, and non-collagenous proteins. Absence of neural, lymphatic, or vascular supply makes regeneration of articular cartilage notoriously difficult. Additionally, articular cartilage has a unique zonal architecture that comprises the superficial, middle, and deep zones, each with distinct cellular phenotype, matrix composition, and anisotropy (Becerra et al. 2010; Toh et al. 2016a).

Articular cartilage injuries, commonly caused by sports and recreational activities, are a major risk factor for development of osteoarthritis (OA), an inflammatory and degenerative joint disease characterized by the degradation of the articular cartilage, subchondral bone, meniscus, and ligaments, together with the formation of painful osteophytes. OA is the most common form of arthritis that affects 250 million people worldwide, affecting mobility, ability to continue working, and quality of life, and the number of cases is increasing due to aging and obesity epidemic (Ge et al. 2006; Murray et al. 2012).

Current treatment modalities for articular cartilage repair include microfracture, abrasion, drilling, osteochondral grafting, perichondral grafting, autologous chondrocyte implantation (ACI) and more recently tissue-engineered grafting (matrix-associated ACI, MACI) (Marcacci et al. 2013). Indeed, many of these methods help to repair the tissue and reduce pain in affected patients to some degree. However, these techniques usually require open surgery and multiple steps in the procedure (Jiang et al. 2011). In terms of clinical outcomes, current techniques often generate inferior fibrocartilage repair and do not regenerate the native zonal architecture of the articular cartilage. Tissue engineering is an emerging integrative strategy aims to reconstruct and regenerate a tissue both structurally and functionally by use of biomaterials as scaffolds, in combination with cells and bioactive signaling molecules incorporated to modulate cellular processes including cell migration, proliferation, differentiation, and matrix synthesis (Langer and Vacanti 1993). On this note, recent efforts have been directed at exploring biomaterial scaffolds, particularly hydrogels, for applications in cartilage tissue engineering and repair (Toh et al. 2011). It is envisaged that ease of clinical application and functional improvements in cartilage repair could be achieved through the development of innovative biomaterial systems such as the injectable hydrogels (Fig. 1) (Toh et al. 2011).

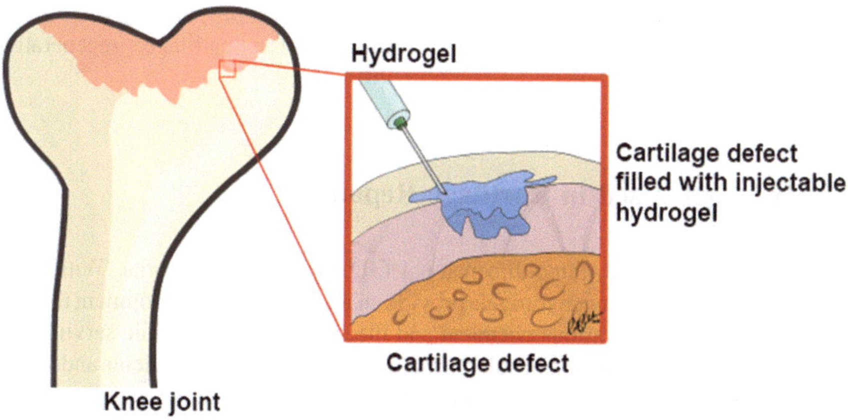

Fig. 1 Utilizing an injectable hydrogel for cartilage repair

2 Scaffold Design in Cartilage Tissue Engineering

The criteria and design parameters of a biomaterial scaffold are important considerations in cartilage tissue engineering (Toh et al. 2011). An ideal scaffold in cartilage tissue engineering should satisfy several essential criteria: It should be (1) biocompatible to minimize any adverse tissue reactions but supports cell growth and overall tissue formation; (2) biodegradable with a favorable resorption rate to provide initial structural support for cell growth and later gradually degraded to facilitate timely tissue remodeling; (3) having the suitable porosity and interconnectivity to enable cell migration and efficient exchange of nutrients and waste; (4) able to serve as a mechanical framework to localize the cells and to support tissue growth under native mechanical loading of the articular cartilage; (5) able to meet the needs of the surgical procedures with either pre-formed polymeric scaffolds for implantation or injectable biomaterials in minimally invasive procedures.

To date, numerous scaffolding biomaterials, based on natural and synthetic polymers, have been explored, either by themselves or in combination with the delivery of cells and/or bioactive molecules, in a variety of forms including hydrogels, sponges, and fibrous meshes for cartilage tissue engineering and regeneration (Chung and Burdick 2008). Of these various material structures, hydrogels are the emerging material of choice and are essentially water-swollen polymer networks cross-linked by either covalent or physical means (Hoffman 2002). Hydrogels are particularly attractive as they can be homogeneously mixed with cells and/or biomolecules, injected in minimally invasive fashion and be able to conform to any size and shape of the defect. Latest strategies have employed chemical modifications of the natural and synthetic polymers and combination to generate hybrid hydrogel scaffolds with combined properties for cartilage tissue engineering. We will focus on the development of hydrogels for cartilage tissue engineering. We will also discuss the materials, modifications, and polymerization mechanisms employed in hydrogel formation for cartilage regenerative applications.

3 Hydrogels Used in Cartilage Repair

Hydrogels are hydrated, three-dimensional (3D) polymeric networks. Within the scope of tissue engineering, hydrogels provide an ideal 3D microenvironment that can signal to cells through various chemical and mechanical signals, while serving as a permeable matrix for diffusion of soluble factors, and exchange of nutrients and waste. In the context of cartilage tissue engineering, hydrogels mimic the native cartilage microenvironment in many aspects. They provide a high water content 3D environment, promote pro-chondrogenic spherical morphology of encapsulated cells, and foster cell–matrix interactions and entrapment of secreted ECM. Additionally, the 3D hydrogel microenvironment can be decorated with adhesive ligands such as arginylglycylaspartic acid (RGD) (Yao et al. 2016; Liu et al. 2010a, b) or loaded with biomolecules such as transforming growth factor (TGF)-β1 (Choi et al. 2015) to enhance cell viability and chondrogenesis, respectively. In recent years, injectable hydrogels have attracted much attention, as they offer advantages of ease of application, minimal invasiveness, and void-filling property. Furthermore, the minimally invasive procedure reduces operating costs and eliminates the risk of infection and tissue morbidity.

Since the ECM of cartilage is mainly composed of proteoglycans and collagens, natural polymers are commonly used in the fabrication of hydrogels for cartilage tissue engineering. Among the natural polymers including agarose (Gadjanski et al. 2013), alginate (Yao et al. 2016), chitosan (Choi et al. 2015), hyaluronic acid (Burdick and Prestwich 2011; Toh et al. 2010, 2012), pullulan (Chen et al. 2016), chondroitin sulfate (Chen et al. 2016), collagen (Choi et al. 2014), dextran (Geng et al. 2012), gelatin (Wang et al. 2014) and fibrin (Ho et al. 2010) have been widely studied in cartilage tissue engineering. Notably, hyaluronic acid injections, also known as viscosupplementation, have been widely used by orthopaedic surgeons in a minimally invasive technique for the treatment of knee OA (Percope de Andrade et al. 2015). Natural hydrogels have inherent advantages such as bioactivity, biocompatibility and biodegradability, but may suffer issues of batch variation and difficulty in control of material degradation and mechanical properties. Natural biodegradable polymer hydrogels and their derivatives have been successfully employed to carry cells and support cell adhesion and tissue formation. They resemble native extracellular matrices that are non-toxic and do not cause immunologic reactions. These matrices are utilized for cell adhesion, growth, proliferation or differentiation and support tissue formation and matrix deposition to fill the defect site. However, most of the natural hydrogels have poor mechanical properties and modifications by chemical or physical means are often required to improve the mechanical properties of the scaffold.

Synthetic polymers have the greatest advantages of defined chemical composition and minimum batch variability. Synthetic hydrogels based on polymers such as poly(ethylene glycol) (Liu et al. 2010a, b), poly(ethylene oxide) (Sharma et al. 2007), poly(ethylene glycol fumarate) (Park et al. 2009a, b), poly(vinyl alcohol) (Nie et al. 2015), and poly(lactic acid) (Basu et al. 2016) have been explored in cartilage tissue engineering. However, synthetic polymers often require modifications to enhance their biocompatibility and cell–cell–matrix interactions, including cell adhesion and spreading, degradability and matrix remodeling. In view of these limitations, there have been significant efforts in the past decade to modify natural and synthetic polymers as well as to develop hybrid composite hydrogels with combined properties of natural and synthetic polymers to fulfill the urgent needs of cartilage tissue engineering.

3.1 *Natural Hydrogels*

Natural hydrogels are often utilized for cell delivery during tissue regeneration, due very much to their innate characteristics and similarity to the native tissue ECM (Toh et al. 2015). Collagen (Choi et al. 2014), gelatin (Wang et al. 2014), chitosan (Choi et al. 2015), hyaluronic acid (Burdick and Prestwich 2011), alginate (Yao et al. 2016), and agarose (Gadjanski et al. 2013) are some of the commonly used natural polymers for hydrogel fabrication. Alternatively, hydrogels made from decellularized tissue matrices are attractive as they mimic closely the native cartilage ECM that is composed of numerous types of matrix proteins (Cheng et al. 2013a, b; Burnsed et al. 2016).

3.1.1 Collagen

Collagen has been widely studied for applications in cartilage tissue engineering as they are the main components that constitute the cartilage ECM and have been found to exert bioactive effects on chondrogenic differentiation of MSCs. For instance, type II collagen hydrogels have been found to have the potential to initiate mesenchymal stem cell (MSC) chondrogenesis independently of exogenous transforming growth factor (TGF)-β1 (Bosnakovski et al. 2006). Similarly, type I collagen hydrogel has been demonstrated to induce chondrogenic differentiation of MSCs without the need of exogenous growth factors (Zhang et al. 2012).

However, collagen-based hydrogels have poor mechanical properties. To overcome this problem, collagen is commonly modified using methods such as physical or chemical modification, addition of cross-linking agents, and blending with other polymers (Schuurman et al. 2013). On this note, collagen hydrogels are commonly cross-linked by agents such as genipin (Mu et al. 2013; Jeng et al. 2012) and transglutaminases (Jeng et al. 2012) to improve the mechanical properties of the scaffold as well as to confer additional beneficial properties (e.g., anti-inflammation) for cartilage regeneration (Lima et al. 2009). Alternatively, collagen could be blended with other polymers to form composite hydrogel scaffolds with desirable biochemical and biomechanical properties. Kontturi et al. (2014) have developed an in situ forming type II collagen/HA composite hydrogel scaffolds for cartilage regeneration. In that study, non-toxic chemical cross-linker poly(ethylene glycol) ether tetrasuccinimidyl glutarate (4SPEG) was used to cross-link type II collagen hydrogels to form strong and mechanically stable hydrogels. The composite type II collagen/HA hydrogels incorporated with TGF-β_1 was able to promote cell viability, chondrogenic gene expression, and GAG synthesis of encapsulated chondrocytes (Kontturi et al. 2014). In another study, Dinescu et al. (2013) fabricated a type I collagen–sericin (Coll-SS) hydrogel with the addition of hyaluronic acid (HA) or chondroitin sulfate (CS) for application in cartilage tissue engineering. In that study, composite Coll-SS hydrogels incorporated with 10% HA and 5% CS displayed the best biological performances in terms of cell viability, proliferation, morphology, and distribution of adipose-derived stem cells (Dinescu et al. 2014). Recently, Zheng and colleagues also reported a novel in situ synthesized nano-hydroxyapatite/collagen/alginate hydrogels for osteochondral tissue engineering. In that study, nano-hydroxyapatite reinforced collagen–alginate hydrogel was prepared by in situ synthesis of nano-hydroxyapatite in collagen I gel followed by addition of alginate and calcium chloride. The multi-phased composite hydrogels not only showed improved mechanical properties but also enhanced cell viability and proliferation, and upregulated expression of cartilaginous markers (type II collagen and aggrecan) by the encapsulated chondrocytes (Zheng et al. 2014).

3.1.2 Gelatin

Gelatin is the hydrolyzed form of collagen that has motifs such as RGD (Arg-Gly-Asp) (Zhao et al. 2013) which are known to promote cell attachment of a wide variety of cell types (Lim et al. 2012, 2013b; Nichol et al. 2010; Wang et al. 2014). It is a natural, biocompatible, biodegradable protein polymer has been extensively studied for cartilage regeneration (Lim et al. 2012). However, gelatin is seldom used directly in cartilage tissue engineering because of its poor mechanical properties (Chan 2015; Zhao et al. 2013). Various strategies such as chemical or physical modifications, cross-linking with specific agents, and blending with other polymers have been frequently employed to improve the properties of gelatin (Wang et al. 2014; Yue et al. 2015; Mazaki et al. 2014; Miljkovic et al. 2009; Han et al. 2015; Levett et al. 2014a, b).

Recently, Wang et al. (2014) developed an injectable gelatin hydrogel system for cartilage tissue engineering. In that study, gelatin–hydroxyphenylpropionic acid (gelatin-HPA) hydrogels were formed using oxidative coupling of hydroxyphenylpropionic acid (HPA) catalyzed by horseradish peroxidase (HRP) and hydrogen peroxide (H_2O_2). By adjusting the concentrations of the gel precursor and/or H_2O_2, the stiffness of the gelatin-HPA gels could be easily tuned. It was found that the cellular functions of chondrocytes such as proliferation, biosynthesis of collagen, and GAG were strongly modulated by the stiffness of hydrogels. The significant impact of hydrogel stiffness on in vitro chondrocyte functions was further demonstrated in vivo in rabbit articular cartilage defect model where the hydrogels of medium stiffness (1000 Pa) seeded with chondrocytes were able to promote enhanced cartilage repair, compared to other hydrogels at low (570 Pa) or high stiffness (2750 Pa) (Wang et al. 2014).

For photopolymerization, gelatin has been modified to include methacrylate groups, and by means of adjusting the polymer precursor concentration, and the UV light intensity, a wide range of properties can be achieved (Yue et al. 2015). In view of potential UV light-induced toxicity, a novel photocross-linkable furfurylamine-conjugated gelatin (gelatin-FA) was developed that allowed rapid photopolymerization by visible light with rose bengal as the light sensitizer. Gelatin-FA hydrogel was subsequently employed as a scaffold for bone marrow MSCs and demonstrated promising efficacy in repair of osteochondral defects, generating hyaline-like cartilage and regenerated subchondral bone at 12-week post-operation (Mazaki et al. 2014). Other investigators have combined gelatin with other polymers including chitosan (Miljkovic et al. 2009), polyethylene glycol (Han et al. 2015), and hyaluronic acid (Levett et al. 2014b) to make hybrid scaffolds with improved composition and mechanical properties for cartilage tissue engineering.

3.1.3 Chitosan

Chitosan is a linear cationic polysaccharide that is anti-bacterial, biocompatible, and biodegradable and has been widely used in the tissue engineering of various tissue including skin, bone, and cartilage. Various strategies such as chemical or physical modifications, cross-linking with specific agents, and blending with other polymers have been employed to improve the properties of chitosan (Hoemann et al. 2007; Martins et al. 2014; Gittens et al. 2016; Wang and Stegemann 2011; Wang and Tsai 2013).

Thermogelling chitosan hydrogels are commonly cross-linked by glycerophosphate (GP) and have demonstrated applications for bone and cartilage tissue engineering (Hoemann et al. 2007; Martins et al. 2014). Hoemann and colleagues showed in a rabbit model that chitosan-GP hydrogels applied in conjunction with bone marrow stimulation elicited a more hyaline and integrated repair tissue with subchondral bone and with improvements in histological score compared to bone marrow stimulation by drilling alone (Hoemann et al. 2007). Other agents including genipin and glyoxal have also been investigated to develop injectable hydrogels of chitosan and composites with other polymers such collagen and hyaluronic acid for various biomedical applications including cartilage tissue engineering (Gittens et al. 2016; Wang and Stegemann 2011; Wang and Tsai 2013; Jalani et al. 2015).

Various chemical modifications (e.g., thiols, methacrylates, maleimides) of chitosan have been reported (Wu et al. 2009; Matsumoto et al. 2016; Choi et al. 2014; Kim et al. 2015). Recently, an injectable and photopolymerizable methacrylated chitosan (MeGC) prepared by free radical polymerization in the presence of riboflavin under visible blue light have been developed (Choi et al. 2014). This system overcomes the issues associated with the potential toxicity and side effects of UV irradiation and toxic initiators. They showed that these hydrogels supported the proliferation and cartilage matrix production by encapsulated chondrocytes and MSCs. Incorporation of type II collagen or chondroitin sulfate in the hydrogel matrix further enhanced cellular condensation and chondrogenesis. Kim et al. (2015) further showed a bio-functional hydrogel for specific use in cartilage regeneration by conjugating TGF-β1 to MeGC hydrogels impregnated with cartilage-specific type II collagen (Kim et al. 2015). TGF-β1 was delivered from MeGC hydrogels in a controlled manner with reduced burst release by chemically conjugating the protein to MeGC. The hydrogel system did not compromise viability of encapsulated human synovium-derived MSCs. Type II collagen impregnation and TGF-β1 delivery significantly enhanced cellular aggregation and deposition of cartilaginous ECM by the encapsulated cells, compared with pure MeGC hydrogels. This report showed the successful engineering of a bio-functional hydrogel with a specific microenvironment tailored to promote chondrogenesis that might prove efficacious for treatment of cartilage defects.

3.1.4 Hyaluronic Acid

Hyaluronic acid (HA) is one of the most extensively studied natural polymers for cartilage tissue engineering (Kim et al. 2011). HA is a linear polysaccharide that consists of alternating units of a repeating disaccharide, β-1,4-D-glucuronic acid-β-1,3-*N*-acetyl-D-glucosamine. It is an highly hydrated, polyanionic, non-sulfated GAG macromolecule with a wide range of molecular weights (100,000–8,000,000 Da) that exist throughout the body from vitreous of the eye to ECM of skin and cartilage tissues and is involved in cellular signaling, tissue morphogenesis, matrix organization, and wound repair (Dicker et al. 2014).

Due to its consistency, biocompatibility, viscoelasticity, limited immunogenicity, and hydrophilic properties, HA is a strong candidate for biomedical applications (Burdick and Prestwich 2011). Chemical modifications of HA to include moieties (e.g., thiols, methacrylates, tyramines) have been frequently performed to improve the structural and mechanical properties of HA for various biomedical applications including cartilage regenerative applications (Liu et al. 2006; Kurisawa et al. 2005; Nettles et al. 2004).

For photopolymerization, HA macromers can be modified to include reactive methacrylate groups to form methacrylated HA (MeHA) and by varying the molecular weight, macromer concentration and cross-linking density, a wide range of properties can be achieved. Studies performed on chondrocytes and MSCs supported the use of photocross-linkable MeHA hydrogels for cartilage tissue engineering (Nettles et al. 2004; Erickson et al. 2009). For instance, photocross-linkable MeHA hydrogels promoted the retention of chondrocytic phenotype and matrix synthesis of encapsulated porcine chondrocytes and enhanced cell infiltration and tissue repair in rabbit osteochondral defect model at 2-week post-surgery (Nettles et al. 2004). In another study by Erickson and co-workers, it was found that variations in HA macromer concentration (1, 2, and 5%), that resulted in different network cross-linking degrees and mechanical properties, exerted profound effects on MSC chondrogenesis. Notably, increasing HA macromer concentration promoted chondrogenesis and matrix synthesis, but generated functionally inferior constructs due to limited matrix distribution throughout the construct (Erickson et al. 2009). In recent years, methacrylation of polymers as gelatin, chitosan, and chondroitin sulfate have also been performed, and these methacrylated derivatives have been blended with MeHA to generate hydrogels with diverse properties for cartilage repair applications (Hayami et al. 2016; Levett et al. 2014a).

Liu and co-workers first reported thiol modification of hyaluronic acid and gelatin that allows cross-linking by addition of poly(ethylene glycol) diacrylate (PEGDA) to form an injectable HA/Gelatin composite hydrogels for osteochondral defect repair (Liu et al. 2006). In that study, osteochondral defects repaired using HA/Gelatin hydrogels seeded with MSCs showed firm translucent cartilage with superior integration to the adjacent normal cartilage at 12 weeks. On contrary, defects repair with only cells showed hyaline-like cartilage at the periphery but the center was filled with fibrous tissue (Liu et al. 2006). Using the same HA-based hydrogel system (Toh et al. 2010; Toh and Cao 2014), Toh and colleagues encapsulated chondrogenically committed cells derived from human embryonic stem cells (hESCs) in hydrogels to form tissue-engineered cartilage constructs. These cartilaginous constructs showed rich deposition of type II collagen and GAG. Following transplantation for 12 weeks, osteochondral defects repaired with the tissue-engineered cartilage constructs showed hyaline cartilage with good structural integration with adjacent cartilage and underlying subchondral bone (Toh et al. 2010).

More recently, an injectable HA hydrogel utilizing enzyme-mediated in situ cross-linking was described, of which coupling of tyramine to HA produced HA-tyramine (HA-Tyr) (Kurisawa et al. 2005). Cross-linking to form hydrogel was induced by addition of HRP and H_2O_2 to HA-Tyr solution in the presence or absence of cells or soluble factors (Lee et al. 2009; Ren et al. 2015; Toh et al. 2012). HA-Tyr hydrogels have been shown to support chondrocytes and MSCs in chondrogenesis and cartilage tissue formation (Ren et al. 2015; Toh et al. 2012). In a study by Toh and co-workers, it was found that by varying the cross-linking degree of HA-Tyr hydrogels, different cartilage subtypes (i.e., hyaline, hypertrophic, fibrocartilage) could be generated through control of MSC condensation during chondrogenesis. Notably, lower cross-linking degree promoted the highest extent of condensation of MSCs and was shown to enhance chondrogenesis toward hyaline cartilage formation, while higher degrees of cross-linking directed MSC differentiation toward fibrocartilage and fibrous tissue formation (Toh et al. 2012). The similar chemical modification was performed with polymers such as pullulan (Chen et al. 2016), chondroitin sulfate (Chen et al. 2016), dextran (Jin et al. 2010), and these tyramine-conjugated derivatives have been blended with HA-Tyr to generate composite hydrogels for cartilage tissue engineering (Jin et al. 2010).

3.1.5 Alginate

Alginate is a natural polyanionic polysaccharide extracted from brown algae and commonly cross-linked with bivalent cations (e.g., calcium, zinc, strontium) to form ionically cross-linked hydrogels. Calcium alginate beads and hydrogels have been used to propagate chondrocytes in cultures and to induce stem cell differentiation. Alginate is commonly loaded with soluble factors, modified with adhesion peptides (e.g., RGD) or blended with other polymers to form composite hydrogels for cartilage tissue engineering applications (Yao et al. 2016; Zheng et al. 2014; Little et al. 2014; Park and Lee 2014).

In a recent study, it was found that alginate gel encapsulation of human-induced pluripotent stem cells (hiPSCs) cultured under TGF-β3 supplemented chondrogenic differentiation condition promoted cartilage formation with upregulated expression of chondrogenic markers including Sox-9, type II collagen, and aggrecan. Upon further implantation in rat osteochondral defects, chondrogenically induced alginate-hiPSC constructs induced significantly better quality of cartilage repair than the control defects, and implanted hiPSCs were still detected in the repaired cartilage (Ko et al. 2014). Bone marrow stimulation by microfracture is a common surgical technique for cartilage repair. Baba and colleagues investigated the use of ultrapurified alginate (UPAL) gel as an adjuvant scaffold in combination with bone marrow stimulation in a rabbit osteochondral defect model (Baba et al. 2015). Notably, it was found that the combination of bone marrow stimulation augmented by UPAL gel significantly improved the macroscopic and histological scores, as well as GAG content of the repair tissue compared to those untreated or treated with bone marrow stimulation at 16-week post-operation (Baba et al. 2015). This study demonstrates the feasibility of a one-step, minimally invasive cartilage repair procedure that omits the need for harvesting donor cells.

3.1.6 Agarose

Agarose is a linear polysaccharide that is soluble in water and solidifies when cooled. Agarose hydrogels have been widely used for study of chondrocyte and MSC response to mechanical loading as they are able to transmit mechanical forces readily to cells during compression while allowing adequate diffusion of soluble factors and exchange of nutrients and wastes (Schmitt et al. 2012; Byers et al. 2008; Karim and Hall 2016; Kaupp et al. 2012; Huang et al. 2009). In a recent study, Karim and colleagues showed that chondrocytes encapsulated within soft agarose gels (0.2% w/v) displayed abnormal morphology and increased clustering compared to that of cells seeded in stiff agarose gels (2% w/v) (Karim and Hall 2016). Separately, Huang and colleagues showed that MSCs seeded at high density of

60 million cells/ml agarose hydrogel and exposed to transient application of TGF-β3 exhibited strong anabolic chondrogenic response and generated mechanically stable cartilage constructs with equilibrium modulus of ~200 kPa by the end of 7-week culture (Huang et al. 2009). Agarose is also frequently blended with other polymers such as gellan (Amici et al. 2000), polyethylene glycol diacrylate (PEGDA) (Rennerfeldt et al. 2013), and poly(acrylic acids) microgels (Tunesi et al. 2015) to generate semi-interpenetrating networks (IPN) with improved composition, architectural, and mechanical properties for cartilage tissue engineering as well as drug delivery.

3.1.7 Native Tissue ECM-Derived Hydrogels

Native tissue-derived ECM and derivatives have been commonly investigated for tissue repair and regeneration (Zhang et al. 2016; Mercuri et al. 2013; Wu et al. 2015; Visser et al. 2015). In particular, decellularized matrices from cells and tissues have attracted attention as they comprise of tissue-specific matrix proteins, and growth factors and receptors that offer unique advantages of biomimetic tissue specificity and bioactivity (Fu et al. 2011; Toh et al. 2014; Wu et al. 2015; Zhang et al. 2016). Recently, an injectable ECM-derived hydrogel material from porcine meniscus was developed for meniscus repair application. In that study, hydrogel was obtained by decellularization and enzymatic digestion of the meniscus. The meniscus-derived ECM hydrogel exhibited good cellular compatibility and promoted the growth of chondrocytes and fibroblasts encapsulated in the hydrogel for two weeks. Further, subcutaneous implantation revealed good tissue compatibility of the tissue-derived ECM hydrogel, thus supporting the use of this injectable meniscus-derived ECM hydrogel for meniscus repair and regeneration (Wu et al. 2015). Visser and colleagues investigated the generation of ECM hydrogels from several tissue-derived matrices (Visser et al. 2015). Accordingly, equine cartilage, meniscus, and tendon tissue were decellularized, enzymatically digested, and functionalized with methacrylamide groups that enable photocross-linking. It was found that these tissue ECM hydrogels were supportive of chondrocyte growth and MSC differentiation to chondrocytes (Visser et al. 2015). Importantly, this study demonstrates that native tissues can be processed into cross-linkable hydrogels for tissue engineering applications. Nevertheless, future work would need to assess the potential of these tissue-derived ECM hydrogels for cartilage repair in an animal model.

3.2 Synthetic Hydrogels

Unlike natural hydrogels, synthetic hydrogels have defined composition with minimal batch variability and are readily available and versatile matrices for tissue engineering (Toh and Loh 2014; Toh et al. 2015). To date, several synthetic polymers have been explored for applications in cartilage tissue engineering and regeneration (Chung and Burdick 2008).

3.2.1 Poly(Ethylene Glycol)

Among the synthetic polymers, poly(ethylene glycol) (PEG) is one of the most commonly used polymers in cartilage tissue engineering, as it is relatively inert and highly versatile for modifications (DeForest and Anseth 2012). Of note, PEG can be easily modified to include chemical moieties, hydrolyzable units, and bioactive ECM, peptides and growth factors to enhance cartilage tissue growth and matrix remodeling (Sridhar et al. 2014; Lutolf et al. 2003; Park et al. 2004; Anderson et al. 2011; Zhang et al. 2015; Liu et al. 2010a, b). Often, PEG hydrogels are considered as a moldable bioinert platform to introduce specific cues for study of their effects on cell functions (DeForest and Anseth 2012).

For instance, PEG hydrogels can be modified through a thiol–ene photopolymerization scheme to incorporate matrix metalloproteinase (MMP)-degradable peptides. In doing so, cell-degradable PEG-peptide hydrogels are created (Lutolf et al. 2003; Park et al. 2004; Anderson et al. 2011). It has been found that MMP-degradable hydrogels enhanced migration of dermal fibroblasts (Lutolf et al. 2003), promoted survival and ECM deposition of encapsulated chondrocytes (Park et al. 2004), and multilineage differentiation of encapsulated MSCs (Anderson et al. 2011). Separately, other peptides such as RGD and collagen mimetic peptide (CMP) have also been incorporated into PEG or poly(ethylene oxide) diacrylate (PEODA) hydrogels and found to promote ECM synthesis and differentiation of encapsulated chondrocytes and MSCs (Lee et al. 2008; Liu et al. 2010a, b; Zhang et al. 2015).

3.2.2 Poly(Ethylene Glycol Fumarate)

Poly(ethylene glycol fumarate)-based hydrogels are commonly used as carriers for growth factors and cells including chondrocytes and MSCs for cartilage tissue engineering (Dadsetan et al. 2007; Holland et al. 2005; Park et al. 2009a). Of note, oligo(poly(ethylene glycol) fumarate) (OPF) has unsaturated double bonds along its macromolecular chain that allows for tailored structure and properties of the hydrogels fabricated (Jo et al. 2001). Using the porcine model, it has been found that rehydrated freeze-dried OPF hydrogel enhanced repair of osteochondral defects compared to the control (no scaffold). There was significantly increased percentage of neotissue filling with hyaline-like cartilage making up $\sim$40% of neotissue at 4 months (Hui et al. 2013). When implanted with bone marrow MSCs, OPF hydrogel was found to create a conducive environment for cell infiltration and proliferation that repair the defects with 99% defect filling and 84% hyaline-like cartilage formation at 4 months (Lim et al. 2013a). These studies provide a clinically feasible procedure utilizing OPF hydrogels for cartilage repair.

3.3 Composite Hydrogels and Hybrids

Hydrogels have been interesting for tissue engineering due to their biocompatibility, high water content, porosity, and permeability of molecules. However, in most cases due to complex structural nature of the native tissues, one type of material could not adequately provide the appropriate environment for cartilage regeneration. A potential solution to this limitation is to introduce another functional component or material into the system to create composite hydrogels and hybrids.

Composite hydrogel scaffolds utilizing both natural and synthetic polymers could be fabricated by different techniques to have a combination of mechanical and biochemical properties. In a study by Wright and colleagues, electrospun and salt-leached scaffolds were embedded into chitosan hydrogel to determine the potential for cartilage tissue regeneration. Polylactide/polycaprolactone embedded chitosan hydrogels displayed good mechanical properties that resemble native cartilage. Additionally, primary chondrocytes exhibited high production of type II collagen and proteoglycans within the hydrogel scaffold. Consequently, these hydrogels have been reported to be promising for osteoarthritis treatment (Wright et al. 2014).

Thermo-sensitive chitosan-PNIPAAm copolymers prepared by graft polymerization of PNIPAAm into chitosan using ceric ammonium nitrate showed similar sol–gel transition properties as PNIPAAm (Cho et al. 2004). When injected with MSCs into the rabbit bladder wall, cartilage formation by chondrogenic differentiation of MSCs was detected following implantation for 14 weeks. Park and colleagues also reportedly grafted pluronic onto chitosan using EDC/NHS chemistry to synthesize a thermo-sensitive chitosan–pluronic hydrogel (Park et al. 2009b) that gelled at 25 °C and had a storage modulus of 10^4 Pa similar to the stiffness of cartilage tissue. Cell culture studies further showed that the thermo-sensitive hydrogel could promote chondrocyte proliferation and enhance GAG synthesis. RGD (Arg-Gly-Asp) could also be conjugated onto the chitosan–pluronic copolymers by coupling the carboxyl group in the peptide with the residual amine group in the copolymers, and in doing so, the viability and proliferation of chondrocytes as well as ECM expression were markedly enhanced (Park et al. 2008). Zhang and co-workers synthesized an amphiphilic block copolymer, poly (ε-caprolactone-*co*-lactide)-PEG-poly(ε-caprolactone-*co*-lactide) and found that the block copolymer formed a sol–gel system, and the transition temperature could be tuned between 26 and 40 °C (Zhang et al. 2010). Subsequently, they immobilized RGD into either hydrophobic poly(ε-caprolactone-*co*-lactide) (PCLA) blocks or hydrophilic PEG blocks and observed that rat chondrocytes preferred to grow on the thermogel of RGD in hydrophilic blocks rather than those in hydrophobic blocks, highlighting the influence of immobilizing sites of RGD peptides in amphiphilic polymers on cell adhesion. Park et al. encapsulated tonsil-derived MSCs into a thermogelling system of PEG-poly(L-alanine-co-L-phenyl alanine) copolymers in 3D culture and treated in culture medium supplemented with adipogenic, osteogenic, or chondrogenic factors to induce differentiation (Park et al. 2014). Interestingly, encapsulated MSCs in this thermogel system preferentially underwent in vitro chondrogenesis with high expressions of type II collagen and GAG and demonstrated cartilage-like tissue formation following subcutaneous implantation in vivo. It was suggested that the stiffness of the thermogels can provide biomechanical cues to guide stem cell differentiation.

In efforts to improve the mechanical properties of hydrogels, Mirahmadi and colleagues incorporated silk fibers into thermo-sensitive chitosan/glycerophosphate hydrogels and observed that the silk fiber-reinforced chitosan hydrogels not only showed enhanced stiffness, but also supported the retention of chondrogenic phenotype by the encapsulated chondrocytes (Mirahmadi et al. 2013). In another study by Wan et al. (2015), amino-diethoxypropane was added into alginate and the modified alginate was able to form hydrogel with chitosan (Wan et al. 2015). When seeded with bone marrow mesenchymal stromal cells (BMSCs) and implanted into a rabbit knee cartilage defect, high levels of GAG and gene expression of cartilage markers including aggrecan, type II collagen, and Sox-9 were detected in the reparative tissues after 12 weeks. Compared to control defect group, BMSC-laden alginate–chitosan hydrogel group showed enhanced cartilaginous tissue formation and matrix deposition.

Besides chitosan, other polysaccharides such as alginate and hyaluronic acid (HA) have been utilized for cartilage regeneration. Park and co-workers reportedly developed a rigid–flexible block copolymer thermogel based on self-assembling ionic complex between (+)-charged amphiphilic copolymers (polyalanine–PLX–polyalanine) and (−)-charged HA (Park et al. 2012). It was found that the temperature-sensitive sol-to-gel transition of the complex aqueous solution facilitated encapsulation of chondrocytes and provided a compatible environment for the cells similar to a biomimetic 3D culture system. Furthermore, the long-range nanofibrous structure of the thermogel played an important role in cell proliferation and protein expression.

Promising advancement with the use of polypeptides and proteins in cartilage tissue engineering is also noted in recent years. For instance, Kim and colleagues directly coupled gelatin to monocarboxylated pluronic to synthesize a gelatin/pluronic thermo-sensitive polymer system that showed reversible sol–gel transition behavior at around 37 °C (Kim et al. 2010). The gelatin/pluronic hydrogel with the addition of gelatin supported higher viability and proliferation of chondrocytes than that in control pluronic group. In another study, an in situ thermal gelling polypeptide (polyalanine–poloxamer–polyalanine block copolymer) that displayed sol–gel transition at 37 °C was investigated for 3D culturing of chondrocytes (Choi et al. 2010). Interestingly, it was found that the β-sheet structure of the polyalanine and the fibrous structure and stiffness of the hydrogel could regulate proliferation and protein expression of the encapsulated chondrocytes (Choi et al. 2011). In addition, methoxy PEG-PCL di-block copolymers were found to have a sol–gel phase transition at body temperature. Kwon et al. (2013) investigated the potential use of methoxy PEG-PCL hydrogel as an in situ-forming scaffold for seeding with chondrocytes in cartilage regeneration (Kwon et al. 2013). After injection into mice, the hydrogel formed an interconnected pore structure to support the growth, proliferation and differentiation of encapsulated chondrocytes, thereby enabling significant cartilage tissue growth over time in vivo.

Yan et al. (2016) reported the design of novel poly(L-glutamic acid) injectable hydrogels made from the self-cross-linking of adipic dihydrazide (ADH)-modified poly(L-glutamic acid) (PLGA-ADH) and aldehyde-modified poly(L-glutamic acid) (PLGA-CHO) (Yan et al. 2016). The authors investigated their potential in cartilage tissue engineering. Both the hydrazide modification degree of PLGA-ADH and oxidation degree of PLGA-CHO can be adjusted by the amount of activators and sodium periodate, respectively. Results showed that the solid content of the hydrogels, $-NH_2/CHO$ molar ratio, and oxidation degree of PLGA-CHO greatly affect the gelation time, equilibrium swelling, mechanical properties, microscopic morphology, and in vitro degradation of the hydrogels. When encapsulated with chondrocytes, these injectable hydrogels showed good cytocompatibility with high cell viability. These results suggest that the PLGA hydrogel has potential as an injectable cell delivery carrier for cartilage regeneration and could serve as a new biomaterial for tissue engineering. Transdermal photopolymerization, which is a minimally invasive method for implantation, was used to subcutaneously place a mixture of polymer and isolated chondrocytes to regenerate cartilage tissue in vivo. The potential use of transdermal photopolymerization for minimally invasive subcutaneous implantation of hydrogels and chondrocytes for in vivo cartilage regeneration has been demonstrated (Elisseeff et al. 1999). Semi-interpenetrating networks of varying proportions of poly(ethylene oxide)dimethacrylate and poly (ethylene oxide) and bovine chondrocytes were implanted in athymic mice. Four mice (12 implants) were harvested at 2, 4, and 7 weeks. Chondrocytes survived implantation and photopolymerization and formed neocartilage containing 1.5–2.9% wet weight collagen and 4–7% GAG. Thirty-five percent of the total collagen was type II collagen.

To enhance the chondroinductivity of hydrogel scaffolds, decellularized tissue ECMs and/or microparticles have been a popular approach and have been used either by themselves or incorporated into hydrogel matrices to form composite hydrogels (Burnsed et al. 2016; Beck et al. 2016). Recently, Beck and colleagues incorporated devitalized cartilage microparticles into photocross-linkable methacrylated hyaluronic acid hydrogel to create a chondroinductive hydrogel composite. It was found that devitalized cartilage microparticles incorporated in the hydrogel improved the rheological properties of the gel and enhanced MSC chondrogenesis without the need of TGF-β3 (Beck et al. 2016). In another study, Stocco et al. (2014) combined decellularized Wharton's jelly and polyvinyl alcohol (PVA) hydrogel to make a hybrid composite material for cartilage regeneration. When seeded with chondrocytes, the hybrid hydrogel scaffold supported cell colonization and proliferation. This study suggests the combination of as composite hydrogels composed of Wharton's jelly and PVA hydrogels for cartilage restoration (Stocco et al. 2014).

3.4 *Multifunctional Hydrogels*

During cartilage injury, there will be prevailing processes of inflammation and oxidative stress that make the tissue environment highly deleterious for cell survival. Inflammation is associated with production of proinflammatory cytokines and proteases including interleukin (IL)-1α/β, interferon (IFN)-γ, and tumor necrosis factor (TNF)-α, together with MMPs involved in matrix degradation. Oxidative stress is largely associated with the elevated production of reactive oxygen species (ROS) that can damage the cellular components (proteins, lipids and DNA) and result in chondrocyte senescence and apoptosis (Toh and Loh 2014; Toh et al. 2015, 2016b).

To enhance cartilage regeneration, various strategies of functionalizing or loading the injectable hydrogel with protective agents have been recently developed. These agents include antioxidants, anti-inflammatory factors, and receptor blocking agents against inflammation and oxidative stress. Sheu and co-workers developed an injectable hydrogel composed of oxidized hyaluronic acid (Oxi-HA) and resveratrol (Res). In that study, Oxi-HA-Res hydrogel demonstrated good cytocompatibility with chondrocytes, facilitated ECM synthesis and also ameliorated lipopolysaccharide (LPS)-induced inflammation and damage (Sheu et al. 2013). Resveratrol, found commonly in grapes, is well-known for its health benefits including antioxidant, anti-inflammatory, anti-aging properties (Kuršvietienė et al. 2016). Similarly, ferulic acid is a naturally available phenolic phytochemical with excellent antioxidant and anti-inflammatory properties. Injectable thermo-sensitive chitosan/gelatin/glycerophosphate hydrogel incorporated with ferulic acid attenuated H_2O_2-induced oxidative stress of nucleus pulposus cells, with decreased apoptosis, and upregulated expression levels of aggrecan and type II collagen, and downregulated level of MMP-3 (Cheng et al. 2011, 2013a, b). The development of injectable hydrogels that can modulate an injurious cartilage tissue environment for enhanced cell survival, proliferation, and matrix synthesis is likely to improve the treatment outcome.

4 Conclusions and Perspectives

This chapter has provided valuable insights into the applications of hydrogels for treatment of cartilage injuries and osteoarthritis. In recent years, promising advances have been made in the hydrogel design and fabrication techniques to enhance survival and chondrogenesis of chondrocytes and MSCs delivered in injectable hydrogels. While studies of cell–material interactions have provided useful insights into the design of hydrogels that would support chondrogenesis and matrix formation of MSCs and chondrocytes, engineering the cartilage tissue that reproduces the properties of the native cartilage remains a challenge.

As the field continually explores new materials, modifications and cross-linking agents, deciphering the role of polymers, functional groups and moieties, and their effects on cell survival, morphogenesis, and functions may aid in the progress and development toward in vivo regeneration of cartilage tissue with native levels of matrix organization, composition, and mechanics.

As mentioned, the state and composition of the joint environment subjected to oxidative stress and inflammation during injury are likely to affect the survival of the cells and these are important considerations when delivering cells in injectable hydrogels. On this note, modification/functionalization of hydrogels to have additional beneficial properties of anti-inflammation and/or antioxidative stress is likely to enhance the survival and functions of chondrocytes and MSCs in cartilage repair. The controllable release of biomolecules, cell-responsiveness, and modulation of the tissue environment are some of the capabilities that have yet to be fully explored in injectable hydrogels for functional cartilage regeneration. These advances in injectable hydrogels would certainly guide the development of therapies for treatment of cartilage injuries and osteoarthritis.

Acknowledgements This work was supported by National University of Singapore, National University Healthcare System, Ministry of Education, and Agency for Science Technology and Research (A*STAR), Singapore.

Disclosure The author indicates no potential conflict of interest.

References

Amici E, Clark AH, Normand V, Johnson NB (2000) Interpenetrating network formation in gellan–agarose gel composites. Biomacromol 1:721–729

Anderson SB, Lin CC, Kuntzler DV, Anseth KS (2011) The performance of human mesenchymal stem cells encapsulated in cell-degradable polymer-peptide hydrogels. Biomaterials 32:3564–3574

Baba R, Onodera T, Momma D, Matsuoka M, Hontani K, Elmorsy S, Endo K, Todoh M, Tadano S, Iwasaki N (2015) A novel bone marrow stimulation technique augmented by administration of ultrapurified alginate gel enhances osteochondral repair in a rabbit model. Tissue Eng Part C Methods 21:1263–1273

Basu A, Kunduru KR, Doppalapudi S, Domb AJ, Khan W (2016) Poly(lactic acid) based hydrogels. Adv Drug Deliv Rev 107:192–205

Becerra J, Andrades JA, Guerado E, Zamora-Navas P, Lopez-Puertas JM, Reddi AH (2010) Articular cartilage: structure and regeneration. Tissue Eng Part B Rev 16:617–627

Beck EC, Barragan M, Libeer TB, Kieweg SL, Converse GL, Hopkins RA, Berkland CJ, Detamore MS (2016) Chondroinduction from naturally derived cartilage matrix: a comparison between devitalized and decellularized cartilage encapsulated in hydrogel pastes. Tissue Eng Part A 22:665–679

Bosnakovski D, Mizuno M, Kim G, Takagi S, Okumura M, Fujinaga T (2006) Chondrogenic differentiation of bovine bone marrow mesenchymal stem cells (MSCs) in different hydrogels: influence of collagen type II extracellular matrix on MSC chondrogenesis. Biotechnol Bioeng 93:1152–1163

Burdick JA, Prestwich GD (2011) Hyaluronic acid hydrogels for biomedical applications. Adv Mater 23:41–56

Burnsed OA, Schwartz Z, Marchand KO, Hyzy SL, Olivares-Navarrete R, Boyan BD (2016) Hydrogels derived from cartilage matrices promote induction of human mesenchymal stem cell chondrogenic differentiation. Acta Biomater

Byers BA, Mauck RL, Chiang IE, Tuan RS (2008) Transient exposure to transforming growth factor beta 3 under serum-free conditions enhances the biomechanical and biochemical maturation of tissue-engineered cartilage. Tissue Eng Part A 14:1821–1834

Chan PPY (2015) Hydrogels with tuneable properties. Methods Mol Biol 1340:121–132

Chen F, Yu S, Liu B, Ni Y, Yu C, Su Y, Zhu X, Yu X, Zhou Y, Yan D (2016) An injectable enzymatically crosslinked carboxymethylated pullulan/chondroitin sulfate hydrogel for cartilage tissue engineering. Sci Rep 6:20014

Cheng NC, Estes BT, Young TH, Guilak F (2013a) Genipin-crosslinked cartilage-derived matrix as a scaffold for human adipose-derived stem cell chondrogenesis. Tissue Eng Part A 19:484–496

Cheng YH, Yang SH, Lin FH (2011) Thermosensitive chitosan-gelatin-glycerol phosphate hydrogel as a controlled release system of ferulic acid for nucleus pulposus regeneration. Biomaterials 32:6953–6961

Cheng YH, Yang SH, Liu CC, Gefen A, Lin FH (2013b) Thermosensitive hydrogel made of ferulic acid-gelatin and chitosan glycerophosphate. Carbohydr Polym 92:1512–1519

Cho JH, Kim SH, Park KD, Jung MC, Yang WI, Han SW, Noh JY, Lee JW (2004) Chondrogenic differentiation of human mesenchymal stem cells using a thermosensitive poly(*N*-isopropylacrylamide) and water-soluble chitosan copolymer. Biomaterials 25:5743–5751

Choi BG, Park MH, Cho SH, Joo MK, Oh HJ, Kim EH, Park K, Han DK, Jeong B (2010) In situ thermal gelling polypeptide for chondrocytes 3D culture. Biomaterials 31:9266–9272

Choi BG, Park MH, Cho S, Joo MK, Oh HJ, Kim EH, Park K, Han DK, Jeong B (2011) Thermal gelling polyalanine-poloxamine-polyalanine aqueous solution for chondrocytes 3D culture: initial concentration effect. Soft Matter 7:456–462

Choi B, Kim S, Lin B, Wu BM, Lee M (2014) Cartilaginous extracellular matrix-modified chitosan hydrogels for cartilage tissue engineering. ACS Appl Mater Interfaces 6:20110–20121

Choi B, Kim S, Fan J, Kowalski T, Petrigliano F, Evseenko D, Lee M (2015) Covalently conjugated transforming growth factor-beta1 in modular chitosan hydrogels for the effective treatment of articular cartilage defects. Biomater Sci 3:742–752

Chung C, Burdick JA (2008) Engineering cartilage tissue. Adv Drug Deliv Rev 60:243–262

Dadsetan M, Szatkowski JP, Yaszemski MJ, Lu L (2007) Characterization of photo-cross-linked oligo[poly(ethylene glycol) fumarate] hydrogels for cartilage tissue engineering. Biomacromol 8:1702–1709

DeForest CA, Anseth KS (2012) Advances in bioactive hydrogels to probe and direct cell fate. Annu Rev Chem Biomol Eng 3:421–444

Dicker KT, Gurski LA, Pradhan-Bhatt S, Witt RL, Farach-Carson MC, Jia X (2014) Hyaluronan: a simple polysaccharide with diverse biological functions. Acta Biomater 10:1558–1570

Dinescu S, Gălăţeanu B, Albu M, Lungu A, Radu E, Hermenean A (2014) Biocompatibility assessment of novel collagen-sericin scaffolds improved with hyaluronic acid and chondroitin sulfate for cartilage regeneration. Biomed Res Int 2013:1–9

Elisseeff J, Anseth KS, Sims D, McIntosh W, Randolph M, Yaremchuk M, Langer R (1999) Transdermal photopolymerization of poly(ethylene oxide)-based injectable hydrogels for tissue-engineered cartilage. Plast Reconstr Surg 104:1014–1022

Erickson IE, Huang AH, Sengupta S, Kestle S, Burdick JA, Mauck RL (2009) Macromer density influences mesenchymal stem cell chondrogenesis and maturation in photocrosslinked hyaluronic acid hydrogels. Osteoarthr Cartil 17:1639–1648

Fu X, Toh WS, Liu H, Lu K, Li M, Cao T (2011) Establishment of clinically compliant human embryonic stem cells in an autologous feeder-free system. Tissue Eng Part C Methods 17:927–937

Gadjanski I, Yodmuang S, Spiller K, Bhumiratana S, Vunjak-Novakovic G (2013) Supplementation of exogenous adenosine 5′-triphosphate enhances mechanical properties of 3D cell-agarose constructs for cartilage tissue engineering. Tissue Eng Part A 19:2188–2200

Ge Z, Hu Y, Heng BC, Yang Z, Ouyang H, Lee EH, Cao T (2006) Osteoarthritis and therapy. Arthritis Rheum 55:493–500

Geng X, Mo X, Fan L, Yin A, Fang J (2012) Hierarchically designed injectable hydrogel from oxidized dextran, amino gelatin and 4-arm poly(ethylene glycol)-acrylate for tissue engineering application. J Mater Chem 22:25130

Gittens J, Haleem AM, Grenier S, Smyth NA, Hannon CP, Ross KA, Torzilli PA, Kennedy JG (2016) Use of novel chitosan hydrogels for chemical tissue bonding of autologous chondral transplants. J Orthop Res 34:1139–1146

Han F, Yang X, Zhao J, Zhao Y, Yuan X (2015) Photocrosslinked layered gelatin-chitosan hydrogel with graded compositions for osteochondral defect repair. J Mater Sci Mater Med 26:160

Hayami JW, Waldman SD, Amsden BG (2016) Chondrocyte generation of cartilage-like tissue following photoencapsulation in methacrylated polysaccharide solution blends. Macromol Biosci 16:1083–1095

Ho ST, Cool SM, Hui JH, Hutmacher DW (2010) The influence of fibrin based hydrogels on the chondrogenic differentiation of human bone marrow stromal cells. Biomaterials 31:38–47

Hoemann CD, Sun J, McKee MD, Chevrier A, Rossomacha E, Rivard GE, Hurtig M, Buschmann MD (2007) Chitosan-glycerol phosphate/blood implants elicit hyaline cartilage repair integrated with porous subchondral bone in microdrilled rabbit defects. Osteoarthr Cartil 15:78–89

Hoffman AS (2002) Hydrogels for biomedical applications. Adv Drug Deliv Rev 54:3–12

Holland TA, Tabata Y, Mikos AG (2005) Dual growth factor delivery from degradable oligo(poly (ethylene glycol) fumarate) hydrogel scaffolds for cartilage tissue engineering. J Control Release 101:111–125

Huang AH, Stein A, Tuan RS, Mauck RL (2009) Transient exposure to transforming growth factor beta 3 improves the mechanical properties of mesenchymal stem cell-laden cartilage constructs in a density-dependent manner. Tissue Eng Part A 15:3461–3472

Hui JH, Ren XF, Afizah MH, Chian KS, Mikos AG (2013) Oligo[poly(ethylene glycol)fumarate] hydrogel enhances osteochondral repair in porcine femoral condyle defects. Clin Orthop Relat Res 471:1174–1185

Jalani G, Rosenzweig DH, Makhoul G, Abdalla S, Cecere R, Vetrone F, Haglund L, Cerruti M (2015) Tough, in-situ thermogelling, injectable hydrogels for biomedical applications. Macromol Biosci 15:473–480

Jeng L, Olsen BR, Spector M (2012) Engineering endostatin-expressing cartilaginous constructs using injectable biopolymer hydrogels. Acta Biomater 8:2203–2212

Jiang YZ, Zhang SF, Qi YY, Wang LL, Ouyang HW (2011) Cell transplantation for articular cartilage defects: principles of past, present, and future practice. Cell Transplant 20:593–607

Jin R, Teixeira LS, Dijkstra PJ, van Blitterswijk CA, Karperien M, Feijen J (2010) Enzymatically-crosslinked injectable hydrogels based on biomimetic dextran-hyaluronic acid conjugates for cartilage tissue engineering. Biomaterials 31:3103–3113

Jo S, Shin H, Shung AK, Fisher JP, Mikos AG (2001) Synthesis and characterization of oligo(poly (ethylene glycol) fumarate) macromer. Macromolecules 34:2839–2844

Karim A, Hall AC (2017) Chondrocyte morphology in stiff and soft agarose gels and the influence of foetal calf serum. J Cell Physiol 232:1041–1052

Kaupp JA, Weber JF, Waldman SD (2012) Mechanical stimulation of chondrocyte-agarose hydrogels. J Vis Exp e4229

Kim DH, Heo SJ, Shin JW, Mun CW, Park KM, Park KD, Jee KS (2010) Preparation of thermosensitive gelatin-pluronic copolymer for cartilage tissue engineering. Macromol Res 18:387–391

Kim IL, Mauck RL, Burdick JA (2011) Hydrogel design for cartilage tissue engineering: a case study with hyaluronic acid. Biomaterials 32:8771–8782

Kim J, Lin B, Kim S, Choi B, Evseenko D, Lee M (2015) TGF-beta1 conjugated chitosan collagen hydrogels induce chondrogenic differentiation of human synovium-derived stem cells. J Biol Eng 9:1

Ko JY, Kim KI, Park S, Im GI (2014) In vitro chondrogenesis and in vivo repair of osteochondral defect with human induced pluripotent stem cells. Biomaterials 35:3571–3581

Kontturi LS, Järvinen E, Muhonen V, Collin EC, Pandit AS, Kiviranta I, Yliperttula M, Urtti A (2014) An injectable, in situ forming type II collagen/hyaluronic acid hydrogel vehicle for chondrocyte delivery in cartilage tissue engineering. Drug Deliv Transl Res 4:149–158

Kurisawa M, Chung J, Yang Y, Gao S, Uyama H (2005) Injectable biodegradable hydrogels composed of hyaluronic acid–tyramine conjugates for drug delivery and tissue engineering. Chem Commun 2005:4312–4314

Kuršvietienė L, Stanevičienė I, Mongirdienė A, Bernatonienė J (2016) Multiplicity of effects and health benefits of resveratrol. Medicina 52:148–155

Kwon JS, Yoon SM, Kwon DY, Kim DY, Tai GZ, Jin LM, Song B, Lee B, Kim JH, Han DK, Min BH, Kim MS (2013) Injectable in situ-forming hydrogel for cartilage tissue engineering. J Mater Chem B 1:3314–3321

Langer R, Vacanti JP (1993) Tissue engineering. Science 260:920–926

Lee F, Chung JE, Kurisawa M (2009) An injectable hyaluronic acid–tyramine hydrogel system for protein delivery. J Controlled Release 134:186–193

Lee HJ, Yu C, Chansakul T, Hwang NS, Varghese S, Yu SM, Elisseeff JH (2008) Enhanced chondrogenesis of mesenchymal stem cells in collagen mimetic peptide-mediated microenvironment. Tissue Eng Part A 14:1843–1851

Levett PA, Melchels FP, Schrobback K, Hutmacher DW, Malda J, Klein TJ (2014a) A biomimetic extracellular matrix for cartilage tissue engineering centered on photocurable gelatin, hyaluronic acid and chondroitin sulfate. Acta Biomater 10:214–223

Levett PA, Hutmacher DW, Malda J, Klein TJ (2014b) Hyaluronic acid enhances the mechanical properties of tissue-engineered cartilage constructs. PLoS ONE 9:e113216

Lim CT, Ren X, Afizah MH, Tarigan-Panjaitan S, Yang Z, Wu Y, Chian KS, Mikos AG, Hui JH (2013a) Repair of osteochondral defects with rehydrated freeze-dried oligo[poly(ethylene glycol) fumarate] hydrogels seeded with bone marrow mesenchymal stem cells in a porcine model. Tissue Eng Part A 19:1852–1861

Lim TC, Toh WS, Wang LS, Kurisawa M, Spector M (2012) The effect of injectable gelatin-hydroxyphenylpropionic acid hydrogel matrices on the proliferation, migration, differentiation and oxidative stress resistance of adult neural stem cells. Biomaterials 33:3446–3455

Lim TC, Rokkappanavar S, Toh WS, Wang LS, Kurisawa M, Spector M (2013b) Chemotactic recruitment of adult neural progenitor cells into multifunctional hydrogels providing sustained SDF-1α release and compatible structural support. FASEB J 27:1023–1033

Lima EG, Tan AR, Tai T, Marra KG, DeFail A, Ateshian GA, Hung CT (2009) Genipin enhances the mechanical properties of tissue engineered cartilage and protects against inflammatory degradation when used as a medium supplement. J Biomed Mater Res A 91:692–700

Little CJ, Kulyk WM, Chen X (2014) The Effect of Chondroitin Sulphate and Hyaluronic Acid on Chondrocytes Cultured within a Fibrin-Alginate Hydrogel. J Funct Biomater 5:197–210

Liu SQ, Tian Q, Hedrick JL, Hui JH, Ee PL, Yang YY (2010a) Biomimetic hydrogels for chondrogenic differentiation of human mesenchymal stem cells to neocartilage. Biomaterials 31:7298–7307

Liu SQ, Tian Q, Wang L, Hedrick JL, Hui JH, Yang YY, Ee PL (2010b) Injectable biodegradable poly(ethylene glycol)/RGD peptide hybrid hydrogels for in vitro chondrogenesis of human mesenchymal stem cells. Macromol Rapid Commun 31:1148–1154

Liu Y, Shu XZ, Prestwich GD (2006) Osteochondral defect repair with autologous bone marrow-derived mesenchymal stem cells in an injectable, in situ, cross-linked synthetic extracellular matrix. Tissue Eng 12:3405–3416

Lutolf MP, Raeber GP, Zisch AH, Tirelli N, Hubbell JA (2003) Cell-responsive synthetic hydrogels. Adv Mater 15:888–892

Marcacci M, Filardo G, Kon E (2013) Treatment of cartilage lesions: what works and why? Injury 44(Suppl 1):S11–S15

Martins EA, Michelacci YM, Baccarin RY, Cogliati B, Silva LC (2014) Evaluation of chitosan-GP hydrogel biocompatibility in osteochondral defects: an experimental approach. BMC Vet Res 10:197

Matsumoto M, Udomsinprasert W, Laengee P, Honsawek S, Patarakul K, Chirachanchai S (2016) A water-based chitosan-maleimide precursor for bioconjugation: an example of a rapid pathway for an in situ injectable adhesive gel. Macromol Rapid Commun 37:1618–1622

Mazaki T, Shiozaki Y, Yamane K, Yoshida A, Nakamura M, Yoshida Y, Zhou D, Kitajima T, Tanaka M, Ito Y, Ozaki T, Matsukawa A (2014) A novel, visible light-induced, rapidly cross-linkable gelatin scaffold for osteochondral tissue engineering. Sci Rep 4:4457

Mercuri JJ, Patnaik S, Dion G, Gill SS, Liao J, Simionescu DT (2013) Regenerative potential of decellularized porcine nucleus pulposus hydrogel scaffolds: stem cell differentiation, matrix remodeling, and biocompatibility studies. Tissue Eng Part A 19:952–966

Miljkovic ND, Lin YC, Cherubino M, Minteer D, Marra KG (2009) A novel injectable hydrogel in combination with a surgical sealant in a rat knee osteochondral defect model. Knee Surg Sports Traumatol Arthrosc 17:1326–1331

Mirahmadi F, Tafazzoli-Shadpour M, Shokrgozar MA, Bonakdar S (2013) Enhanced mechanical properties of thermosensitive chitosan hydrogel by silk fibers for cartilage tissue engineering. Mater Sci Eng C-Mater Biol Appl 33:4786–4794

Mu C, Zhang K, Lin W, Li D (2013) Ring-opening polymerization of genipin and its long-range crosslinking effect on collagen hydrogel. J Biomed Mater Res Part A 101A:385–393

Murray CJ, Vos T, Lozano R, Naghavi M, Flaxman AD, Michaud C et al (2012) Disability-adjusted life years (DALYs) for 291 diseases and injuries in 21 regions, 1990–2010: a systematic analysis for the Global Burden of Disease Study 2010. Lancet 380:2197–2223

Nettles DL, Vail TP, Morgan MT, Grinstaff MW, Setton LA (2004) Photocrosslinkable hyaluronan as a scaffold for articular cartilage repair. Ann Biomed Eng 32:391–397

Nichol JW, Koshy ST, Bae H, Hwang CM, Yamanlar S, Khademhosseini A (2010) Cell-laden microengineered gelatin methacrylate hydrogels. Biomaterials 31:5536–5544

Nie L, Zhang G, Hou R, Xu H, Li Y, Fu J (2015) Controllable promotion of chondrocyte adhesion and growth on PVA hydrogels by controlled release of TGF-beta1 from porous PLGA microspheres. Colloids Surf B Biointerfaces 125:51–57

Park H, Lee KY (2014) Cartilage regeneration using biodegradable oxidized alginate/hyaluronate hydrogels. J Biomed Mater Res A 102:4519–4525

Park KM, Joung YK, Park KD, Lee SY, Lee MC (2008) RGD-conjugated chitosan-Pluronic hydrogels as a cell supported scaffold for articular cartilage regeneration. Macromol Res 16:517–523

Park H, Temenoff JS, Tabata Y, Caplan AI, Raphael RM, Jansen JA, Mikos AG (2009a) Effect of dual growth factor delivery on chondrogenic differentiation of rabbit marrow mesenchymal stem cells encapsulated in injectable hydrogel composites. J Biomed Mater Res A 88:889–897

Park KM, Lee SY, Joung YK, Na JS, Lee MC, Park KD (2009b) Thermosensitive chitosan-Pluronic hydrogel as an injectable cell delivery carrier for cartilage regeneration. Acta Biomater 5:1956–1965

Park MH, Choi BG, Jeong B (2012) Complexation-induced biomimetic long range fibrous orientation in a rigid-flexible block copolymer thermogel. Adv Func Mater 22:5118–5125

Park MH, Yu Y, Moon HJ, Ko DY, Kim HS, Lee H, Ryu KH, Jeong B (2014) 3D culture of tonsil-derived mesenchymal stem cells in poly(ethylene glycol)-Poly(L-alanine-co-L-phenyl alanine) Thermogel. Adv Healthcare Mater 3:1782–1791

Park Y, Lutolf MP, Hubbell JA, Hunziker EB, Wong M (2004) Bovine primary chondrocyte culture in synthetic matrix metalloproteinase-sensitive poly(ethylene glycol)-based hydrogels as a scaffold for cartilage repair. Tissue Eng 10:515–522

Percope de Andrade MA, Campos TV, Abreu-E-Silva GM (2015) Supplementary methods in the nonsurgical treatment of osteoarthritis. Arthroscopy 31:785–792

Ren CD, Gao S, Kurisawa M, Ying JY (2015) Cartilage synthesis in hyaluronic acid-tyramine constructs. J Mater Chem B 3:1942–1956

Rennerfeldt DA, Renth AN, Talata Z, Gehrke SH, Detamore MS (2013) Tuning mechanical performance of poly(ethylene glycol) and agarose interpenetrating network hydrogels for cartilage tissue engineering. Biomaterials 34:8241–8257

Schmitt JF, See KH, Yang Z, Hui JH, Lee EH (2012) Sequential differentiation of mesenchymal stem cells in an agarose scaffold promotes a physis-like zonal alignment of chondrocytes. J Orthop Res 30:1753–1759

Schuurman W, Levett PA, Pot MW, van Weeren PR, Dhert WJA, Hutmacher DW, Melchels FP, Klein TJ, Malda J (2013) Gelatin-methacrylamide hydrogels as potential bio-materials for fabrication of tissue-engineered cartilage constructs. Macromol Biosci 13:551–561

Sharma B, Williams CG, Khan M, Manson P, Elisseeff JH (2007) In vivo chondrogenesis of mesenchymal stem cells in a photopolymerized hydrogel. Plast Reconstr Surg 119:112–120

Sheu SY, Chen WS, Sun JS, Lin FH, Wu T (2013) Biological characterization of oxidized hyaluronic acid/resveratrol hydrogel for cartilage tissue engineering. J Biomed Mater Res A 101:3457–3466

Sridhar BV, Doyle NR, Randolph MA, Anseth KS (2014) Covalently tethered TGF-beta1 with encapsulated chondrocytes in a PEG hydrogel system enhances extracellular matrix production. J Biomed Mater Res A 102:4464–4472

Stocco E, Barbon S, Dalzoppo D, Lora S, Macchi V, Paolo Parnigotto P, Grandi C (2014) In vitro assessment of a novel composite scaffold for articular cartilage restoration. Italian J Anat Embryol 119:188

Toh WS, Loh XJ (2014) Advances in hydrogel delivery systems for tissue regeneration. Mater Sci Eng C 45:690–697

Toh WS, Lee EH, Guo XM, Chan JK, Yeow CH, Choo AB, Cao T (2010) Cartilage repair using hyaluronan hydrogel-encapsulated human embryonic stem cell-derived chondrogenic cells. Biomaterials 31:6968–6980

Toh WS, Spector M, Lee EH, Cao T (2011) Biomaterial-mediated delivery of microenvironmental cues for repair and regeneration of articular cartilage. Mol Pharm 8:994–1001

Toh WS, Lim TC, Kurisawa M, Spector M (2012) Modulation of mesenchymal stem cell chondrogenesis in a tunable hyaluronic acid hydrogel microenvironment. Biomaterials 33:3835–3845

Toh WS, Foldager CB, Pei M, Hui JH (2014) Advances in mesenchymal stem cell-based strategies for cartilage repair and regeneration. Stem Cell Rev 10:686–696

Toh WS, Toh YC, Loh XJ (2015) Hydrogels for stem cell fate control and delivery in regenerative medicine. In: Loh JX (ed) In-situ gelling polymers. Springer, Singapore, pp. 187–214. https://doi.org/10.1007/978-981-287-152-7_8

Toh WS, Cao T (2014) Derivation of chondrogenic cells from human embryonic stem cells for cartilage tissue engineering. Methods Mol Biol 1307:263–279

Toh WS, Foldager CB, Hui JH, Olsen BR, Spector M (2016a) Exploiting stem cell-extracellular matrix interactions for cartilage regeneration: a focus on basement membrane molecules. Curr Stem Cell Res Ther 11:618–625

Toh WS, Brittberg M, Farr J, Foldager CB, Gomoll AH, Hui JH, Richardson JB, Roberts S, Spector M (2016b) Cellular senescence in aging and osteoarthritis. Acta Orthop 87:6–14. https://doi.org/10.1080/17453674.2016.1235087

Tunesi M, Prina E, Munarin F, Rodilossi S, Albani D, Petrini P, Giordano C (2015) Cross-linked poly(acrylic acids) microgels and agarose as semi-interpenetrating networks for resveratrol release. J Mater Sci Mater Med 26:5328

Visser J, Levett PA, te Moller NC, Besems J, Boere KW, van Rijen MH, de Grauw JC, Dhert WJ, van Weeren PR, Malda J (2015) Crosslinkable hydrogels derived from cartilage, meniscus, and tendon tissue. Tissue Eng Part A 21:1195–1206

Wan W, Li Q, Gao H, Ge L, Liu Y, Zhong W, Ouyang J, Xing M (2015) BMSCs laden injectable amino-diethoxypropane modified alginate-chitosan hydrogel for hyaline cartilage reconstruction. J Mater Chem B 3:1990–2005

Wang L, Stegemann JP (2011) Glyoxal crosslinking of cell-seeded chitosan/collagen hydrogels for bone regeneration. Acta Biomater 7:2410–2417

Wang LS, Du C, Toh WS, Wan ACA, Gao SJ, Kurisawa M (2014) Modulation of chondrocyte functions and stiffness-dependent cartilage repair using an injectable enzymatically crosslinked hydrogel with tunable mechanical properties. Biomaterials 35:2207–2217

Wang PY, Tsai WB (2013) Modulation of the proliferation and matrix synthesis of chondrocytes by dynamic compression on genipin-crosslinked chitosan/collagen scaffolds. J Biomater Sci Polym Ed 24:507–519

Wright LD, McKeon-Fischer KD, Cui Z, Nair LS, Freeman JW (2014) PDLA/PLLA and PDLA/PCL nanofibers with a chitosan-based hydrogel in composite scaffolds for tissue engineered cartilage. Tissue Eng Regen Med 8:946–954

Wu J, Ding Q, Dutta A, Wang Y, Huang YH, Weng H, Tang L, Hong Y (2015) An injectable extracellular matrix derived hydrogel for meniscus repair and regeneration. Acta Biomater 16:49–59

Wu ZM, Zhang XG, Zheng C, Li CX, Zhang SM, Dong RN, Yu DM (2009) Disulfide-crosslinked chitosan hydrogel for cell viability and controlled protein release. Eur J Pharm Sci 37:198–206

Yan S, Zhang X, Zhang K, Di H, Feng L, Li G, Fang J, Cui L, Chen X, Yin J (2016) Injectable in situ forming poly(L-glutamic acid) hydrogels for cartilage tissue engineering. J Mater Chem B 4:947–961

Yao Y, Zeng L, Huang Y (2016) The enhancement of chondrogenesis of ATDC5 cells in RGD-immobilized microcavitary alginate hydrogels. J Biomater Appl 31:92–101

Yue K, Trujillo-de Santiago G, Alvarez MM, Tamayol A, Annabi N, Khademhosseini A (2015) Synthesis, properties, and biomedical applications of gelatin methacryloyl (GelMA) hydrogels. Biomaterials 73:254–271

Zhang J, Mujeeb A, Du Y, Lin J, Ge Z (2015) Probing cell-matrix interactions in RGD-decorated macroporous poly (ethylene glycol) hydrogels for 3D chondrocyte culture. Biomed Mater 10:035016

Zhang L, Yuan T, Guo L, Zhang X (2012) An in vitro study of collagen hydrogel to induce the chondrogenic differentiation of mesenchymal stem cells. J Biomed Mater Res A 100:2717–2725

Zhang S, Lu Q, Cao T, Toh WS (2016) Adipose tissue and extracellular matrix development by injectable decellularized adipose matrix loaded with basic fibroblast growth factor. Plast Reconstr Surg 137:1171–1180

Zhang Z, Lai Y, Yu L, Ding J (2010) Effects of immobilizing sites of RGD peptides in amphiphilic block copolymers on efficacy of cell adhesion. Biomaterials 31:7873–7882

Zhao W, Jin X, Cong Y, Liu Y, Fu J (2013) Degradable natural polymer hydrogels for articular cartilage tissue engineering. J Chem Technol Biotechnol 88:327–339

Zheng L, Jiang X, Chen X, Fan H, Zhang X (2014) Evaluation of novel in situ synthesized nano-hydroxyapatite/collagen/alginate hydrogels for osteochondral tissue engineering. Biomed Mater 9:065004

Chapter 13
DNA-Based Hydrogels: An Approach for Multifunctional Bioapplications

Diana Costa, Artur J. M. Valente and João Queiroz

Abstract DNA-based networks have attracted significant interest in the last decades, due to its hydrophilicity, biocompatibility and stimuli responsiveness. These characteristics make them very suitable for a variety of applications in the biomedical field. In this context, relevant advances on the design and formulation of DNA-based systems as technological devices to be used in clinical applications have been accomplished. In the last few years, particular attention has been focused on the plasmid DNA (*p*DNA) hydrogels. Biocompatible *p*DNA gel networks were synthesized by a cross-linking reaction. In order to enhance transfection efficiency and targeting of the systems, transferrin has been included in the protocol of hydrogels preparation. All developed carriers are photodegradable which opens the possibility for the sustained and controlled delivery of different plasmids and anticancer drugs. The cancer treatment approach based on the combination of chemotherapy and specific gene delivery demonstrated to possess stronger ability to weaken the growth and proliferation of tumour cells. The effect is enhanced when transferrin is present in the pDNA hydrogels. This finding is a great achievement and instigates further research focused on the generation of new vectors for the delivery of biopharmaceuticals contributing for the evolution of cancer therapy.

Keywords Hydrogels · Cross-linked systems · pDNA · Controlled release
Gene delivery · Co-delivery · Cancer therapy

D. Costa (✉) · J. Queiroz
CICS-UBI—Health Sciences Research Centre, University of Beira Interior,
Av. Infante D. Henrique, 6200-506 Covilhã, Portugal
e-mail: dcosta@fcsaude.ubi.pt

A. J. M. Valente
Department of Chemistry, University of Coimbra, 3004-535 Coimbra, Portugal

V. K. Thakur and M. K. Thakur (eds.), *Hydrogels*, Gels Horizons: From Science to Smart Materials, https://doi.org/10.1007/978-981-10-6077-9_13

1 Introduction

Polymer gels are cross-linked polymer networks inflated with a solvent, such as water. Hydrogels show singular physical and chemical characteristics and structures that can swell in water presenting a three-dimensional matrix in virtue of covalent or non-covalent bonding (chemical and physical gels, respectively). Their intrinsic characteristics such as the degree of ionization of the gel, the swelling profile and the specific mesh size can be tailored to optimize the use of hydrogels in the various categories of the technological field. Responsive gels, networks that can change their equilibrium swelling degree due to the changes in the surrounding environment, have several uses in the biomedical area (controlled drug/gene delivery, ocular devices, biological assays and biomimetics), agricultural, cosmetics to industrial (Gehrke 1993; Gehrke and Lee 1990; Peppas 1986; Thakur and Thakur 2014, 2015). The chemical and physical properties of networks can be modified to enhance their performance in the controlled release of bioactive agents (Peppas 1986; Peppas et al. 2000). For example, the formulation of temperature-responsive gels (Jeong et al. 2002), pH-responsive matrices (Torres-Lugo et al. 2002), glucose-responsive networks (Podual et al. 2000) and tuning of the gel swelling pattern (Peppas and Langer 2004) were engineered to enhance the sustained/controlled drug release area. When gel systems are formulated to be biodegradable, these vectors gained interesting properties which make them suitable for delivery applications (Chen et al 2013; Klinger et al. 2012; Lee et al 2013). Biodegradable networks seem to be very adequate carriers for the release of DNA, as the release rate can be largely controlled and the cross-linking density demonstrated to be a parameter that allows for efficient modulation (Costa et al. 2012, 2014a). To be used in therapeutic applications, these systems should gather a set of requirements, such as its biocompatibility, and possess an adequate size and shape, they must be able to load and release biocargos, and they should protect them against nucleases elimination. Moreover, they must exhibit a good biodistribution along with decreased side effects and inflammation, leading to a controlled and sustained delivery of the incorporated biomolecules (Al-Dosari and Gao 2009; Vercauteren et al. 2012). Additionally, when the release profile is controlled, relevant progresses can be made, namely in gene delivery, with a major effect in the higher extent and duration of transgene expression and lower toxicity to non-target cells (Gojgini et al. 2011; Noga et al. 2012). Concerning cancer therapy, gene delivery protocols based on hydrogels gave rise to impressive achievements (Han et al. 2011; Xu et al. 2012). Various fascinating works demonstrated the efficacy of cancer therapy using the p53 gene, a tumour suppressor gene involved in cellular pathways such as DNA repair, maintenance of the cell cycle and apoptosis (programmed cell death) (Costa et al. 2012; Gaspar et al. 2011; Li et al. 2011a; Sasaki et al. 2012). Conventional treatments in cancer clinic, including radiotherapy and chemotherapy, are frequently ineffective, and/or their use instigates a set of undesirable secondary effects; chemotherapy, for instance, can be accompanied by the loss of sensitivity of cancer cells to anticancer drugs, a multidrug resistance effect. Therefore, gene delivery emerges as an advanced and exciting tool to face this serious and deadly disease. This kind of treatment can be used in an individual manner or it can be combined with radiotherapy or chemotherapy, since specific genes have the ability to sensitize cancer cells to radiation or drugs improving its therapeutic effect. In particular, in the literature one can find a variety of reports showing the release genes and pharmaceutics to the same cells (Chen et al. 2011, 2012; Costa et al. 2012; Li et al. 2011b), with consequent demonstration of the synergetic effect resulting from its combination in the cancer treatment.

This chapter describes our research in the biomaterials area, focused on the materials design, and concerns the development of novel hydrogel systems as innovative platforms to operate therapeutically in order to instigate the translation of hydrogels design into bioapplications. Our findings represent a significant basis for the creation/engineering of novel biopharmaceutical vehicles for advanced controlled release and targeted delivery strategies offering innovative solutions for the, always growing, clinical demands.

2 DNA-Based Hydrogels

Deoxyribonucleic acid is an amphiphilic bio-polylectrolyte present in the cell nuclei, being responsible for the preservation of genetic content of life (Nelson and Cox 2000). The amphiphilic character of DNA is due to the existence of hydrophobic bases and carbohydrates and phosphate which confer the hydrophilic character of DNA; the latter is characterized by a pK_a a ca. 7, while the former allows the self-assembly of DNA in a double helix structure.

DNA compaction is a process of paramount importance once it allows the facilitated transport through the cellular membrane and thus to more easily reach the nucleus of cells. In the last decade or so, several authors have focused their work on developing efficient routes to find efficient ways for DNA compaction. For example, Horn and Peterson have found that histone proteins are able to compact DNA in living cells (Horn and Peterson 2002). In in vitro conditions, there are several studies focusing on this issue by using different approaches: by using multivalent ions (Sen and Cothers 1986), cationic surfactants or polyamines, such as, e.g. poly (ethylenimine). While multivalent ions promote DNA condensation by charge neutralization and screening effect, the interaction mechanism behind the use of surfactants and polyamines is somewhat complex. In the case of cationic surfactants (e.g. alkyltrimethylammonium bromides), the interaction occurs at pre-micelle concentrations and is characterized by a cooperative association process where both hydrophobic and hydrophilic parts of surfactant might play an important role. Polyamines are known as efficient delivery vectors; furthermore, they show ability to bind DNA by promoting its condensation. These two important features make these systems quite useful for in vivo and in vitro applications (Cao et al. 2013; Móran et al. 2009; Kircheis et al. 2001).

Bearing in mind that strong interactions occurring between DNA and the above-mentioned compounds, it could be anticipated that, in suitable concentrations, physical DNA hydrogels could be prepared and thus to contribute for the development of new vectors for the sustained and controlled release of DNA, the development of a biomatrix for drug release or both (i.e. the co-transport of DNA and drug). In 2006, Costa et al. (2006a) developed physical DNA hydrogels by using hydrophobic-modified cellulose-based polymers; these hydrogels have shown a complex behaviour characterized by well-defined phase transitions and interesting rheological properties (quite different from the pristine polymers); surprisingly, it was found that the effect of hydrophobic interactions in the gel formation is very minor or even neglected (Costa et al. 2006a). Following these achievements, *ss*- or *ds*DNA-based hydrogels were prepared by electrostatic interactions using cationic polyelectrolytes, surfactants (e.g. hexadecyltrimethylammonium bromide—CTAB) or proteins (e.g. lysozyme and protamine sulphate) (Móran et al. 2009, 2011; Tang et al. 2009). It should be highlighted that surfactant–DNA gels have shown high versatility due, in particular, to the reversibility of the gel formation and, consequently, to the DNA compaction/decompaction. A different approach for the development of DNA hydrogels is based on DNA supramolecular interactions to obtain DNA hydrogels. In this framework, Um et al. used enzymes, as a catalyst, to obtain DNA hydrogels formed by the self-assembly of DNA building blocks (Um et al. 2006). Following the same approach, Cheng et al. synthesized external stimulus (pH)-responsive DNA hydrogels; the pH-responsive DNA hydrogel is formed by the assembly of previously formed Y-shaped DNA structures (Cheng et al. 2009). Following this approach, thermal-responsive physically DNA hydrogels have been prepared some years later (Xing et al. 2011). In a similar way, tetravalent DNA nanostars can be stabilized by lowering the temperature being able, after that, to form highly thermodynamically stabilized. This unconventional behaviour occurs as a consequence of the high freedom degree nanstars's arms; such unique property allows to modify the hydrogel properties by changing the nanostar architecture (Rovigatti et al. 2014). By using a different strategy, Kahn et al. used gold-coated surfaces as a matrix for the assembly of DNA-based copolymers; they showed that the functionalization of copolymers, such as acrylamide, with, for example, G-rich DNA tethers is possible to modify the stiffness of the obtained hydrogel (Kahn et al. 2015). Recently, a supramolecular polypeptide–DNA hydrogel for in situ 3D bioprinting was developed (Li et al. 2015). These authors take advantage of physical and mechanical properties of the gel to print stable structures, keeping their shape and function, up to millimetre scale. Astonishing live cellular applications can be tested by three-dimensional cell printing (Li et al. 2015).

A different aspect of DNA hydrogels is related with chemically cross-linked hydrogels. Tanaka had a pioneer role on this issue through the synthesis of covalent DNA hydrogels (Amiya and Tanaka 1987). Following the work of Tanaka's group, Horkay et al. synthesized cross-linked DNA gels, using ethylene glycol diglycidyl ether (EGDE) as cross-linker, and the effect of non-associated salts with mono- and di-valent cations on the mechanical and osmotic properties of gels has been investigated (Horkay and Basser 2004). The effect of EGDE concentration on the rheological properties of cross-linked DNA hydrogels was further studied by Topuz and Okay, who found that at low EGDE concentrations the hydrogels become responsive to thermal stimulus (Topuz and Okay 2008).

One decade ago, one of us has devoted efforts to deeply study the mechanism of interaction between different co-solutes and DNA hydrogels as well as to develop stimuli-responsive chemically DNA-based hydrogels. Free radical aqueous polymerization of DNA from salmon testes, using EGDE as cross-linker and TEMED as initiator (Costa et al. 2006a). The mechanism of cross-linking involves the guanine nitrogen atom of the DNA and the more substituted C (or the least hindered end) of the bifunctional epoxy-based cross-linker. Rheological and swelling behaviour have been assessed in detail. It has been found that these gels show a shear thinning behaviour. Concerning the swelling, DNA gels behave differently when in equilibrium with salts or cationic surfactants. In general, the swelling degree (Q) of gels decreases by increasing the ionic strength (either by increasing the concentration or the ionic valence) in the gel's surrounding media. This can be justified by the electrostatic screening effect. On the other hand, the Q of these gels shows two different regimes in the presence of surfactants. At concentrations below the critical micelle concentration (*cmc*), or critical aggregation concentration (*cac*), a slight deswelling is noticed; however, at concentrations above the *cmc* (or *cac*) the collapse is significant. This is in agreement with the effect of the surfactant chain length; i.e. Q decreases with an increase of the surfactant chain length (Costa et al. 2006b, 2007).

Following these findings, it was our scope to develop DNA gel particles for the simultaneous delivery of DNA and drugs (in particular, highly hydrophobic drugs). In order to achieve that, plasmid DNA (pcDNA3-FLAG-p53) microgels were prepared by using the inverse microemulsion polymerization method (Costa et al. 2012, 2014b). By playing with the concentration of EGDE, acting as cross-linker, it was possible to obtain light-responsive gels; in other words, it was possible to synthesize microparticles that after UV radiation suffer degradation with consequent release of DNA (without losing its properties). The effect of light on DNA gels and DNA gels previously loaded with doxorubicin (a fluorescent anticancer drug) was tested with quite promising results (Costa et al. 2014b). Additionally, spherical DNA-based nanogels were synthesized for the first time by using polyamines (e.g. PEI) as *p*DNA condensation agent. These nanogels showed very low size distribution, high loading capacity and interesting release kinetics of simultaneous delivery of pDNA and drugs (Costa et al. 2015a). Very recently, DNA gel particles have been synthesized by using cationic surfactants (Costa et al. 2015b). However, these particles show an important drawback: their cytotoxicity. In order to overcome that, the particles were coated with polyethylenimine (PEI) (Costa et al. 2015b). This strategy will allow to address the cytotoxicity issue without compromising the transport properties. These formulated vectors and their potentialities to be used as delivery matrices will be discussed in the following section.

3 Polyethylenimine Coated Surfactant–DNA Gels

DNA strongly interacts with positively charged surfactants, and from their association, two phenomena can result: DNA compaction and condensation (Estévez-Torres and Baigl 2011; García et al. 2014). Generally, a common phase diagram representing this associative behaviour includes one phase concentrated in both charged agents and the other diluted in the cationic entity and the DNA. Due to the opposite charges, the electrostatic attraction is responsible for this situation, along with the increase in the entropy of the system arising from the release of the respective counterions (Costa et al. 2006b). At the critical aggregation concentration (cac) or at surfactant concentrations above this value, precipitation may occur. It will be dependent on the concentration of the polyelectrolyte and the individual characteristics of each system (Dias et al. 2000).

In the biomedical field, the application of surfactants presents a major obstacle due to their displayed cytotoxicity. To overcome this drawback and enhance the biocompatibility of surfactant-based release vehicles, various approaches have been designed. Some of them are focused in the alteration of the surfactant's head group (Jadhav et al. 2008) or hydrocarbon chain (Courrier et al. 2003) and reducing the surfactant amount that induces DNA compaction. In line with this, our group has developed plasmid pcDNA3-FLAG-p53/surfactant nanoparticles that have been coated with a densely charged and biocompatible cationic amine, polyethylenimine (PEI), to improve the biocompatibility of the plasmid DNA (pDNA) carriers. Following this idea, we formed pDNA nanosystems by dropwise addition of the aqueous pDNA solution to the hexyl, octyldecyl, dodecyl or hexadecyl trimethyl ammonium bromide (C_6TAB, C_8TAB, C_{12}TAB or C_{16}TAB, respectively) solutions. Nanoparticles were synthesized for all systems studied, except for pDNA-C_6TAB where the short hydrophobic chain (therefore, a high solubility) prevented a strong binding of C_6TAB molecules to pDNA and the surfactant was not able to fully neutralize the negative charges of pDNA. Contrary, cationic amphiphiles with longer alkyl tails efficiently compacted pDNA molecules resulting in the formation of particles. The resulting particles present sizes ranging from 620 to 210 nm, approximately, and this parameter varies with the surfactant tail length (Costa et al. 2015b). A trend can be ascribed: small particles can be formed when surfactants with longer hydrocarbon tails are used. The explanation can be related with the stronger hydrophobic interaction between longer chain surfactants and stronger associative binding and, therefore, with their great potential for DNA condensation. In the case of C_6TAB, the interactions of hydrophobic nature between the chains are weak and not sufficient to the pDNA charge neutralization, suggesting another mechanism of binding (Husale et al. 2008). On the contrary, C_8, C_{12} and C_{16}TAB amphiphiles are all able to interact with pDNA promoting its condensation which results in the formation of nanoparticles, with pDNA-C_{16}TAB particles being the smallest ones. Summarizing, it was found that by increasing the surfactant tail length, small pDNA-based nanoparticles can be formed.

Information concerning the cytotoxicity of the designed particles comes from the 3-[4,5-dimethyl-thiazol-2-yl]-2,5-diphenyltetrazolium bromide (MTT) assay on fibroblast (NDHF) cells. This colorimetric assay shows that all systems exhibited significant cytotoxicity to NDHF cells, with higher effect as surfactant chain increases. The same kind of pattern was found for HeLa cells, and thus, these particles may be cytotoxic, therefore not suitable for biomedical applications. To overcome this inconvenient, the coating of nanoparticles with polyethylenimine (PEI) appears to be an interesting approach. The biocompatibility of the PEI/CnTAB/pDNA systems has been considerably improved, as shown in Fig. 1. PEI, a positively charged polyamine, has a higher charge density, and by electrostatic interactions, it can strongly bind to pDNA leading to its compaction (Korolev et al. 2010). This property of PEI to bind to pDNA forming small nanoparticles has been greatly explored, and a variety of nucleic acid-based vehicles have been developed and their potentiality for therapeutic uses evaluated (Chen et al. 2014; Nawwab Al-Deen et al. 2014). p53-C_nTAB (n = 8, 12 and 16) particles have been coated with PEI; no relevant interaction between PEI and pDNA has been observed for molar ratios below 4.8. Additionally, the systems remain cytotoxic. According to this profile, molar ratios of 4.8, 6 and 10 have been chosen to further studies. For all p53-C_nTAB-PEI systems developed, scanning electron microscopy images (Fig. 2) show spherical or oval particles. Along with this information, size and zeta potential parameters were analysed by means of dynamic light scattering analysis. The particles have sizes ranging from 500 to 90 nm, approximately, which make them suitable for cell uptake and internalization. As observed earlier, smaller-sized pDNA nanoparticles appear in the systems where surfactants with longer chains are

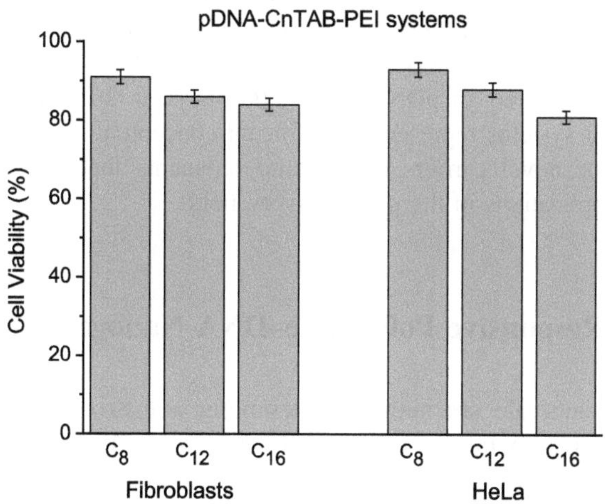

Fig. 1 Cytotoxicity profile of plasmid DNA-CnTAB-PEI nanoparticles with an *N*/*P* ratio of 10 on human fibroblasts and HeLa cells after 2 days of incubation, as measured by the MTT assay. The percentage viability is expressed relative to control cells. The data were obtained by averaging the results from three experiments. The respective errors were determined as below 0.05%. The error bars in the figure represent the standard deviations (i.e. ±SD) from the three experiments

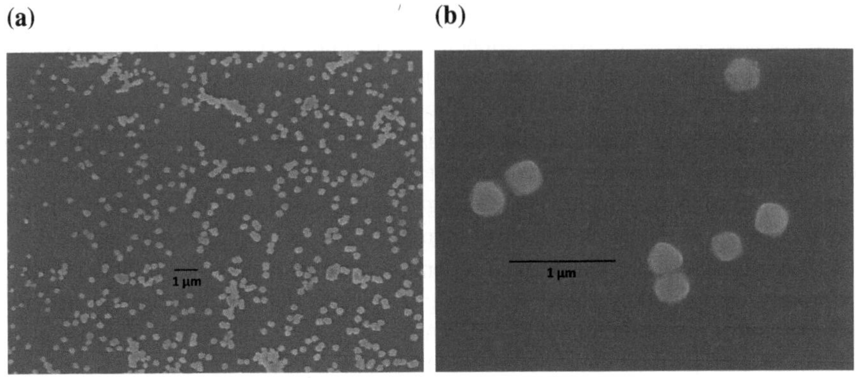

Fig. 2 Scanning electron micrographs of p53-C_8TAB-PEI nanoparticles, *N*/*P* = 4.8 (**a**), and p53-C_{12}TAB-PEI nanoparticles, *N*/*P* = 6 (**b**)

employed. However, the most important finding is the dramatic reduction in the particle size that has been achieved by PEI coating, this fact becoming more relevant for the C_{16}TAB (Costa et al. 2015b). The alteration of particle size with PEI nitrogens to pDNA phosphates (*N*/*P* ratio) is very mild with only a minor reduction as *N*/*P* ratio increases. A different situation was found on particle surface charges when particles were coated with PEI and the *N*/*P* ratio was varied. It seems that pDNA negative charges have been fully neutralized, as positive values have been found for p53-C_nTAB-PEI particles.

Furthermore, the pDNA-C_8TAB-PEI particles present an interesting dissolution pattern while those including C_{12}TAB and C_{16}TAB display a swelling/deswelling behaviour that can be deeply investigated for the delivery of pDNA in a controlled manner. Additionally, both the surfactant tail length and the *N*/*P* ratio can be used as controlling parameters of pDNA release (Costa et al. 2015b). As a whole, the p53-C_nTAB-PEI systems represent a significant conceptual advance in the design and synthesis of novel carriers for specific bioagents that can find innovative therapeutical applications in the gene delivery field.

4 Stimuli-Responsive Polyamine–DNA Nanogels

Hydrogels at nanoscale are networks presenting a matrix in three dimensions adequate for the incorporation of biological components. To be used in therapeutic protocols, these hydrogels should, among other requirements, encapsulate biomolecules efficiently (e.g. DNA) and ensure the controlled and sustained delivery of the therapeutic dosage. Considering cancer therapy, it is desirable that nanogels ensure the efficient target of genes with anticancer effect to the tumoral tissue. Once this location is reached, they should bind to the target cells and instigate transgene expression. Several strategies have been explored for gene release at a specific target, as for example, the use of ligands that can bind cells, receptor-mediated endocytosis (Kircheis et al. 2001) and proteins or peptides with nuclear localization signals (NLS) to facilitate nucleus targeting (Bremner et al. 2001; Ogris et al. 2001). Considering these approaches, the gene expression and the in vivo applications gained relevant enhancement (Chen et al. 2014; Ogris et al. 2001). Our team contributed to this topic developing chemically engineered polyamine gels through the combination of pVAX1-*LacZ* or pcDNA3-FLAG-p53 (p53) condensation and cross-linking reaction with ethylene glycol diglycidyl ether (EGDE). The polyamines protamine sulphate (PS) and polyethylenimine (PEI) were considered to induce pDNA condensation into nanosized particles. The plasmids pVAX1-*LacZ* and pcDNA3-FLAG-p53 have been condensed by PS and PEI, respectively, and thereafter cross-linked with EGDE (Costa et al. 2015b). This novel strategy involving the combined use of polyamine condensation ability and cross-linker gel synthesis gives rise to nanoparticles of reduced size. Following this protocol, PS-pVAX1-*LacZ* and PEI-p53 nanogels with 0.08, 0.1 and 0.15% cross-linker densities were formulated. Additionally, the protein transferrin has been also encapsulated in PEI-p53 nanogels by imbibition (Costa et al. 2015c). The obtained networks exhibit a spherical or oval shape with diameter sizes ranging from 80 to 300 nm, and so they are convenient to be used in gene release situations. In order to get further information on these gel vehicles, the mean values of size and zeta potential were acquired by dynamic light scattering (Table 1). The smaller sizes have been confirmed. The cross-linker density is a parameter that also deeply influences the size of the gels, with highest cross-linked PS or PEI systems showing the smallest sizes. Transferrin (Tf) does not seem to dictate the size of hydrogels. Moreover, in general, PEI gels show smaller sizes in comparison with the correspondent cross-linked PS. Polyamines, due to their positive charges and high

Table 1 Characterization of protamine sulphate (PS) pVAX1-*LacZ* and polyethylenimine (PEI) pcDNA3-FLAG-p53 cross-linked nanogels at various *N/P* ratios (4.8, 6 and 10) and cross-linker densities. The values of the different parameters were calculated with the data obtained from three independent measurements (mean ± SD, $n = 3$)

System	*N/P*	Particle size (nm)	Zeta potential (mV)	Viability (%)
pDNA-PS-0.08% EGDE	1	127 ± 3	41 ± 0.2	89
pDNA-PS-0.1% EGDE	1	100 ± 1	40 ± 0.5	87
pDNA-PS-0.15% EGDE	1	89 ± 1	45 ± 0.1	83
pDNA-PEI-0.08% EGDE	4.8	107 ± 1	48 ± 0.4	89
	6	105 ± 1	55 ± 0.1	87
	10	109 ± 2	69 ± 0.3	87
pDNA-PEI-0.1% EGDE	4.8	90 ± 0.9	47 ± 0.4	87
	6	90 ± 0.3	58 ± 0.2	87
	10	91 ± 0.8	67 ± 0.5	85
pDNA-PEI-0.15% EGDE	4.8	81 ± 0.2	48 ± 0.3	82
	6	86 ± 0.4	57 ± 0.6	80
	10	83 ± 0.1	69 ± 0.9	80

charge density, strongly interact with DNA mainly by electrostatic interactions. PEI is, thus, considered the most efficient pDNA-compacting molecule (Jorge et al. 2014), what may justify the sizes of PEI-p53 carriers. Information from gel size has been correlated with the surface charges of the system. All PS hydrogels have positive zeta potential values ranging from +40 to +45 mV, approximately. High positive values were found for all PEI-p53 networks, demonstrating that *N*/*P* ratio is a relevant parameter controlling the size. As the PEI content in the PEI-p53 systems increases, a significant increment on surface charges can be observed. Concerning the presence of the protein in PEI-p53 systems, it was found a decrease in the zeta potential values. This fact can be attributed to its shielding ability in PEI/DNA formulations (Kircheis et al. 2001). Furthermore, when transferrin is present in PEI-p53 vectors, the positive charges can be shielded and this may lead to a lowering of the unspecific interactions, what may improve the targeted payload release after systemic application (Kircheis et al. 2001).

The most interesting characteristic of the formulated nanogels is their photodegradable profile under ultraviolet light exposure (Costa et al. 2015c). As these systems are photodegradable, therapeutic cargos encapsulated in their matrices can be delivered in a controlled and sustained manner, leading to enhanced therapeutic effects in cancer cells.

4.1 Delivery of Genes and Anticancer Drugs

The great drug and gene loading capacity of the developed PEI-p53 and PEI-Tf-p53 hydrogels allow for the investigation of the co-delivery strategy in cancer therapy. The p53-based plasmid and the drugs doxorubicin (DOX), epirubicin (EPIR) and paclitaxel (PTX) were considered, and the performance of the designed nanogels as delivery vectors was evaluated (Costa et al. 2015c). Due to the photodegradation process, they can suffer in UV conditions, and the p53 gene and anticancer drugs

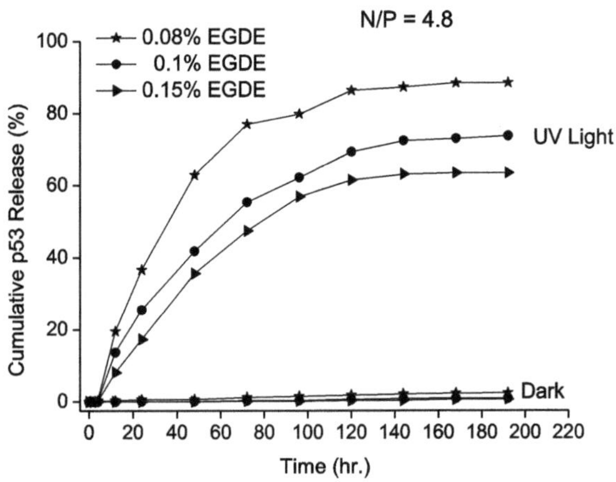

Fig. 3 Cumulative release of pcDNA3-FLAG-p53 gene from 0.08, 0.1 and 0.15% EGDE cross-linked PEI p53 nanogels at *N*/*P* ratio of 4.8, 6 and 10 after their irradiation with UV light (400 nm) and in the dark conditions as a function of time

can be simultaneously released. Figure 3 presents the release curve of the p53 gene from PEI-p53 gels, *N*/*P* ratio of 4.8, after their irradiation with UV light. Higher levels of p53 can be delivered from PEI-p53 networks, with cross-linker density controlling the release extent. In the dark, no relevant p53 plasmid content can be delivered from PEI matrices. The same finding was observed for the systems with *N*/*P* ratios of 6 and 10. Conventional drugs applied in chemotherapy, DOX, EPIR and PTX, can also be efficiently released from PEI nanovectors under photodegradation, for all *N*/*P* ratios considered (Costa et al. 2015c). Figure 4 summarizes the release behaviour of EPIR and PTX at *N*/*P* ratios of 6 and 10, respectively. The effect of transferrin in the drug release profile is noticeable; Tf retards and decreases the drug amount that is delivered from the nanogels. The dual encapsulation of Tf and anticancer drug into the nanosystems induces an initial time lag in the release curve and lowers the maximum drug-released value. Moreover, a

Fig. 4 Cumulative release of epirubicin (EPIR), (**a**), and paclitaxel (PTX), (**b**), from 0.08, 0.1 and 0.15% EGDE cross-linked PEI p53 nanogels at *N*/*P* ratio of 6 and 10, respectively, after their irradiation with UV light (400 nm)

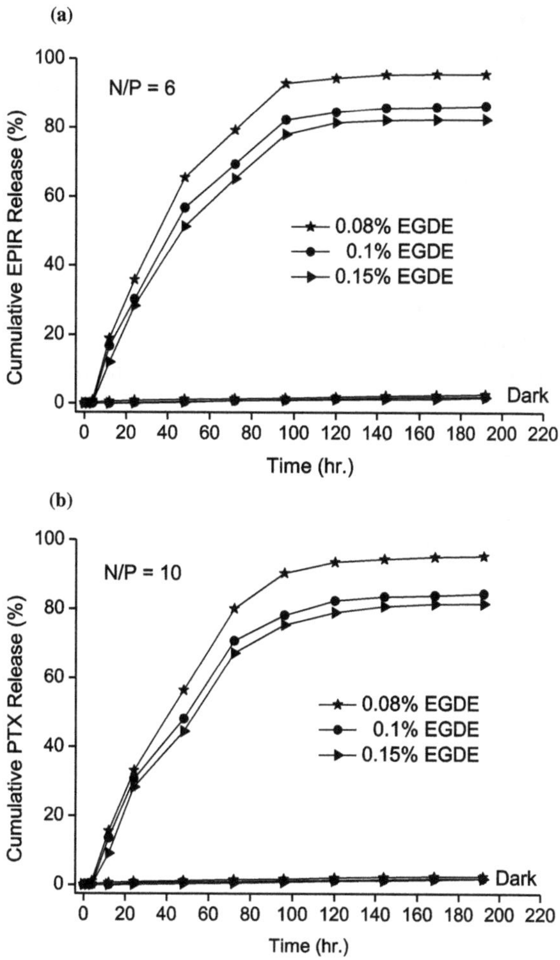

low content of Tf (<10%) can be released from PEI-Tf-p53 hydrogels after UV irradiation, a fact most probably related with its higher molecular weight (76,000 Da) which turns into a hard task of its transport through the PEI-p53 network. Transferrin can improve the cell transfection phenomenon through the binding to a receptor expressed on the surface of most cancer cells (Daniels et al. 2012), and this may facilitate the uptake of PEI-Tf-p53 vectors into endosomes. The determination of the p53 protein expression in breast MCF-7 cancer cells (Fig. 5), after mediated transfection with PEI-p53 and PEI-Tf-p53 systems, at *N*/*P* ratios of 4.8, 6 and 10, highlights these statements. Before the transfection studies, the PEI-p53 and PEI-Tf-p53 nanogels were ultraviolet light irradiated (400 nm) and the resulting network photodisruption allows for the delivery of the incorporated p53 plasmid. The release of p53 plasmid and the gene expression in the cell nucleus through a transfection process mediated by PEI-p53 vehicles, with/without the presence of transferrin, have been accomplished (Costa et al. 2015c). Furthermore, it was found that depending on PEI-p53 gel vehicle, transfection occurs by slightly different mechanisms. PEI-p53 gels, at *N*/*P* ratios of 4.8, 6 and 10, are able to surpass the cellular barrier and release pDNA into the site of cytosol. The entry of foreign pDNA into the nucleus is an inefficient process and the only possible mechanism known involves cell proliferation where mitotic cell activity results in the opening of the cell nucleus. Tumour cells present a high rate of cell division which favours both transfection and p53 expression, with the consequent and desirable protein production. On the other hand, the inclusion of Tf into PEI-p53 gels permits the receptor-mediated endocytosis pathway to occur. Following this, the crossing of cell membrane, an always difficult internalization stage, may be facilitated. Gene delivery and expression rates can be greatly improved by the combination of this effect induced by Tf and the recognized endosomolytic capacity of PEI. Formulated PEI-p53 gels at *N*/*P* ratio of 6 are the most suitable to be further

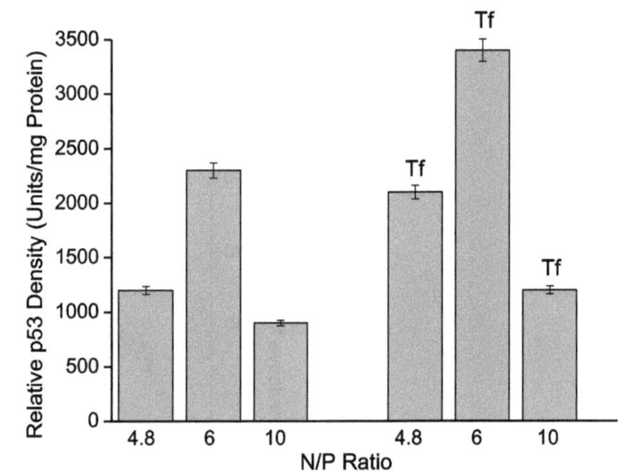

Fig. 5 Quantification of p53 protein expression in MCF-7 cells using a pan-p53 ELISA kit for 0.08% EGDE cross-linked PEI p53 and PEI-Tf p53 nanogels at *N*/*P* ratios of 4.8, 6 and 10. The data were obtained by calculating the average of three independent experiments. The respective errors were determined and were below 0.05%

exploited for p53 gene release into cancer cells, as these complexes mediate the expression of higher levels of p53 protein, in comparison with similar designed vehicles at *N*/*P* ratios of 4.8 and 10, the latter presenting the lowest values (Costa et al. 2015c).

4.2 Cancer Gene Therapy

Applied to cancer therapy, the co-delivery of p53 gene and anticancer drugs can result in a considerable improvement in the treatment efficacy. To investigate this issue, an in vitro growth assay on breast cancer cells, MCF-7, mediated by the use of PEI or PEI-Tf dual systems (Costa et al. 2015c) was performed. The co-delivery of gene and anticancer drug promoted by PEI and PEI-Tf vectors and its consequence on the rate of cancer cells growth, relative to that displayed by control cells, was followed for 8 days. The results are summarized in Fig. 6. The PEI-p53 mediated cell growth was dramatically decreased, when compared to that of control cells. Moreover, and as expected, a more significant growth inhibition can be detected when MCF-7 cells are transfected with PEI-p53-anticancer drug vectors. The treatment of cancer cells with both p53 gene and anticancer drug leads to a strong suppressor effect in the growth/proliferation of MCF-7 cells instigating apoptosis. The PEI-p53-EPIR-based carriers appear to be the most effective of all studied vectors in inducing the MCF-7 cells growth inhibition. The loading of transferrin into PEI-p53-anticancer drug systems reveals to be quite relevant; due to the targeting capacity, the PEI co-delivery nanosystems show great potential in decreasing the growth of cancer cells (Costa et al. 2015c). Furthermore, this inhibition effect promoted by PEI-p53 matrices has been confirmed by an MTT-based analysis. The results are shown in Fig. 7. The treatment of cancer cells MCF-7 with

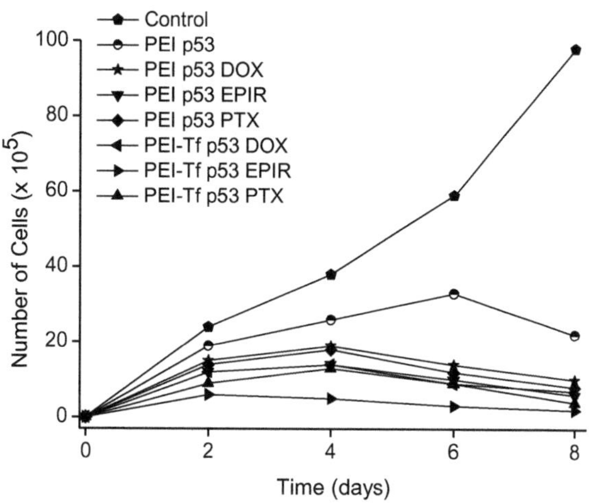

Fig. 6 Cell growth assay in MCF-7 cells after transfection with 0.08% EGDE cross-linked PEI p53, PEI p53 DOX, PEI p53 EPIR, PEI p53 PTX, PEI-Tf p53 DOX, PEI-Tf p53 EPIR and PEI-Tf p53 PTX nanogels at *N*/*P* ratios of 6, compared with the in vitro growth of untreated cells. The data were obtained by calculating the average of three experiments ($P < 0.05$)

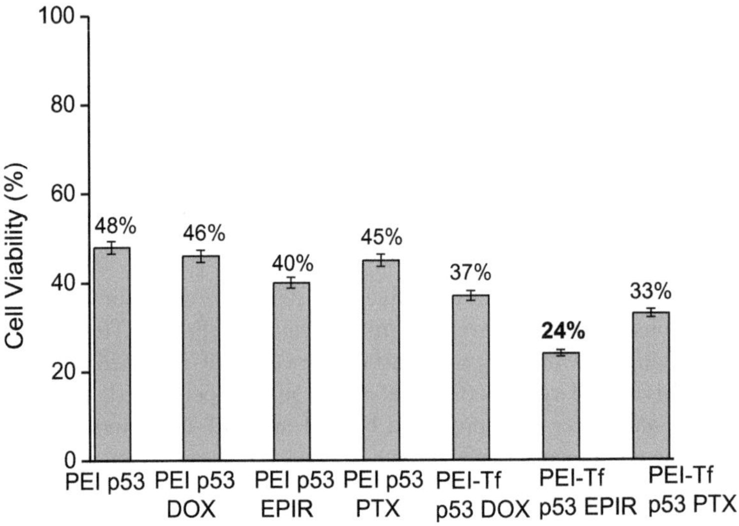

Fig. 7 Viability of MCF-7 cells after transfection with the EGDE cross-linked PEI p53 vectors and EGDE cross-linked PEI-Tf p53, after 8 days measured by MTT assay. Viability of MCF-7 cells after treatment with DOX, EPIR and PTX is also presented. Per cent viability is expressed relative to control cells. The data were obtained by calculating the average of three experiments ($P < 0.05$)

the developed p53-based pDNA-PEI networks results in the inhibition of tumoral growth. The mediated transfection by pDNA-PEI system leads to the transgene expression and consequently to the p53 protein production. Due to the p53 protein regulatory function related to damaged DNA, introducing p53 gene plasmid into cancer cells induces their apoptosis. The combination of gene therapy, by the use of PEI-p53 nanogels, and chemotherapy is more efficient in promoting an inhibition effect on tumour growth than the individual application of a p53 encoding plasmid or the free anticancer drug. The synergetic effect achieved by the simultaneous delivery of gene and drug to the same cancer cells reveals to be an excellent therapy to control tumour growth and proliferation. Additionally, this effect is even more pronounced when PEI-p53 vector with EPIR encapsulated is considered (Costa et al. 2015c). Also, larger apoptosis can be observed when transferrin is incorporated into PEI-Tf-p53-anticancer drug nanocarriers. Therefore, the formulated bifunctional gene-/drug-based PEI gels are tunable matrices that allow for efficient and controlled drug and gene delivery giving rise to a synergistic effect that enhances cancer therapy, greatly contributing for clinical translation.

5 Concluding Remarks

This chapter highlights the current developments in the formulation of DNA-based hydrogels exhibiting tunable skills and advanced performances towards innovative bioapplications. The progresses accomplished by our group in this particular area, by using plasmid DNA vectors, have been presented. This represents our contribution to the scientific field dealing with engineering tailored DNA networks for the delivery of biopharmaceuticals that can operate therapeutically in a broad range of biomedical issues. In the light of this, the dual release of genes and anticancer drugs, alone or as a bifunctional system, is a powerful tool enhancing the treatment of deadly diseases, such as cancer. The associated synergetic effect promotes further evolution in the design of new engineered delivery systems that can demonstrate impressive technological and biomedical utility, instigating clinical translation.

Although significant studies have been conducted aiming the targeted and controlled release of plasmid DNA, there are still many issues to address. Therefore, the future of biopharmaceutical investigation holds for the creation of original devices with tunable properties, controlled delivery profile and desirable biological interactions with body elements intensifying the application of DNA systems to the many puzzling demands still searching for solution in the biomedical field. The impact of this knowledge on bio-multifunctional applications will be tremendous, and it will, surely, contribute to the development of advanced therapies for the most deadly diseases.

References

Al-Dosari MS, Gao X (2009) Nonviral gene delivery: principle, limitations and recent progress. Am. Ass. Pharm. Sci. J. 11:1–74

Amiya T, Tanaka T (1987) Phase transition in crosslinked gels of natural polymers. Macromolecules 20:1162–1165

Bremner KH, Seymour LW, Pouton CW (2001) Harnessing nuclear localization pathways for transgene delivery. Curr Opin Mol Ther 3:170–177

Cao M, Deng X, Su S, Zhang F, Xiao X, Hu Q, Fu Y, Yang BB, Wu Y, Sheng W, Zeng Y (2013) Protamine sulfate-nanodiamond hybrid nanoparticles as a vector for MiR-203 restoration in esophageal carcinoma cells. Nanoscale 5:12120–12125

Chen G, Zheng L, Liu S, He X (2011) rAd-p53 enhances the sensitivity of human gastric cancer cells to chemotherapy. World J Gastroenterol 17:4289–4297

Chen S, Zhao D, Li F, Zhuo R, Cheng S (2012) Co-delivery of genes and drugs with nanostructured calcium carbonate for cancer therapy. RSC Adv 2:1820–1826

Chen W, Zheng M, Meng F, Cheng R, Deng C, Feijen J, Zhong Z (2013) In situ forming reduction-sensitive degradable nanogels for facile loading and triggered intracellular release of proteins. Biomacromolecules 14:1214–1222

Chen Z, Zhang L, He Y, Li Y (2014) Sandwich-type Au-PEI/DNA/PEI-Dexa nanocomplex for nucleus-targeted gene delivery in vitro and in vivo. ACS Appl Mater Interfaces 6:14196–14206

Cheng EJ, Xing YZ, Chen P, Yang Y, Sun YW, Zhou DJ, Xu LJ, Fan QH, Liu DS (2009) A pH-triggered, fast-responding DNA hydrogel. Angew Chem 121:7796–7799

Costa D, Santos S, Antunes FE, Miguel MG, Lindman B (2006a) Some novel aspects of DNA physical and chemical gels. Arkivoc iv:161–172
Costa D, Hansson P, Schneider S, Miguel MG, Lindman B (2006b) Interaction between covalent DNA gels and a cationic surfactant. Biomacromolecules 7:1090–1095
Costa D, Miguel MG, Lindman B (2007) Responsive polymer gels: double-stranded versus single-stranded DNA. J. Phys. Chem. B 111:10886–10896
Costa D, Valente AJM, Miguel MG, Queiroz J (2012) Plasmid DNA microgels for a therapeutical strategy combining the delivery of genes and anticancer drugs. Macromol Biosci 9:1243–1252
Costa D, Valente AJM, Miguel MG, Queiroz J (2014a) Plasmid DNA hydrogels for biomedical applications. Adv Colloid Interface Sci 205C:264–270
Costa D, Valente AJM, Miguel MG, Queiroz J (2014b) Plasmid DNA microgels for drug/gene co-delivery: a promising approach for cancer therapy. Colloids Surf A 442:181–190
Costa D, Valente AJM, Queiroz J (2015a) Plasmid DNA nanogels as photoresponsive materials for multifunctional bio-applications. J Biotechnol 20:98–104
Costa D, Briscoe W, Queiroz J (2015b) Polyethylenimine coated plasmid DNA-surfactant complexes as potential gene delivery systems. Colloids Surf B 133:156–163
Costa D, Valente AJM, Queiroz J (2015c) Stimuli-responsive polyamine-DNA blend nanogels for co-delivery in cancer therapy. Colloids and Surf B 132:194–201
Courrier HM, Krafft MP, Butz N, Porté C, Frossard N, Rémy-Krsitensen A, Méli Y, Pons F, Vandamme TF (2003) Evaluation of cytotoxicity of new semi-fluorinated amphiphiles derived from dimorpholinophosphate. Biomaterials 24:689–696
Daniels TR, Bernabeu E, Rodríguez JA, Patel S, Kozman M, Chiappetta DA, Holler E, Ljubimova JY, Helguera G, Penichet ML (2012) Transferrin receptors and the targeted delivery of therapeutic agents against cancer. Biochim Biophys Acta 1820:291–317
Dias R, Mel'nikov S, Lindman B, Miguel MG (2000) DNA phase behavior in the presence of oppositely charged surfactants. Langmuir 16:9577–9583
Estévez-Torres A, Baigl D (2011) DNA compaction: fundamentals and applications. Soft Matter 7:6746–6756
García JP, Marrón E, Martín VI, Moyá ML, Lopez-Correjo P (2014) Conformational changes of DNA in the presence of 12-s-12 gemini surfactants (s = 2 and 10). Role of the spacer's length in the interaction surfactant-polynucleotide. Colloids Surf B 118:90–100
Gaspar VM, Correia IJ, Sousa A, Silva F, Paquete CM, Queiroz JA, Sousa F (2011) Nanoparticle mediated delivery of pure p53 supercoiled plasmid DNA for gene therapy. J Control Release 156:212–222
Gehrke SH (1993) Synthesis, equilibrium swelling, kinetics, permeability and applications of environmentally responsive gels. Adv Polym Sci 110:81–144
Gehrke SH, Lee PI (1990) Hydrogels for drug delivery systems. Specialized drug delivery systems. Marcel Dekker, New York
Gojgini S, Tokatlian T, Segura T (2011) Utilizing cell-matrix interactions to modulate gene transfer to stem cells inside hyaluronic acid hydrogels. Mol Pharm 8:1582–1591
Han HD, Mora EM, Roh JW, Nishimura M, Lee SJ, Stone RL, Bar-Eli M, Lopez-Berestein G, Sood AK (2011) Chitosan hydrogels for localized gene silencing. Cancer Biol Ther 11:839–845
Horkay F, Basser P (2004) Osmotic observations on chemically cross-linked DNA gels in physiological salt solutions. Biomacromolecules 5:232–237
Horn PJ, Peterson CL (2002) Molecular biology: chromatin higher order folding: wrapping up transcription. Science 297:1824–1827
Husale S, Grange W, Karle M, Bürgi S, Hegner M (2008) Interaction of cationic surfactants with DNA: a single molecule study. Nucleic Acids Res 36:1443–1449
Jadhav V, Maiti S, Dasgupta A, Das PK, Dias R, Miguel MG, Lindman B (2008) Effect of the head-group geometry of the amino acid-based cationic surfactants on interaction with plasmid DNA. Biomacromolecules 9:1852–1859
Jeong B, Kim S, Bae Y (2002) Thermosensitive sol-gel reversible hydrogels. Adv Drug Deliv Rev 54:37–51

Jorge AF, Pereira RFP, Nunes SCC, Valente AJM, Dias R, Pais AACC (2014) Interpreting the rich behavior of ternary DNA-PEI-Fe (III) complexes. Biomacromolecules 15:478–491

Kahn JS, Trifonov A, Guo W, Fan C, Willner I (2015) Integration of switchable DNA-based hydrogels with surfaces by the hybridization chain reaction. Nano Lett 15:7773–7778

Kircheis R, Wightman L, Schreiber A, Robitza B, Rössler V, Kursa M, Wagner E (2001) Polyethylenimine/DNA complexes shielded by transferrin target gene expression to tumors after systemic application. Gene Ther 8:28–40

Klinger D, Aschenbrenmer EM, Weiss CK, Landfester K (2012) Enzimatically degradable nanogels by inverse miniemulsion copolymerization of acrylamide with dextran methacrylates as crosslinkers. Polym Chem 3:204–216

Korolev N, Lyubartsev A, Nordenskiold L (2010) Cation-induced polyelectrolyte-polyelectrolyte attraction in solutions of DNA and nucleosome core particles. Adv Colloid Interface Sci 158:32–47

Lee J, Jeong D, Seo S, Na K (2013) Biodegradable nanogel based on all-trans retinoic acid/ pullulan conjugate for anti-cancer drug delivery. J Pharm Investig 43:63–69

Li Y, Hua S, Xiao W, Wang H, Luo X, Li C, Cheng S, Zhang X, Zhuo R (2011a) Dual vectors of anti-cancer drugs and genes based on pH sensitive micelles self-assembled from hybrid polypeptide copolymers. J Mater Chem 21:3100–3106

Li H, Lakshmikanth T, Garofalo C, Enge M, Spinnler C, Anichini A, Szeekely L, Kärren K, Carbone E, Selivanova G (2011b) Pharmacological activation of p53 triggers anticancer innate immune response through induction of ULBP2. Cell Cycle 10:3346–3358

Li C, Faulkner-Jones A, Dun AR, Jin J, Chen P, Xing Y, Yang Z, Li Z, Shu W, Liu D, Duncan R (2015) Rapid formation of a supramolecular polypeptide-DNA hydrogel for in situ three-dimensional multilayer bioprinting. Angew Chem 54:3957–3961

Móran MC, Baptista FR, Ramalho A, Miguel MG, Lindman B (2009) DNA gel nanoparticles: preparation and controlling the size. Soft Matter 5:2538–2542

Móran MC, Miguel MG, Lindman B (2011) DNA gel particles from single and double-tail surfactants: supramolecular assemblies and release characteristics. Soft Matter 7:2001–2010

Nawwab Al-Deen FM, Selomulya C, Kong YY, Xiang SD, Ma C, Coppel RL, Plebanski M (2014) Design of magnetic polyplexes taken up efficiently by dendritic cell for enhanced DNA vaccine delivery. Gene Ther 21:212–218

Nelson DL, Cox MM (2000) Lenhinger: principles of biochemistry. Wroth Publishers, New York

Noga M, Edinger D, Rödl W, Wagner E, Winter G, Besheer A (2012) Controlled shielding and deshielding of gene delivery polyplexes using hydroxyethyl starch (HES) and alpha-amylase. J Control Release 159:92–103

Ogris M, Carlisle RC, Bettinger T, Seymour LW (2001) Melittin enables efficient vesicular escape and enhanced nuclear access of nonviral gene delivery vectors. J Biol Chem 276:47550–47555

Peppas NA (1986) Hydrogels in medicine and pharmacy. CRC Press, Florida

Peppas NA, Langer R (2004) Origins and development of biomedical engineering within chemical engineering. AIChE J 50:536–546

Peppas NA, Bures P, Leobandung W, Ichikawa H (2000) Hydrogels in pharmaceutical formulations. Eur J Pharm Biopharm 50:27–46

Podual K, Doyle FJ, Peppas NA (2000) Dynamic behaviour of glucose-oxidase-containing microparticles of poly(ethylene glycol)-grafted cationic hydrogels in an environment of changing pH. Biomaterials 21:1439–1450

Rovigatti L, Smallenburg F, Romano F, Sciortino F (2014) Gels of DNA nanostars never crystallize. ACS Nano 8:3567–3574

Sasaki Y, Oshima Y, Koyama R, Tamura M, Kashima L, Idogawa M, Yamashita T, Toyota M, Imai K, Shinomura Y, Tokino T (2012) A novel approach to cancer treatment using structural hybrids of the p53 gene family. Cancer Gene Ther 19:749–756

Sen D, Cothers DM (1986) Condensation of chromatin: role of multivalent cations. Biochemistry 25:1495–1503

Tang HW, Duan XR, Feng XL, Liu LB, Wang S, Li YL, Zhu DB (2009) Fluorescent DNA-poly (phenylenevinylene) hybrid hydrogels for monitoring drug release. Chem Commun 6:641–643

Thakur VK, Thakur MK (2014) Recent trends in hydrogels based on psyllium polysaccharide: a review. J Clean Prod 82:1–15
Thakur VK, Thakur MK (2015) Recent advances in green hydrogels from lignin: a review. Int J Biol Macromolec 72:834–847
Topuz F, Okay O (2008) Rheological behavior of responsive DNA hydrogels. Macromolecules 41:8847–8854
Torres-Lugo M, Garcia M, Record R, Peppas NA (2002) Physicochemical behavior and cytotoxic effects of p(methacrylic acid-g-ethylene glycol) nanospheres for oral delivery of proteins. J Control Release 80:197–205
Um SH, Lee JB, Park N, Kwon SY, Umbach CC, Luo D (2006) Enzyme-catalyzed assembly of DNA hydrogel. Nat Mater 5:797–801
Vercauteren D, Rejman J, Martens TF, Demeester J, De Smedt SC, Braeckmans K (2012) On the cellular processing of non-viral nanomedicines for nucleic acid delivery: mechanisms and methods. J Control Release 161:566–581
Xing Y, Cheng E, Yang Y, Chen P, Zhang T, Sun Y, Yang Z, Liu D (2011) Self-assembled DNA hydrogels with designable thermal and enzymatic responsiveness. Adv Mater 23:1117–1121
Xu L, Li X, Takemura T, Hanagata N, Wu G, Chou LL (2012) Genotoxicity and molecular response of silver nanoparticle (NP)-based hydrogel. J Nanobiotechnol 10:11

Chapter 14
Stem Cell Culture on Polymer Hydrogels

Akon Higuchi, Hsing-Fen Li, S. Suresh Kumar, Abdullah A. Alarfaj and Murugan A. Munusamy

Abstract The fate of stem cell differentiation is guided by several different factors of the stem cell microenvironment, such as cell culture biomaterial elasticity (physical cues) and cell–biomaterial interactions (biological cues). Mimicking the stem cell microenvironment using polymer hydrogels with optimal elasticities is an excellent strategy for stem cell expansion and differentiation. This chapter describes poly(vinyl alcohol) (PVA) hydrogels grafted with several nanosegments that are designed for the culture and differentiation of human hematopoietic and progenitor cells (hHSPCs), human amniotic fluid stem cells (hAFSCs), and human pluripotent stem cells (hPSCs). The elasticity of the cell culture hydrogels can regulate stem cell adhesion overall, as well as cell phenotype, focal adhesions, and morphology, especially in 2-D culture conditions. The mechano-sensing of cell culture biomaterials by stem cells is typically regulated by integrin-mediated focal adhesion signaling. PVA hydrogels having a storage modulus (E') of 12–30 kPa were found to be efficient materials for ex vivo hHSPC expansion. We also developed PVA hydrogels grafted with oligopeptides derived from vitronectin (PVA-oligoVN hydrogels), which can be produced to have a variety of stiffnesses, for the xeno-free culture of hPSCs. The ideal stiffness of the PVA-oligoVN hydrogels for hPSC culture was found to be 25.3 kPa. A high concentration of oligoVN (500–1500 μg/mL) should be used to prepare the PVA-oligoVN hydrogels to achieve a sufficient

A. Higuchi (✉) · H.-F. Li
Department of Chemical and Materials Engineering, National Central University, 300 Jhongli, Taoyuan 32001, Taiwan
e-mail: higuchi@ncu.edu.tw

A. Higuchi
Nano Medical Engineering Laboratory, RIKEN, 2-1, Hirosawa, Wako, Saitama 351-0198, Japan

A. Higuchi · A. A. Alarfaj · M. A. Munusamy
Department of Botany and Microbiology, College of Science, King Saud University, 2455, Riyadh 11451, Saudi Arabia

S. Suresh Kumar
Department of Medical Microbiology and Parasitology, Universities Putra Malaysia, 43400 Serdang, Selangor, Malaysia

V. K. Thakur and M. K. Thakur (eds.), *Hydrogels*, Gels Horizons: From Science to Smart Materials, https://doi.org/10.1007/978-981-10-6077-9_14

oligoVN surface density to maintain hPSC pluripotency. Optimized stiffness (physical cues) and cell-binding moiety surface density (biological cues) are the key factors for designing hydrogel-based cell culture materials for supporting hPSC pluripotency in xeno-free culture conditions.

Keywords Hydrogels · Stem cells · Elasticity · Differentiation
Pluripotency · Nanosegment · Oligopeptide · Poly(vinyl alcohol)

1 Introduction

Humans can sustain injuries and lose organs and tissue because of diseases, birth defects, and accidents. Stem cells, such as human mesenchymal stem cells (hMSCs), human induced pluripotent stem cells (hiPSCs), and human embryonic stem cells (hESCs), are valuable cell sources in translational medicine.

The stem cell characteristics and differentiation fates are governed by the microenvironment of the stem cells. Therefore, it is a reasonable strategy to develop biomaterials, such as hydrogels, nanofibers, and microcarriers, to mimic the stem cell microenvironments; these materials could be used to maintain pluripotency for stem cell expansion or to promote stem cell differentiation into specific desired lineages in vitro.

Several biological cues, such as soluble factors (i.e., growth factors, hormones, inhibitors, and agonists) and extracellular matrices (ECMs), strongly affect stem cell pluripotency and differentiation fate (Fig. 1) (Higuchi et al. 2011b, 2012, 2013). However, recent research suggests that the physical cues provided by cell culture biomaterials (i.e., topography, elasticity, and cyclic biomaterial stretching) also affect stem cell differentiation (Fig. 1) (Higuchi et al. 2015b; Engler et al. 2006; Murphy et al. 2014; Wen et al. 2014).

It was reported that human bone marrow stem cell (hBMSC) differentiation into specific tissue lineages could be efficiently induced when the hBMSCs were cultivated on collagen-coated polyacrylamide hydrogels with a stiffness that was similar to the tissue of interest (Wen et al. 2014; Murphy et al. 2014; Engler et al. 2006). Softer hydrogels, with an elasticity close to that of the brain (e.g., 0.3 kPa) promoted cells to display neuronal morphologies and early markers of neural cells (P-NFH, β-III tubulin), whereas stiffer hydrogels (10 kPa) tended to promote the differentiation of cells into muscle-guided hBMSCs expressing myogenic markers (MyoD) (Engler et al. 2006). Furthermore, even stiffer hydrogels (approximately 35 kPa), which were similar in elasticity to collagenous bones, tended to induce hBMSCs into cells expressing early markers (Runx2) of osteoblasts (Engler et al. 2006).

Several other investigators also noted that cell culture biomaterial (hydrogel) elasticity is an important factor in the differentiation of hMSCs in two-dimensional (2-D) culture conditions (Lanniel et al. 2011; Park et al. 2011a, b), as the elasticity of the cell culture biomaterial guides hMSC differentiation fate. It should be noted

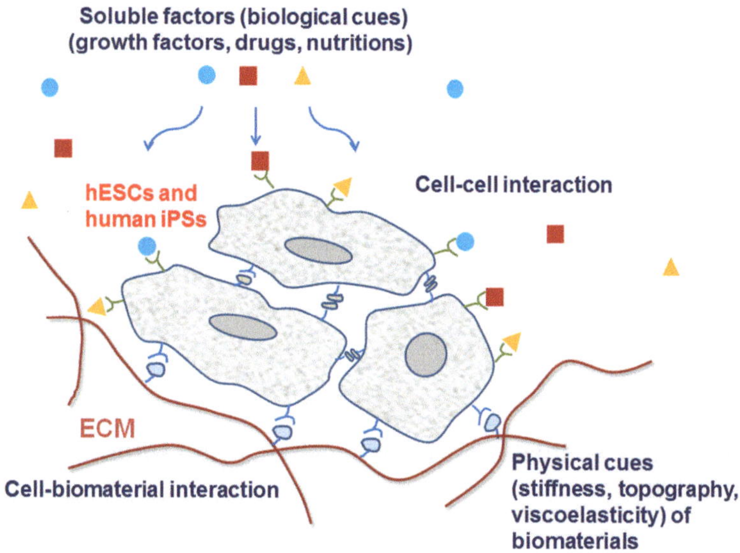

Fig. 1 Microenvironment of human stem cells is regulated by the following factors: (1) physical factors, e.g., stiffness, of the tissues or cell culture matrices, (2) interactions among molecules biologically active with cells, e.g., ECMs and biomaterials; (3) cell–cell interactions; and (4) several soluble factors, such as growth factors, inhibitors, nutrients, and other bioactive molecules (Higuchi et al. 2011a, b). Copyright 2011. Adapted with permission from the American Chemical Society

that some studies reported conflicting results and have offered different, but intriguing, ideas for explaining the effect of cell culture biomaterial elasticity on hMSC differentiation fate (Higuchi et al. 2013; Trappmann et al. 2012; Huebsch et al. 2010). While Engler's landmark report (Engler et al. 2006) showed that the elasticity of cell culture hydrogels guides the differentiation fate of hBMSCs, this was true only under limited conditions, in which hBMSCs were cultured on collagen-coated polyacrylamide hydrogels in 2-D culture conditions. Collagen type I can diffuse into soft polyacrylamide hydrogels, whereas it cannot diffuse into stiff hydrogels deeply. Hydrogels with different elasticities yield different ECM anchoring densities on the hydrogels; subsequently, different mechanical feedback is provided from the ECM to the stem cells.

The mechanical feedback seems to allow stem cells to either differentiate into specific cell lineages or maintain their stemness. The differentiation fate and pluripotency of ESCs has also been shown to be dictated by the cell culture microenvironment (Higuchi et al. 2011b). Even without leukemia inhibitory factor (LIF) in the culture medium, mouse ESCs (mESCs) maintained their pluripotency when they were cultivated on soft hydrogels (0.6 kPa) with an elasticity similar to the intrinsic elasticity of the mESCs; in contrast, the mESCs could not maintain their pluripotency when they were cultured on stiff tissue culture polystyrene plates (TCPS, 3 GPa) coated with collagen type I (Chowdhury et al. 2010).

Valuable research investigating the effects of cell culture biomaterial elasticity on the differentiation fate and pluripotency of stem cells has been conducted (Higuchi et al. 2015a, b; Engler et al. 2006; Murphy et al. 2014; Wen et al. 2014). However, little is known of how cell culture hydrogel elasticity affects the pluripotency and differentiation fate of (a) human hematopoietic stem and progenitor cells (hHSPCs) (Kumar et al. 2013), (b) human amniotic fluid stem cells (hAFSCs) (Wang et al. 2015), and (c) human pluripotent stem cells (hPSCs) (Higuchi et al. 2015a). In this chapter, we discuss (1) the culture of hHSPCs on hydrogels having optimal elasticity, (2) the maintenance of hAFSC pluripotency on hydrogels having optimal elasticity, and (3) the xeno-free culture of hPSCs on hydrogels having optimal elasticity.

2 hHSPCs Expansion on Hydrogels Having Optimal Elasticity

hHSPCs are multipotent cells that have the potential to generate all types of mature blood cells as well as the capacity for self-renewal (Mendez-Ferrer et al. 2010; Xie et al. 2009; Higuchi et al. 2009). Intravenous infusion of hHSPCs has been performed to cure patients having malignant diseases or hematological disorders after radiation and/or chemotherapy (Remberger et al. 2011; Copelan 2006; Doran et al. 2009). Transplantation of hHPSCs into patients has been performed via transplanting bone marrow, umbilical cord blood (UCB), and hHSPCs isolated from peripheral blood (Chua et al. 2007; Yamamoto et al. 2011).

In particular, UCB is considered a valuable source of hHPSCs because UCB transplantation carries a lower risk of graft-versus-host disease (GVHD) compared with that of other hHPSC sources, such as peripheral blood and bone marrow (Park et al. 2011a; Rocha et al. 2004; Kishore et al. 2011). In addition, donors do not have any risks for side effects when UCB is collected to obtain hHPSCs. However, side effects in donors are reported in some cases when hHSPCs are obtained from bone marrow or peripheral blood, and death occurs in approximately one out of every 4,000 bone marrow donors (Copelan 2006). The small volume (60–180 mL) that is generally obtained from a single donor of UCB limits the UCB transplantation treatment to pediatric patients (Higuchi et al. 2009). Approximately 1.7×10^7 $CD34^+$ cells/kg is expected to be used for transplanting UCB into a patient (Kumar et al. 2013). Therefore, the transplantation of UCB must be restricted to children weighing less than or equal to 20 kg (Fujimoto et al. 2007; Franke et al. 2007). The disadvantage of UCB transplantation is the low dose of hHSPCs, which leads to slower engraftment and higher chances of engraftment failure than those resulting from bone marrow transplantation. Therefore, several efforts have been directed toward the expansion of hHSPCs ex vivo not only to improve engraftment time and to reduce graft failure but also to develop this therapy for adult patients (Fujimoto et al. 2007; Franke et al. 2007).

In this chapter, we discuss on the ex vivo expansion of hHSPCs from UCB through culture on hydrogels grafted with various nanosegments. Several researchers have already developed cell culture materials for the ex vivo expansion of hHSPCs by mimicking the microenvironment of bone marrow (Mendez-Ferrer et al. 2010; Chen et al. 2012; Kumar et al. 2013; Calvi et al. 2003; Di Maggio et al. 2011).

One idea is to develop surface-modified electrospun nanofibers for the expansion of hHSPCs. Chua et al. investigated hHSPC expansion on aminated, hydroxylated, carboxylated, or nonmodified films and nanofibers (Chua et al. 2006). The aminated films and nanofibers led to the highest hHSPC expansion (180- and 200-fold, respectively). In addition, while the aminated films led to less hHSPC expansion than the aminated nanofibers, it should be noted that the difference did not appear to be significant (Chua et al. 2006).

Bone marrow microenvironments contain several ECM components that are crucial for maintaining the stemness of hHSPCs. ECMs mediate the bonds between cell adhesion molecules (CAMs) and hHSPCs, which are extensively important for the signal transduction involved in hematopoiesis and other interactions of hHSPCs in the bone marrow (Higuchi et al. 2009; Feng et al. 2006). Feng et al. reported the ex vivo expansion of hHSPCs on polyethylene terephthalate (PET) mesh with immobilized fibronectin (FN) and collagen (COL) (Feng et al. 2006). The hHSPCs cultivated on the PET mesh with immobilized FN expanded at the highest rate. FN is reported to be the preferred ECM component compared with other ECM components for the ex vivo expansion of hHSPCs (Feng et al. 2006).

CS1 (connecting segment-1, EILDVPST) and arginine-glycine-aspartic acid (RGD) motifs are known to be valuable adhesion domains in fibronectin that bind to surface receptors of hematopoietic progenitors. hHSPCs derived from UCB were expanded on PET films grafted with CS1 and RGD (GRGDSPC) oligopeptides, as reported by Jiang et al. (2006). The hHPSCs on the CS1 peptide-grafted PET films expanded the most, and hHPSCs (600 $CD34^+$ cells) expanded ex vivo for 10 days on the CS1 peptide-grafted PET films were successively engrafted into mice. However, the hHSPCs cultured on the PET film grafted with the RGD peptide expanded less efficiently. These results suggested that having optimal adhesion peptides (e.g., CS-1) on cell culture biomaterials could extensively influence the expansion of hHSPCs isolated from UCB (Jiang et al. 2006).

The effect of nanosegment species (e.g., RGDS, RGES, CS1, FN, and polyamine) immobilized on polystyrene (PSt) plates on the ex vivo expansion of hHSPCs isolated from UCB has been systematically investigated by our group (Chen et al. 2012).

hHSPCs cultivated on a PSt surface with immobilized CS1 exhibited higher-fold expansion and more pluripotent colony-forming units (CFUs) (i.e., CFU-granulocytes, erythroid cells, macrophages, and megakaryocytes [CFU-GEMM]) than did hHSPCs cultured on a PSt surface with immobilized polyamine, RGD, and FN (Chua et al. 2006). However, the elasticity of the cell culture biomaterials used in this study was approximately 3 GPa, and this investigation did not study the effect of the biomaterial stiffness on the expansion of the cultured hHSPCs.

In this chapter, we discuss the physical effect of cell culture hydrogel elasticity on the stemness, proliferation, and fate of hHSPCs cultured on synthetic hydrogels consisting of poly(vinyl alcohol-co-itaconic acid) (PVA) grafted with CS1 and FN. These hydrogels were selected because the elasticity of the hydrogels could be easily controlled by crosslinking time (intensity) with glutaraldehyde. The stiffness of the PVA hydrogels could be changed using the same polymeric main chain with different degrees of crosslinking. CS1 and FN were grafted with the carboxylic acid group of PVA via NHS (carbodiimide reaction using N-hydroxysuccinimide) and EDC (*N*-(3-dimethylaminopropyl)-*N*′-ethylcarbodiimide hydrochloride) in an aqueous solution. The PVA hydrogels were transparent with or without immobilized CS1 and FN, which allowed observation of hHSPCs on the PVA hydrogels via optical microscopy, much as cells are observed on conventional TCPS dishes. The goal of this study was to discuss the optimal stiffness of PVA hydrogels with immobilized CS1 and FN for the ex vivo expansion of hHSPCs isolated from UCB (Kumar et al. 2013).

2.1 Characterization of PVA Hydrogels Immobilized CS1 and FN

PVA hydrogels with immobilized CS1 and FN and different elasticities were prepared by varying the crosslinking time in the applied glutaraldehyde solution. These gels were used to investigate the optimal hydrogel elasticity for the expansion of hHSPCs when they were cultured on hydrogels containing specific immobilized cell-binding domains or proteins (e.g., CS1 and FN) (Fig. 2) (Kumar et al. 2013). PVA hydrogels with a thickness of 1.4–1.6 μm under dry conditions were prepared in this study. The thickness of the PVA hydrogels in aqueous conditions was also measured (2.2–8.9 μm thick) and is summarized in Table 1. Thick PVA hydrogels (approximately 25 μm thick) were prepared to measure the E' (storage modulus) of the gels using a rheometer, each of which is also listed in Table 1 (Kumar et al. 2013). The stiffest PVA hydrogel (PVA-48) had an E'= 30.4 kPa, whereas the softest PVA hydrogel (PVA-0.5) had an E'= 3.7 kPa, nearly eightfold less than that of PVA-48. PVA-*X* (e.g., PVA-0.5) indicates a PVA hydrogel crosslinked for *X* hours (e.g., 0.5 h). The PVA hydrogels with immobilized CS1 (PVA-*X*-CS1) and FN (PVA-*X*-FN) were measured to have E' values identical to those of the corresponding PVA hydrogels (PVA-*X*), within the experimental errors. This was because the amount of immobilized CS1 and FN was small; therefore, the grafted CS1 and FN segments did not contribute to the storage moduli of the bulk PVA hydrogels.

We attempted to quantify the number of immobilized nanosegments (i.e., CS1 and FN) on the PVA hydrogels with immobilized CS1 and FN and different

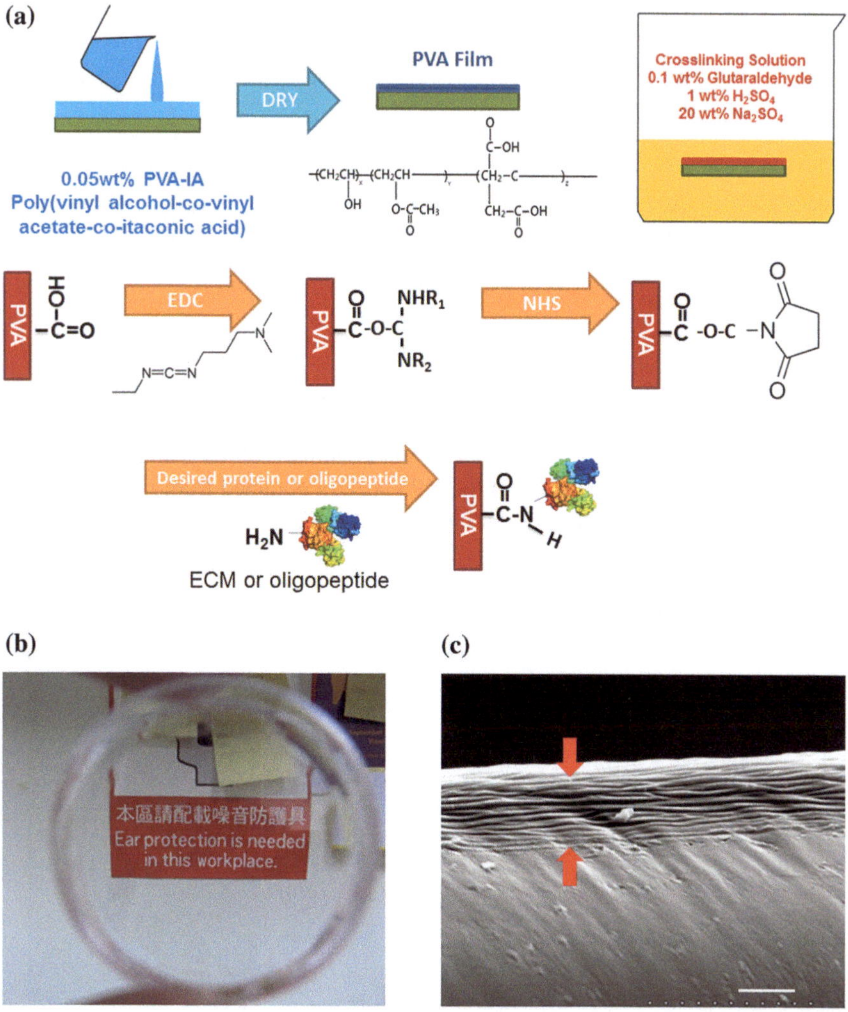

Fig. 2 Preparation of PVA hydrogels with immobilized CS1 and FN. **a** Reaction scheme of PVA hydrogels with immobilized CS1 and FN. **b** Macroscopic image of PVA-6-CS1 hydrogels. **c** Cross-sectional SEM image of PVA-6-CS1 hydrogels. The bar indicates 10 μm (Kumar et al. 2013). Copyright 2013. Adapted with permission from Elsevier Ltd.

elasticities. However, we failed to measure the absolute quantity of immobilized CS1 and FN on the PVA hydrogels using colorimetric (e.g., micro-BCA), chemical reaction, and titration methods in this study. Furthermore, it is known that the enzyme-linked immunosorbent assay (ELISA) cannot measure the absolute quantity of nanosegments immobilized on the biomaterials. Therefore, XPS analysis was performed to quantify the CS1 and FN nanosegments present on the PVA

Table 1 Characteristics of PVA hydrogels (Kumar et al. 2013). Copyright 2013. Adapted with permission from Elsevier Ltd.

Dishes	Thickness on dry condition (μm)	Thickness on wet condition (μm)	*E*′ (kPa)	*H* (%)
PVA-0.5	1.4–1.6	8.9	3.7	83
PVA-1	1.4–1.6	5.1	10.3	71
PVA-6	1.4–1.6	3.6	12.2	59
PVA-24	1.4–1.6	2.3	25.3	36
PVA-48	1.4–1.6	2.2	30.4	31
TCPS	2000	2000	1,200.0	0.2

hydrogels. The C1 s peaks of the XPS spectra obtained from PVA hydrogels with and without immobilized CS1 or FN are shown in Fig. 3 (Kumar et al. 2013).

C–H and C–C bonding (285 eV), O–C=O bonding (289 eV), and C–N bonding (286 eV) were clearly identified in the XPS spectra obtained from the PVA-*X*-CS1 and PVA-*X*-FN hydrogels (excluding PVA-0.5-CS1), whereas C–H and C–C

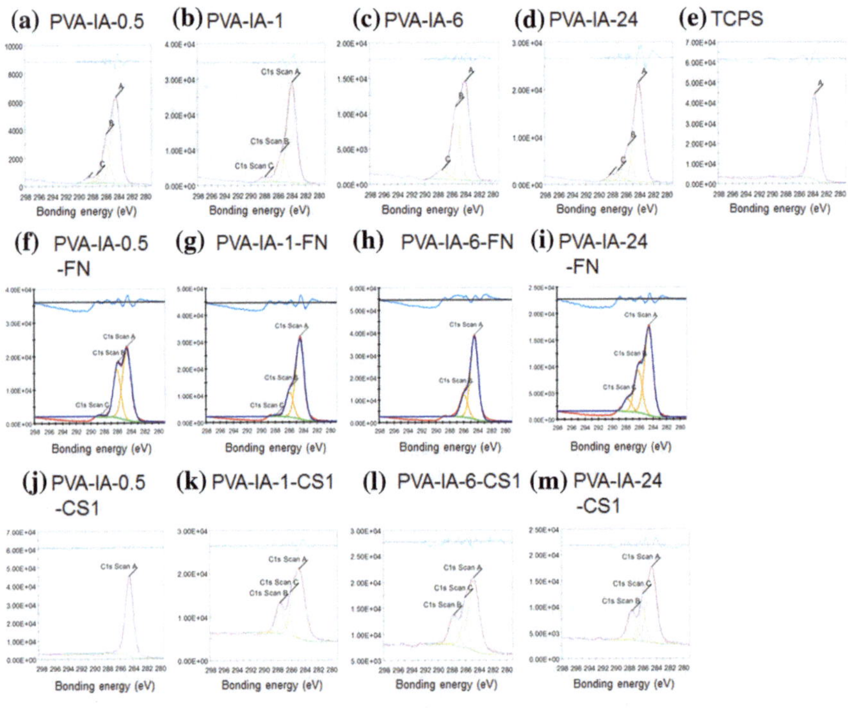

Fig. 3 XPS analysis of PVA hydrogels with immobilized CS1 and FN. C1 s spectra of *a*-*d* nonmodified PVA-*X* hydrogels, *f*-*i* PVA-*X*-FN hydrogels, *j*-*m* PVA-*X*-CS1 hydrogels, and **e** TCPS plates, where *X* is the reaction time *h* (Kumar et al. 2013). Copyright 2013. Adapted with permission from Elsevier Ltd.

bonding (285.0 eV) were mainly detected in the XPS spectra from TCPS and nonmodified PVA-*X* hydrogels (Fig. 3) (Kumar et al. 2013). These results indicated that CS1 and FN are securely bound to the PVA hydrogels, except for PVA-0.5-CS1.

The N1s peaks of the XPS spectra obtained from PVA-*X*-CS1, PVA-*X*-FN, and nonmodified PVA hydrogels were measured and are shown in Fig. 4 (Kumar et al. 2013). An N1s peak at 399 eV was broadly detected in the PVA-*X*-CS1 and PVA-*X*-FN hydrogel spectra, whereas only a trace N1s peak was observed in the non-modified PVA hydrogel spectra. This is because PVA hydrogels do not contain nitrogen molecules, and the nitrogen atom can be attributed to the CS1 oligopeptides or FN protein. The atomic ratios of N/C of the PVA-*X*-CS1, PVA-*X*-FN, and PVA hydrogels after having reacted with EDC/NHS (PVA-EDC) were analyzed using the XPS spectra and are summarized in Fig. 5 (Kumar et al. 2013). A higher density of CS1 than FN was immobilized on the PVA hydrogels when the same crosslinking time (*X*) was used ($p < 0.01$), except for PVA-0.5-CS1 and PVA-0.5-FN. Approximately 2.5- to 3-fold more CS1 than FN was immobilized on the PVA-24-CS1 and PVA-48-CS1 hydrogels. Furthermore, the atomic ratios of N/C on the PVA-24-FN and PVA-48-FN gels were higher than those of the

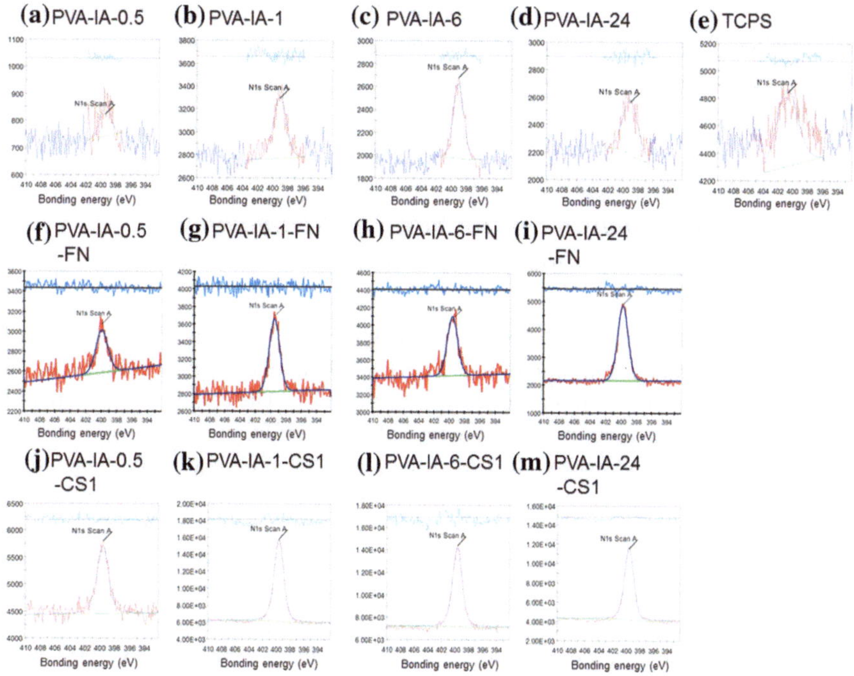

Fig. 4 XPS analysis of PVA hydrogels with immobilized CS1 and FN. N1s spectra of **a-d** nonmodified PVA-*X* hydrogels, **f-i** PVA-*X*-FN hydrogels, **j-m** PVA-*X*-CS1 hydrogels, and **e** TCPS plates, where *X* is the reaction time **h** (Kumar et al. 2013). Copyright 2013. Adapted with permission from Elsevier Ltd.

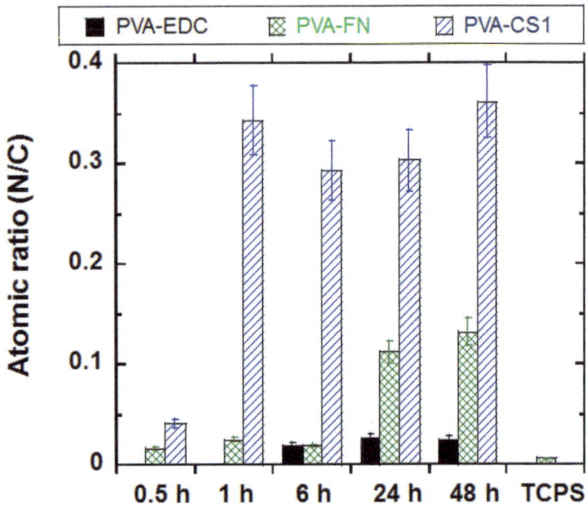

Fig. 5 XPS analysis of PVA hydrogels with immobilized CS1 and FN. The atomic ratios of nitrogen to carbon (N/C) in PVA hydrogels activated by EDC/NHS (PVA-*X*-EDC, black column, left), PVA-*X*-FN hydrogels (green column, middle), and PVA-*X*-CS1 hydrogels (blue column, right), where *X* is the reaction time (h) (Kumar et al. 2013). Copyright 2013. Adapted with permission from Elsevier Ltd.

PVA-0.5-FN, PVA-1-FN, and PVA-6-FN gels ($p < 0.01$). These results indicated that grafting FN onto PVA hydrogels is more unmanageable than grafting CS1 onto PVA hydrogels. This is because the FN molecule is larger than CS1, which leads to a less efficient reaction. The amounts of CS1 immobilized on the surface of PVA-1-CS1, PVA-6-CS1, PVA-24-CS1, and PVA-48-CS1 hydrogels appeared to be almost the same, corresponding to 2.5–3 times the surface density of FN immobilized on the PVA-24-FN and PVA-48-FN hydrogels (Kumar et al. 2013).

2.2 *Ex Vivo Expansion of hHSPCs in Nonmodified and Surface-Modified PVA Hydrogels*

hHSPCs purified from UCB were expanded for ten days on TCPS and PVA hydrogels with and without immobilized CS1 and FN, and the morphologies of the cultivated cells are shown in Fig. 6 (Kumar et al. 2013). The cell numbers increased after ten days of cultivation. The numbers of two hHSPC populations were analyzed using flow cytometry before and after hHSPC expansion, including $CD34^{+}CD45^{+}CD^{SS}$ cells that were analyzed using the method described by ISHAGE guidelines (Keeney et al. 1998; Gori et al. 2012; Higuchi et al. 2010), and $CD34^{+}CD38^{-}$ cells (Mortera-Blanco et al. 2011; Gori et al. 2012; Flores-Guzman et al. 2013; Rodriguez-Pardo and Vernot 2013; Salati et al. 2013; Roy et al. 2012; Holmes et al. 2012). The cells measured using the ISHAGE guidelines were determined to be hematopoietic progenitor cells, whereas the $CD34^{+}CD38^{-}$ cells were considered hematopoietic stem cells (Keeney et al. 1998). The flow cytometry scattergrams of the $CD34^{+}$ cells in UCB, which correspond to $CD34^{+}CD38^{-}$ cells (a), and the flow cytometry scattergrams of the UCB cells analyzed using ISHAGE guidelines (b and c) are shown in Fig. 7A (Kumar et al. 2013). The fold expansion

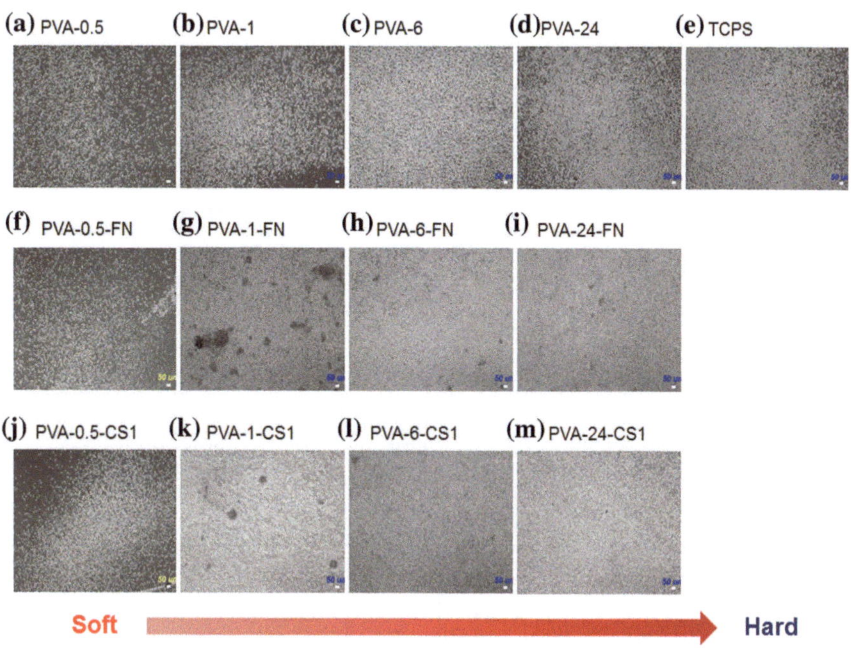

Fig. 6 Morphologies of cells cultivated on **a** PVA-0.5 hydrogels, **b** PVA-1 hydrogels, **c** PVA-6 hydrogels, **d** PVA-24 hydrogels, **e** TCPS plates, **f** PVA-0.5-FN hydrogels, **g** PVA-1-FN hydrogels, **h** PVA-6-FN hydrogels, **i** PVA-24-FN hydrogels, **j** PVA-0.5-CS1 hydrogels, **k** PVA-1-CS1 hydrogels, **l** PVA-6-CS1 hydrogels, and **m** PVA-24-CS1 hydrogels after 10 days of cultivation. The bar indicates 50 μm (Kumar et al. 2013). Copyright 2013. Adapted with permission from Elsevier Ltd.

of hHSPCs was analyzed using the protocol shown in Fig. 7a(a–c); these cells were cultured on TCPS dishes and nonmodified PVA hydrogels with different elasticities for ten days (Fig. 7b) (Kumar et al. 2013). The fold expansions of the analyzed $CD34^{+}CD38^{-}$ hHSPCs cells and those analyzed using the ISHAGE guidelines (Keeney et al. 1998) were observed to be the same, within experimental error. Therefore, $CD34^{+}CD38^{-}$ cell numbers were used in subsequent experiments to analyze the fold expansion of hHSPCs after $CD34^{+}$ cells isolated from UCB were cultured on TCPS and PVA hydrogels with and without immobilized CS1 and FN; the results are shown in Fig. 7b, c (Kumar et al. 2013). The expansion of hHSPCs cultured on nonmodified PVA hydrogels was found to increase with increasing PVA hydrogel elasticity (Fig. 7b). Although the E' of the PVA-48 hydrogels was 30.4 kPa, the fold expansion of hHSPCs on PVA-48 hydrogels was found to be almost equal to that of hHSPCs cultured on TCPS, which has the stiffest Young's modulus, i.e., 3 GPa.

The viability of hHSPCs cultured on PVA hydrogels with or without immobilized CS1 and FN was 60–75%, as determined by flow cytometry using a dye (7-AAD) exclusion method. There were no significant differences in the viability of

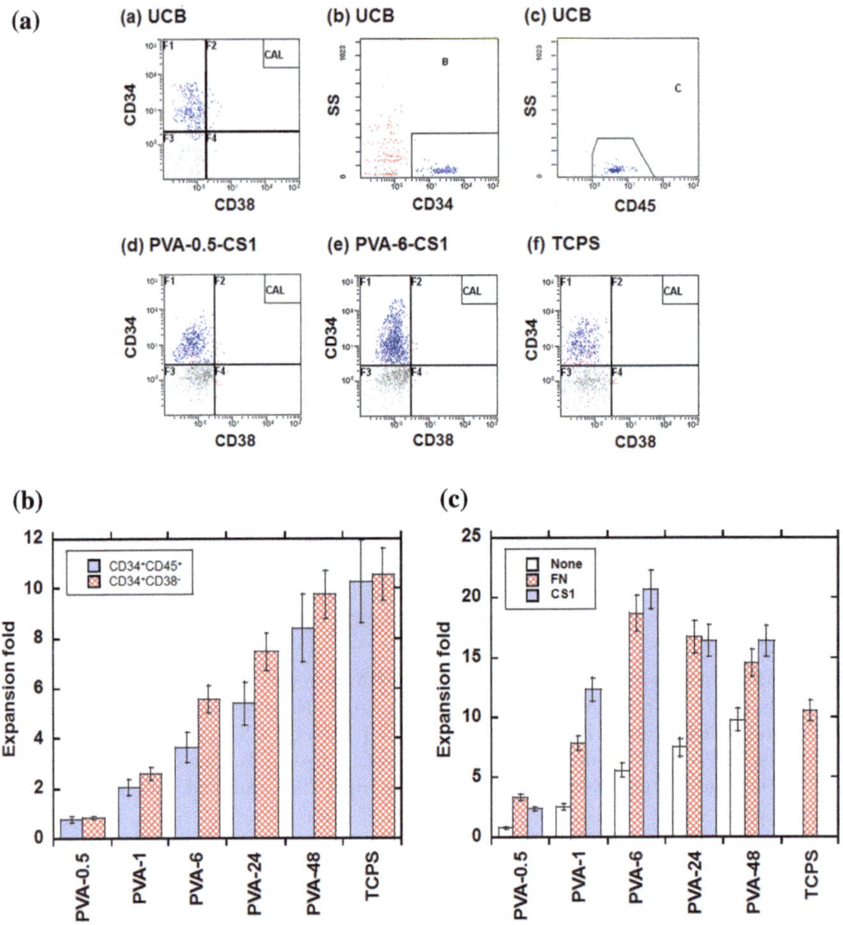

Fig. 7 hHSPC culture on PVA hydrogels with immobilized CS1 and FN. **a** (*a*, *d*, *e*, *f*) CD38 and CD34 expression analyzed by flow cytometry, (*b*) CD34 and side-scattering (SS) expression, and (*c*) CD45 and SS expression, obtained from hHSPCs prior to expansion (*a*, *b*, *c*) and after ten days of cultivation on (*d*) PVA-0.5-CS1 hydrogels, (*e*) PVA-6-CS1 hydrogels, and (*f*) TCPS plates. **b** Expansion (fold) of hHSPCs analyzed by $CD34^{+}CD45^{+}$ cells (blue column, left) and by $CD34^{+}CD38^{-}$ cells (red column, right) after the expansion of hHSPCs for ten days on several PVA hydrogels. **c** Expansion (fold) of hHSPCs ($CD34^{+}CD38^{-}$ cells) after the expansion of hHSPCs for ten days on PVA hydrogels (white column, left), PVA-FN hydrogels (red column, middle), PVA-CS1 hydrogels (blue column, right), and TCPS plates (Kumar et al. 2013). Copyright 2013. Adapted with permission from Elsevier Ltd.

hHSPCs cultured on nonmodified and surface-modified PVA hydrogels ($p > 0.05$). Only the number of viable hHSPCs was evaluated in this study.

The expansion of hHSPCs cultivated on PVA-6-CS1, PVA-FN-6, PVA-24-CS1, PVA-FN24, PVA-48-CS1, and PVA-48 hydrogels was higher than the expansion of hHSPCs cultured on TCPS ($p < 0.05$). In particular, although there was no statistically significant difference, the expansion of hHSPCs cultured on PVA-*X*-CS1 hydrogels ($X = 6$–48) was found to be higher than that of hHSPCs cultured on PVA-*X*-FN hydrogels that had been prepared with the same crosslinking times ($p > 0.05$). PVA-6-CS1 and PVA-6-FN hydrogels, with an E' of 12.2 kPa, were selected as the optimal hydrogels for high-expansion hHSPC cultures in this study.

From Engler's hypothesis, it is generally believed that the optimal method of differentiating stem cells into specific lineages in vitro is to culture the stem cells on materials having the same elasticity as the tissue of interest (Engler et al. 2006). If we follow Engler's hypothesis, the optimal elasticity of hydrogels cultured for hHSPC expansion should coincide with the elasticity of bone marrow, which was reported to be 150 Pa (Higuchi et al. 2013). However, our present results indicate that using softer PVA hydrogels grafted with CS1 and FN (i.e., PVA-1-CS1 and PVA-1-FN, having E' values of 10.3 kPa, and PVA-0.5-CS1 and PVA-0.5-FN, having E' values of 3.7 kPa) for hHSPC expansion is less effective than using moderately stiff PVA hydrogels with immobilized CS1 and FN (i.e., PVA-6-CS1 and PVA-6-FN, having an E' of 12.2 kPa). In summary, we noted that Engler's hypothesis is not applicable for hHSPC expansion on PVA hydrogels with and without immobilized CS1 and FN.

XPS analysis indicated that the surface densities of CS1 on PVA-1-CS1, PVA-6-CS1, PVA-24-CS1, and PVA-48-CS1 were nearly equal. It should be mentioned that the surface density of FN on PVA-6-FN was much less than that on PVA-24-FN and PVA-48-FN. This result suggests that the key factor rendering the PVA-6-CS1 and PVA-6-FN hydrogels optimal for hHSPC expansion was the elasticity of these hydrogels rather than the surface densities of FN and CS1 (Kumar et al. 2013).

2.3 *Analysis of hHSPCs Expanded on Nonmodified and Surface-Modified PVA Hydrogels via Colony-Forming Assays*

The quality of the hHSPCs was characterized through colony-forming unit (CFU) assays after the hHSPCs were expanded on TCPS and PVA hydrogels with and without immobilized CS1 and FN. The macroscopic [(a–e)] and microscopic [(f–j)] morphologies of the colonies cultured on the methylcellulose gels for 14 days using the same hHSPC seeding densities (450 cells/dish) are shown in Fig. 8 (Kumar et al. 2013). Colonies originating from the hHSPCs could be clearly observed in each dish. The number of each type of colony (i.e., CFU-GEMM, CFU-GM, CFU-M, CFU-G, BFU-E, and CFU-E) in each dish was analyzed for hHSPCs cultured on TCPS and PVA hydrogels with and without immobilized CS1 and FN (Fig. 9) (Kumar et al. 2013). It was found that the total numbers of colonies generated from hHSPCs cultured on each PVA hydrogel were higher than the total numbers of colonies generated from hHSPCs cultured on TCPS ($p < 0.05$).

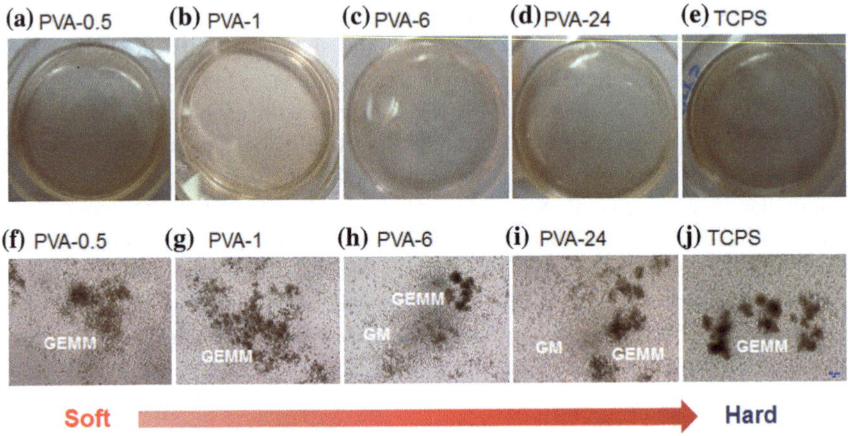

Fig. 8 Macroscopic and microscopic images of colonies on methylcellulose gels generated from hHSPCs after expansion for 14 days on PVA-0.5 hydrogels **a** and **f**, PVA-1 hydrogels **b** and **g**, PVA-6 hydrogels **c** and **h**, PVA-24 hydrogels **d** and **i**, and TCPS plates **e** and **j** at a seeding density of 450 hHSPCs/dish (Kumar et al. 2013). Copyright 2013. Adapted with permission from Elsevier Ltd.

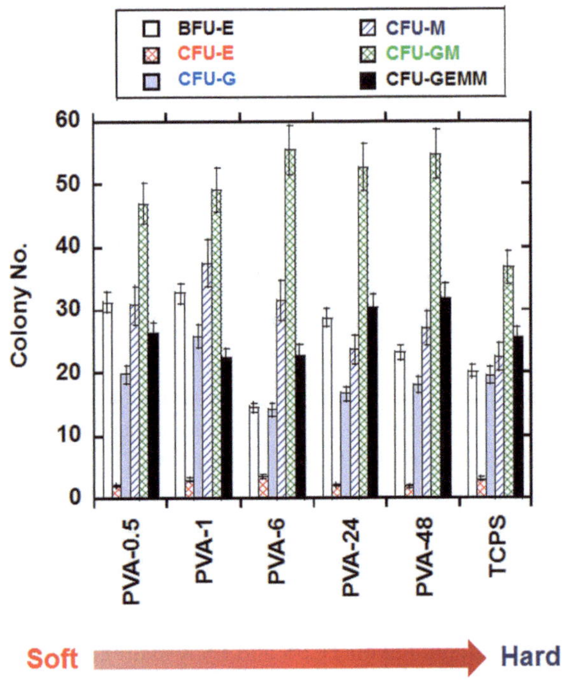

Fig. 9 CFU numbers (CFU-GEMM, CFU-GM, CFU-M, CFU-G, CFU-E, and BFU-E colonies from the right column to left column) originating from hHSPCs after expansion for 14 days on PVA hydrogels and TCPS plates at a seeding density of 450 hHSPCs/dish (Kumar et al. 2013). Copyright 2013. Adapted with permission from Elsevier Ltd.

CFU-GEMMs and CFU-GMs, the numbers of pluripotent colonies, are valuable indices for evaluating hHSPC expansion efficiency. Figure 10 represents the numbers of CFU-GEMMs and CFU-GMs produced by hHSPCs cultivated on TCPS and PVA hydrogels with and without immobilized CS1 and FN (Kumar et al. 2013). The numbers of both CFU-GEMMs and CFU-GMs generated from the hHSPCs cultured on PVA-6-FN, PVA-6-CS1, PVA-24-FN, PVA-24-CS1, PVA-48-FN, and PVA-48-CS1 hydrogels were higher than those cultivated on TCPS ($p < 0.05$), which was similar to the trend observed in the fold expansion of the hHSPCs. The number of CFU-GMs arising from hHSPCs cultivated on PVA-6-CS1 hydrogels was found to be the highest among all PVA-*X*-CS1 hydrogels evaluated in this study (<0.05), whereas the numbers of CFU-GEMMs and CFU-GMs produced from hHSPCs cultivated on PVA-6-FN hydrogels were slightly less than those produced from hHSPCs cultivated on PVA-24-FN and PVA-48-FN hydrogels ($p < 0.05$). These results can probably be attributed to the low surface density of FN on PVA-6-FN hydrogels compared with that on PVA-24-FN and PVA-48-FN hydrogels (Fig. 5), whereas the surface density of CS1 on PVA-6-CA1 hydrogels was almost the same as that on PVA-24-CS1 and PVA-48-CS1 hydrogels. We were not able to enhance the surface density of CS1 or FN on the soft PVA hydrogels (e.g., CS1 immobilized on PVA-0.5 hydrogels and FN immobilized on PVA-0.5, PVA-1, and PVA-6 hydrogels), even though we increased the concentration of CS1 and FN in the reaction solution as well as the reaction time (e.g., 48–60 h). It was found that the physical properties of appropriate cell culture materials (e.g., hydrogels) guided the expansion of hHSPCs and the formation of pluripotent colonies when the elasticity (E') ranges from 12 kPa (PVA-6-CS1 and PVA-6-FN) to 25 kPa (PVA-24-CS1 and PVA-24-FN) and when

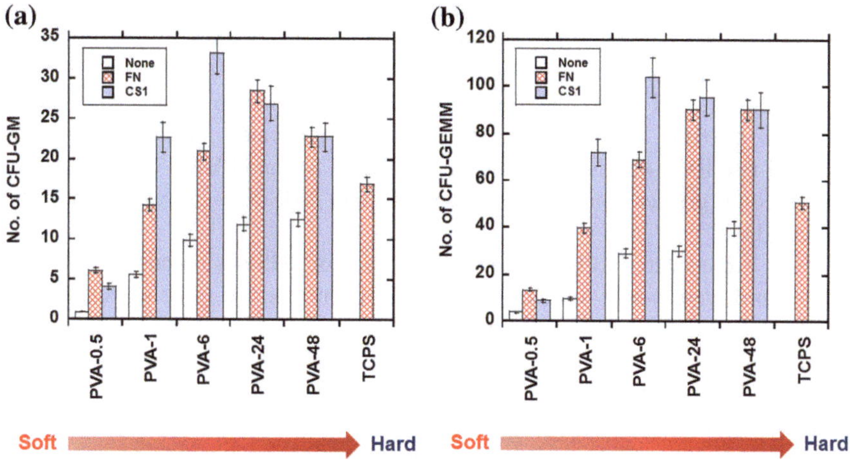

Fig. 10 Numbers of CFU-GEMM and CFU-GM colonies originating from hHSPCs after expansion for 14 days on PVA hydrogels (white column, left), PVA-FN hydrogels (red column, middle), PVA-CS1 hydrogels (blue column, right), and TCPS plates at a seeding density of 450 hHSPCs/dish (Kumar et al. 2013). Copyright 2013. Adapted with permission from Elsevier Ltd.

the cell culture materials have the optimal surface density of cell-binding molecules, such as CS1 and FN (Kumar et al. 2013).

2.4 hHSPC Culture on Hydrogels Having Optimal Elasticity

CS1 is one of the most valuable domains of FN and bind to surface receptors on hematopoietic progenitors in early stages (Kerst et al. 1993). The VLA-4 integrin receptor, $\alpha_4\beta_1$, is known to bind to CS1 and appears on hHSPCs in early stages (Kerst et al. 1993). The signal transduction pathways generated by the binding of FN and CS1 to the VLA-4 integrin receptor on hHSPCs exert key effects on the expansion (Figs. 7B, C) and stemness maintenance of HSPCs (Fig. 10).

The total number of hHSPC colonies produced after hHSPC expansion is calculated from both the fold expansion of hHSPCs (Fig. 7B, C) and the number of colonies per the initial seeding density (Fig. 10, 450 HSPCs/dish). Total CFU expansion is defined as follows:

$$\text{Total CFU expansion} = (\text{hHSPC fold expansion}) \times (\text{Number of CFUs per 450 hHSPCs}) \quad (1)$$

If the total CFU expansion is higher than that of UCB without expansion, then the hHSPCs generated after expansion were more able to produce CFUs than the initial UCB hHSPCs before expansion, indicating the utility of the hHSPC expansion.

Total CFU expansion calculated via hHSPC expansion (Fig. 7) and the CFU assay (Fig. 10) of hHSPCs after expansion on PVA hydrogels with and without immobilized CS1 and FN are summarized in Fig. 11 (Kumar et al. 2013). The hHSPCs cultivated on the hydrogels, except the PVA-0.5, PVA-0.5-FN, PVA-0.5-CS1, and PVA-1 hydrogels, exhibited a higher expansion than that of UCB ($p < 0.01$). In particular, the hHSPCs cultivated on the PVA-6-FN, PVA-24-FN, PVA-48-FN, PVA-6-CS1, PVA-24-CS1, and PVA-48-CS1 hydrogels expressed a greater total CFU expansion than hHSPCs cultivated on TCPS ($p < 0.01$).

UCB is considered to be an important source of hHSPCs because much lower risk of graft-versus-host disease (GVHD) has been reported to be associated with UCB transplantation than with peripheral blood or bone marrow transplantation (Chua et al. 2006; Higuchi et al. 2009; Feng et al. 2006; Jiang et al. 2006). In addition, donors have no risk for side effects when UCB is collected for hHSPC transplantation, whereas several side effects for donors have been reported when hHSPCs are obtained from bone marrow or peripheral blood (Copelan 2006). A total of 1.7×10^7 $CD34^+$ cells/kg are reportedly used for UCB transplantation (Chua et al. 2006). As such, UCB transplantation is restricted to use in children weighing less than or equal to 20 kg (Fujimoto et al. 2007; Franke et al. 2007). The main disadvantage of UCB transplantation is the low cell number involved,

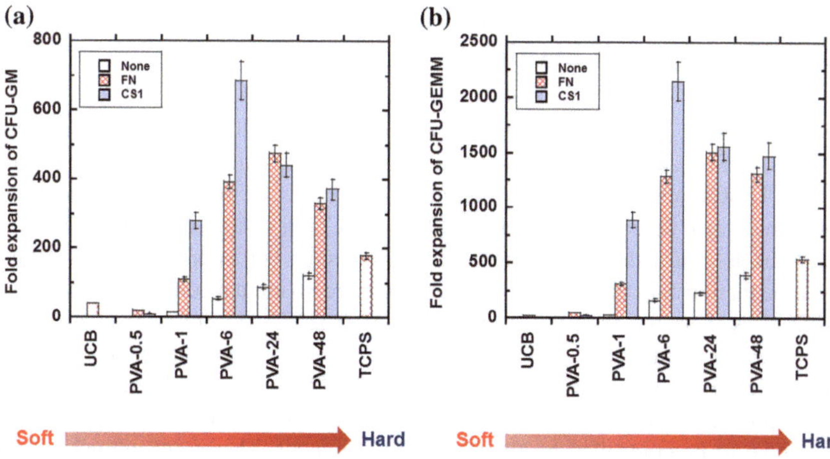

Fig. 11 CFU fold expansion (CFU-GEMM and CFU-GM colonies) of fresh hHSPCs isolated from UCB using the Ficoll–Paque method followed by MACS (UCB), or hHSPCs after expansion for 14 days on PVA-*X* hydrogels (white column, left), PVA-*X*-FN hydrogels (red column, middle), PVA-*X*-CS1 hydrogels (blue column, left), and TCPS plates at a seeding density of 450 hHSPCs/dish, where *X* is the reaction time (h) (Kumar et al. 2013). Copyright 2013. Adapted with permission from Elsevier Ltd.

which leads to prolonged engraftment times and higher percentages of engraftment failure compared with those of the bone marrow transplantation. Several ideas have been considered for hHSPC expansion to reduce the required engraftment time and decrease the engraftment failure rate, particularly for the development of UCB transplantation for adult patients (Fujimoto et al. 2007; Franke et al. 2007). Toward this goal, several investigators have studied the development of cell culture materials for ex vivo hHSPC expansion through mimicking the microenvironment of bone marrow (Chua et al. 2006; Feng et al. 2006; Jiang et al. 2006; Mendez-Ferrer et al. 2010; Di Maggio et al. 2011). However, the effect of and the optimal stiffness of cell culture materials have not yet been investigated in the context of hHSPC expansion. Therefore, the effect of the elasticity of PVA hydrogels with immobilized CS1 and FN on hHSPC expansion was investigated in this study. The elasticity of the bone marrow microenvironment is reported to be 150 kPa (Higuchi et al. 2013), whereas the optimal elasticity of PVA hydrogels for hHSPC expansion was determined to be 12–30 kPa in this study. Engler's idea (Engler et al. 2006) that stem cells tend to differentiate into specific tissue lineages most efficiently when the stem cells are cultivated on materials having an elasticity similar to the tissue of interest does not explain the ex vivo hHSPC expansion observed in this study. Furthermore, the hHSPCs cultivated on PVA hydrogels with immobilized CS1 and FN were found to expand more efficiently and produce higher numbers of multipotent CFU-GEMMs and CFU-GMs than hHSPCs cultivated on nonmodified PVA hydrogels (Kumar et al. 2013).

3 Pluripotent Maintenance of HAFSCs on Hydrogels Having Optimal Elasticity

MSCs are easily obtainable stem cells that can be used for clinical applications. hAFSCs, which are stem cells obtained from amniotic fluid, are pluripotent fetal cells. hAFSCs are capable of differentiating into cells representing the three embryonic germ layers (Zheng et al. 2008; De Coppi et al. 2007; Higuchi et al. 2011a, b). hAFSCs should be a more promising stem cell source for translational medicine than hESCs and hiPSCs because there are no ethical concerns about using hAFSCs, as there are for hESCs. Furthermore, hAFSCs do not raise concerns of xenogeneic contamination generated from the use of a feeder layer (e.g., mouse embryonic fibroblasts, MEFs), as do hESCs and hiPSCs. Stem cell properties, such as pluripotency maintenance and appropriate differentiation into desired lineages, are known to be controlled by their microenvironment. Therefore, mimicking the microenvironments of the stem cells using natural polymers, such as ECMs, should be a good strategy. This strategy may promote the production of the large numbers of stem cells and specifically differentiated cells that are in demand for translational medicine (Dellatore et al. 2008). The goal of this chapter is to discuss (a) which ECM-derived oligopeptides or ECMs immobilized onto the hydrogels are efficient for the maintenance of hAFSC pluripotency and (b) the optimal hydrogel elasticity for the maintenance of hAFSC pluripotency.

3.1 Culture on hAFSCs on Hydrogels Having Optimal Elasticity

hAFSCs were cultivated on (a) TCPS, (b) TCPS plates coated with CELLstart, (c) PVA hydrogels grafted with several oligopeptides derived from ECMs (PVA-oligoECM), (d) PVA hydrogels grafted with several ECMs (PVA-ECM) and nonmodified PVA in expansion medium. The ECMs used in this chapter are vitronectin (VN), fibronectin (FN), and collagen type I (COL). The oligoECMs used in this chapter are oligoVN (KGGPQVTRGDVFTMP), oligoFN (KGGAVTGRGDSPASS), COL-A (GTPGPQGIAGQRGVV), and COL-B ($(RADA)_4$GGDGEA). Figure 12 shows the morphology of hAFSCs cultivated for seven days on (a) PVA hydrogels and PVA-COL hydrogels having elasticities of $E' = 25$ kPa (24 h crosslinking time), 12 kPa (6 h crosslinking time), 11 kPa (4 h crosslinking time), and 10 kPa (2 h crosslinking time) and (b) TCPS and TCPS dishes coated with CELLstart with an elasticity of 3 GPa (Wang et al. 2015). hAFSCs were not able to be cultivated on PVA hydrogels with and without immobilized COL when the PVA hydrogels had an E' less than 12 kPa (PVA-2-COL, PVA-2 and PVA-4), whereas hAFSCs could be cultured and expanded on PVA hydrogels with and without immobilized COL when the E' of the PVA hydrogels was more than 12 kPa (i.e., PVA-24-COL, PVA-24, PVA-12-COL,

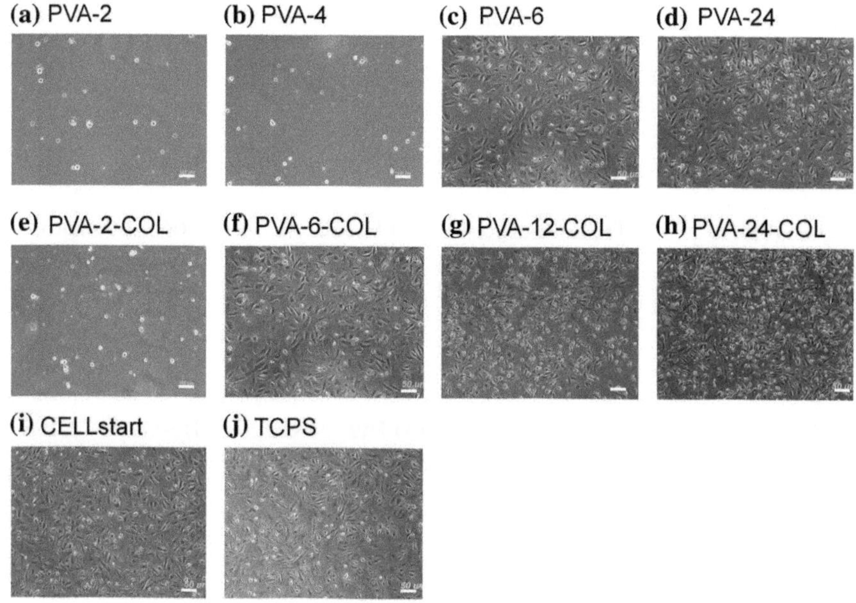

Fig. 12 Morphologies of hAFSCs cultivated on **a** PVA-2 hydrogels, **b** PVA-4 hydrogels, **c** PVA-6 hydrogels, **d** PVA-24 hydrogels, **e** PVA-2-COL hydrogels, **f** PVA-6-COL hydrogels, **g** PVA-12-COL hydrogels, **h** PVA-24-COL hydrogels, **i** TCPS plates coated with CELLstart, and **j** TCPS plates for seven days. The bar represents 50 μm (Wang et al. 2015). Copyright 2015. Adapted with permission from the Royal Society of Chemistry

PVA-6-COL, and PVA-6). The stiffest surface, TCPS plates coated with CELLstart (3 GPa elasticity) and TCPS, allowed the culture and expansion of hAFSCs. Therefore, relatively stiff PVA hydrogels with several immobilized ECM-derived oligopeptides and ECMs (PVA-24-oligoECM and PVA-24-ECM), relatively soft PVA hydrogels with several immobilized ECM-derived oligopeptides and ECMs (PVA-6-oligoECM and PVA-6-ECM), and the stiffest plates (TCPS dishes coated with CELLstart and TCPS) were chosen to study the doubling time, expression of differentiation genes, and expression of pluripotent genes or proteins of hAFSCs in this study.

Figure 13 depicts the morphology of hAFSCs cultivated for seven days on relatively stiff (PVA-24) and soft (PVA-6) PVA hydrogels with several immobilized ECM-derived oligopeptides (oligoVN, oligoFN, COL-A, and COL-B) and ECMs (VN, FN, and COL) in expansion medium (Wang et al. 2015). hAFSCs could expand and proliferate on any of these PVA hydrogels with and without immobilized ECM-derived oligopeptides and ECMs.

Figure 14 shows the doubling time of hAFSCs cultured on TCPS dishes coated with CELLstart, TCPS, and PVA hydrogels with and without immobilized ECM-derived oligopeptides and ECMs (Wang et al. 2015). The doubling time of the hAFSCs was found to be approximately 6 days when the hAFSCs were

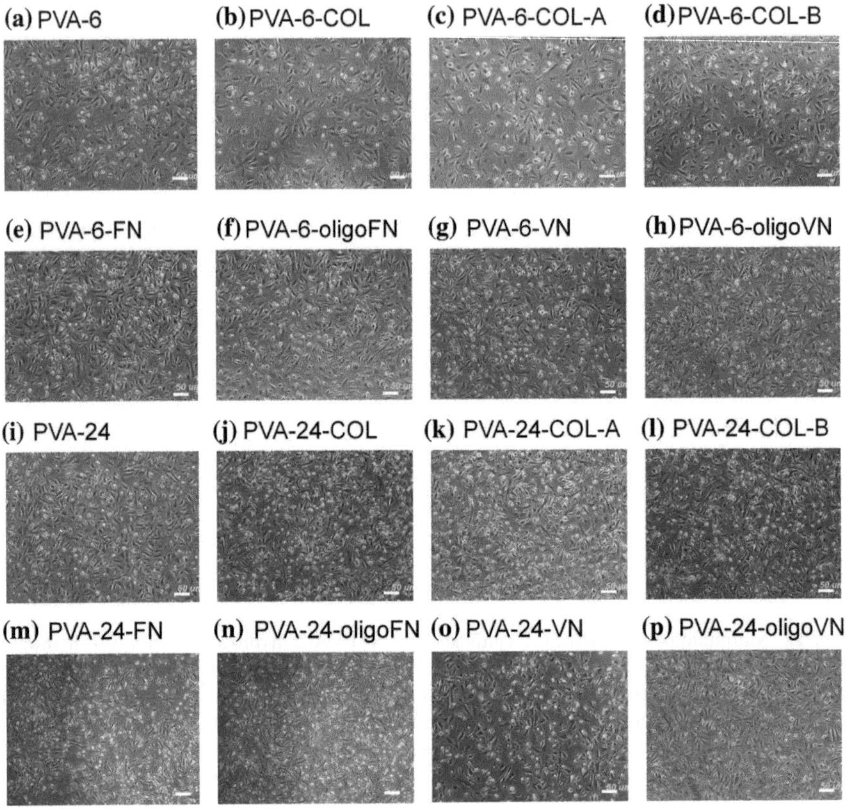

Fig. 13 Morphologies of hAFSCs cultivated on **a** PVA-6 hydrogels, **b** PVA-6-COL hydrogels, **c** PVA-6-COL-A hydrogels, **d** PVA-6-COL-B hydrogels, **e** PVA-6-FN hydrogels, **f** PVA-6-oligoFN hydrogels, **g** PVA-6-VN hydrogels, **h** PVA-6-oligoVN hydrogels, **i** PVA-24 hydrogels, **j** PVA-24-COL hydrogels, **k** PVA-24-COL-A hydrogels, **l** PVA-24-COL-B hydrogels, **m** PVA-24-FN hydrogels, **n** PVA-24-oligoFN hydrogels, **o** PVA-24-VN hydrogels, and **p** PVA-6-oligoVN hydrogels for seven days. The bar represents 50 μm (Wang et al. 2015). Copyright 2015. Adapted with permission from the Royal Society of Chemistry

cultivated on relatively stiff PVA-24 hydrogels with and without immobilized ECM-derived oligopeptides and ECMs or on the stiff surface of TCPS and TCPS dishes coated with CELLstart. This doubling time is shorter than that of cells cultivated on the relatively soft PVA-6 hydrogels with and without immobilized ECM-derived oligopeptides and ECMs in expansion medium. The effect of different ECM-derived oligopeptides and ECMs present on PVA hydrogels on the doubling time of the hAFSCs was not clear. The soft surface of the PVA-6 hydrogels led to slow hAFSC growth. Other researchers have also observed slower stem cell growth when the cells were cultivated on softer cell culture materials (Higuchi et al. 2013; Winer et al. 2009). For example, hMSCs cultivated on soft polyacrylamide hydrogels (PAAm, 250 Pa) stopped growing, as reported by Winter

Fig. 14 Doubling time of hAFSCs cultivated on stiff PVA-24 hydrogels (red column) and soft PVA-6 hydrogels (gray column) with immobilized ECMs (FN, VN, and COL) or ECM-derived oligopeptides (oligoFN, oligoVN, COL-A, and COL-B), on TCPS plates coated with CELLstart (blue bar), and on TCPS plates (blue bar) (Wang et al. 2015). Copyright 2015. Adapted with permission from the Royal Society of Chemistry

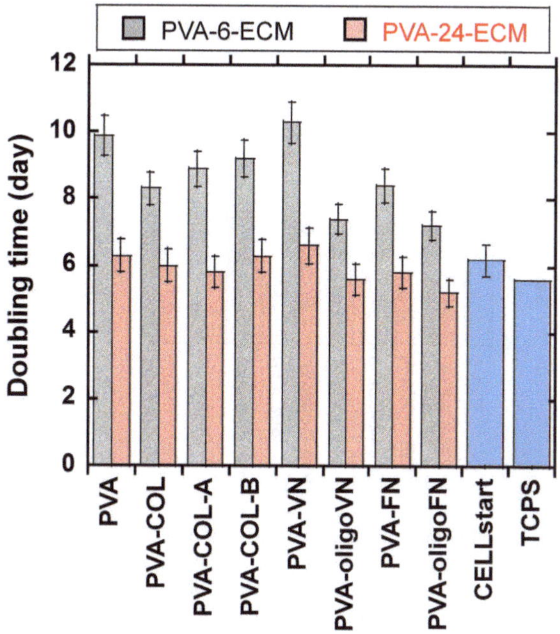

et al. (Higuchi et al. 2013; Winer et al. 2009). These resting hMSCs began to grow again when they were shifted onto stiff hydrogels. Therefore, hMSCs on soft PAAm hydrogels remain quiescent but have the potential to reinitiate growth or to differentiate upon transfer to stiffer materials (Wang et al. 2015; Higuchi et al. 2013; Winer et al. 2009).

3.2 The Gene Expression of Pluripotency and Differentiation in hAFSCs Cultured on PVA Hydrogels

We investigated whether slow or fast stem cell growth maintains their pluripotency or facilitates their differentiation. Therefore, hAFSC gene expression relating to differentiation and pluripotency was investigated when hAFSCs were cultivated on stiff and soft PVA hydrogels with and without immobilized ECM-derived oligopeptides and ECMs in expansion medium.

Figure 15 describes the expression of genes related to pluripotency, *Nanog, Sox2,* and *Oct4*, in hAFSCs cultivated on PVA-6-oligoECM (Fig. 15a), PVA-6-ECM (Fig. 15a), PVA-24-oligoECM (Fig. 15b), and PVA-24-ECM (Fig. 15b) hydrogels, as well as TCPS plates coated with CELLstart and TCPS plates (Fig. 15a, b) (Wang et al. 2015). The pluripotency genes of hAFSCs cultivated on the stiffest TCPS plates coated with and without CELLstart expressed the

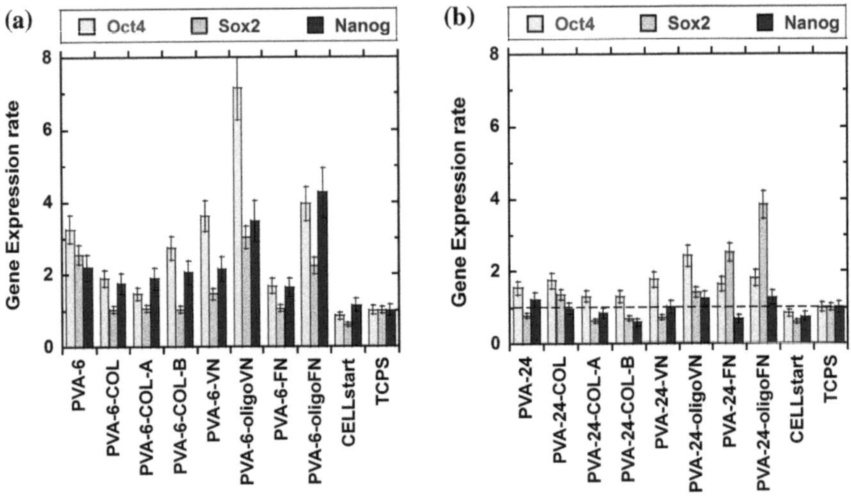

Fig. 15 Pluripotency gene expression (red bar, *Nanog*; black bar, *Sox2*; green bar, *Oct4*) in hAFSCs cultivated on soft PVA-6 hydrogels **a** and stiff PVA-24 hydrogels **b** with immobilized ECMs (FN, VN, and COL) or ECM-derived oligopeptides (oligoFN, oligoVN, COL-A, and COL-B) and on TCPS plates coated with CELLstart and TCPS plates after seven days of culture (Wang et al. 2015). Copyright 2015. Adapted with permission from the Royal Society of Chemistry

lowest levels of the genes investigated in this study. The hAFSCs cultivated on the relatively soft PVA-6-oligoECM and PVA-6-ECM hydrogels showed higher gene expression of *Nanog* and *Oct4* (pluripotency genes) compared with that of hAFSCs cultivated on the stiffest surfaces, TCPS dishes coated with CELLstart and TCPS plates ($p < 0.05$). Furthermore, hAFSCs cultivated on the soft PVA-6-oligoECM hydrogels and PVA-6-ECM hydrogels expressed higher levels of pluripotency genes (*Nanog* and *Oct4*) than hAFSCs on the stiff PVA-24-oligoECM and PVA-24-ECM hydrogels.

When the pluripotency gene expression of hAFSCs cultivated on PVA hydrogels immobilized with ECM-derived oligopeptides and ECMs was analyzed, we found no significant differences in the gene expression of hAFSCs on PVA hydrogels with immobilized collagen-derived oligopeptides (COL-A and COL-B) and collagen type 1. However, hAFSCs cultivated on PVA hydrogels with immobilized oligoFN and oligoVN were found to express higher levels of pluripotency genes (*Nanog, Sox2*, and *Oct4*) than hAFSCs cultivated on PVA hydrogels with immobilized fibronectin and vitronectin, respectively ($p < 0.05$). These results suggest that ECM-derived oligopeptides, such as oligo FN and oligoVN, efficiently act as bioactive nanosegments immobilized on PVA hydrogels to maintain hAFSC pluripotency.

We proposed $I_{pluripotency}$, the index of pluripotency gene expression, which is defined as the sum of the pluripotency gene expression levels of hAFSCs, calculated from the relative gene expression of hAFSCs cultivated on TCPS plates:

$$I_{pluripotency} = I_{Nanog} + I_{Sox2} + I_{Oct4} \tag{2}$$

where I_{Nanog}, I_{Sox2}, and I_{Oct4}, are the expression levels of *Nanog*, *Sox2*, and *Oct4*, respectively, which are calculated to be relative to the expression of the same genes of hAFSCs cultivated on TCPS plates. Figure 16 depicts $I_{pluripotency}$ of hAFSCs cultivated on PVA-24 and PVA-6 hydrogels with and without immobilized ECM-derived oligopeptides and ECMs (Wang et al. 2015). hAFSCs cultivated on the soft PVA-6 hydrogels with immobilized oligoFN (PVA-6-oligoFN) and oligoVN (PVA-6-oligoVN) expressed higher $I_{pluripotency}$ values than hAFSCs cultivated on the stiff PVA-24 hydrogels with and without immobilized ECM-derived oligopeptides or ECMs. Furthermore, hAFSCs grown on the stiff PVA-6-oligoFN and PVA-6-oligoVN showed higher $I_{pluripotency}$ values than those grown on PVA-6 hydrogels with immobilized ECM or ECM-derived oligopeptides other than oligoVN and oligoFN. It should be noted that hAFSCs cultivated on nonmodified PVA-6 hydrogels expressed relatively high $I_{pluripotency}$ values among the PVA hydrogels investigated in this study, although hAFSCs cultivated on PVA-6-oligoVN hydrogels showed higher the $I_{pluripotency}$ values compared to hAFSCs cultivated on nonmodified PVA-6 hydrogels ($p < 0.05$).

The expression of typical differentiation genes of the three germ layers, *Runx2* (mesoderm), *Sox17* (endoderm), and *Nestin* (ectoderm), in hAFSCs was evaluated

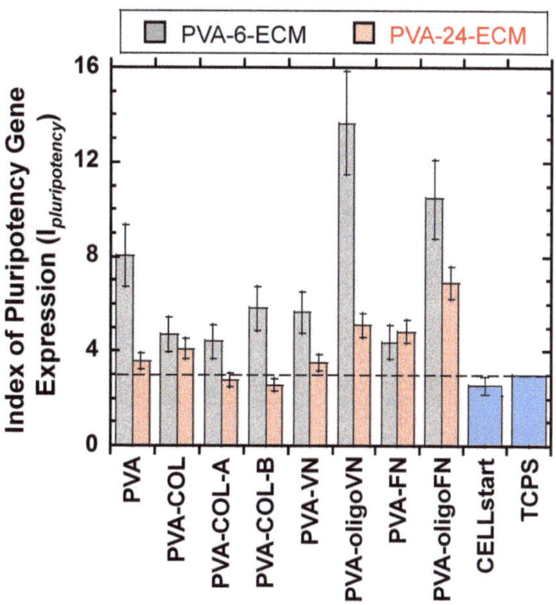

Fig. 16 $I_{pluripotency}$ (index of pluripotency gene expression) of hAFSCs cultivated on stiff PVA-24 hydrogels (red column) and soft PVA-6 hydrogels (black column) with immobilized ECMs (FN, VN, and COL) or ECM-derived oligopeptides (oligoFN, oligoVN, COL-A, and COL-B) and on TCPS plates coated with CELLstart and TCPS plates for seven days (Wang et al. 2015). Copyright 2015. Adapted with permission from the Royal Society of Chemistry

when hAFSCs were cultivated on CELLstart-coated TCPS plates, TCPS, PVA-24 hydrogels, and PVA-6 hydrogels in expansion medium; the results are shown in Fig. 17 (Wang et al. 2015). hAFSCs grown on soft PVA-6 hydrogels with or without ECM-derived oligopeptides and ECMs exhibited more than two-fold higher expression of *Nestin* than hAFSCs grown on TCPS plates. Soft PVA-6-FN, PVA-6-oligoVN, and PVA-6-COL hydrogels seem to be preferable for hAFSCs to spontaneously differentiate into early neural cells, indicated by the high levels of *Nestin* expressed by the hAFSCs.

hAFSCs grown on PVA-6-oligoVN showed high *Sox17* expression compared with that of hAFSCs cultured on PVA with or without immobilized ECM-derived oligopeptides and ECMs and compared with hAFSCs cultured on CELLstart-coated TCPS plates and TCPS. hAFSCs cultured on PVA-6-oligoFN and PVA-6-oligoVN showed higher expression of *Runx2* than did hAFSCs grown on PVA-6 with and without immobilized ECM-derived oligopeptides and ECMs. However, hAFSCs cultivated on the stiff PVA-24-oligoECM and PVA-24-ECM hydrogels exhibited higher expression of *Sox17* and *Runx2* than did hAFSCs cultured on soft PVA-6-oligoECM and PVA-6-ECM hydrogels or stiff TCPS dishes coated with CELLstart and TCPS plates. PVA-oligoFN, PVA-24-FN, PVA-24-oligoVN, and PVA-24-COL-B hydrogels, as well as PVA-6-oligoFN hydrogels, seem to be preferable cell culture materials for spontaneously inducing early stages of osteoblast differentiation, as indicated by the expression of *Runx2*. Furthermore, PVA-24-oligoFN, PVA-24-oligoVN, and PVA-24-COL hydrogels, as well as

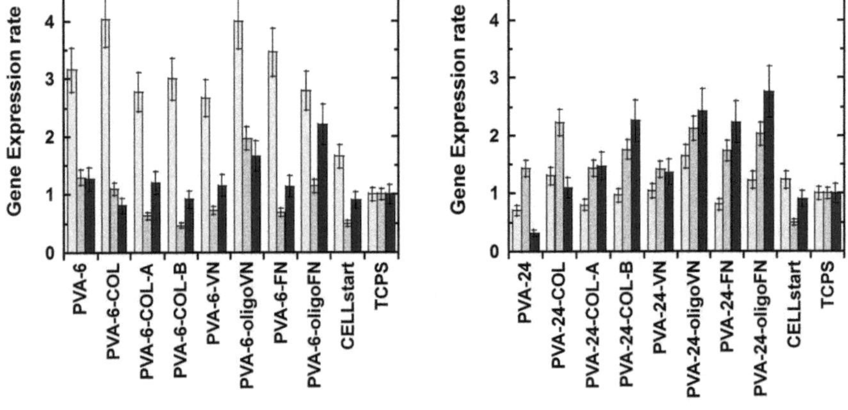

Fig. 17 Expression of differentiation genes (red bar, *Runx2*; black bar, *Sox17*; green bar, *Nestin*) in hAFSCs cultivated on soft PVA-6 hydrogels **a** and stiff PVA-24 hydrogels **b** with immobilized ECMs (FN, VN, and COL) or ECM-derived oligopeptides (oligoFN, oligoVN, COL-A, and COL-B) and on TCPS plates coated with CELLstart and TCPS plates for seven days (Wang et al. 2015). Copyright 2015. Adapted with permission from the Royal Society of Chemistry

PVA-6-oligoVN hydrogels, seem to be preferable cell culture materials for spontaneously inducing early stages of endoderm differentiation, as indicated by the expression of *Sox17*.

Here, we propose $I_{\text{differentiation}}$, the index of differentiation gene expression, which is defined as the sum of the differentiation gene expression levels of hAFSCs, calculated from the relative gene expression of hAFSCs cultivated on TCPS plates:

$$I_{\text{differentiation}} = I_{\text{Runx2}} + I_{\text{Sox17}} + I_{\text{Nestin}} \tag{3}$$

where I_{Runx2}, I_{Sox17}, and I_{Nestin} are the expression levels of *Runx2*, *Sox17,* and *Nestin,* respectively, which are calculated to be relative to the expression of the same genes of hAFSCs cultivated on TCPS plates. Figure 18 depicts $I_{\text{differentiation}}$ of hAFSCs cultivated on PVA-24 and PVA-6 hydrogels with and without immobilized ECM-derived oligopeptides and ECMs. hAFSCs grown on stiff and soft PVA-oligoECM and PVA-ECM hydrogels exhibited higher values of $I_{\text{differentiation}}$ than hAFSCs cultured on PVA-24 hydrogels, TCPS plates coated with CELLstart, and TCPS plates (Wang et al. 2015). The $I_{\text{differentiation}}$ of hAFSCs cultivated on PVA-24-oligoECM and PVA-24-ECM hydrogels was almost equal to the $I_{\text{differentiation}}$ of hAFSCs cultured on PVA-6-oligoECM and PVA-6-ECM hydrogels with each ECM-derived oligopeptide or ECM, within experimental error ($p > 0.05$). This is because of compensation via the expression of different genes. Stiff PVA-24-oligoECM and PVA-24-ECM hydrogels extensively induce the spontaneous differentiation of hAFSCs into early osteoblasts (expression of *Runx2*) and early endoderm cells (expression of *Sox17*); in contrast, soft PVA-6-oligoECM

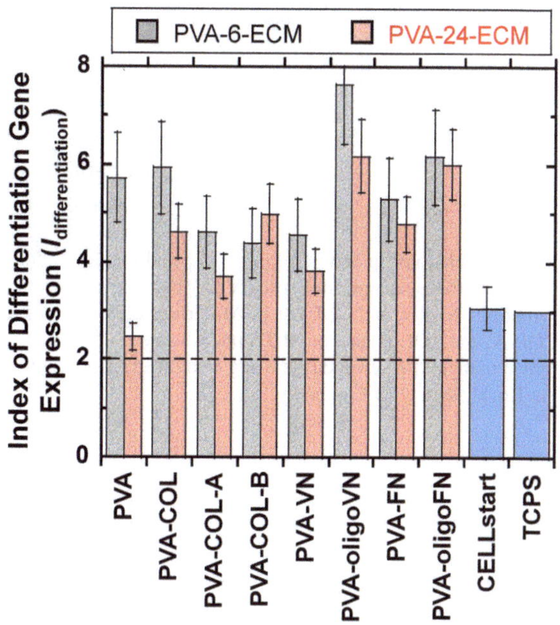

Fig. 18 $I_{\text{differentiation}}$ (index of differentiation gene expression) of hAFSCs cultivated on stiff PVA-24 hydrogels (red column) and soft PVA-6 hydrogels (black column) with immobilized ECMs (FN, VN, and COL) or ECM-derived oligopeptides (oligoFN, oligoVN, COL-A, and COL-B) and on TCPS plates coated with CELLstart and TCPS plates for seven days (Wang et al. 2015). Copyright 2015. Adapted with permission from the Royal Society of Chemistry

and PVA-6-ECM hydrogels extensively induce the spontaneous differentiation of hAFSCs into early neural cells (expression of *Nestin*).

The relationship between $I_{pluripotency}$ and $I_{differentiation}$ was examined for hAFSCs grown on PVA-oligoECM and PVA-ECM hydrogels in expansion medium (Fig. 19) (Wang et al. 2015). $I_{differentiation}$ increased with increasing $I_{pluripotency}$, with a linearity of 0.80. This result may be unexpected because the pluripotency gene expression of hiPSCs and hESCs would be downregulated upon the induction of differentiation (Takahashi et al. 2007). The unexpected result shown in Fig. 19 is considered as follows: The cells used in the experiments were hAFSCs, not hiPSCs or hESCs; hAFSCs, which tend to maintain high stemness on optimal hydrogels having nanosegments, are a heterogeneous population of cells that includes cells that spontaneously differentiated from the hAFSCs. This phenomenon is generally found in cultures of heterogeneous stem cell populations, such as hAFSCs, in expansion medium (Wang et al. 2015).

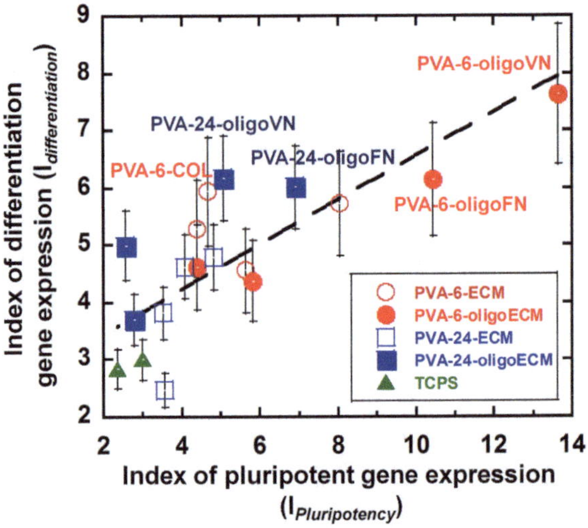

Fig. 19 Relationship between $I_{differentiation}$ and $I_{pluripotency}$ for hAFSCs cultivated on stiff PVA-24 hydrogels (blue square) and soft PVA-6 hydrogels (red circle) with immobilized ECMs (open symbols, FN, VN, and COL) or ECM-derived oligopeptides (closed symbols, oligoFN, oligoVN, COL-A, and COL-B) and on TCPS plates coated with CELLstart (green triangle) and TCPS plates (green triangle) for seven days (Wang et al. 2015). Copyright 2015. Adapted with permission from the Royal Society of Chemistry

3.3 The Differentiation and Pluripotency Proteins Expressed on HAFSCs Cultivated on PVA-OligoECM and PVA-ECM Hydrogels

The differentiation and pluripotency of hAFSCs cultivated on PVA-oligoECM and PVA-ECM hydrogels were investigated from gene expression levels determined in the previous sections (Figs. 15, 16, 17, 18, and 19). Here, we investigated the expression of pluripotency and differentiation surface markers of hAFSCs grown on PVA-oligoECM and PVA-ECM hydrogels.

Figures 20 and 21 represent the expression levels of the surface markers SSEA4 and Sox2 (pluripotency markers), the differentiation protein marker AFP (hepatocytes, endoderm), and β-III tubulin (neural cells, ectoderm) on hAFSCs cultivated on the TCPS plates, stiff PVA-24 hydrogels (PVA-24-oligoFN, PVA-24-oligoVN, PVA-24-FN, PVA-24-VN, and PVA-24), and soft PVA-6 hydrogels (PVA-6-oligoFN, PVA-6-oligoVN, PVA-6-FN, PVA-6-VN, and PVA-6) (Wang et al. 2015). These surface markers were evaluated because oligoFN and oligoVN

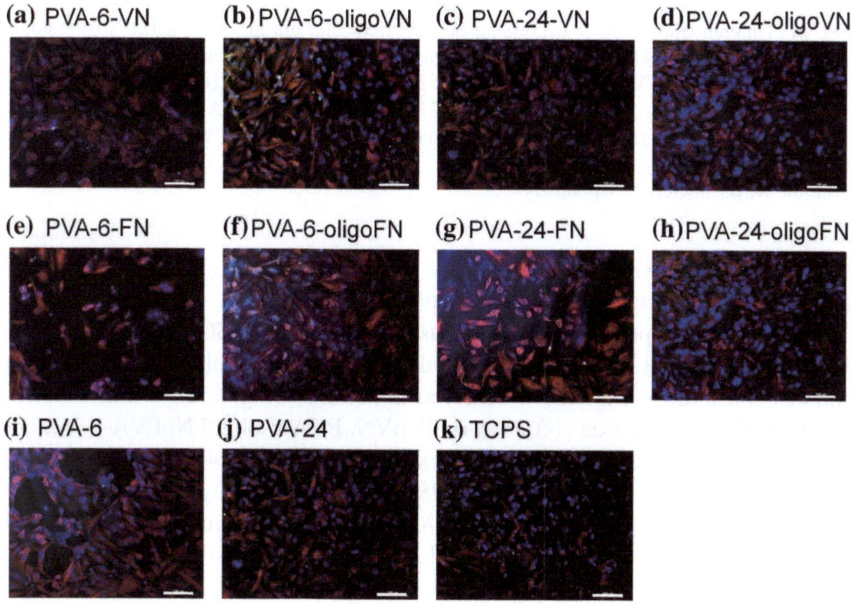

Fig. 20 Differentiation marker expression of β-III tubulin (red) and pluripotency marker expression of Sox2 (green) in hAFSCs cultivated on **a** PVA-6-VN hydrogels, **b** PVA-6-oligoVN hydrogels, **c** PVA-24-VN hydrogels, **d** PVA-24-oligoVN hydrogels, **e** PVA-6-FN hydrogels, **f** PVA-6-oligoFN hydrogels, **g** PVA-24-FN hydrogels, **h** PVA-24-oligoFN hydrogels, **i** PVA-6 hydrogels, **j** PVA-24 hydrogels, and TCPS plates **k** after seven days of culture. Nuclei were stained with Hoechst 33,342 (blue) (Wang et al. 2015). Copyright 2015. Adapted with permission from the Royal Society of Chemistry

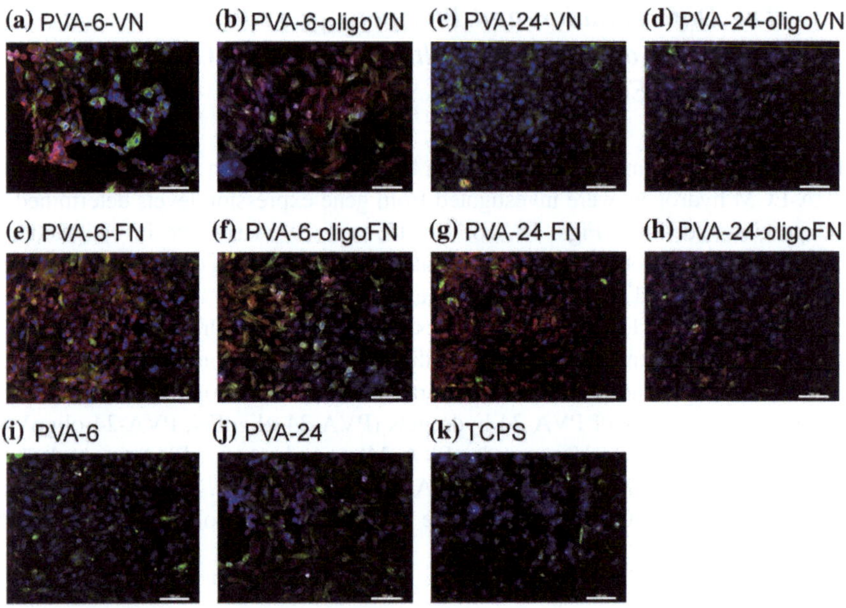

Fig. 21 Differentiation marker expression of AFP (red) and pluripotency marker expression of SSEA4 (green) in hAFSCs cultivated on (**a**) PVA-6-VN hydrogels, (**b**) PVA-6-oligoVN hydrogels, (**c**) PVA-24-VN hydrogels, (**d**) PVA-24-oligoVN hydrogels, (**e**) PVA-6-FN hydrogels, (**f**) PVA-6-oligoFN hydrogels, (**g**) PVA-24-FN hydrogels, (**h**) PVA-24-oligoFN hydrogels, (**i**) PVA-6 hydrogels, (**j**) PVA-24 hydrogels, and TCPS plates (**k**) for seven days. Nuclei were stained with Hoechst 33,342 (blue) (Wang et al. 2015). Copyright 2015. Adapted with permission from the Royal Society of Chemistry

tend to maintain hAFSC pluripotency as well as trigger the spontaneous differentiation of hAFSCs when immobilized on PVA hydrogels.

Slightly high expression of Sox2 was observed in hAFSCs cultivated on soft PVA-6-oligoFN and PVA-6-oligoVN hydrogels (Fig. 20), which was similar to the earlier trend in *Sox2* expression observed in hAFSCs (Fig. 15a). We also noticed that hAFSCs cultivated on soft PVA-6-oligoVN, PVA-6-oligoFN, PVA-6-VN, and PVA-6-FN hydrogels exhibited high expression of SSEA4 (Fig. 21). hAFSCs grown on the stiffest materials, the TCPS plates, showed only the faintest expression of SSEA4 and Sox2. The soft PVA-6 hydrogels with immobilized oligoVN and oligoFN were found to be favorable for supporting hAFSC pluripotency, as indicated by the surface marker expression of SSEA4 and Sox 2; these results showed a trend similar to that observed in the gene expression of *Nanog, Sox2*, and *Oct4* in hAFSCs (Fig. 15a).

hAFSCs grown on stiff and soft PVA hydrogels with and without immobilized oligoECMs and ECMs exhibited similar expression of β-III tubulin (Fig. 20). High expression of AFP was found in hAFSCs grown on soft PVA-6, PVA-6-oligoVN, PVA-6-oligoFN, PVA-6-VN, and PVA-6-FN hydrogels, whereas hAFSCs displayed only faint expression of AFP when they were cultivated on stiff TCPS plates and PVA-24 hydrogels with and without immobilized oligoECMs and ECMs (Fig. 21). Soft PVA-6-oligoVN hydrogels seem to be favorable cell culture biomaterials for the induction of spontaneous hAFSC differentiation into early-stage hepatocytes due to the high expression of AFP (Fig. 21) as well as the high expression of Sox17 (Fig. 17a) in the hAFSCs (Wang et al. 2015).

3.4 Comparison of the Present Study with Previously Published Studies

Engler et al. presented a hypothesis for the effect of elasticity of cell culture matrices on the spontaneous differentiation of hBMSCs into some cell lineages (i.e., early-stage neural cells, muscle cells, and osteoblasts) depending on the elasticities of the matrices (Engler et al. 2006); when hBMSCs were cultivated in expansion medium containing low serum concentrations, stiff matrices with elasticity similar to that of collagenous bone induced hBMSCs toward early osteoblast differentiation. Softer matrices having elasticity similar to that of the brain (0.3 kPa) induced the differentiation of hBMSCs into early neural cell lineages, whereas stiffer matrices (10 kPa) mimicking muscles spontaneously induced hBMSCs to differentiate into early myoblasts (Engler et al. 2006; Higuchi et al. 2013; 2015b). The effect of PVA hydrogel elasticity on the differentiation and pluripotency of hAFSCs was investigated in this study. hAFSCs spontaneously differentiated into early neural cells on soft PVA-6 hydrogels with immobilized oligoECMs and ECMs, which were identified by the expression of *Nestin*, whereas hAFSCs grown on stiff PVA-24 hydrogels with immobilized oligoECMs and ECMs spontaneously differentiated into early-stage osteoblasts, which were identified by the expression of *Runx2*, similar to the results reported by Engler (Engler et al. 2006; Higuchi et al. 2013; 2015b). However, hAFSCs grown on the stiffest surface, TCPS plates, did not spontaneously differentiate into osteoblasts, but instead exhibited the lowest differentiation and pluripotency in this study, as indicated by the evaluation of both gene and surface marker expression.

Tse and Engler made PAAm hydrogels using photopolymerization and a gradient-patterned photomask, which had radial elastic modulus gradients from 1 to 14 kPa and 1 kPa/mm (Higuchi et al. 2013; 2015b; Tse and Engler 2011). hBMSCs were observed to migrate from softer hydrogels into stiffer hydrogels in expansion medium and subsequently differentiate into contractile myogenic cells. In contrast, hBMSCs expressing neuronal markers tended to stay on soft sites of the PAAm hydrogels (Tse and Engler 2011). The soft cell culture hydrogels were suggested to induce hBMSCs to differentiate toward neuronal lineages in expansion medium (Engler et al. 2006; Higuchi et al. 2013; 2015b). These results are consistent with those found in this study; hAFSCs grown on soft PVA-6 hydrogels with immobilized oligoECMs and ECMs preferred to maintain pluripotency and to express

Nestin compared with hAFSCs cultivated on stiff TCPS plates and PVA-24 hydrogels. However, PVA hydrogels having an E' of less than 11 kPa were found to be too soft to support hAFSC proliferation in this study, despite having several immobilized oligoECMs and ECMs (Fig. 12).

Gilbert et al. evaluated whether the elastic modulus of cell culture materials serves an important factor in the differentiation or self-renewal of muscle stem cells (MuSCs) during muscle regeneration (Gilbert et al. 2010; Higuchi et al. 2013; 2015b). They prepared crosslinked poly(ethylene glycol) (PEG) hydrogels with immobilized laminin that had different elastic moduli (2–42 kPa). The MuSCs were found to expand more on soft PEG hydrogels than on stiff TCPS plates (Gilbert et al. 2010). In addition, the MuSCs showed higher cell survival when grown on soft PEG hydrogels. Furthermore, it was found that the MuSCs grown on soft PEG hydrogels expressed less myogenin (differentiation marker) than the cells grown on stiff TCPS plates (Gilbert et al. 2010; Higuchi et al. 2013; 2015b). Soft cell culture materials seemed to increase the number of MuSCs by restricting MuSC differentiation as well as by enhancing cell viability (Gilbert et al. 2010). In contrast, our studies showed that hAFSCs had high pluripotent gene expression and expressed several differentiation genes when grown on soft PVA-6 hydrogel dishes grafted with oligoECMs and ECMs (Figs. 15A and 17A) (Wang et al. 2015). These conflicting results may be related to the heterogeneous population of hAFSCs, as MuSCs are not a heterogeneous cell population.

Saha et al. prepared hydrogels having an interpenetrating polymer network with oligopeptides containing the cell-binding sequence RGD and elasticity ranging from 10 Pa to 10 kPa (Higuchi et al. 2013; Saha et al. 2008). The hydrogels required elastic moduli higher than 0.1 kPa for the expansion of rat neural stem cells. The expression of the neural marker β-III tubulin was found to be the highest in the rat neural stem cells cultured on the hydrogels with an elastic modulus of 0.5 kPa, which is similar to the physiological elasticity of brain tissue (Higuchi et al. 2013; Saha et al. 2008). The differentiation of rat neural stem cells into neurons was preferentially induced on softer hydrogels in the differentiation medium. On the contrary, differentiation into glial cells was preferentially induced on stiffer hydrogels (Higuchi et al. 2013; Saha et al. 2008). These results indicated that physical cues (i.e., hydrogel stiffness) are key factors involved in the differentiation and proliferation of stem cells, which was also demonstrated by the spontaneous differentiation and pluripotency of hAFSCs grown on PVA hydrogels of different stiffness in this study.

The soft PVA-6-oligoVN hydrogels could support hAFSC pluripotency, as indicated by high pluripotency gene expression (Figs. 15 and 16) and high surface marker expression (Figs. 20 and 21). The amino acid sequence of the oligoVN chosen in this study has been used for cultivating hESCs and hiPSCs that maintain their pluripotency, as reported in the literature (Higuchi et al. 2015a; Melkoumian et al. 2010; Higuchi et al. 2014a). hPSCs were reported to have been expanded for over ten passages on the plates with immobilized oligoVN without differentiating, whereas hPSCs could not be expanded on plates with immobilized peptides obtained from fibronectin and laminin because they differentiated (Higuchi et al.

2014a; 2015a; Melkoumian et al. 2010) Therefore, oligoVN tends to maintain stem cell pluripotency, for both hPSCs and hAFSCs. Therefore, both the physical cues (e.g., low elasticity) and biochemical cues (e.g., specific sequence of oligopeptides such as oligoVN) of PVA-6-oligoVN seem to contribute to better maintenance of hAFSC pluripotency than the other cell culture hydrogels evaluated in this study (Wang et al. 2015).

4 Xeno-Free Culture of HPSCs on Hydrogels with Optimal Elasticity

hPSCs should be cultured on biomaterials in chemically defined and xeno-free conditions if they are to be considered for use in regenerative medicine. Currently, hPSC cultures for maintaining pluripotency are typically performed using the following conditions: (a) cultivation on feeder layers of MEFs or human fibroblasts or (b) cultivation on Geltrex or Matrigel (Higuchi et al. 2011b, 2014b); Geltrex and Matrigel are produced from an extract of Engelbreth-Holm-Swarm mice. Therefore, the above typical culture methods of hPSCs are undefined and contain components of xeno origin, which hinder the clinical usage of hPSCs.

There is an increasing demand for the development of cell culture materials for hPSC culture on xeno-free and feeder-free conditions for the clinical usage of hPSCs. Some cell culture biomaterials have been proposed for culturing hPSCs and maintaining their pluripotency under xeno-free conditions. These biomaterial designs rely on the immobilization of biological cues, such as oligoECMs (Park et al. 2015; Chen et al. 2014; Wu et al. 2014; Fan et al. 2014; Deng et al. 2013; Pennington et al. 2015; Lin et al. 2014; Higuchi et al. 2014a), ECMs (Jonas et al. 2013; Liu et al. 2014; Lu et al. 2014; Rodin et al. 2010; Miyazaki et al. 2012; Tsutsui et al. 2011), and heparin-mimicking polymers (Brafman et al. 2010; Zhang et al. 2013; Villa-Diaz et al. 2010; Qian et al. 2014; Nandivada et al. 2011), on cell culture plates.

Cell culture plates with immobilized ECMs, such as fibronectin (CELLstart), laminin (laminin-322, laminin-521, and laminin-511), and vitronectin, have resulted in good hPSC proliferation in chemically defined media (Jonas et al. 2013; Liu et al. 2014; Lu et al. 2014; Rodin et al. 2010; Miyazaki et al. 2012; Tsutsui et al. 2011). Cell culture biomaterials grafted with oligoECM have also been found to support the pluripotency of hPSCs in chemically defined media (Park et al. 2015; Chen et al. 2014; Wu et al. 2014; Fan et al. 2014; Deng et al. 2013; Pennington et al. 2015; Lin et al. 2014; Higuchi et al. 2014a, b). Synthetic polymer plates, such as (a) copoly[2-(acryloyloxyethyl)] trimethylammonium-co-2-(diethylamino)ethyl acrylate] (Zhang et al. 2013), (b) aminopropylmethacrylamide (Irwin et al. 2011), (c) poly[2-(methacryloyloxy](ethyldimethyl-(3-sulfopropyl) ammonium hydroxide (Nandivada et al. 2011; Qian et al. 2014; Villa-Diaz et al. 2010), and (d) poly (methylvinylether-alt-maleic anhydride) (Brafman et al. 2010), have also been reported to be feasible cell culture plates for hPSC culture in chemically defined medium.

The biological cues and physical cues, such as stiffness of the cell culture materials, have been suggested to guide stem cell pluripotency and differentiation fate (Engler et al. 2006; Higuchi et al. 2013). Several previous studies investigated the effect of the cell culture hydrogel stiffness on the differentiation fate and pluripotency of mouse ESCs and hBMSCs (Engler et al. 2006; Higuchi et al. 2013; Kumar et al. 2013; Chowdhury et al. 2010). However, the effect of the cell culture hydrogel stiffness on the expansion of hESCs has not yet been reported, aside from our study (Higuchi et al. 2015a). Therefore, we prepared PVA hydrogels with immobilized oligoVN to investigate the physical effect of hydrogel stiffness on the expansion of hiPSCs and hESCs. The PVA hydrogels were controlled to have different stiffnesses by regulating the crosslinking time. OligoVN was covalently bound to the carboxylic acid site of the PVA hydrogels using an EDC/NHS reaction. The goal of this study was to determine the optimal stiffness of PVA-oligoVN hydrogels for the long-term (10–20 passages) culture of hESCs and hiPSCs in xeno-free conditions (Higuchi et al. 2015a).

4.1 Cultivation of hESCs and hiPSCs on PVA hydrogels Having Optimal Stiffness

hPSCs were cultivated on PVA-oligoVN hydrogels with different stiffnesses to investigate the effect of PVA-oligoVN hydrogel stiffness on hPSC proliferation. For the preliminary screening experiments, we investigated hESC attachment and morphology on PVA-oligoVN hydrogels prepared using different conditions; hESCs exhibit high attachment ratios and well-formed colonies when they are cultured on adequate materials. The morphology of WA09 cells (hESCs) cultivated on PVA-oligoVN hydrogels having different stiffnesses (PVA-48-500, PVA-24-500, PVA-12-500, PVA-6-500, and PVA-1-500 hydrogels) as well as on plates coated with Synthemax II (Synthemax II plates) at passage 1 is shown in Fig. 22 (Higuchi et al. 2015a). PVA-*X*-*Y* hydrogels are defined as PVA hydrogels crosslinked for *X* hours and grafted with oligoVN using a reaction solution containing *Y* μg/mL of oligoVN. While the WA09 cells could not adhere well to the soft PVA-1-500 hydrogels, WA09 cells could attach to the PVA hydrogels with stiffnesses higher than 12–15 kPa (i.e., the PVA-48-500, PVA-24-500, PVA-12-500, and PVA-6-500 hydrogels). It is found that PVA-oligoVN hydrogels need to have a minimum stiffness to allow hESCs to attach to the surface.

The attachment ratio of hPSCs cultivated on Matrigels, Synthemax II plates, and PVA-oligoVN hydrogels was investigated at passages 1–5, and the attachment ratios at passage 3 are described in Fig. 21 (Higuchi et al. 2015a). High attachment ratios of HPS0077 (hiPSCs) and WA09 cells (hESCs) were observed on PVA-24-500 hydrogels and Matrigels. Low or moderate attachment ratios of hPSCs

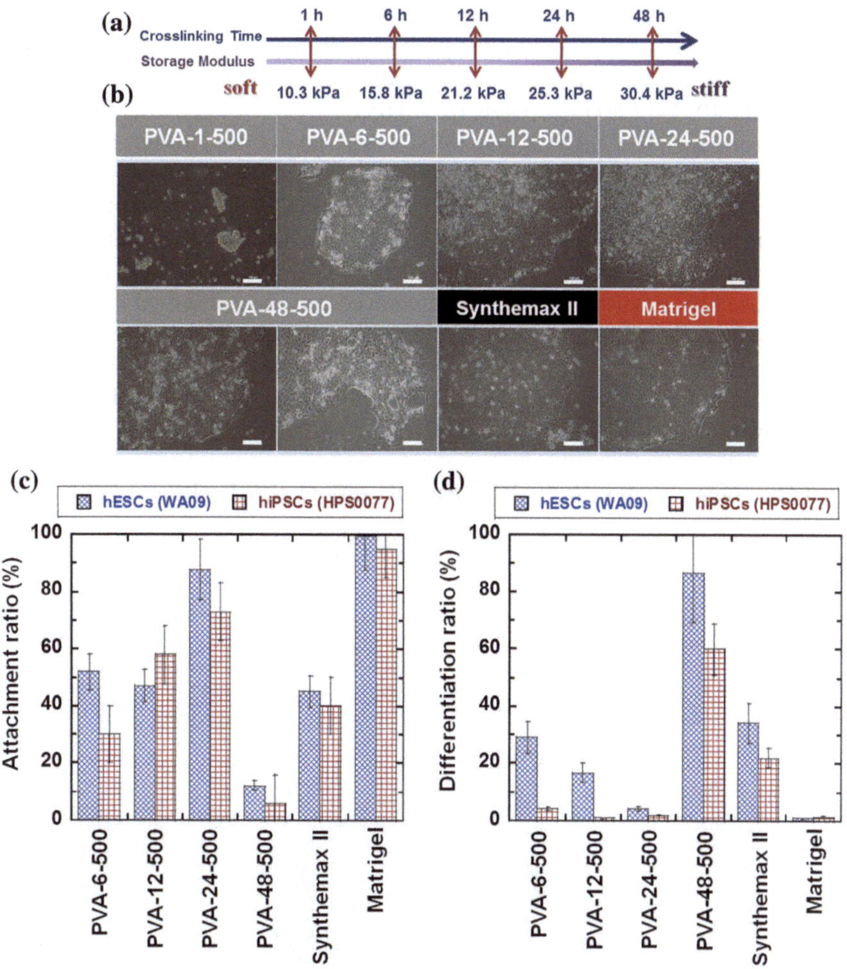

Fig. 22 hPSC cultivation on PVA-oligoVN hydrogels having optimal elasticity. **a** Stiffness (storage modulus) can be controlled by the crosslinking intensity (time) of PVA-oligoVN hydrogels. **b** Picture of WA09 cells (hESCs) cultivated on PVA-oligoVN hydrogels having different stiffnesses (PVA-1-500, PVA-6-500, PVA-12-500, PVA-24-500, and PVA-48-500 hydrogels), Matrigels, and Synthemax II plates at passage 1. The bar represents 100 μm. **c** Attachment ratio of hPSCs (red column, HPS0077 cells [hiPSCs] and blue column, WA09 cells [hESCs]) on PVA-oligoVN hydrogels with different stiffnesses, Matrigels, and Synthemax II plates at passage 3. **d** Differentiation ratio of hPSCs (red column, HPS0077 cells [hiPSCs] and blue column, WA09 cells [hESCs]) on PVA-oligoVN hydrogels with different stiffnesses, Matrigels, and Synthemax II plates at passage 3 (Higuchi et al. 2015a). Adapted with permission from a Creative Commons Attribution License

were observed on the Synthemax II plates and PVA-48-500, PVA-12-500, and PVA-6-500 hydrogels. Those PVA-oligoVN hydrogels that provided the optimal elasticity (25 kPa, PVA-24-500 hydrogels) resulted in the highest hPSC attachment ratios among the PVA-oligoVN hydrogels. hPSCs cultured on Synthemax II plates exhibited a lower attachment ratio than hPSCs cultured on PVA-24-500 hydrogels ($p < 0.05$) and an approximately equal attachment ratio compared to hPSCs cultured on the PVA-12-500 hydrogels ($p > 0.05$).

The pluripotency of hESCs and hiPSCs was investigated using colony morphology and live staining of alkaline phosphatase. Figure 23 displays the morphology of (a) completely differentiated, (b) partially differentiated, and (c) pluripotent WA09 cells (c) (Higuchi et al. 2015a). The completely differentiated cells did not show alkaline phosphatase (ALP) activity. On the other hand, alkaline phosphatase activity of the partially differentiated cells was observed in the colony center, while cells on the colony edge did not show ALP activity. The pluripotent cells showed excellent colony morphology and ALP activity. The differentiation ratio was investigated for whole hPSC colonies on the plates and evaluated using the following equation:

$$\begin{aligned} \text{Differentiation ratio}\,(\%\,) = {} & \%\,\text{of partially differentiated cells} \times 0.5\,(\%) \\ & + \%\text{ of differentiated cells} \times 1.0\,(\%) \end{aligned} \tag{4}$$

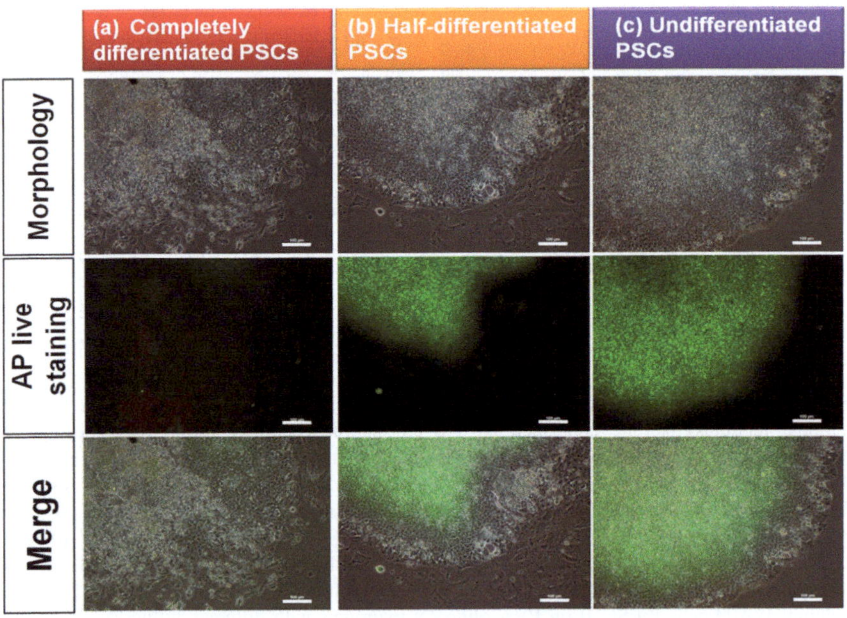

Fig. 23 Definition of completely differentiated hPSCs **a**, partially differentiated hPSCs **b**, and undifferentiated (pluripotent) hPSCs **c**. **a** No ALP (alkali phosphatase) activity was observed in completely differentiated cells. **b** ALP activity was observed in the center of the colonies, whereas the edges of the partially differentiated cell colonies did not show ALP activity. **c** The undifferentiated (pluripotent) cells displayed good colony morphologies and ALP activity in whole cells (Higuchi et al. 2015a). Adapted with permission from a Creative Commons Attribution License

The differentiation ratios of hPSCs cultivated on Matrigels, Synthemax II plates, and PVA-oligoVN hydrogels were investigated at passages 1–5, and the results at passage 3 are described in Fig. 22. hPSCs cultured on Matrigels and PVA-24-500 hydrogels showed low differentiation ratios (Higuchi et al. 2015a). On the other hand, hPSCs cultured on Synthemax II plates and PVA-48-500 hydrogels exhibited high differentiation ratios. These results indicate that the cell culture hydrogels should have an optimal stiffness (e.g., 25.3 kPa) to maintain hPSC pluripotency when they are cultivated on PVA-oligoVN hydrogels. PVA-24-500 hydrogels provided higher hPSC pluripotency (lower differentiation ratio) and attachment ratios than the Synthemax II plates (commercial dishes) ($p < 0.05$) (Higuchi et al. 2015a).

4.2 *Effect of Surface Density of Biological Cues (OligoVN) on HPSC Proliferation*

PVA-oligoVN hydrogels with a stiffness of 25.3 kPa (crosslinking time = 24 h) were selected as the optimal cell culture material having optimal stiffness of those described in the previous section. In the following experiments, we evaluated the effect of the oligoVN surface density of PVA-oligoVN hydrogels on hPSC proliferation. The concentration of oligoVN in the reaction solution can control the surface density of oligoVN on PVA-24-EDC hydrogels. Figure 24a shows the morphologies of WA09 cells cultured on PVA-oligoVN gels having different oligoVN surface densities (PVA-24-1500, PVA-24-1000, PVA-24-500, PVA-24-250, PVA-24 h-100, and PVA-24 h-50 hydrogels), as well as on Matrigels and Synthemax II plates at passage 1 (Higuchi et al. 2015a). WA09 cells detached easily from the PVA-oligoVN hydrogels with an oligoVN concentration < 500 μg/mL (PVA-24-250, PVA-24-100, and PVA-24-50 hydrogels). WA09 cells could not be cultivated on PVA-24-50 gels after two passages. Figure 24b shows the attachment ratio of HPS0077 cells (hiPSCs) and WA09 cells (hESCs) on PVA-oligoVN hydrogels prepared with 100–1500 μg/mL of oligoVN in the reaction solution, as well as on Matrigels and Synthemax II plates at passage 3 (Higuchi et al. 2015a). The attachment ratios of HPS007 and WA09 cells increased with increasing oligoVN concentrations up to 500 μg/mL. The attachment ratios of HPS007 and WA09 cells on the PVA-24-500, PVA-24-1000, and PVA-24-1500 hydrogels showed no significant differences ($p > 0.05$). Furthermore, HPS007 and WA09 cells cultured on the PVA-24-1500 hydrogels showed higher attachment ratios than the cells cultured on the Synthemax II plates ($p < 0.05$), but slightly less than the cells cultured on Matrigels ($p < 0.05$).

We evaluated the differentiation ratios of HPS007 and WA09 cells grown on PVA-oligoVN hydrogels prepared with 100–1500 μg/mL of oligoVN in the reaction solution, as well as on Matrigels and Synthemax II plates at passage 3 to investigate the hPSC pluripotency; the results are displayed in Fig. 24c

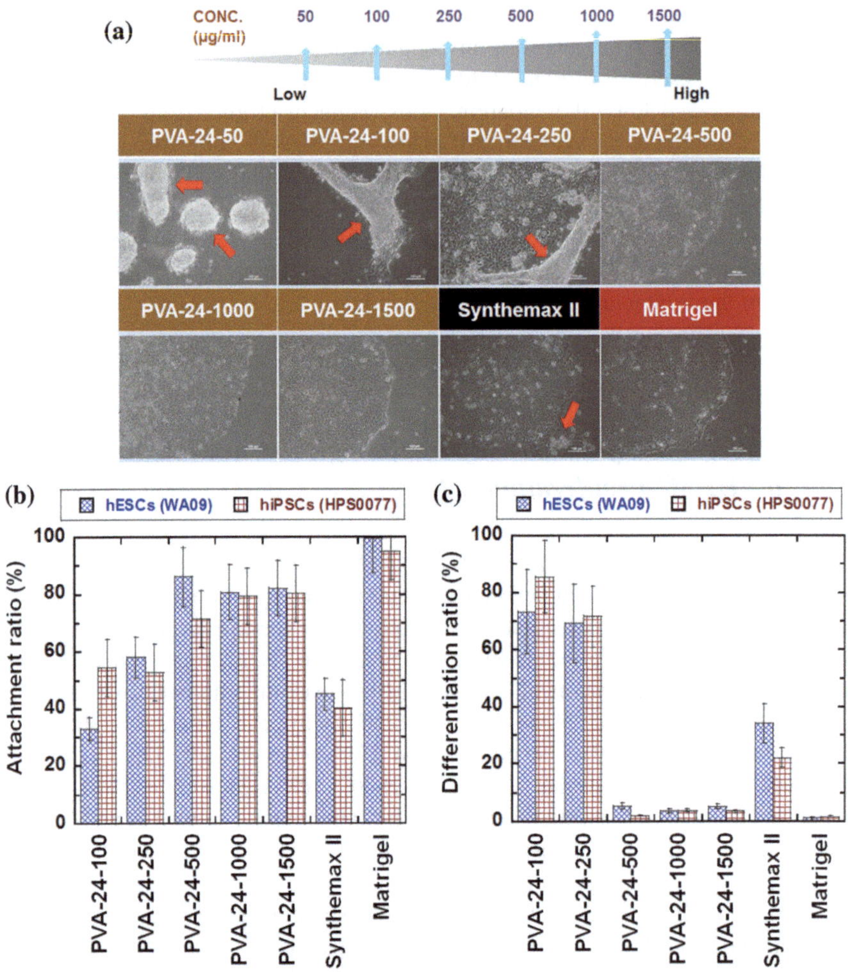

Fig. 24 hPSC cultivation on PVA-oligoVN hydrogels having several surface densities of oligoVN. **a** Pictures of WA09 (hESCs) cultivated on PVA-oligoVN hydrogels having different surface densities of oligoVN (PVA-24-50, PVA-24-100, PVA-24-250, PVA-24-500, PVA-24-1000, and PVA-24-1500 hydrogels), Matrigels, and Synthemax II plates at passage 1. The red arrows indicate detached cells. The bar represents 100 μm. **b** Attachment ratio of hPSCs (red column, HPS0077 cells [hiPSCs] and blue column, WA09 cells [hESCs]) on PVA-oligoVN hydrogels having different surface densities of oligoVN, Matrigels, and Synthemax II plates at passage 3. **c** Differentiation ratio of hPSCs (red column, HPS0077 cells [hiPSCs] and blue column, WA09 cells [hESCs]) on PVA-oligoVN hydrogels having different surface densities of oligoVN, Matrigels, and Synthemax II plates at passage 3 (Higuchi et al. 2015a). Adapted with permission from a Creative Commons Attribution License

(Higuchi et al. 2015a). hPSCs on the PVA-24-500, PVA-24-1000, and PVA-24-1500 hydrogels, as well as on the Matrigels, showed very low differentiation ratios, whereas hPSCs on PVA-24-100 and PVA-24-250 hydrogels, as well as on

Synthemax II plates, displayed high differentiation ratios. It was found that there is a minimum surface density of oligoVN (500 μg/mL) on the PVA-oligoVN hydrogels required to maintain hPSC pluripotency, with a low differentiation ratio, as well as to achieve a high attachment ratio. The active layer of Synthemax dishes is reported to be polyacrylate grafted with oligoVN of the same amino acid sequence as was prepared for the PVA-oligoVN hydrogels in this study (Melkoumian et al. 2010). High concentrations, such as 1 mM (1590 μg/mL), of oligoVN in the reaction solution was also necessary to produce the Synthemax plates (Melkoumian et al. 2010). Therefore, it is reasonable that high concentrations of oligoVN in the reaction solution are required to maintain hPSC pluripotency on the surface of PVA-oligoVN hydrogels (Higuchi et al. 2015a).

4.3 hPSC Culture for Long Period Under Xeno-Free Culture Conditions

In the previous section, PVA-24-500, PVA-24-1000, and PVA-24-1500 hydrogels were found to be good materials for hPSC culture; these materials have optimal biological cues (optimal surface density of oligoVN) and good physical cues (optimal stiffness). Based on these results, we performed long-term (20 passages) hPSC culture using PVA-24-1000 hydrogels, Synthemax II dishes and Matrigels in Essential 8 medium (xeno-free culture medium).

Figure 25 shows the differentiation ratios, attachment ratios, and expansion rates of HPS0077 (hiPSCs) and WA09 (hESCs) cells cultivated on PVA-24-1000 hydrogels, Matrigels and Synthemax II plates for 1–20 passages (Higuchi et al. 2015a). HPS0077 and WA09 cells on the PVA-24-1000 hydrogels showed almost the same fold expansion as the cells cultivated on commercial dishes (Synthemax II), but they exhibited slightly worse expansion than cells cultured on Matrigels ($p < 0.05$). HPS0077 and WA09 cells cultured on the PVA-24-1000 hydrogels exhibited slightly higher attachment ratios compared with the cells on Synthemax II plates over 20 passages, but the difference was not significant ($p > 0.05$). However, it should be noted that HPS0077 and WA09 cells on the PVA-24-1000 hydrogels showed significantly higher attachment ratios than cells on Synthemax II plates ($p < 0.05$) during the early passages (i.e., <5 passages). The attachment ratios of HPS0077 and WA09 cells cultured on Matrigels were more than 80% after 20 passages, which was extensively greater than the attachment ratios of cells on the Synthemax II plates and PVA-24-1000 hydrogels ($p < 0.05$). HPS0077 and WA09 cells on Matrigels and PVA-24-1000 hydrogels showed lower differentiation ratios than the cells on Synthemax II plates ($p < 0.05$), indicating that hPSCs could maintain pluripotency on the PVA-24-1000 hydrogels as well as on Matrigels for a long time (e.g., >20 passages).

These results suggest that hESCs and hiPSCs can be cultured on PVA-24-1000 hydrogels under xeno-free and feeder-free conditions with better performance than on commercially available plates (Synthemax II), although hPSCs cultivated on

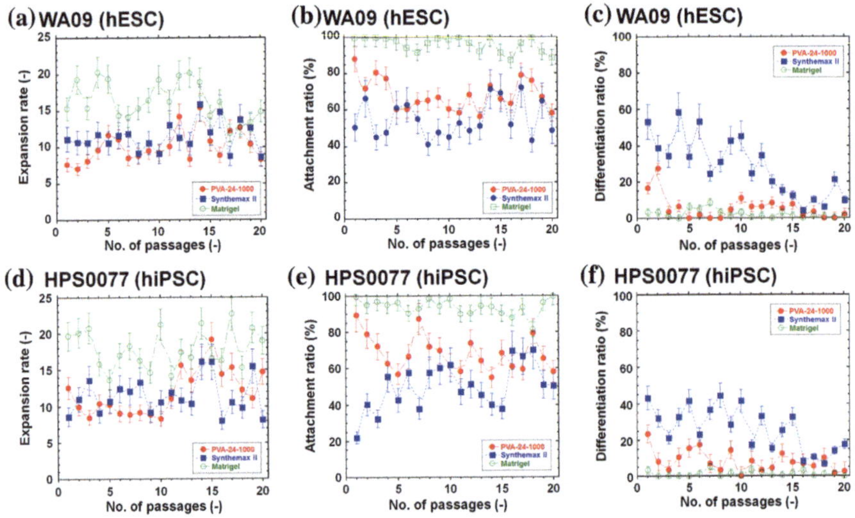

Fig. 25 hPSCs cultivated on PVA-oligoVN hydrogels having an optimal elasticity for long-term, xeno-free culturing. **a** Expansion rate, **b** attachment ratio, and **c** differentiation ratio of WA09 cells (hESCs) cultured on PVA-24-1000 hydrogels (closed red circle), Matrigels (open green circle), and Synthemax II plates (closed blue square) for 1–20 passages. **d** Expansion rate, **e** attachment ratio, and **f** differentiation ratio of HPS0077 cells (hiPSCs) cultured on PVA-24-1000 hydrogels (closed red circle), Matrigels (open green circle), and Synthemax II plates (closed blue square) for 1–20 passages. (Higuchi et al. 2015a). Adapted with permission from a Creative Commons Attribution License

Matrigels showed a slightly better attachment ratio and expansion rate. However, Matrigel is not a xeno-free material; hESC and hiPSC cultures on PVA-24-1000 hydrogels are xeno-free, which is required for clinical applications. The hESC and hiPSC cultures on PVA-24-1000 hydrogels showed better performance (i.e., lower differentiation ratio) than the commercial dishes, i.e., Synthemax II; thus, PVA-24 h-1000 hydrogels allow hPSCs to maintain higher levels of pluripotency than do Synthemax II plates. This finding was further evaluated in the experiments described below.

HPS0077 and WA09 cells used in the experiments shown in Figs. 22, 23, 24, and 25 were expanded on Matrigels for several passages in advance to allow them to adjust to feeder-free conditions. These procedures are necessary for the cells to transition from cell culture on MEFs to cell culture on the materials in feeder-free conditions. WA09 cells were also shifted from cell culture on MEFs to cell culture on Synthemax II and PVA-24-1000 hydrogels, starting with WA09 cells cultured on MEFs and not cultured on Matrigels in advance. The results are described in Fig. 26 (Higuchi et al. 2015a). The WA09 cells transferred directly from MEFs to the Synthemax II plates showed differentiated cells at passage 1, whereas the WA09 cells transferred from MEFs to PVA-24-1000 hydrogels were found to maintain pluripotency. These data show trends similar to those in Figs. 25c, f, where hPSCs tended to differentiate on Synthemax II plates compared to hPSCs on PVA-24-1000 hydrogels.

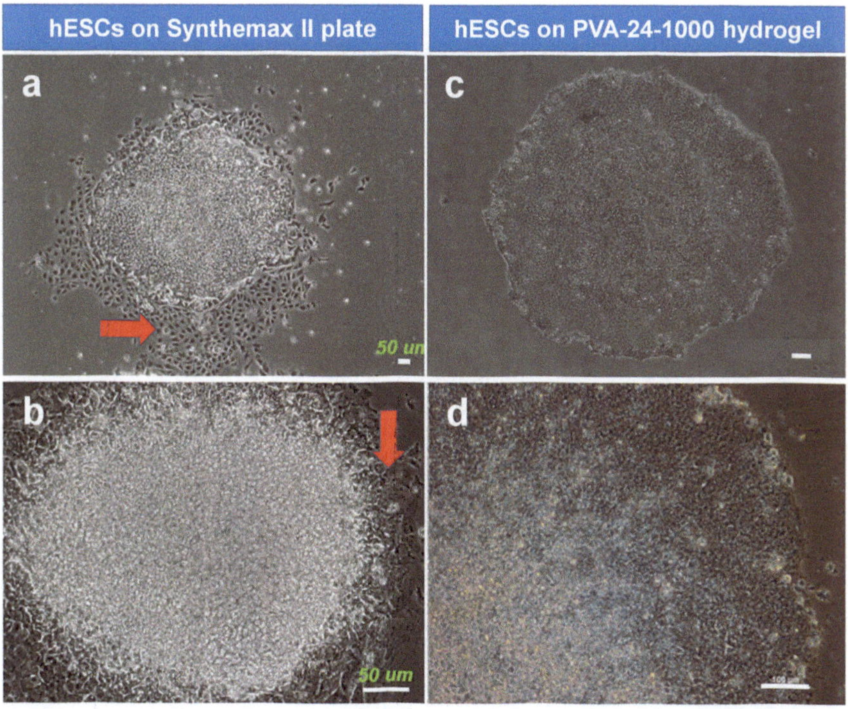

Fig. 26 Comparison of hESCs cultivated on PVA-oligoVN hydrogels and Synthemax II plates. The morphologies of WA09 cells (hESCs) cultivated on Synthemax II plates **a**, **b** and PVA-24-1000 hydrogels **c**, **d** at passage 1 after the WA09 cells were transferred from cultivation on MEFs to cultivation on Synthemax II plates or PVA-24-1000 dishes. Red arrows indicate differentiated cells. The bar represents 50 μm **a**, **b** and 100 μm **c**, **d** (Higuchi et al. 2015a). Adapted with permission from a Creative Commons Attribution License

The pluripotency of HPS0077 and WA09 cells was investigated by immunostaining for pluripotency proteins SSEA-4, Tra-1-81, Sox2, and Oct3/4 after the cells on PVA-24-1000 hydrogels were passaged 20 times; the results are shown in Fig. 27 (Higuchi et al. 2015a). The pluripotent proteins (SSEA-4, Tra-1-81, Sox2, and Oct3/4) were expressed on HPS0077 and WA09 cells cultivated on PVA-24-1000 hydrogels in Essential 8 xeno-free culture medium for 20 passages (Higuchi et al. 2015a).

4.4 Differentiation of hESCs and hiPSCs Cultured on PVA Hydrogels

We evaluated the ability of HPS0077 and WA09 cells to differentiate into cells derived from the three germ layers in vitro (embryoid body (EB) formation method) and in vivo (teratoma formation method) to evaluate their pluripotency after being

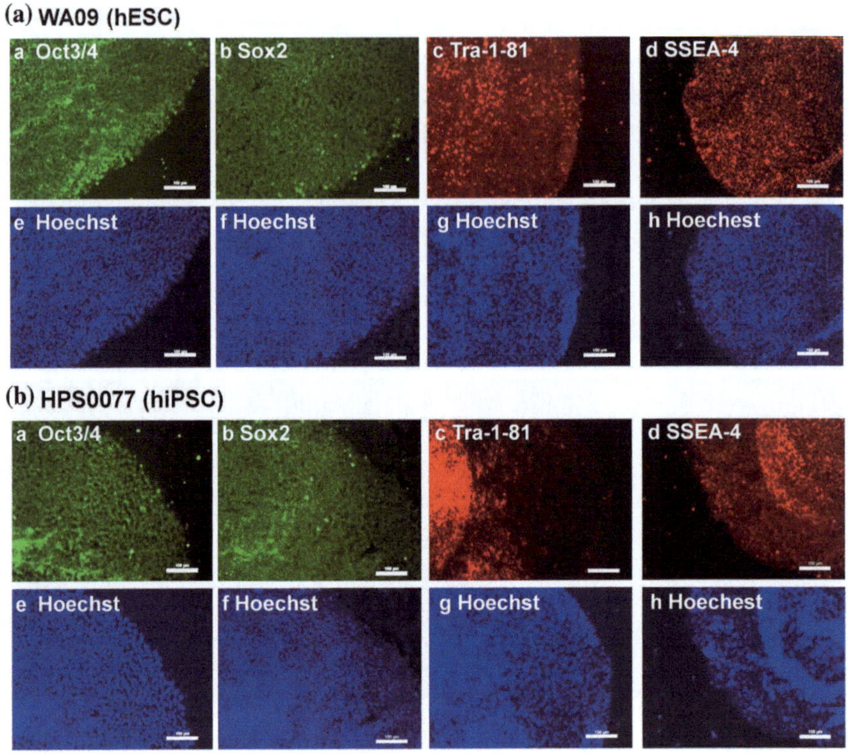

Fig. 27 Expression of pluripotency proteins of hiPSCs and hESCs cultivated on PVA-oligoVN hydrogels. **a** Expression of pluripotency proteins on WA09 cells (hESCs) measured by immunostaining after cultivation on PVA-24-1000 hydrogels in xeno-free culture conditions for 20 passages. (*a*) Oct3/4, (b) Sox2, (*c*) Tra-1-81, (*d*) SSEA-4, and (*e–h*) Hoechst staining of hESCs used in (*a–d*). The bar represents 100 μm. **b** Expression of pluripotency proteins on HPS0077 cells (hiPSCs) measured by immunostaining after cultivation on PVA-24-1000 hydrogels in xeno-free conditions for 20 passages. (*a*) Oct3/4, (*b*) Sox2, (*c*) Tra-1-81, (*d*) SSEA-4, and (*e–h*) Hoechst staining of hESCs used in (*a–d*). The bar represents 100 μm (Higuchi et al. 2015a). Adapted with permission from a Creative Commons Attribution License

cultured on PVA-24-1000 hydrogels in xeno-free conditions for 20 passages. Figure 28a shows EBs generated from WA09 and HPS0077 cells after a suspension culture of hPSCs on ultra-low-binding plates (Higuchi et al. 2015a). Figure 28b, c shows the cells from EBs prepared from WA09 to HPS0077 cells, respectively, and immunostained with glial fibrillary acidic protein (GFAP, ectoderm), β-III tubulin (ectoderm), smooth muscle actin (SMA, mesoderm), and alpha-fetoprotein (AFP, endoderm) (Higuchi et al. 2015a). It was found that both WA09 and HPS0077 cells could differentiate into cells derived from the three germ layers, as indicated by the expression of GFAP, β-III tubulin, SMA, and AFP. Therefore, the hiPSCs and hESCs not only maintained their pluripotency after being cultivated on

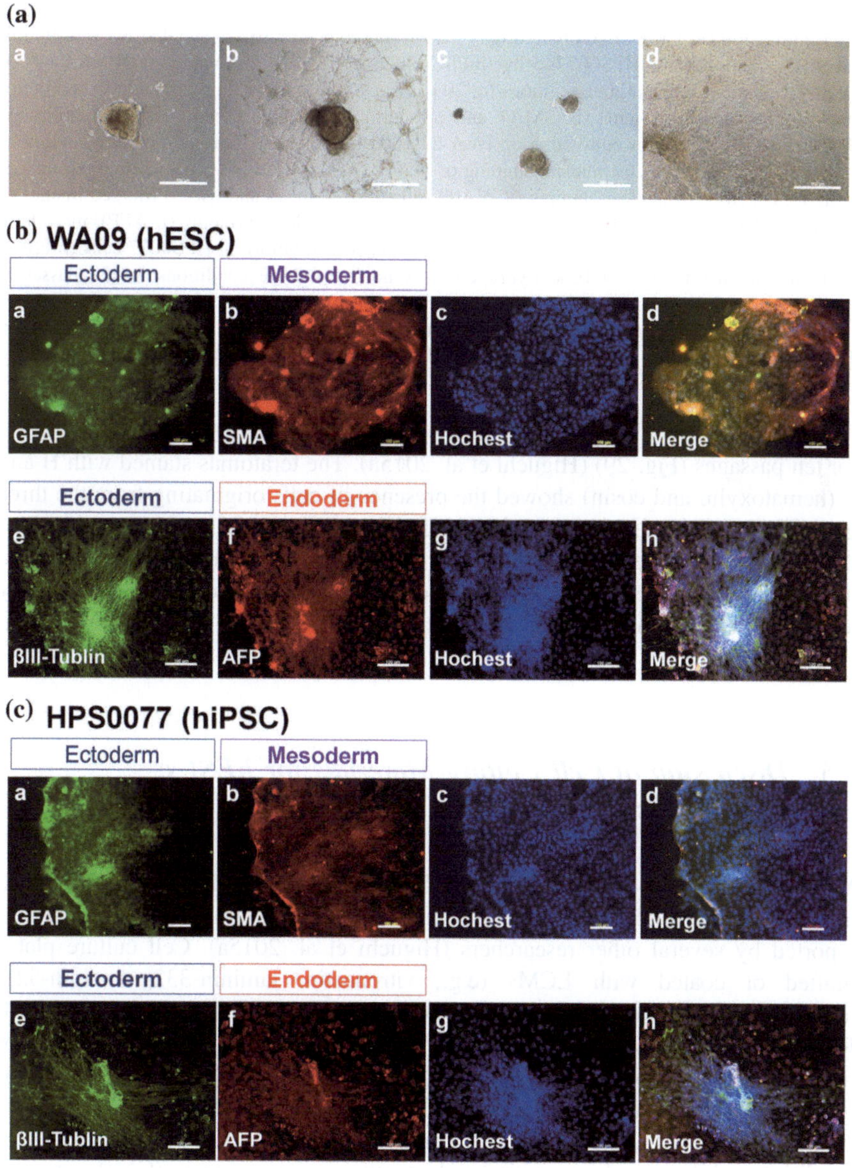

PVA-24-1000 hydrogels under xeno-free conditions but also maintain the ability to differentiate into cells derived from the three germ layers in vitro.

Using the teratoma formation assay, the ability of WA09 cells to differentiate into cells derived from the three germ layers was investigated in vivo. WA09 cells were transplanted into SCID (non-obese diabetic/severe combined immunodeficiency) mice to produce teratomas after being cultured on PVA-24-1000 hydrogels

◂**Fig. 28** Expression of differentiation proteins in embryoid bodies (EBs) generated from hPSCs after cultivation on PVA-oligoVN hydrogels. **a** Morphologies of EBs generated by WA09 cells (hESCs, a and b) and HPS0077 cells (hiPSCs, c and d) after cultivation on PVA-24-1000 hydrogels in xeno-free culture conditions for 20 passages. **b** Immunostaining of endoderm proteins (f, AFP), mesoderm proteins (b, SMA), and ectoderm proteins (a, GFAP; e, β-III tubulin) on WA09 cells (hESCs) after cultivation on PVA-24-1000 hydrogels in xeno-free culture conditions for 20 passages. (c) Hoechst nuclear staining of WA09 cells utilized in (*a*, *b*). (d) Merged images of (*a–c*). (*g*) Hoechst nuclear staining of WA09 cells utilized in (*e*) and (*f*). (*h*) Merged images of (*e–g*). The bar represents 100 μm. **c** Immunostaining of endoderm proteins (f, AFP), mesoderm proteins (b, SMA), and ectoderm proteins (a, GFAP; e, β-III tubulin) on HPS0077 cells (hiPSCs) after cultivation on PVA-24-1000 hydrogels in xeno-free culture conditions for 20 passages. (c) Hoechst nuclear staining of HPS0077 cells utilized in (*a*) and (*b*). (*d*) Merged images of (*a–c*). (*g*) Hoechst nuclear staining of HPS0077 cells utilized in (*e*) and (*f*). (*h*) Merged images of (*e–g*). The bar represents 100 μm (Higuchi et al. 2015a). Adapted with permission from a Creative Commons Attribution License

for ten passages (Fig. 29) (Higuchi et al. 2015a). The teratomas stained with H and E (hematoxylin and eosin) showed the presence of cells originating from the three germ layers (neurons (ectoderm), chondrocytes (mesoderm), osteoblasts (mesoderm), and enterocytes (endoderm)). Therefore, hESCs cultivated on PVA-24-1000 hydrogels for long periods (10–20 passages) are able to maintain their pluripotency and can differentiate into cells of the three germ layers in vivo and in vitro (Higuchi et al. 2015a).

4.5 Discussion of Cell Culture Materials for hESCs and hiPSCs

Various matrices for culturing hPSCs have been investigated in chemically defined and feeder-free conditions. Table 2 summarizes cell culture matrices for hPSCs reported by several other researchers (Higuchi et al. 2015a). Cell culture plates grafted or coated with ECMs (e.g., vitronectin, laminin-332, laminin-521, laminin-511, and fibronectin) (Jonas et al. 2013; Liu et al. 2014; Lu et al. 2014; Rodin et al. 2010; Miyazaki et al. 2012; Tsutsui et al. 2011) and ECM-derived oligopeptides (Park et al. 2015; Chen et al. 2014; Wu et al. 2014; Fan et al. 2014; Deng et al. 2013; Pennington et al. 2015; Lin et al. 2014; Higuchi et al. 2014a) have been extensively reported. Because generating human ECMs under xeno-free conditions while complying with good manufacturing practices (GMPs) is relatively expensive, the development of cell culture plates with immobilized ECM-derived oligopeptides could be a reasonable alternative. To date, totally synthetic polymeric materials have been prepared as hPSC culture substrates, including (a) poly(methyl vinyl ether-alt-maleic anhydride), or PMVE-alt-MA, (b) poly[2-(methacryloyloxy)ethyl dimethyl-(3-sulfopropyl)ammoniumhydroxide], or PMEDSAH, and (c) aminopropylmethacrylamide, or APMAAm (Brafman et al.

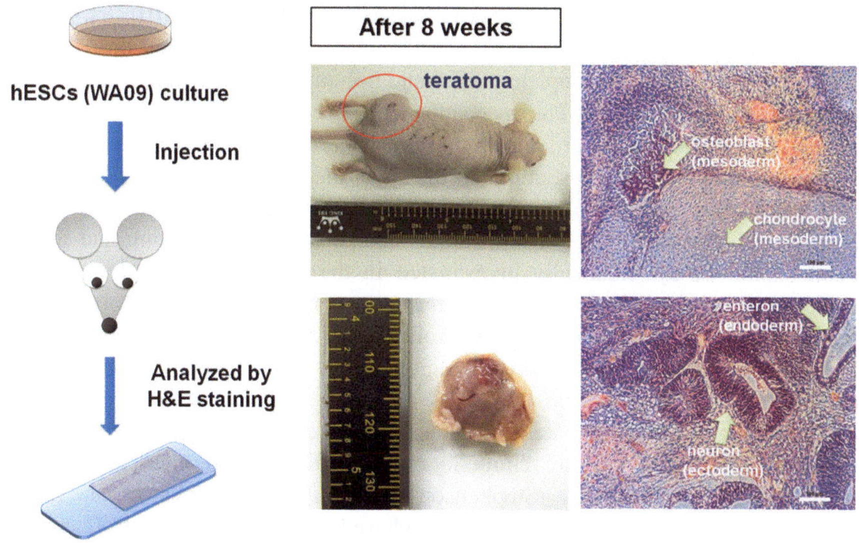

Fig. 29 Differentiation potential of hESCs (WA09) demonstrated by teratoma formation *in vivo* after cultivation on PVA-24-1000 hydrogels for 20 passages. Enterocytes (endoderm), neurons (ectoderm), chondrocytes (mesoderm), and osteoblasts (mesoderm) were generated in the teratomas. The bar represents 100 μm (Higuchi et al. 2015a). Adapted with permission from a Creative Commons Attribution License

Table 2 Coating and substrates for feeder-free hPSC culture in a defined medium[a] (Higuchi et al. 2015a). Adapted 2015. Reproduced with permission a Creative Commons Attribution License

(a) *Coating materials of ECMs*
Gelatin, fibronectin, laminin, **laminin-332**, **laminin-511**, **laminin-521**, **vitronectin**, **recombinant vitronectin**
(b) *Polysaccharide substrates*
Chitin/alginate, cellulose, positively charged cellulose
(c) *Oligopeptides for immobilization on substrates*
Poly-D-lysine, cyclic RGD, pronectin, **oligovitronectin**
(d) *Chimera protein for immobilization on substrates*
E-cadherin chimera
(e) *Synthetic polymer of substrates*[b]
MEASAH, PMVE-alt-MA[39], PMEDSAH[40–43], APMAAm[44], Copoly(AEtMA-co-DEAEA)[45], Poly-3,4-dihydroxy-L-phenyl-alanine

[a]Bold biomolecules are typically used

[b]PMVE-alt-MA, poly(methyl vinyl ether-alt-maleic anhydride); PMEDSAH, poly[2-(methacryloyloxy(ethyl dimethyl))-(3-sulfopropyl) ammoniumhydroxide]; APMAAm, aminopropylmethacrylamide; Copoly(AEtMA-co-DEAEA), copoly[2-(acryloyloxyethyl) trimethylammonium-co-2-(diethylamino)ethyl acrylate]

2010; Nandivada et al. 2011; Qian et al. 2014; Villa-Diaz et al. 2010; Irwin et al. 2011; Zhang et al. 2013). However, hPSCs are generally cultivated on Matrigels prior to being grown on completely synthetic polymer materials, and Matrigels are xeno-containing materials. The mechanism of successful hPSC culture on completely synthetic polymer materials is generally considered to involve heparin-mimicking polymers, although the exact mechanism of hPSC proliferation on completely synthetic polymer materials is still under investigation. The mechanism of how synthetic polymer materials enable hPSC proliferation and attachment should be systematically investigated in the future. Another drawback of using completely synthetic polymer materials is that currently these polymers are not commercially available, as only professional chemists can synthesize these polymers; in contrast, ECM-derived oligopeptides and ECMs are commercially available.

Currently, only coating materials can be obtained commercially for hPSC culture matrices; these materials are chimeric proteins, ECMs and matrices made of ECM-derived oligopeptides (Stephenson et al. 2012; Nagaoka et al. 2010). Synthetic polymer materials cannot be purchased commercially for hPSC culture. Therefore, the development of reliable synthetic polymer materials that can support hPSC expansion with high pluripotency and high reproducibility is required.

The PVA-24-1000 hydrogels include the hPSC-binding site of oligoVN that binds to integrins $\alpha_V\beta_5$ and $\alpha_V\beta_3$ (Higuchi et al. 2014b). This is why hPSCs grown on MEFs can be safely transferred onto PVA-24-1000 hydrogels without differentiating, whereas hPSCs shifted from MEFs onto Synthemax II plates will differentiate significantly (Fig. 26). However, it should be noted that the sequence of the oligoVN cell-binding domain on the Synthemax II plates is the same as that used on PVA-24-1000 hydrogels (Melkoumian et al. 2010).

The effect of cell culture material stiffness on the differentiation and pluripotency of stem cells (i.e., stem cell fate) has mainly been studied using adult stem cells, such as hBMSCs and hHSPCs (Engler et al. 2006; Higuchi et al. 2013; Kumar et al. 2013). However, we suggest that finely tuning the stiffness of the cell culture matrices is an extremely important factor for the expansion and attachment of hESCs and hiPSCs. hPSCs grown on the stiffest substrates (E'= 30 kPa) were induced to differentiate after 5 days of cultivation, whereas only a small number of hPSCs colonies attached to the softest substrates (E'= 10–20 kPa). hPSCs cultivated on the PVA-oligo hydrogels with adequate stiffness (E'= 25 kPa) could maintain their pluripotency and proliferate for over 10–20 passages under feeder-free and xeno-free conditions. It was reported that hPSCs could be cultured on Synthemax II plates for 10–20 passages under xeno-free conditions (Melkoumian et al. 2010). However, in this study, hPSCs were found to differentiate more on Synthemax II plates than on PVA-24 h-1000 hydrogels and Matrigels. This can probably be explained by the extremely stiff base of the Synthemax II cell culture plates, as Synthemax II is coated on TCPS (approximately 3 GPa modulus). If the base of the Synthemax II cell culture plates was selected to be a softer material than TCPS, then hPSCs should maintain more pluripotency on the Synthemax II coated on the soft material.

hPSCs can be cultured for a long period on Matrigels and maintain their pluripotency because Matrigels are composed of several growth factors and ECMs that support the pluripotency of hPSCs. Although Matrigels are loaded on the extremely stiff surface of TCPS, the top layer of Matrigel on the plate is expected to produce a somewhat soft hydrogel (in comparison with TCPS alone) for hPSCs, which enables hPSCs to grow and maintain their pluripotency during long-term culture.

hESCs were reported to be cultured on polyacrylamide hydrogels with immobilized glycosaminoglycan, which had elastic moduli of 0.7, 3, and 10 kPa (Musah et al. 2012). In this study, hESCs could be grown on only relatively stiff hydrogels (10 kPa), which is similar to the results of the present study indicating that very soft hydrogels (0.7–3 kPa) cannot allow hESCs to attach to the surface. Unfortunately, their study of hESCs cultured on glycosaminoglycan-binding polyacrylamide hydrogels did not investigate hESCs culture on hydrogels stiffer than 10–12 kPa; our results indicated that the optimal stiffness of the hydrogels was approximately 25 kPa. However, their study suggests that hPSCs are able to receive mechanical information from cell culture hydrogels using glycosaminoglycan engagement that may activate specific signaling pathways relating to pluripotency.

Hydrogels with heparin-mimicking sites and several different moduli were prepared by Chang et al. by changing the crosslinking intensity (55, 140, and 345 kPa) (Chang et al. 2013). hESCs could attach slightly to the hydrogels with a low stiffness (55 kPa), whereas hESCs could attach to the hydrogels with a moderate stiffness (140 kPa); these cells would spontaneously differentiate. hESCs showed excellent attachment and could maintain their pluripotency on the most rigid hydrogels (345 kPa) for more than 20 passages. Although these hydrogels do not possess binding sites for hPSCs, FGF-2-binding moieties originating from heparin-mimicking moieties may contribute to the hPSCs binding to the hydrogels. In particular, FGF-2 could bind to a lesser extent to their softer hydrogels due to the lower density of heparin-mimicking sites. The minimum stiffness requirement of hydrogels for the attachment of hPSCs indicates the existence of a minimum density of FGF-2-binding moieties for hPSC binding. In our study, PVA-24-1000 hydrogels having a modulus of E'= 25 kPa were found to be the optimal materials for maintaining long-term cultures of pluripotent hPSCs. Both our present study and Chang's study showed that there is an optimal hydrogel stiffness that supports the long-term culture of hPSCs. The difference of optimal stiffness between our present study and Chang's study is probably due to the different hydrogel materials used.

We should consider optimizing the stiffness (physical cues) of the materials used for hPSC expansion and fine-tuning the surface density of cell-binding moieties (biological cues) to be important steps of designing cell culture materials for hPSC culture in feeder-free and xeno-free conditions. PVA hydrogels with concentrations of oligoVN higher than 500 μg/mL and an optimal stiffness of 25 kPa were the optimal cell culture materials for hPSC expansion in this study. The PVA-24-1500, PVA-24-1000, and PVA-24-500 hydrogels exhibited lower differentiation ratios and excellent attachment ratios for hPSCs compared with those of the Synthemax II plates (commercial plates). Because Synthemax II plates are prepared by coating bioactive polymers on stiff TCPS plates, Synthemax II plates do not possess the optimal stiffness for hPSC culture, whereas the PVA-oligoVN hydrogels can easily be made with the optimal stiffness for hPSC culture by controlling the crosslinking intensity. hPSCs could be cultivated on PVA-24-1000 hydrogels for many passages (10–20 passages) while maintaining their pluripotency and their ability to differentiate into cells of the three germ layers both in vitro and in vivo (Higuchi et al. 2015a).

5 Conclusion and Future Perspective

This research studied the effect of hydrogel elasticity on the culture of hHSPCs, hAFSCs, and hPSCs. PVA hydrogels having an E' of 12–30 kPa were found to be efficient cell culture materials for hHSPC expansion ex vivo. Among the PVA hydrogels and TCPS plates investigated in this study, the PVA-6-FN and PVA-6-CS1 hydrogels were found to be the best cell culture materials for hHSPCs exhibiting high expansion and capable of generating many multipotent CFU-GMs and CFU-GEMMs. The XPS analysis suggested that the surface density of FN on PVA-6-FN gels was extensively less than that on PVA-24-FN and PVA-48-FN hydrogels. Furthermore, hHSPCs cultured on PVA-1-CS1 hydrogels expansion slightly less and generated fewer CFU-GEMMs and CFU-GMs than hHSPCs cultured on PVA-6-CS1, PVA-24-CS1, and PVA-48-CS1 hydrogels, although the surface density of CS1 on PVA-1-CS1 was almost equal to that on PVA-6-CS1, PVA-24-CS1, and PVA-48-CS1 hydrogels, as determined via XPS analysis. The presentation of the specific CS1 cell-binding domains to hHSPCs at high concentrations relative to the density of FN should be an essential point for the signal transduction pathways promoting hHSPC expansion and maintaining hHSPC stemness. Therefore, both biological cues (i.e., surface density of CS1 and FN) and physical cues (i.e., cell culture material elasticity) are key factors in determining the success of efficient hHSPC expansion and stemness maintenance ex vivo.

hAFSCs were cultivated on stiff and soft PVA hydrogels with and without immobilized ECM-derived oligopeptides and ECMs to investigate the spontaneous differentiation and pluripotency of stem cells in expansion medium. It is found to be important for the PVA hydrogels with and without immobilized ECM-derived oligopeptides and ECMs to possess a minimum storage modulus ($E' > 12$ kPa) to support hAFSC expansion. The soft PVA-6-oligoECM and PVA-6-ECM hydrogels ($E' = 12$ kPa) were found to be optimal for culturing hAFSCs with high pluripotency, as evaluated by protein and gene expression analyses, compared with the relatively stiff PVA-24-oligoECM and PVA-24-ECM hydrogels ($E' = 25$ kPa), as well as with TCPS coated with CELLstart (3 GPa of elastic modulus) and TCPS plates. Soft PVA-6-oligoVN and PVA-6-oligoFN hydrogels were found to be the most optimal cell culture plates for supporting hAFSC pluripotency and spontaneous differentiation into early-stage neural cells, as determined from high *Nestin* expression. hAFSCs cultured on relatively stiff PVA-24-oligoFN, PVA-24-FN, PVA-24-oligoVN, and PVA-24-COL-B hydrogels spontaneously differentiated into early osteoblasts, as determined from high *Runx2* expression. We found that hAFSCs showing high levels of pluripotency genes (*Nanog*, *Sox2,* and *Oct4*) on PVA-oligoECM and PVA-ECM hydrogels also expressed high levels of genes relating to the three early germ layers (*Nestin, Runx2*, and *Sox17*) in expansion medium. This evidence indicates that the hAFSC population of stem cells is inhomogeneous. Therefore, hAFSCs contain stem cells that spontaneously differentiate as well as those with high pluripotency. Both physical cues (stiffness) and biochemical cues (cell adhesive oligopeptides) of cell culture materials are key factors for supporting the stemness of hAFSCs.

We also developed PVA-oligoVN hydrogels with a variety of stiffnesses for the xeno-free culture of hESCs and hiPSCs, supporting their pluripotency for more than 20 passages. The best stiffness of the PVA-oligoVN hydrogels was found to be 25.3 kPa. Furthermore, high concentrations of oligoVN (500–1500 μg/mL) should be used in the preparation of PVA-oligoVN hydrogels to ensure a sufficient surface density of oligoVN on the hydrogels for maintaining hPSC pluripotency. Optimizing the stiffness (physical cues) and fine-tuning the surface density of the cell-binding moieties (biological cues) of hydrogels are the key factors for designing hPSC culture materials that will support pluripotency in xeno-free conditions.

Acknowledgements This research was partially supported by the Ministry of Science and Technology, Taiwan, under grant number 104-2221-E-008-107-MY3. This work was also supported by the LandSeed Hospital project (NCU-LSH-105-A-001) and the Cathay General Hospital Project (105CGH-NCU-A3). A Grant-in-Aid for Scientific Research (15K06591) from the Ministry of Education, Culture, Sports, Science, and Technology of Japan is also acknowledged. The Deanship of Scientific Research, College of Science Research Centre, King Saud University, Kingdom of Saudi Arabia, is also acknowledged.

References

Brafman DA, Chang CW, Fernandez A, Willert K, Varghese S, Chien S (2010) Long-term human pluripotent stem cell self-renewal on synthetic polymer surfaces. Biomaterials 31:9135–9144

Calvi LM, Adams GB, Weibrecht KW, Weber JM, Olson DP, Knight MC, Martin RP, Schipani E, Divieti P, Bringhurst FR, Milner LA, Kronenberg HM, Scadden DT (2003) Osteoblastic cells regulate the haematopoietic stem cell niche. Nature 425:841–846

Chang CW, Hwang Y, Brafman D, Hagan T, Phung C, Varghese S (2013) Engineering cell-material interfaces for long-term expansion of human pluripotent stem cells. Biomaterials 34:912–921

Chen LY, Chang Y, Shiao JS, Ling QD, Chang Y, Chen YH, Chen DC, Hsu ST, Lee HHC, Higuchi A (2012) Effect of the surface density of nanosegments immobilized on culture dishes on ex vivo expansion of hematopoietic stem and progenitor cells from umbilical cord blood. Acta Biomater 8:1749–1758

Chen X, Prowse AB, Jia Z, Tellier H, Munro TP, Gray PP, Monteiro MJ (2014) Thermoresponsive worms for expansion and release of human embryonic stem cells. Biomacromol 15:844–855

Chowdhury F, Li Y, Poh YC, Yokohama-Tamaki T, Wang N, Tanaka TS (2010) Soft substrates promote homogeneous self-renewal of embryonic stem cells via downregulating cell-matrix tractions. PLoS ONE 5:e15655
Chua KN, Chai C, Lee PC, Ramakrishna S, Leong KW, Mao HQ (2007) Functional nanofiber scaffolds with different spacers modulate adhesion and expansion of cryopreserved umbilical cord blood hematopoietic stem/progenitor cells. Exp Hematol 35:771–781
Chua KN, Chai C, Lee PC, Tang YN, Ramakrishna S, Leong KW, Mao HQ (2006) Surface-aminated electrospun nanofibers enhance adhesion and expansion of human umbilical cord blood hematopoietic stem/progenitor cells. Biomaterials 27:6043–6051
Copelan EA (2006) Hematopoietic stem-cell transplantation. New Eng J Med 354:1813–1826
De Coppi P, Bartsch G, Siddiqui MM, Xu T, Santos CC, Perin L, Mostoslavsky G, Serre AC, Snyder EY, Yoo JJ, Furth ME, Soker S, Atala A (2007) Isolation of amniotic stem cell lines with potential for therapy. Nat Biotechnol 25:100–106
Dellatore SM, Garcia AS, Miller WM (2008) Mimicking stem cell niches to increase stem cell expansion. Curr Opin Biotechnol 19:534–540
Deng Y, Zhang X, Zhao X, Li Q, Ye Z, Li Z, Liu Y, Zhou Y, Ma H, Pan G, Pei D, Fang J, Wei S (2013) Long-term self-renewal of human pluripotent stem cells on peptide-decorated poly (OEGMA-co-HEMA) brushes under fully defined conditions. Acta Biomater 9:8840–8850
Di Maggio N, Piccinini E, Jaworski M, Trumpp A, Wendt DJ, Martin I (2011) Toward modeling the bone marrow niche using scaffold-based 3D culture systems. Biomaterials 32:321–329
Doran MR, Markway BD, Aird IA, Rowlands AS, George PA, Nielsen LK, Cooper-White JJ (2009) Surface-bound stem cell factor and the promotion of hematopoietic cell expansion. Biomaterials 30:4047–4052
Engler AJ, Sen S, Sweeney HL, Discher DE (2006) Matrix elasticity directs stem cell lineage specification. Cell 126:677–689
Fan Y, Hsiung M, Cheng C, Tzanakakis ES (2014) Facile engineering of xeno-free microcarriers for the scalable cultivation of human pluripotent stem cells in stirred suspension. Tissue Eng Part A 20:588–599
Feng Q, Chai C, Jiang XS, Leong KW, Mao HQ (2006) Expansion of engrafting human hematopoietic stem/progenitor cells in three-dimensional scaffolds with surface-immobilized fibronectin. J Biom Mater Res A 78:781–791
Flores-Guzman P, Fernandez-Sanchez V, Valencia-Plata I, Arriaga-Pizano L, Alarcon-Santos G, Mayani H (2013) Comparative in vitro analysis of different hematopoietic cell populations from human cord blood: in search of the best option for clinically oriented ex vivo cell expansion. Transfusion 53:668–678
Franke K, Pompe T, Bornhauser M, Werner C (2007) Engineered matrix coatings to modulate the adhesion of CD133 + human hematopoietic progenitor cells. Biomaterials 28:836–843
Fujimoto N, Fujita S, Tsuji T, Toguchida J, Ida K, Suginami H, Iwata H (2007) Microencapsulated feeder cells as a source of soluble factors for expansion of CD34(+) hematopoietic stem cells. Biomaterials 28:4795–4805
Gilbert PM, Havenstrite KL, Magnusson KE, Sacco A, Leonardi NA, Kraft P, Nguyen NK, Thrun S, Lutolf MP, Blau HM (2010) Substrate elasticity regulates skeletal muscle stem cell self-renewal in culture. Science 329:1078–1081
Gori JL, Chandrasekaran D, Kowalski JP, Adair JE, Beard BC, D'Souza SL, Kiem HP (2012) Efficient generation, purification, and expansion of CD34(+) hematopoietic progenitor cells from nonhuman primate-induced pluripotent stem cells. Blood 120:e35–e44
Higuchi A, Huang SC, Shen PY, Ling QD, Zhao JK, Chang Y, Wang HC, Bing JT, Hsu ST (2011a) Differentiation ability of amniotic fluid-derived stem cells cultured on extracellular matrix-immobilized surface. Curr Nanosci 7:893–901
Higuchi A, Kao SH, Ling QD, Chen YM, Li HF, Alarfaj AA, Munusamy MA, Murugan K, Chang SC, Lee HC, Hsu ST, Kumar SS, Umezawa A (2015a) Long-term xeno-free culture of human pluripotent stem cells on hydrogels with optimal elasticity Sci Rep 5:18136

Higuchi A, Lin FL, Cheng YK, Kao TC, Kumar SS, Ling QD, Hou CH, Chen DC, Hsu ST, Wu GJ (2014a) Preparation of induced pluripotent stem cells on dishes grafted on oligopeptide under feeder-free conditions. J Taiwan Inst Chem Eng 45:295–301

Higuchi A, Ling QD, Chang Y, Hsu ST, Umezawa A (2013) Physical cues of biomaterials guide stem cell differentiation fate. Chem Rev 113:3297–3328

Higuchi A, Ling QD, Hsu ST, Umezawa A (2012) Biomimetic cell culture proteins as extracellular matrices for stem cell differentiation. Chem Rev 112:4507–4540

Higuchi A, Ling QD, Ko YA, Chang Y, Umezawa A (2011b) Biomaterials for the feeder-free culture of human embryonic stem cells and induced pluripotent stem cells. Chem Rev 111:3021–3035

Higuchi A, Ling QD, Kumar S, Munusamy M, Alarfajj AA, Umezawa A, Wu GJ (2014b) Design of polymeric materials for culturing human pluripotent stem cells: progress toward feeder-free and xeno-free culturing. Prog Polym Sci 39:1348–1374

Higuchi A, Ling QD, Kumar SS, Chang Y, Alarfaj AA, Munusamy MA, Murugan K, Hsu ST, Umezawa A (2015b) Physical cues of cell culture materials lead the direction of differentiation lineages of pluripotent stem cells. J Mater Chem B 3:8032–8058

Higuchi A, Yang ST, Li PT, Chang Y, Tsai EM, Chen YH, Chen YJ, Wang HC, Hsu ST (2009) Polymeric materials for ex vivo expansion of hematopoietic progenitor and stem cells. Polym Rev 49:181–200

Higuchi A, Yang ST, Li PT, Tamai M, Tagawa Y, Chang Y, Chang Y, Ling QD, Hsu ST (2010) Direct ex vivo expansion of hematopoietic stem cells from umbilical cord blood on membranes. J Membr Sci 351:104–111

Holmes T, Yan F, Ko KH, Nordon R, Song E, O'Brien TA, Dolnikov A (2012) Ex vivo expansion of cord blood progenitors impairs their short-term and long-term repopulating activity associated with transcriptional dysregulation of signalling networks. Cell Prolif 45:266–278

Huebsch N, Arany PR, Mao AS, Shvartsman D, Ali OA, Bencherif SA, Rivera-Feliciano J, Mooney DJ (2010) Harnessing traction-mediated manipulation of the cell/matrix interface to control stem-cell fate. Nat Mater 9:518–526

Irwin EF, Gupta R, Dashti DC, Healy KE (2011) Engineered polymer-media interfaces for the long-term self-renewal of human embryonic stem cells. Biomaterials 32:6912–6919

Jiang XS, Chai C, Zhang Y, Zhuo RX, Mao HQ, Leong KW (2006) Surface-immobilization of adhesion peptides on substrate for ex vivo expansion of cryopreserved umbilical cord blood CD34 + cells. Biomaterials 27:2723–2732

Jonas SJ, Alva HA, Richardson W, Sherman SP, Galic Z, Pyle AD, Dunn B (2013) A spatially and chemically defined platform for the uniform growth of human pluripotent stem cells. Mater Sci Eng, C 33:234–241

Keeney M, Chin-Yee I, Weir K, Popma J, Nayar R, Sutherland DR (1998) Single platform flow cytometric absolute CD34 + cell counts based on the ISHAGE guidelines. Cytometry 34:61–70

Kerst JM, Sanders JB, Slaper-Cortenbach IC, Doorakkers MC, Hooibrink B, van Oers RH, von dem Borne AE, van der Schoot CE (1993) Alpha 4 beta 1 and alpha 5 beta 1 are differentially expressed during myelopoiesis and mediate the adherence of human CD34 + cells to fibronectin in an activation-dependent way. Blood 81:344–351

Kishore V, Eliason JF, Matthew HW (2011) Covalently immobilized glycosaminoglycans enhance megakaryocyte progenitor expansion and platelet release. J Biomed Mater Res A 96:682–692

Kumar SS, Hsiao JH, Ling QD, Dulinska-Molak I, Chen GP, Chang Y, Chang Y, Chen YH, Chen DC, Hsu ST, Higuchi A (2013) The combined influence of substrate elasticity and surface-grafted molecules on the ex vivo expansion of hematopoietic stem and progenitor cells. Biomaterials 34:7632–7644

Lanniel M, Huq E, Allen S, Buttery L, Williams PM, Alexander MR (2011) Substrate induced differentiation of human mesenchymal stem cells on hydrogels with modified surface chemistry and controlled modulus. Soft Matter 7:6501–6514

Lin PY, Hung SH, Yang YC, Liao LC, Hsieh YC, Yen HJ, Lu HE, Lee MS, Chu IM, Hwang SM (2014) A synthetic peptide-acrylate surface for production of insulin-producing cells from human embryonic stem cells. Stem Cells Dev 23:372–379

Liu L, Yoshioka M, Nakajima M, Ogasawara A, Liu J, Hasegawa K, Li S, Zou J, Nakatsuji N, Kamei K, Chen Y (2014) Nanofibrous gelatin substrates for long-term expansion of human pluripotent stem cells. Biomaterials 35:6259–6267

Lu HF, Chai C, Lim TC, Leong MF, Lim JK, Gao S, Lim KL, Wan AC (2014) A defined xeno-free and feeder-free culture system for the derivation, expansion and direct differentiation of transgene-free patient-specific induced pluripotent stem cells. Biomaterials 35:2816–2826

Melkoumian Z, Weber JL, Weber DM, Fadeev AG, Zhou Y, Dolley-Sonneville P, Yang J, Qiu L, Priest CA, Shogbon C, Martin AW, Nelson J, West P, Beltzer JP, Pal S, Brandenberger R (2010) Synthetic peptide-acrylate surfaces for long-term self-renewal and cardiomyocyte differentiation of human embryonic stem cells. Nat Biotech 28:606–610

Mendez-Ferrer S, Michurina TV, Ferraro F, Mazloom AR, Macarthur BD, Lira SA, Scadden DT, Ma'ayan A, Enikolopov GN, Frenette PS (2010) Mesenchymal and haematopoietic stem cells form a unique bone marrow niche. Nature 466:829–834

Miyazaki T, Futaki S, Suemori H, Taniguchi Y, Yamada M, Kawasaki M, Hayashi M, Kumagai H, Nakatsuji N, Sekiguchi K, Kawase E (2012) Laminin E8 fragments support efficient adhesion and expansion of dissociated human pluripotent stem cells. Nat Commun 3:1236

Mortera-Blanco T, Mantalaris A, Bismarck A, Aqel N, Panoskaltsis N (2011) Long-term cytokine-free expansion of cord blood mononuclear cells in three-dimensional scaffolds. Biomaterials 32:9263–9270

Murphy WL, McDevitt TC, Engler AJ (2014) Materials as stem cell regulators. Nat Mater 13: 547–557

Musah S, Morin SA, Wrighton PJ, Zwick DB, Jin S, Kiessling LL (2012) Glycosaminoglycan-binding hydrogels enable mechanical control of human pluripotent stem cell self-renewal. ACS Nano 6:10168–10177

Nagaoka M, Si-Tayeb K, Akaike T, Duncan SA (2010) Culture of human pluripotent stem cells using completely defined conditions on a recombinant E-cadherin substratum. BMC Dev Biol 10:60

Nandivada H, Villa-Diaz LG, O'Shea KS, Smith GD, Krebsbach PH, Lahann J (2011) Fabrication of synthetic polymer coatings and their use in feeder-free culture of human embryonic stem cells. Nat Protoc 6:1037–1043

Park DH, Lee JH, Borlongan CV, Sanberg PR, Chung YG, Cho TH (2011a) Transplantation of umbilical cord blood stem cells for treating spinal cord injury. Stem Cell Rev 7:181–194

Park HJ, Yang K, Kim MJ, Jang J, Lee M, Kim DW, Lee H, Cho SW (2015) Bio-inspired oligovitronectin-grafted surface for enhanced self-renewal and long-term maintenance of human pluripotent stem cells under feeder-free conditions. Biomaterials 50:127–139

Park JS, Chu JS, Tsou AD, Diop R, Tang ZY, Wang AJ, Li S (2011b) The effect of matrix stiffness on the differentiation of mesenchymal stem cells in response to TGF-beta. Biomaterials 32:3921–3930

Pennington BO, Clegg DO, Melkoumian ZK, Hikita ST (2015) Defined culture of human embryonic stem cells and xeno-free derivation of retinal pigmented epithelial cells on a novel, synthetic substrate. Stem Cells Transl Med 4:165–177

Qian X, Villa-Diaz LG, Kumar R, Lahann J, Krebsbach PH (2014) Enhancement of the propagation of human embryonic stem cells by modifications in the gel architecture of PMEDSAH polymer coatings. Biomaterials 35:9581–9590

Remberger M, Mattsson J, Olsson R, Ringden O (2011) Second allogeneic hematopoietic stem cell transplantation: a treatment for graft failure. Clin Transplantation 25:E68–E76

Rocha V, Labopin M, Sanz G, Arcese W, Schwerdtfeger R, Bosi A, Jacobsen N, Ruutu T, de Lima M, Finke J, Frassoni F, Gluckman E (2004) Transplants of umbilical-cord blood or bone marrow from unrelated donors in adults with acute leukemia. New Eng J Med 351:2276–2285

Rodin S, Domogatskaya A, Strom S, Hansson EM, Chien KR, Inzunza J, Hovatta O, Tryggvason K (2010) Long-term self-renewal of human pluripotent stem cells on human recombinant laminin-511. Nat Biotech 28:611–615

Rodriguez-Pardo VM, Vernot JP (2013) Mesenchymal stem cells promote a primitive phenotype CD34 + c-kit + in human cord blood-derived hematopoietic stem cells during ex vivo expansion. Cell Mol Biol Lett 18:11–33

Roy S, Tripathy M, Mathur N, Jain A, Mukhopadhyay A (2012) Hypoxia improves expansion potential of human cord blood-derived hematopoietic stem cells and marrow repopulation efficiency. Eur J Haematol 88:396–405

Saha K, Keung AJ, Irwin EF, Li Y, Little L, Schaffer DV, Healy KE (2008) Substrate modulus directs neural stem cell behavior. Biophy J 95:4426–4438

Salati S, Lisignoli G, Manferdini C, Pennucci V, Zini R, Bianchi E, Norfo R, Facchini A, Ferrari S, Manfredini R (2013) Co-culture of hematopoietic stem/progenitor cells with human osteblasts favours mono/macrophage differentiation at the expense of the erythroid lineage. PLoS ONE 8: e53496

Stephenson E, Jacquet L, Miere C, Wood V, Kadeva N, Cornwell G, Codognotto S, Dajani Y, Braude P, Ilic D (2012) Derivation and propagation of human embryonic stem cell lines from frozen embryos in an animal product-free environment. Nat Protoc 7:1366–1381

Takahashi K, Tanabe K, Ohnuki M, Narita M, Ichisaka T, Tomoda K, Yamanaka S (2007) Induction of pluripotent stem cells from adult human fibroblasts by defined factors. Cell 131:861–872

Trappmann B, Gautrot JE, Connelly JT, Strange DG, Li Y, Oyen ML, Cohen Stuart MA, Boehm H, Li B, Vogel V, Spatz JP, Watt FM, Huck WT (2012) Extracellular-matrix tethering regulates stem-cell fate. Nat Mater 11:642–649

Tse JR, Engler AJ (2011) Stiffness gradients mimicking in vivo tissue variation regulate mesenchymal stem cell fate. PLoS ONE 6:e15978

Tsutsui H, Valamehr B, Hindoyan A, Qiao R, Ding X, Guo S, Witte ON, Liu X, Ho CM, Wu H (2011) An optimized small molecule inhibitor cocktail supports long-term maintenance of human embryonic stem cells. Nat Commun 2:167

Villa-Diaz LG, Nandivada H, Ding J, Nogueira-de-Souza NC, Krebsbach PH, O'Shea KS, Lahann J, Smith GD (2010) Synthetic polymer coatings for long-term growth of human embryonic stem cells. Nat Biotech 28:581–583

Wang PY, Lee HHC, Higuchi A, Ling QD, Lin HR, Li HF, Kumar SS, Chang Y, Alarfaj AA, Munusamy MA, Chen DC, Hsu ST, Wang HC, Hsiao HY, Wu GJ (2015) Pluripotency maintenance of amniotic fluid-derived stem cells cultured on biomaterials. J Mater Chem B 3:3858–3869

Wen JH, Vincent LG, Fuhrmann A, Choi YS, Hribar KC, Taylor-Weiner H, Chen S, Engler AJ (2014) Interplay of matrix stiffness and protein tethering in stem cell differentiation. Nat Mater 13:979–987

Winer JP, Janmey PA, McCormick ME, Funaki M (2009) Bone marrow-derived human mesenchymal stem cells become quiescent on soft substrates but remain responsive to chemical or mechanical stimuli. Tissue Eng Part A 15:147–154

Wu S, Johansson J, Damdimopoulou P, Shahsavani M, Falk A, Hovatta O, Rising A (2014) Spider silk for xeno-free long-term self-renewal and differentiation of human pluripotent stem cells. Biomaterials 35:8496–8502

Xie Y, Yin T, Wiegraebe W, He XC, Miller D, Stark D, Perko K, Alexander R, Schwartz J, Grindley JC, Park J, Haug JS, Wunderlich JP, Li H, Zhang S, Johnson T, Feldman RA, Li L (2009) Detection of functional haematopoietic stem cell niche using real-time imaging. Nature 457:97–101

Yamamoto S, Ikeda H, Toyama D, Hayashi M, Akiyama K, Suzuki M, Tanaka Y, Watanabe T, Fujimoto Y, Hosaki I, Nishihira H, Isoyama K (2011) Quality of long-term cryopreserved umbilical cord blood units for hematopoietic cell transplantation. Int J Hematol 93:99–105

Zhang R, Mjoseng HK, Hoeve MA, Bauer NG, Pells S, Besseling R, Velugotla S, Tourniaire G, Kishen RE, Tsenkina Y, Armit C, Duffy CR, Helfen M, Edenhofer F, de Sousa PA, Bradley M (2013) A thermoresponsive and chemically defined hydrogel for long-term culture of human embryonic stem cells. Nat Commun 4:1335

Zheng YB, Gao ZL, Xie C, Zhu HP, Peng L, Chen JH, Chong YT (2008) Characterization and hepatogenic differentiation of mesenchymal stem cells from human amniotic fluid and human bone marrow: a comparative study. Cell Biol Int 32:1439–1448

Chapter 15
Various Functional and Stimuli-Responsive Hydrogel Based on Polyaspartamides

Bo Wang and Ji-Heung Kim

Abstract Polyaspartamides are amide derivatives of biodegradable poly(aspartic acid), the thermal polycondensate of aspartic acid, which have been investigated as carriers for macromolecular prodrugs, platforms for many functional polymers and gels, and macromolecular nano-assemblies for drug delivery system. In this chapter, various different stimuli-responsive (temperature, pH, ion, redox, CO_2, etc.), physical and chemical, hydrogel systems based on polyaspartamides will be reviewed with interests for the development of various smart biomaterials and industrial polymeric gels.

Keywords Polyaspartamide · Hydrogel · Poly(aspartic acid) · Stimuli-responsive Biomaterials

Covalent or non-covalent bonding (hydrogen-bonding interactions, metal-ligand coordination bonding, or guest–host interactions) can form soft three-dimensional polymeric network hydrogels, which can be specified as chemical hydrogels and physical hydrogels. Hydrogels prepared from both natural and man-made polymers have persistently attracted interests for drug encapsulation, contact lenses, biosensors, etc., and also such hydrogels have become of great interest to the field of "tissue engineering" as matrices for repair and regeneration of various organs and tissue (Domb et al. 1998; Schmaljohann 2006; Ballauff and Lu 2007; Thakur and Thakur 2014, 2015). Hydrogels are cross-linked network of hydrophilic or water-soluble polymer that may absorb water from 10 to 20% (an arbitrary lower limit) to more than hundreds of times of their dry weight. By using chemical and physical cross-linking processes, the corresponding gel can be prepared from a variety of hydrophilic polymers. More recently, stimuli-responsive polymers have attracted great attention for use in biotechnology and medicines. The stimuli-responsive and smart hydrogels are capable to undergo reversible volume

---

B. Wang · J.-H. Kim (✉)
Department of Chemical Engineering, Sungkyunkwan University,
Suwon 16419, Republic of Korea
e-mail: kimjh@skku.edu

V. K. Thakur and M. K. Thakur (eds.), *Hydrogels*, Gels Horizons: From Science to Smart Materials, https://doi.org/10.1007/978-981-10-6077-9_15

change in response to a specific stimulus, such as light, temperature, ionic strength, pH, and electric field.

Poly(aspartic acid), abbreviated as PASP, is one of biodegradable and water-soluble poly(amino acid)s which possesses carboxylic acid groups on the side chain. PASP has been considered as an attractive alternative of conventional non-biodegradable polymers in various industrial and biomedical fields. Polyaspartamides are generally termed for all kinds of amide derivatives of poly (aspartic acid), which has been widely investigated as carriers for macromolecular prodrugs, platforms for many functional polymers or gels, and macromolecular nano-assemblies for drug delivery system (DDS) due to its excellent performances such as inherent biodegradability, water-soluble and biocompatible nature, and multi-point drug attachment. Poly(*N*-2-hydroxyethyl-D,L-aspartamide), abbreviated as PHEA, is one example polymer which has hydroxyl pendent groups on the polyamide backbone. By the modification reaction of precursor polymer, polysuccinimide (PSI), typically by aminolysis reaction, or via a secondary reaction with the –COOH or –OH groups of PASP and PHEA, various kinds of biodegradable functional polymers or the cross-linked gel as well as its macromolecular self-assembly structures can be afforded. Many of these polymers show physicochemical features appropriate for the evolution of biological materials in the forms of polymer prodrugs, hydrogels, or nano-materials. On the other hand, stimuli-responsive polymers and hydrogel materials have received a great attention in recent years and showed substantial promise in the fields of cancer drug delivery and regenerative medicines.

To date, various functional polymers and gels based on polyaspartamides have been investigated with interests to identify themselves as novel and promising biomaterials. In this chapter, the research works on the synthesis, properties, and diverse applications of polyaspartamide-based hydrogels will be reviewed including some related works on the poly(aspartic acid).

1 Synthesis of Poly(Aspartic Acid) and Polyaspartamides

The significance of biodegradable polymers and hydrogels is increasingly approved, and widespread researches have been carried out for various different biological and medical applications. Polymers of α-amino acids with the molecular formula of H_2N–CH(R)–COOH are called as poly(amino acid)s or polypeptides. Poly(amino acid)s, containing amide bonds similar to protein structure, have been regarded as good biocompatible material and studied for a variety of biomedical and pharmaceutical applications. Poly(aspartic acid), PASP, is one of water-soluble and acidic poly(amino acid)s, which can be prepared from PSI through chemical alkaline hydrolysis (Nakato et al. 1998; Thombre and Sarwade 2005). PSI, the reactive precursor polymer, can be readily synthesized by the thermal polycondensation of aspartic acid or the ammonium salts of maleic acid and malic acid (Neri et al. 1973; Tomida et al. 1996). Aspartic acid itself is often chemically synthesized from maleic anhydride, or by biological fermentation from glucose and fumaric acid. PASP, obtained by thermal reaction processes, was known to have an approximate molar ratio of 30% α- and 70% β-moieties which was randomly distributed over the macromolecular chain (Fig. 1). Many research works on the methods of synthesis, structural analysis, and potential applications of PASP have been reported (Pivcova et al. 1981, 1982; Wolk et al. 1994; Freeman et al. 1996; Matsubara et al. 1998; Nakato et al. 1998; Tang and Wheeler 2001; Ross et al. 2001; Hiraishi 2015). Low et al. (1996) introduced economically viable poly(aspartic acid) having high purity, non-toxic, and environmentally compatible nature. Some grades of PASP are commercially available currently as dispersants, water treatment chemicals, and soil modifier.

PSI is a reactive aliphatic polyimide, which can be synthesized by polycondensation of aspartic acid monomer via high-temperature bulk or solution processes. Different methods of synthesis and reaction conditions can affect not only the yield but the purity of PSI; thus, the structure and molar mass of the polymer can affect the performance and the application of PASP derivatives. Nakato and coworkers studied systematically on the structure and properties of poly(aspartic acid). They have synthesized PSI at various different reaction conditions, using *o*-phosphoric acid as the catalyst in mesitylene/sulfolane mixed solvent (7:3 by

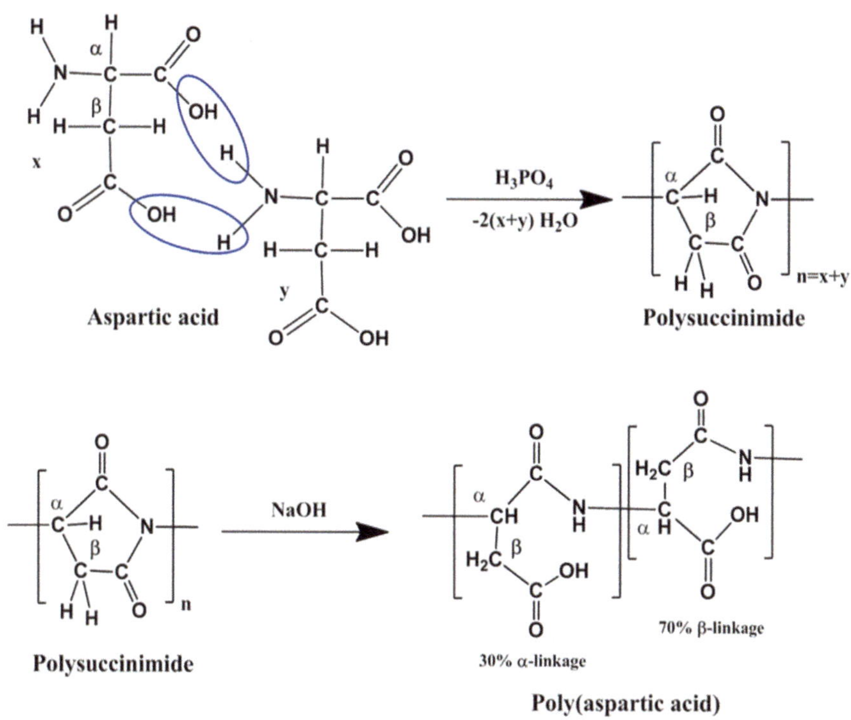

Fig. 1 Reaction scheme for thermal synthesis of PASP

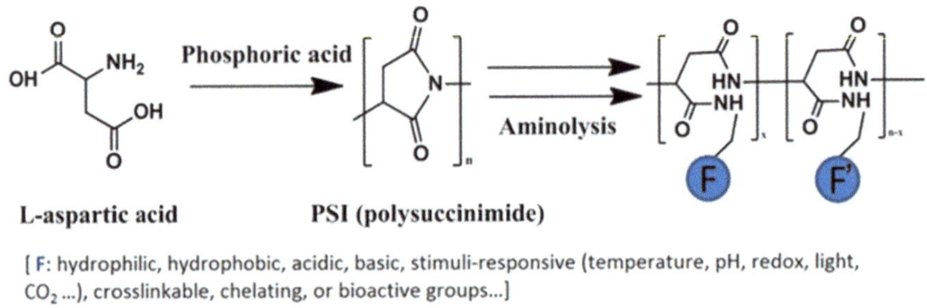

Fig. 2 Reaction scheme for various functional polyaspartamide derivatives by ring-opening aminolysis reaction of PSI

weight) and compared the effect on the PSI molecular weight (Nakato et al. 1998). Also, a high molecular weight PSI was prepared by the condensation polymerization of L-aspartic acid using excess amount of *o*-phosphoric under reduced pressure at high temperature (Tomida et al. 1996; Kim et al. 2011). By varying the reaction conditions (temperature, time, pressure), a wide range of molecular weight was obtained and the pressure (vacuum) control seemed important to give higher molecular weight PSI, up to around 160 K Dalton, as measured by GPC (Kim et al. 2011).

Polyaspartamides are the synthetic protein-like polymers, which have amide backbone and a variety of different side groups. Due to the reactivity of succinimide ring on PSI, polyaspartamides can be easily prepared with various combinations of pendants via aminolysis reaction process (Fig. 2). From the decades of researches, it has been assuredly established that the succinimide rings can be opened quantitatively by alkaline hydrolysis or aminolysis reaction, forming carboxylate or other amide derivatives to provide all different polyaspartamides, respectively. For example, α,β-poly(*N*-2-hydroxyethyl-DL-aspartamide), PHEA, one of typical derivative polymer containing hydroxyl pendants, is prepared simply by ring-opening aminolysis reaction of PSI with ethanolamine. Of course, a secondary polymer reaction upon PASP or one polyaspartamide containing specific functional groups can be used to prepare another polyaspartamide derivative or the related cross-linked gel by means of proper reaction chemistry.

2 Hydrogel Based on the Cross-linked PASPs—Superabsorbent Polymer

In the last decades, biodegradable hydrogels have attracted much attention and have been used for industrial and agricultural applications, such as superabsorbent, metal-chelating agent, soil stabilizer, and also for the applications in biomedical fields involving drug delivery system. Poly(aspartic acid) (PASP), one of poly(amino acid)s with carboxylic pendent groups, is a promising water-soluble and biodegradable polymer. In addition, PASP is also considered as a "green" substitute for poly(acrylic acid) and the copolymers which are traditional non-biodegradable vinyl polymers. Many researchers have introduced PASP-based hydrogels as a representative biodegradable superabsorbent (Yoshimura et al. 2005). PASP exhibits high water absorbency with pH and electrolyte sensitivity in its neutralized and cross-linked form. The superabsorbency of this hydrogel is owing to the polymer network structure of partially ionized and lightly cross-linked polycarboxylate. To convert PASP to the cross-linked gel form, various different methods including chemical and radiation processes have been employed. As a representative chemical approach, diamine compound or the polymeric analogs was used to cross-link PSI chain first, and then the resulting gel was hydrolyzed by adding sodium hydroxide or in alkaline buffer solution to provide the sodium polyaspartate hydrogel. Alternatively, the partially neutralized PASP can be cross-linked by using diepoxy compounds, for instance, ethylene glycol diglycidyl ether (EGDGE) or oligomeric PEG-diepoxide, to provide the corresponding PASP hydrogel (Kim et al. 2002). Umeda et al. (2011) prepared superabsorbent gel from allyl-modified PASP by radical reaction process. The swelling degree of PASP gel would depend on the preparation conditions, more specifically on the nature and amount of cross-linker compound used. Typical swelling ratios of PASP gel in pure water and saline solution are observed in the range of several hundred (g/g) and several tens (g/g), respectively. On the other hand, IPN or semi-IPN type composite gels from sodium polyaspartate and natural alginate, hyaluronate, or konjac glucomannan were prepared using both covalent and an ionic cross-linking method (for example Ca^{2+}). In addition, the swelling behavior and dye adsorption property of PASP hydrogels were examined, where the salt- and pH-sensitivity and hydrolytic degradability of the hydrogels were elucidated (Jeon et al. 2008; Liu and Urban 2010). Recently, pH- and temperature dual stimuli-responsive co-network hydrogel based on PASP was prepared by grafting thermo-sensitive PNIPAAm and used for drug delivery experiment (Némethy et al. 2013). Superswelling gels based on PASP are potentially useful in many applications such as personal care, water treatment, agricultural (water retaining, fertilizer) and biomedical areas.

In addition, Zrinyi and our group reported the synthesis of biodegradable PASP derivatives with various amino acid (L-lycine, leucine, γ-amino butylic acid (GABA), etc.) pendants in order to modify the polymer structure and properties and also to use them as bioactive macromolecules (Kim et al. 2008, 2011). Their amphiphilic analogs were also prepared to examine the nanoparticle formation with the physicochemical properties in aqueous medium. The materials were found to be non-toxic and exhibited cell growth activity, suggesting a potential use in pharmaceutical and cosmetic fields. These novel amino acid-conjugated PASPs could be cross-linked to the corresponding hydrogel using diepoxy compounds under mild condition. The hydrogels possessed swelling ratios of 100–150 g/g in water and 15–25 g/g in PBS solution and showed pH-sensitive and hydrolytic degradable characteristics (Son et al. 2011).

3 Polyaspartamides and the Cross-linked Gel

As mentioned above, polyaspartamide and the copolymeric analogs with a variety of different pendent groups can be obtained from the aminolysis reaction to PSI. For example, hydroxyl pendant groups of PHEA provide the sites for making covalent bond between the polymer and bioactive molecules, further chemical modification, and also cross-linking to form hydrogels. Giammona group in their early researches attached benzoic acid or its derivatives to PHEA by ester bonds in order to study the polymer as a macromolecular drug carrier (Giammona et al. 1987, 1995). In addition, Paclitaxel was attached to PHEA via a succinic spacer, and the physicochemical and biological characterization of this drug conjugate was performed. Another example of using polyaspartamide as a macromolecular prodrug was reported, where polymer-drug conjugate was prepared to improve the solubility in water, bioavailability, and to enhance the delivery of lipophilic estrogen hormone (Koncic et al. 2011).

Chemical hydrogel based on PHEA or PHEA derivatives were prepared either by direct coupling with cross-linking reagents such as glutaldehyde and hexamethylene diisocyanate (HMDI) (Kim et al. 2004a, b) or by cross-linking of PSI, the precursor polymer, with diamine compounds and subsequent aminolysis reaction by using excess ethanolamine. The swelling degrees were moderate with several tens of swelling ratios, much lower ranges compared to ionic PASP gels. The partial functionalization of PHEA with methacrylic anhydride or glycidyl methacrylate was attempted to introduce side chain reactive double bonds with ester groups (Giammona et al. 1999a, b). The resulting polymers were used to prepare the cross-linked gels and microparticles by chemical or photochemical method, where drug molecule could be encapsulated physically during the gel preparation process. The use of these PHEA-based hydrogels as drug delivery carrier and matrices for tissue engineering has been investigated. PHEA or glycidyl-modified PHEA was combined with natural polysaccharides including chitosan, inulin, hyaluronic acid, or dextran to prepare the hybrid gels with modified gel properties and biological suitability (Pitarresi et al. 2003). Meanwhile, our laboratory has recently introduced the preparation of PHEA-silica hybrid gels to improve mechanical property and biocompatibility of PHEA hydrogel. A process of sol–gel was utilized in this work using TEOS and ethoxysilane-conjugated PHEA (Jeon et al. 2009).

4 Stimuli-Responsive Smart Hydrogels

Stimuli-responsive and smart hydrogels have gained considerable attention due to their unique properties such as high water content, flexibility, and biocompatibility. In particular, smart polymers and hydrogels originated from amphiphilic copolymers have attracted great interest because of their promising potential applications in biomedical engineering and pharmaceutics (Domb et al. 1998; Ballauff and Lu 2007; Schmaljohann 2006).

Temperature, magnetic or electric fields, light, and mechanical stress, etc., belong to physical stimuli, which can alter the level of energy sources and change the molecular interactions (Ebara et al. 2014). On the other hand, chemical stimuli, such as pH, CO_2, redox, will change the interactions among polymer chains or between polymer chain and solvent at the molecular level. In some developments, two or more responsive mechanisms to stimuli were combined in one hydrogel system.

4.1 LCST Solution Behavior and Sol–Gel Transition from Polyaspartamide Copolymers

Nowadays, stimuli-responsive hydrophilic polymers have become very attractive for the application in biotechnology and medicine. Among them, thermo-responsive polymers having a lower critical solution temperature (LCST) were widely investigated due to their promising applications in controlled drug delivery, biomimetic actuators, chromatographic separations and tissue and cell engineering. Poly(*N*-isopropylacrylamide) (PNIPAAm) is the representative thermo-responsive polymer, which exhibits a rapid and reversible hydration–dehydration change in response to temperature near 32 °C in water. In addition to PNIPAAm, numerous other temperature-responsive polymers, including poly(vinyl methyl ether), poly (2-isopropyl-2-oxazoline), poly(*N*-vinyl alkyl amide)s, and polyphosphazenes, are known. However, the majority of them are non-biodegradable, thus restricting their applications in biomedical areas.

Poly(amino acid)s, as biocompatible and biodegradable materials, have been widely investigated due to their broad potential applications. For the potential uses in biomedical field, based on the reversible phase transition behavior, various approaches have been reported for the modification of structure of polyaspartamides. For instance, two ω-amino alcohols (C5 and C6) were attached to yield the poly(*N*-substituted α/β asparagines) which displayed a sharp LCST at the temperature ranges of 20–40 °C (Tachibana et al. 2003a, b). The measurement of dynamic light scattering (DLS) showed the particle size of the polymer depending on the temperature in aqueous solution, where the mean size was observed to be less than 10 nm diameter at the temperature below 25 °C, while the size was increased to over 200 nm diameter beyond 30 °C. Kim et al. investigated several amphiphilic copolymer systems including poly(*N*-isopropyl aminoethyl-co-alkyl aspartamide) and poly(*N*-isopropyl aminoethyl-co-hydroxyalkyl aspartamide), which displayed both temperature and pH-responsive LCST behavior in aqueous solution (Moon et al. 2010; Moon and Kim 2006, 2008). The LCSTs were controllable by changing the relative content of each pendent group through altering the hydrophobic and hydrophilic balance. In the former polyaspartamide system, *n*-alkyl chains (C6, C8, or C12) function as hydrophobic groups and *N*-isopropylethylene moiety as hydrophilic and stimuli-responsive groups.

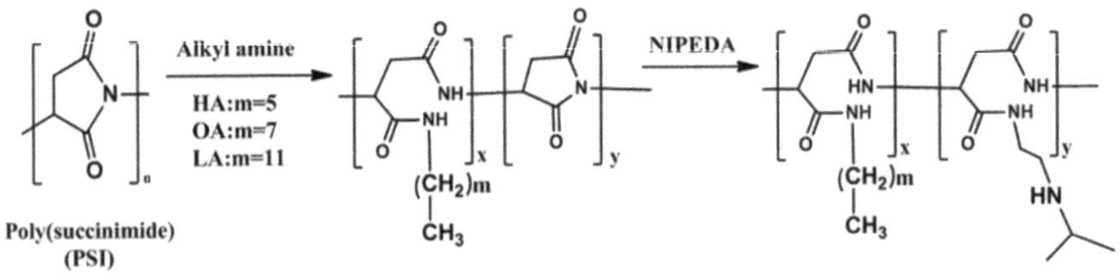

Fig. 3 Synthesis of amphiphilic poly(*N*-isopropyl aminoethyl-co-alkyl aspartamide)s, PolyAspAm(alkyl/NIPEDA)

Figure 3 represents the synthetic scheme for this special amphiphilic copolyaspartamides containing *N*-isopropyl amine and several different alkyl pendants. At the middle composition range, the system exhibited a LCST behavior (20–50 °C) in aqueous solution (Moon et al. 2009). The transition temperature (this system is also pH-dependent due to the ionization equilibrium of the basic isopropyl amine groups) could be regulated by changing the length and content of alkyl graft. The same amphiphilic copolymers formed nanoparticle in aqueous solution as evidenced by their SEM images, and the nanoparticle size increased by increasing temperature due to the aggregation of polymeric particles caused by the increased hydrophobic interactions between pendent groups (Moon and Kim 2008; Moon et al. 2009; Bach et al. 2008). Recently, biodegradable and thermo-sensitive polyaspartamide derivatives bearing pendent aromatic groups were also synthesized and their LCST solution behavior has been characterized (Ma et al. 2014).

Injectable polymeric hydrogels have been investigated for their attractive applications in the biomedical fields such as local drug delivery or therapeutic implant. The polymers, loaded with biologically active molecules or cells in aqueous media, can be converted into gels in situ via physical cross-linking which is triggered externally by temperature, pH, or light. Among them, thermo-responsive hydrogels have been vastly studied, which can experience fast conversion from sol (liquid) to gel phase without any additives at human body temperature. The typical examples are poly(ethylene oxide)-poly(propylene oxide)-poly(ethylene oxide) (Pluronic®) and block-type amphiphilic poly(vinyl ether)s. The behavior of sol–gel switching depends sharply on the composition and concentration of the block copolymer. The majority of the synthesized injectable hydrogels, however, are non-biodegradable, and this may limit their applications in biomedical field.

Amphiphilic block/graft copolymers based on polyaspartamides are potential candidates for injectable in situ hydrogels. As discussed previously, our group prepared several amphiphilic polyaspartamide copolymer systems showing LCST behavior from their dilute aqueous solutions. In a concentrated solution, a poly(*N*-isopropylaminoethyl-co-laurylaspartamide) copolymer system exhibited sol–gel transition as the sol-to-gel phase diagrams are given in Fig. 4. Upon the temperature increase, the transparent solution has changed to non-flowing gel, and the conversion was thermally reversible. In addition, the gel formation occurred near body

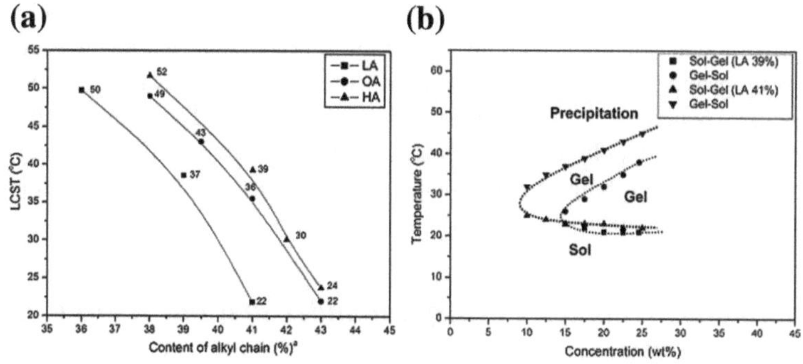

Fig. 4 LCSTs of PolyAspAm(alkyl/NIPEDA) solutions (1 wt%) with different alkyl contents (**a**), sol–gel phase diagrams of amphiphilic PolyAspAm(LA/NIPEDA) solutions at various concentrations (**b**)

temperature, suggesting the potential of this amphiphilic copolymer system as injectable polymer gel for biomedical applications (Moon and Kim 2006; Moon et al. 2009, 2011). Takeuchi et al. (2006) reported sol–gel transition of amphiphilic polyaspartamide derivatives containing both alkyl and hydroxyalkyl groups (Takeuchi et al. 2006). Also, a collection of polyaspartamides with *N*-isopropyl and hydroxyalkyl amide pendants were prepared, and their self-assembly to nanoparticle and temperature-responsive sol–gel phase transition behavior were observed (Gu et al. 2013). Interestingly, hydrophobic and biodegradable PLA, poly(lactic acid), was grafted onto PHEA polymer backbone to provide a novel graft copolymers PHEA-*g*-PLA. At high PLA content, the copolymer exhibited in situ gel forming behavior (Pitarresi et al. 2010). Lu et al. (2013) synthesized two PASP derivatives with complementary functional groups, i.e., hydrazine and aldehyde, and used them to prepare biodegradable injectable hydrogel under physiological conditions (Lu et al. 2013). Sun et al. introduced in situ hydrogel based on acryloyl-modified PHEA. The time for gelation, degree of swelling, gel content, and other physicochemical characteristics were measured to evaluate the feasibility as an injectable in situ gel. The transition from liquid to gel at body temperature was induced by chemical cross-linking by using redox initiator. By modulating the component composition, they could control the gelation time in a broad range from one min to tens of minutes (Sun et al. 2009).

4.2 Thermo-Responsive Polyaspartamides Chemical Gels

A thermo-responsive (also pH-responsive) polyaspartamide hydrogel was prepared from polyaspartamide derivative containing 2-diisopropylamine and hydroxyethyl pendants by using hexamethylene diisocyanate (HMDI) as the cross-linker (Moon and Kim 2010). The LCST solution behavior of this amphiphilic copolymer system and the thermo- and pH-responsive swelling properties of the cross-linked hydrogel were investigated together with the gel morphology and degradation behavior. Kobayashi group also synthesized thermo-responsive polyaspartamides with hydroxyalkyl pendants showing LCST behavior in water (Tachibana et al. 2003a, b). They prepared cross-linked hydrogels from the same series of polymers by using HMDI and discussed their swelling behavior in response to the temperature.

4.3 Photo-cross-linked Polyaspartamides and the Hybrid Gel with Polysaccharides

Compared with conventional methods, there is an advantage of using radiation chemistry in preparing biomaterials because it is possible to obtain pure products (Carenza 1992; Chapiro 1995). Recently, high energy radiation technology was widely employed in various different disciplines including biomaterials technology and biomedicine (Safranj et al. 1995; Morita and Kaetsu 1992; Li et al. 1991). γ-Irradiation is a common method to make cross-linked hydrogels, and it provides a simple, brief, and fast technique to generate sterilized hydrogels (Rosiak et al. 1995). Method of γ-irradiation in an aqueous media can have more virtues relative to irradiation in solid state, which can improve the reaction of polymers and render the cross-linking process easier due to the high chain mobility and high product yield of reactive components from water radiolysis (Rosiak et al. 1995; Matheson et al. 1973). Giammona group synthesized water-soluble copolymer, PHG, from the reaction PHEA with GMA (glycidyl methacrylate), which was cross-linked through gamma irradiation at 0 °C either with or without *N,N'*-methylenebisacrylamide (Giammona et al. 1999a, b) to provide gels. Pitarresi et al. (2002) investigated on the hydrogel preparation from the copolymer PHG by exposing the polymer solution in water to three different irradiation doses of γ-ray source alone or in the presence of poly(ethylene glycol) dimethacrylate (PEGDMA) or diacrylate (PEGDA).

UV irradiation is another typical source to produce hydrogels. Compared with γ-rays, there are some considerable advantages in the UV-induced cross-linking of biocompatible polymers: the facility, selectivity, safety, and low price (Scranton et al. 1997; Ward and Peppas 2001; Mellott et al. 2001; Elisseeff et al. 1997). The functionalized polysaccharides such as dextran or hyaluronic acid with glycidyl methacrylate (GMA) can generate hydrogels by irradiation. Pitarresi et al. (2003, 2007) prepared a series of biodegradable hydrogels by means of photo-cross-linking of above-modified polysaccharide polymers together with methacrylated PHEA. The irradiation at 313 nm provided the cross-linked hydrogels, for which various physiological characteristics and the suitability for drug delivery system were evaluated. Park et al. (2011) investigated biodegradable PHEA gel and the hybrid gels with Pluronic F127, which were generated by photo-cross-linking of those polymers modified with methacryloyl groups. The hybrid gel showed reversible swelling–deswelling behavior and microporous structure which changed as a function of temperature.

4.4 pH-Responsive Polyaspartamides Hydrogels

The ionizable groups with different charges, concentration, and pKa value can affect the swelling degree of ionic polymers (Nakato et al. 1999). In addition, the swelling properties of pH-responsive hydrogels can be influenced by the nature of swelling medium, typical pH, ionic strength, etc., because ionic groups in the polymer backbone can either donate or accept protons according to the change of ambient pH. Among various parameters used to adjust the intelligent system, pH is the most widely used parameter in some parts of human body (Nakato et al. 1999; Mahmoud et al. 1995; Brazel and Peppas 1996). When the pH of surroundings changes, the extent of ionization of pH-responsive hydrogel can substantially vary at a particular pH related to pKa or pKb. Two classical classes of pH-responsive hydrogels are anionic and cationic ones. Anionic hydrogels mostly contain –COOH or $-SO_3H$ groups which ionize at pH above their pKa values (Ebara et al. 2014). These hydrogels will swell in high pH, due to the electrostatic repulsion between the anionic groups along the chain (Zhao et al. 2012; Ramírez-Fuentes et al. 2008). On the other hand, cationic hydrogels typically possess basic amine groups which are charged at low pH (Khare and Peppas 1995; Soppimath et al. 2001). Thus, the swelling behavior of cationic hydrogel is the opposite of the anionic hydrogel. The hydrogel exhibits lower swelling at a high pH value, because the amine groups become comparatively hydrophobic; on the other hand, when the value of pH decreases below the pKa, positive charges are developed along the polymer backbone to cause an increase in swelling. As a kind of polybasic acid polymer, PASP superabsorbent gel should have a pH-sensitive swelling property. Gyenes et al. (2008) prepared PASP gels by chemical cross-linking and subsequent hydrolysis of the precursor PSI, which exhibit a volume phase transition near the pKa value of aspartic acid moiety. Several amino acids such as putrescin, spermine, spermidine, lysine, and cystamine were also used as cross-linker for PSI to produce the biodegradable and biocompatible hydrogels. In order to obtain polymer gels with non-toxicity for biomedical uses, Torma et al. (2010) introduced PASP chemical gel which is prepared by cross-linking using 1,4-diaminobutane. The pH-sensitivity of PASP, of course, will be attributed to the presence of carboxyl groups on the chains of network. PASP gel displayed a significant volume transition at a specific pH near the pKa value of PASP. Giammona group (Giammona et al. 2006) studied a polymer named polyhydroxyethylaspartamide methacrylated (PHM) which was prepared through the reaction of PHEA with methacrylic anhydride. After UV irradiation to stimulate PHM with or without succinic anhydride (SA), the pH-sensitive polyaspartamide hydrogels were formed, with which the pH-dependent swelling and drug release properties were investigated

(Mandracchia et al. 2004; Giammona et al. 2006). "Click" functional PHEA-azide derivative has been synthesized, and its hydrogel was prepared by click reaction with natural alginate or hyaluronate modified with the complementary alkyne functional groups. These biocompatible hydrogels showed outstanding swelling–deswelling behavior as a function of solution pH (Bui et al. 2015). Polyaspartamide derivatives with various cationic pendants such as substituted alkyl amines, pyridine, or imidazole groups exhibit pH-sensitive solubility changes depending on the pKa value of the cationic groups. A series of cationic polyaspartamide derivatives containing *N*,*N*-dimethylaminoethyl and *N*,*N*-diisopropylaminoethyl were synthesized. Their cross-linked hydrogel possessed pH-dependent swelling behavior due to the existence of basic amine groups, where some systems also exhibited thermo-responsive behavior as the examples were already described in previous section (Moon et al. 2006; Moon and Kim 2008, 2010). It is well known that chitosan possessing amine groups, which is rare in polysaccharides, forms pH-responsive hydrogel. Kim et al. (2004a, b) suggested biodegradable pH-sensitive hydrogels based on IPN type composite hydrogel between synthetic PHEA and chitosan, which were obtained using glutaraldehyde as the cross-linker in aqueous media.

4.5 Redox-Responsive Polyaspartamides Hydrogels

Apart from pH and temperature, redox potential has received attention as external trigger. Many redox-responsive polymers and conjugates have been developed especially for the sophisticated drug delivery systems (Meng et al. 2009), and a sol–gel system driven by redox stimuli has recently been reported (Peng et al. 2008; Kawano et al. 2004; Liu et al. 2008). These redox-responsive polymers often possess characteristic disulfide linkages in the main or side chain, and sometimes in the cross-linker. The disulfide bonds, though stable under the normal body temperature, pH and oxidation environment, are prone to be cleaved under a reducing condition through thiol-disulfide exchange reactions. Gyarmati et al. (2013) first reported the redox-responsive PASP hydrogels with the external redox stimuli. The reversible sol–gel behavior was observed both in organic solution (DMF) and in aqueous solution through the interconversion between thiol and disulfide. PASP modified with side chains containing thiol group was covalently cross-linked with 1,4-diaminobutane to provide a non-cleavable network. In addition to pH-responsiveness, the swelling could be tuned reversibly by treating the hydrogel in reducing or oxidizing media. Zrinyi et al. (2013) reported doubly cross-linked PASP hydrogels utilizing two different cross-linkers at the same time; i.e., one is diaminobutane (DAB) and the other is cystamine (CYS). DAB affords stable covalent cross-link, while CYS with disulfide linkage can be disconnected by reduction.

4.6 CO_2-Responsive Polyaspartamides Hydrogels

Carbon dioxide (CO_2) is an important greenhouse gas, which can absorb and emit infrared ray at its two infrared-active vibrational frequencies (Petty 2006). Carbon dioxide is also non-toxic as a pivotal metabolite in cells and contains good biocompatibility and membrane permeability. As an inexpensive and benign gas, CO_2 has been found to be capable to reversibly switch the hydrophilicity of solute, solvent, or surfactants bearing functional group of amidine or guanidine (Jessop et al. 2005, 2010; Phan and Jessop 2009; Liu et al. 2006). In particular, CO_2 can stabilize the intracellular pH value through a series of equilibrium reactions ($CO_2 + H_2O \leftrightarrow H_2CO_3 \leftrightarrow H^+ + HCO_3^-$). Because few researches have been done on the influence of gas molecule to polymer assemblies, a construction of CO_2-responsive polymer system will be quite new and challenging. It has been shown that amidine or guanidine-containing polymers possessed reversibly responsive behavior to CO_2 and some ranges of related applications have also been evaluated. One prospected application of such polymers can be reversible CO_2 absorption and release as a means of segregating CO_2 as a valuable C-1 raw material (Endo et al. 2004; Zhou et al. 2009).

It was expected that polymers containing guanidine or imidazole functional groups with higher pKa values show a rapid CO_2-responsive property. Guanidines are amine compounds with three kinds of nitrogen functions (one "imine" and two "amides") on one carbon atom and able to be protonated when bubbling CO_2 in aqueous solution. Simple bubbling of a noble gas can reverse the process of protonation (Yan and Zhao 2014). Imidazole is formally an amidine, possessing two nitrogen atoms in the "amide" and "imine" functional groups. Using CO_2 as a "green" gas stimulus, Tran et al. (2015a, b) proved a novel strategy of using L-arginine (Larg), a biofunctional molecule containing guanidine group, in order to prepare CO_2-responsive polyaspartamide derivative polymers and gels. The PHEA-Larg hydrogels and PolyAspAm(OA/Larg) nanoparticle underwent reversible shifting by CO_2/N_2 gas purge, in regard to the changes in hydrogel volume and nanoparticles size, respectively (Tran et al. 2015a, b). The same group also developed novel histamine-conjugated polyaspartamides (PHEA-HIS). The hydrogel from PHEA-HIS demonstrated the similar reversible volume transition behavior by CO_2/N_2 treatment.

4.7 Multi-responsive Polyaspartamides Hydrogels

Novel biomaterials, known as "smart" materials, which can respond to the host surrounding are emerging and evolving into elegant materials which can respond to multiple stimuli in order to incarnate better biomimetic processes (Holzapfel et al. 2013). Multi-responsive hydrogels are promising materials for these purposes because they can respond to two or more ambient stimuli such as temperature, light, pH, temperature, redox potential, enzymes, or ionic strength. The changes in smart and functional materials can be volume contraction, swelling, degradation, color change, etc., and the precise response may be manipulated by optimal design of polymer structures (Garcia et al. 2007; Liu et al. 2010). Temperature and pH as the most common and attractive stimuli were employed in order to develop dual-responsive hydrogel, and many of these hydrogels are interpenetrating polymer networks (IPNs) or semi-IPNs-type materials.

Due to the synthetic versatility involved in polyaspartamides, it is possible to synthesize various multifunctional polyaspartamide polymers via proper molecular design and reaction processes. Moon and Kim (2008) have demonstrated a double-responsive (pH and temperature) hydrogel based on polyaspartamide modified with diisopropylamine and hydroxyethyl amine. The LCST phase transition of the copolymer was modulated by hydrophobic–hydrophilic balance and the environmental pH. Furthermore, temperature or medium pH functioned as stimuli to induce volume phase transition through the swelling–deswelling behavior of the hydrogel (see Fig. 5). A series of thermo- and pH-sensitive IPN gels were prepared from PASP and PNIPAAm (Liu et al. 2012). The IPN gels showed enhanced property in their sensitivity to temperature and rates of swelling–-deswelling, by varying the content of the function in PASP hydrogel. In addition, the gel possessed a pH-responsive property due to the presence of carboxyl groups. Gyarmati et al. (2014) studied a redox- and pH-responsive PASP hydrogel which was synthesized by cross-linking of cysteamine-conjugated PSI with 1,4-diaminobutane. Due to the polyelectrolyte character of the network and the thiolation of the polymer, significant changes in the hydrogel swelling were observed by the external pH and also the swelling degree could be modulated reversibly by reducing or oxidizing the hydrogel in aqueous media.

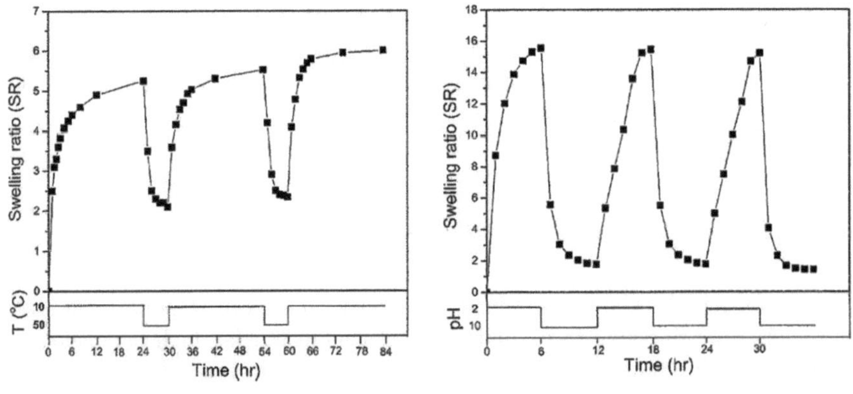

Fig. 5 Reversible swelling curves of polyaspartamide hydrogel as a function of temperature (**a**), and pH of buffer solution (**b**)

5 Applications of Smart Polyaspartamide Hydrogels

5.1 Hydrogels in Drug Delivery Systems

Hydrogels have attracted significant interests for their applications in drug delivery (Lowman et al. 1999; Elvira et al. 2002; Hoffman 2012; Vashist et al. 2014). The benefit of hydrogels for drug delivery application may be the possible sustained release, maintaining a high local drug concentration in the surrounding tissues over a long period. The content of drugs loaded and released depends on the porous structure of hydrogels, capable to be regulated by changing the degree of cross-links and the affinity to water. Another advantage of hydrogels may be less strong interaction with drugs, so that a larger portion of biologically active molecules, especially proteins and peptides, can be released efficiently through hydrogel matrices (Yang et al. 2010). There are several mechanisms of loading drug into hydrogels and releasing therefrom: control by diffusion, control by swelling, control by chemicals, and environmentally responsive release (Place et al. 2009).

Fiorica et al. (2015) prepared a graft copolymer of polyaspartamide with polylactide to provide an in situ gel-like depot and applied this material for the sustained release of sulpiride drug. In this work, it has been proved that the investigated gel depots are cell compatible and able to extend the release time to 7 days for the in vitro release of sulpiride. In addition, rabbits, as an animal model, are used to reveal the profiles of in vivo release and pharmacokinetic parameters. The results suggested that PHEA-g-PLA depots be applicable in clinical applications for a weekly administration of poorly soluble drugs to mankind or animals. Biocompatible hybrid gels based on natural hyaluronic acid and polyaspartamide derivative were used as potential substitutes for the amniotic membrane for epithelial limbal cell delivery (Fiorica et al. 2011).

Stimuli-responsive hydrogel is promising as material for controlled drug delivery because its nature and property will change in response to a particular stimulus. The pH-responsive hydrogels have been widely studied in order to achieve a "switch" in the profile of drug release responding to a gradual change in pH value. Among the various signals employed to modulate intelligent systems, pH is the most widely used since variations in pH exist in different body sites, such as stomach, intestinal tract, vagina, and blood vessels.

Liu et al. (2010) developed a novel controlled release system based on the pH-responsive semi-IPN hydrogel. This hydrogel consisted of poly(aspartic acid) and konjac glucomannan (KGM) mixed with trisodium trimetaphosphate as a non-toxic cross-linker. The pH-dependent swelling of the hydrogel can control 5-Fluorouracil released in vitro. Their research results suggested that the semi-IPN hydrogel be served as a potential drug delivery device, in which the primary objective was in the upper region of small intestine. Recently, zwitterionic polyaspartamide nano-gel was prepared and studied for cancer delivery. They showed remarkable anti-protein adsorption capability and the stability in protein solutions. Doxorubicin (DOX), an anticancer drug, was entrapped into the nano-gel and exhibited release behavior which is accelerated under acidic circumstances. The DOX entrapped in nano-gel was smoothly released into the cancer cells and provided high cytotoxicity (Lu et al. 2014). Like other nonsteroidal agents, anti-inflammatory drug named diflunisal induces gastrointestinal side effects such as hemorrhages and ulceration of stomach or small intestine (Langman et al. 1994). Giammona et al. (1997) developed a novel hydrogel in microparticulate form which is prepared by gamma irradiation of PHEA. When the PHEA hydrogel was digested by enzyme with different concentrations of pepsin or α-chymotrypsin, the degradation reaction did not occur within 24 h. The in vitro release study indicated PHEA hydrogel can be regarded as a system for the delivery of diflunisal. The in vivo studies also revealed that PHEA hydrogel was biocompatible after oral administration to mice, and diflunisal-loaded PHEA microgel particles significantly enhanced the gastric tolerance of drug and bioavailability compared to free diflunisal. In addition, a photo-cross-linked hybrid gel from dextran and polyaspartamide derivatives was used as suitable vector for colon-specific drug delivery (Giammona et al. 1997; Pitarresi et al. 2007; Tripodo et al. 2009). Inulin (INU-SA) derivatives with succinic moiety of varying degrees of substitution were cross-linked with α,β-polyaspartylhydrazide (PAHy) to provide INUPAHy composite hydrogels. Glutathione (GSH) and oxytocin (OT), which are latent agents for colonitis treatment, were entrapped into the INUPAHy hydrogel. The release results, evaluated in simulated gastrointestinal fluids, suggested that these GSH- and OT-loaded INUPAHy hydrogels have useful potential for the oral remedy of inflammatory bowel disease (Mandracchia et al. 2011). The release behavior of a nonsteroidal anti-inflammatory drug (NSAID) and 4-biphenylacetic acid (BPAA) from PHEA hydrogels was conducted at two different pH values (4 and 7.4) using dimyristoyl phosphatidylcholine (DMPC) liposomes (multi-lamellar vesicles, MLV) as a representative biomembrane model to test the drug transfer from loaded hydrogel (Castelli et al. 1997). A further example of controlled drug release was achieved by using thermo-responsive hydrogels. Tachibana et al. (Tachibana et al. 2003a, b) introduced new biodegradable thermo-responsive poly(amino acid) hydrogels for a possible application in the drug delivery, where brilliant blue FCF (BB) was examined as a model drug.

The multi-response hydrogels have been studied most widely for both temperature and pH-sensitivity, which are the most defining parameters in human body. A large number of researches are concentrated on obtaining dual-responsive (pH and temperature) hydrogels with potential applications in drug delivery system (Knipe and Peppas 2014). Némethy et al. (2013) developed multi-responsive co-network hydrogels, PASP-*l*-PNIPAAm, made from pH-responsive PASP and temperature-responsive PNIPAAm. The PASP-*l*-PNIPAAm hydrogel exhibited nonlinear volume transition upon pH and temperature. Specifically, the hydrogel volume decreased under acidic condition or at a temperature range above the volume phase transition temperature (T_{VPT}) of PNIPAAm. On the contrary, the hydrogels swelled in alkaline solutions or at a temperature below the T_{VPT}. The in vitro release of DiclofenacNa (DFS), a widely used NSAID, from hydrogels in PBS at 37 °C and pH = 7.6 after pre-conditioning in an acidic solution, was slower compared to the case in a purely alkaline media. In addition, when the drug-loaded PASP-*l*-PNIPAAm hydrogel was transferred from the stomach to the bowels with the corresponding pH change from 1.2 to 7.6, the release rate increased drastically. These results showed the PASP-*l*-PNIPAAm hydrogel could prevent the adverse effects of the drug in stomach and also protect the drug from highly acidic environment inside stomach. The hydrogel was found to be non-toxic for human epithelial cells (Némethy et al. 2013). pH- and redox-sensitive, disulfide cross-linked PASP hydrogels were developed by Gyarmati et al. (2014). The swelling degrees could be tuned by both pH and oxidation/reduction condition, and the redox process was found to be reversible and repeatable. In the related study by the same research group, drug release behavior of a model drug entrapped in cysteamine-modified poly(aspartic acid) was examined. Firstly, model drug release from disulfide cross-linked PASP-SS-E hydrogels without reducing agent showed that 20% of the molecules being barely released at the time of the whole measurement. After adding a DTT reduction stimulus, the total number of dye release was fivefold more than without DTT. In fact, all of the entrapped drug could be released during the measurement time. Because of the abrupt rise in the swelling, a mass of drugs can be released even before hydrogels was completely dissolved by induction of reduction. The tunable dual-responsive characters of the gels suggest various future applications, such as novel drug delivery as well as sensors, actuators, and various biomedical devices (Gyarmati et al. 2013).

5.2 *Injectable Hydrogels for Biomedical Applications*

Injectable biodegradable hydrogel is becoming more and more attractive in biomedical applications, such as drug/cell delivery and tissue engineering (Li et al. 2012; Huynh et al. 2011). In comparison with classical dose forms, injectable hydrogels can easily be combined with biologically active compounds and injected into a localized site resulting in formulations that can accomplish sustainable or stimulated drug delivery through site-specific control. Furthermore, the hydrogel is compatible with the biological system, lowering the drug administration frequency and surgical invasion (Huynh et al. 2011; Zhao et al. 2011). Injectable hydrogels can be materialized by either physical or chemical cross-linking. Chemically cross-linked hydrogels can be prepared by using various reactions such as photopolymerization, formation of disulfide link, Schiff base, or Michael-addition-type reaction. Physically cross-linked hydrogels are typically formed through sol–gel conversion in response to external stimulation like temperature and pH variations via self-assembly behavior of amphiphilic block or graft copolymers (Chang et al. 2009; Khutoryanskaya et al. 2008; Chiu et al. 2009). Lu et al. (2013) introduced a hydrazone-forming injectable hydrogel by mixing aqueous solutions of two PASP derivatives containing complementary reactive groups, i.e., hydrazine and aldehyde-modified poly(aspartic acid) (PAHy and PAAld). Hydrazone bond is a type of acid-labile linkage, which remains stable at physiological pH but undergoes hydrolytic degradation in acidic conditions. In this work, PASP is modified to possess aldehyde groups with adjustable degree of substitution, as the most important factors to control the density of cross-linking and gelation time. In vitro degradation behavior, DOX-loading and the pH-responsive release behavior were investigated to suggest the potential utility in current biomedical applications. The preparation and in situ gel formation of graft copolymers composed of poly (2-hydroxyethyl aspartamide) and polylactic acid, PHEA-g-PLA, was investigated. At high PLA content, the copolymer exhibited in situ gel forming property. The cell compatibility and in vitro drug release behavior of this injectable gel systems were evaluated (Pitarresi et al. 2010). Moon et al. (2011) developed in situ gel system based on amphiphilic polyaspartamide copolymer, PolyAspAm(LA/NIPEDA), in aqueous solution (25 wt%). The gel formed instantaneously when aqueous solution at 5 °C was injected into an excess amount of PBS (pH 7.4) controlled at 37 °C. The same gel formation in vivo was also performed successfully by subcutaneously injection of the polymer solution into SD rats through syringe needle. In vitro MTT essay confirmed the low cytotoxicity of this material. Furthermore, chlorambucil-loaded in situ gel exhibited a sustained release of drug over the 15 h to suggest potential applications of these novel injectable hydrogels for site-specific delivery of drugs and cells.

5.3 *Scaffold for Tissue Engineering*

Tissue engineering has become a very useful technology for the design of a responsive and living alternative, similar to the function of natural tissue (Langer and Vacanti 1993; Hunt et al. 2014). One of the most popular applications in tissue engineering is for scaffolds which have attracted significant interests as a novel means for cell study in advanced biology by providing 3D tissue culture matrix that supports cell attachment, differentiation and migration, proliferation and guidance of neotissue formation. The main function of the scaffold is to guide new tissue regeneration and to offer proper structural support. Hydrogels with porous, biocompatible properties play a role as space-filling matters that function as delivery channels for bioactive materials and as 3D structures that guide and flourish required cells (Caló and Khutoryanskiy 2015; Place et al. 2009). The first consideration of scaffold materials is biocompatible and mechanical properties (Place et al. 2009). Several synthetic polymers which can constitute hydrogels usefully applied for tissue engineering include poly(ethylene glycol) (PEG), poly(vinyl alcohol) (PVA), poly(2-hydroxyethyl methacrylate) (PHEMA), poly(lactic acid) (PLA), polycaprolactone (PCL), poly(propylene fumarate) (PPF), poly(propylene fumarate-co-ethylene glycol), and polypeptides. Polypeptide, made up of polymers of one or more amino acids, has received tremendous interest over the past ten years, and such poly(amino acid) derivatives have been employed in various fields of biomedical engineering including coatings for therapeutic devices, scaffolds for tissue engineering, and as micro- or nanoparticles for advanced drug delivery.

Pitarresi et al. (2010) have firstly introduced a fibrillar scaffold produced by the electrospinning of PHEA-*g*-PLA solution in mixed *N*,*N*-dimethyl formamide (DMF) and acetone. About 1 μm diameter fibrillar scaffold resisted the chemical degradation in physiological condition up to 15 days. They proposed this novel electrospun scaffold for use in skin regeneration (Pitarresi et al. 2010). Recently, the same group reported heparin-functionalized polyaspartamide-*g*-PLA electrospun scaffold to suggest its potential for use in revascularization. Using human vascular endothelial cells (HVECs), the cell growth and formation of new tissues have been evaluated (Pitarresi et al. 2014).

Gyarmati et al. (2015) introduced supermacroporous PASP chemical hydrogel by a special cryogelation technique in DMSO and subsequent hydrolysis in pH buffer. Due to the biocompatibility, controllable pore structure, and pH-responsive characteristics in addition to the enhanced mechanical property of cryogel, these PASP gels are of great interest as scaffold biomaterials.

Natural polymers, as compared with man-made polymers, have many benefits in their applications in biology and pharmaceutics, due to their abundance, inherent biocompatible and non-toxic nature, and so on. The hybrid gel systems including natural alginate or chitosan with polyaspartamide polymers must have potential as scaffold materials by combining the advantageous properties of both components in future.

5.4 *Self-healing Hydrogel*

Considerable attention have been paid to self-healing, the ability of a material or system to heal spontaneously or repair itself as a function of intelligent materials. Self-healing hydrogels are made commonly by either reversible covalent cross-linking or metal-mediated coordination linkages. When dynamic covalent bonds or non-covalent bonds such as coordination or hydrogen-bonding interaction are used to achieve a reversible system, the self-healing process can be repeated for numerous times (Deng et al. 2012).

In recent years, dopamine (DOPA, 3,4-dihydroxyphenethylamine) and DOPA-containing polymers received a great attention, and the chemistry and related materials development has been extensively pursued. DOPA (with catechol functional group) was known as a major component of adhesive mussel protein and has been proved to be critical in relation to adhesion and polypeptide cross-linking in nature. Many DOPA-functionalized polymer and gel systems have been developed especially for adhesive and surface coating. One of the functions of adhesive DOPA is the extremely powerful ability of metal coordination, which has newly been adopted to prepare self-healing gels (Menyo et al. 2013). Recently, a representative pH-responsive self-healing hydrogel based on DOPA-conjugated polyaspartamide was reported by Wang et al. (2015). By using boric acid (H_3BO_3), the mussel-inspired copolymer exhibited a reversible gelation by pH change, which was based on reversible boron–catechol coordination binding.

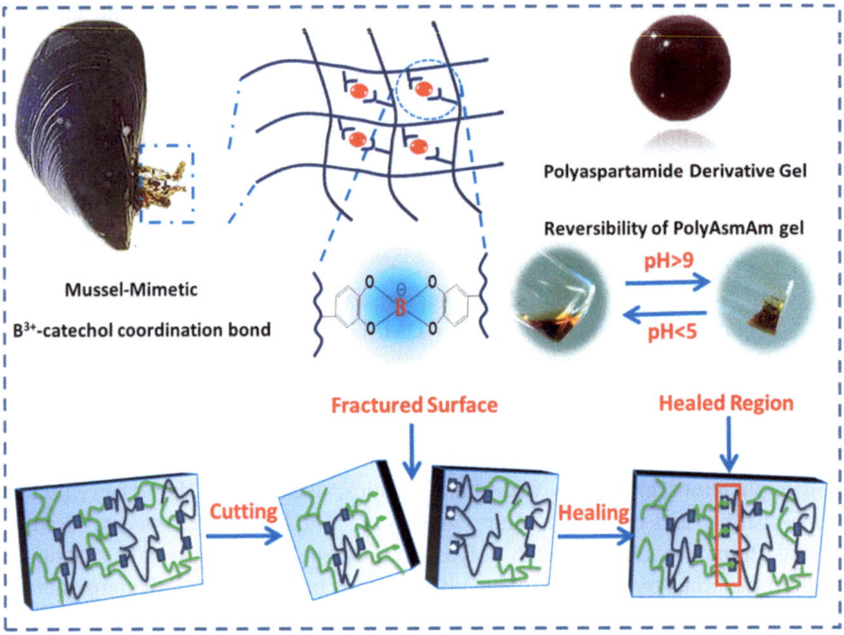

Fig. 6 Schematic illustration of mussel-mimetic pH-responsive self-healing hydrogel from dopamine-conjugated polyaspartamide with boric acid

In addition, the dynamic complexation in boron–catechol coordination can lead to the constitution of a rapid self-healing smart material. Figure 6 represents this self-healing hydrogel system from mussel-inspired polyaspartamide derivative containing catechol pendant groups.

6 Conclusion and Future Perspective

During the last decade, a significant amount of researches has been carried out for the development of biodegradable and biocompatible polymers and hydrogels based on polyaspartamides. As are discussed extensively in this review, polyaspartamide derivatives are versatile and the scope of potential applications and material developments are very broad and comprehensive. Especially many polyaspartamide-based hydrogels for drug delivery devices and scaffolds are promising, even though various novel and functional polyaspartamides are currently under extensive studies and development to find novel applications not only in biomedical but also in environmental and other specialty chemical field. Several advantageous aspects of this sort of polymer will be readily available raw materials, non-toxic and biocompatible nature, environmentally benign, relatively simple synthesis, and easy chemical modifications. In near future, we expect more outstanding achievements in the bio-related materials development based on novel stimuli-responsive and "smart" polyaspartamide derivatives and the hybrid gel systems.

References

Bach QV, Moon JR, Lee DS, Kim JH (2008) Lower critical solution temperature behavior of amphiphilic copolymers based on polyaspartamide derivatives. J Appl Polym Sci 107:509–513

Ballauff M, Lu Y (2007) "Smart" nanoparticles: preparation, characterization and applications. Polymer 48:1815–1823

Brazel CS, Peppas NA (1996) Pulsatile local delivery of thrombolytic and antithrombotic agents using poly(*N*-isopropylacrylamide-co-methacrylic acid) hydrogels. J Control Release 39:57–64

Bui QT, Jeon YS, Um SH, Chung DJ, Kim JH (2015) Preparation of novel hybrid gels from polyaspartamides and natural alginate or hyaluronate by click reaction. J Polym Res 22:27

Caló E, Khutoryanskiy VV (2015) Biomedical applications of hydrogels: a review of patents and commercial products. Eur Polym J 65:252–267

Carenza M (1992) Recent achievements in the use of radiation polymerization and grafting for biomedical applications. Int J Radiat Appl Instrum C Radiat Phys Chem 39:485–493

Castelli F, Pitarresi G, Tomarchio V, Giammona G (1997) Effect of pH on the transfer kinetics of an anti-inflammatory drug from polyaspartamide hydrogels to a lipid model membrane. J Control Release 45:103–111

Chang G, Yu L, Yang Z (2009) A delicate ionizable-group effect on self-assembly and thermogelling of amphiphilic block copolymers in water. Polymer 50:6111–6120

Chapiro A (1995) Radiation chemistry in the field of biomaterials. Radiat Phys Chem 46:159–160

Chiu YL, Chen SC, Su CJ, Hsiao CW, Chen YM, Chen HL, Sung HW (2009) pH-triggered injectable hydrogels prepared from aqueous *N*-palmitoyl chitosan: in vitro characteristics and in vivo biocompatibility. Biomaterials 30:4877–4888

Deng G, Li F, Yu H, Liu F, Liu C, Sun W, Chen Y (2012) Dynamic hydrogels with an environmental adaptive self-healing ability and dual responsive sol-gel transitions. ACS Macro Lett 1:275–279

Domb AJ, Kost J, Wiseman D (1998) Handbook of biodegradable polymers (vol 7). CRC Press

Ebara M, Kotsuchibashi Y, Uto K (2014) Introductory guide to smart biomaterials. Smart Biomaterials. Springer, pp 1–7

Elisseeff J, Anseth K, Langer R, Hrkach JS (1997) Synthesis and characterization of photo-crosslinked polymers based on poly(L-lactic acid-co-L-aspartic acid). Macromolecules 30:2182–2184

Elvira C, Mano JF, San Roman J, Reis RL (2002) Starch-based biodegradable hydrogels with potential biomedical applications as drug delivery systems. Biomaterials 23:1955–1966

Endo T, Nagai D, Monma T, Yamaguchi H, Ochiai B (2004) A novel construction of a reversible fixation-release system of carbon dioxide by amidines and their polymers. Macromolecules 37:2007–2009

Fiorica C, Senior RA, Pitarresi G, Palumbo FS, Giammona G, Deshpande P, MacNeil S (2011) Biocompatible hydrogels based on hyaluronic acid cross-linked with a polyaspartamide derivative as delivery systems for epithelial limbal cells. Int J Pharm 414:104–111

Fiorica C, Palumbo FS, Pitarresi G (2015) In-situ forming gel-like depot of a polyaspartamide-polylactide copolymer for once a week administration of sulpiride. J Pharm Pharmacol 67:78–86

Freeman MB, Paik YH, Swift G (1996) Biodegradability of polycarboxylates: structure-activity studies. Hydrogels Biodegradable Polym Bioapplications 627:118–136

Garcia A, Marquez M, Cai T, Rosario R, Hu Z, Gust D, Rosario R, Hu Z, Gust D, Park CD (2007) Photo-, thermally, and pH-responsive microgels. Langmuir 23:224–229

Giammona G, Carlisi B, Palazzo S (1987) Reaction of α,β-poly(N-hydroxyethyl-DL-aspartamide) with derivatives of carboxylic acids. J Polym Sci Part A Polym Chem 25:2813–2818

Giammona G, Puglisi G, Cavallaro G, Spadaro A, Pitarresi G (1995) Chemical stability and bioavailability of acyclovir coupled to α,β-poly(*N*-2-hydroxyethyl-DL-aspartamide). J Control Release 33:261–271

Giammona G, Pitarresi G, Tomarchio V (1997) A hydrogel based on a polyaspartamide: characterization and evaluation of in-vivo biocompatibility and drug release in the rat. J Pharm Pharmacol 49:1051–1056

Giammona G, Pitarresi G, Cavallaro G, Buscemi S, Saiano F (1999a) New biodegradable hydrogels based on a photocrosslinkable modified polyaspartamide: synthesis and characterization. BBA-Biomembranes 1428:29–38

Giammona G, Pitarresi G, Cavallaro G, Spadaro G (1999b) New biodegradable hydrogels based on an acryloylated polyaspartamide cross-linked by gamma irradiation. J Biomat Sci-Polym Ed 10:969–987

Giammona G, Pitarresi G, Cavallaro G, Mandracchia D (2006) pH-sensitive hydrogel based on a polyaspartamide derivative. J Drug Deliv Sci Tec 16:77–84

Gu X, Wang J, Liu X, Zhao D, Wang Y, Gao H, Wu G (2013) Temperature-responsive drug delivery systems based on polyaspartamides with isopropylamine pendant groups. Soft Matter 9:7267–7273

Gyarmati B, Vajna B, Némethy Á (2013) Redox- and pH-responsive cysteamine-modified poly (aspartic acid) showing a reversible sol-gel transition. Macromol Biosci 13:633–640

Gyarmati B, Némethy Á, Szilágyi A (2014) Reversible response of poly(aspartic acid) hydrogels to external redox and pH stimuli. RSC Adv 4:8764–8771

Gyarmati B, Mészár EZ, Kiss L, MA, László K, Szilágyi A (2015) Supermacroporous chemically cross-linked poly(aspartic acid) hydrogels. Acta Biomater 22:32–38

Gyenes T, Torma V, Gyarmati B, Zrínyi M (2008) Synthesis and swelling properties of novel pH-sensitive poly(aspartic acid) gels. Acta Biomater 4:733–744

Hiraishi T (2015) Poly(aspartic acid) (PAA) hydrolases and PAA biodegradation: current knowledge and impact on applications. Appl Microbiol Biot 1–8

Hoffman AS (2012) Hydrogels for biomedical applications. Adv Drug Deliv Rev 64:18–23

Holzapfel BM, Reichert JC, Schantz JT, Gbureck U, Rackwitz L, Nöth U, Jakob F, Rudert M, Groll J, Hutmacher DW (2013) How smart do biomaterials need to be? A translational science and clinical point of view. Adv Drug Deliv Rev 65:581–603

Huang SJ (1989) 21-Biodegradation. Compr Polym Sci 6:597–605

Hunt JA, Chen R, van Veen T, Bryan N (2014) Hydrogels for tissue engineering and regenerative medicine. J Mater Chem B 2:5319–5338

Huynh CT, Nguyen MK, Lee DS (2011) Injectable block copolymer hydrogels: achievements and future challenges for biomedical applications. Macromolecules 44:6629–6636

Jeon YS, Lei J, Kim JH (2008) Dye adsorption characteristics of alginate/polyaspartate hydrogels. J Ind Eng Chem 14:726–731

Jeon YS, Lei J, Chung DJ, Kim JH (2009) Sol-gel derived organic/inorganic hybrid gels based on poly(2-hydroxyethyl aspartamide) and silica. J Ind Eng Chem 15:544–549

Jessop PG, Heldebrant DJ, Li X, Eckert CA, Liotta CL (2005) Green chem: reversible nonpolar-to-polar solvent. Nature 436:1102

Jessop PG, Phan L, Carrier A, Robinson S, Dürr CJ, Harjani JR (2010) A solvent having switchable hydrophilicity. Green Chem 12:809–814

Kawano S, Fujita N, Shinkai S (2004) A coordination gelator that shows a reversible chromatic change and sol-gel phase-transition behavior upon oxidative/reductive stimuli. J Am Chem Soc 126:8592–8593

Khare AR, Peppas NA (1995) Swelling/deswelling of anionic copolymer gels. Biomaterials 16:559–567

Khutoryanskaya OV, Mayeva ZA, Mun GA, Khutoryanskiy VV (2008) Designing temperature-responsive biocompatible copolymers and hydrogels based on 2-hydroxyethyl (meth) acrylates. Biomacromol 9:3353–3361

Kim JH, Lee JH, Yoon SW (2002) Preparation and swelling behavior of biodegradable superabsorbent gels based on polyaspartic acid. J Ind Eng Chem 8:138–142

Kim JH, Sim SJ, Lee DH, Chung DJ, Kim JH (2004a) Preparation and properties of PHEA/chitosan composite hydrogel. Polym J 36:943–948

Kim JH, Sim SJ, Lee DH, Kim D, Lee YK, Kim JH (2004b) Preparation and properties of biodegradable hydrogel based on glutaraldehyde-crosslinked poly(2-hydroxyethyl aspartamide). J Ind Eng Chem 10:278–282

Kim SI, Min SK, Kim JH (2008) Synthesis and characterization of novel amino acid-conjugated poly(aspartic acid) derivatives. Bull Korean Chem Soc 29:1887–1892

Kim JH, Son CM, Jeon YS, Choe WS (2011) Synthesis and characterization of poly(aspartic acid) derivatives conjugated with various amino acids. J Polym Res 18:881–890

Knipe JM, Peppas NA (2014) Multi-responsive hydrogels for drug delivery and tissue engineering applications. Regenerative Biomat, 57–65

Koncic MZ, Zorc B, Novak P (2011) Macromolecular prodrugs. XIII. Hydrosoluble conjugates of 17β-estradiol and estradiol-17β-valerate with polyaspartamide polymer. Acta Pharm 61:465–472

Langer R, Vacanti J (1993) Tissue engineering. Science 260:920–926

Langman MJS, Weil J, Wainwright P, Lawson DH, Rawlins MD, Logan RF, Colin-Jones DG (1994) Risks of bleeding peptic ulcer associated with individual non-steroidal anti-inflammatory drugs. Lancet 343:1075–1078

Li X, Shen W, Liu C, Nishimoto SI, Kagiya T (1991) Controlled release of 5-fluorouracil or mitomycin-C from polymer matrix: preparation by radiation polymerization and in vivo evaluation of the anticancer drug/polymer composites. Int J Radiat Appl Instrum C-Radiat Phys Chem 38:377–382

Li Y, Rodrigues J, Tomás H (2012) Injectable and biodegradable hydrogels: gelation, biodegradation and biomedical applications. Chem Soc Rev 41:2193–2221

Liu F, Urban MW (2010) Recent advances and challenges in designing stimuli-responsive polymers. Prog Polym Sci 35:3–23

Liu Y, Jessop PG, Cunningham M, Eckert CA, Liotta CL (2006) Switchable surfactants. Science 313:958–960

Liu J, Yan J, Yuan X, Liu K, Peng J, Fang Y (2008) A novel low-molecular-mass gelator with a redox active ferrocenyl group: tuning gel formation by oxidation. J Colloid Interface Sci 318:397–404

Liu C, Chen Y, Chen J (2010) Synthesis and characteristics of pH-sensitive semi-interpenetrating polymer network hydrogels based on konjac glucomannan and poly(aspartic acid) for in vitro drug delivery. Carbohyd Polym 79:500–506

Liu M, Su H, Tan T (2012) Synthesis and properties of thermo- and pH-sensitive poly (*N*-isopropylacrylamide)/polyaspartic acid IPN hydrogels. Carbohydr Polym 87:2425–2431

Low KC, Wheeler AP, Koskan LP (1996) Commercial poly(aspartic acid) and its uses. Adv Chem Ser 248:99–112

Lowman AM, Peppas NA, Hydrogels EM (1999) Encyclopedia of controlled drug delivery (Mathiowitz E, ed), pp 397–418

Lu C, Wang X, Wu G, Wang J, Wang Y, Gao H (2013) An injectable and biodegradable hydrogel based on poly(α, β-aspartic acid) derivatives for localized drug delivery. J Biomed Mater Res A 102:628–638

Lu C, Li B, Liu N, Wu G, Gao H, Ma J (2014) A hydrazone crosslinked zwitterionic polypeptide nanogel as a platform for controlled drug delivery. Rsc Adv 4:50301–50311

Ma Y, Jiang X, Zhuo R (2014) Biodegradable and thermosensitive polyaspartamide derivatives bearing aromatic structures. Mater Lett 121:78–80

Mahmoud EA, Svensson LO, Olsson SE, Mardch PA (1995) Antichlamydal activity of vaginal secretion. Am J Obstet Gynecol 172:1268–1272

Mandracchia D, Pitarresi G, Palumbo FS, Carlisi B, Giammona G (2004) pH-sensitive hydrogel based on a novel photocross-linkable copolymer. Biomacromol 5:1973–1982

Mandracchia D, Denora N, Franco M, Pitarresi G, Giammona G, Trapani G (2011) New biodegradable hydrogels based on Inulin and α, β-polyaspartylhydrazide designed for colonic drug delivery: in vitro release of glutathione and oxytocin. J Biomat Sci Polym Ed 22:313–328

Matheson MS, Mamou A, Silverman J, Rabani J (1973) Reaction of hydroxyl radicals with polyethylene oxide in aqueous solution. J Phys Chem 77:2420

Matsubara K, Nakato T, Tomida M (1998) End group and irregular structure analysis in thermally prepared sodium polyaspartate by ^{1}H and ^{13}C NMR spectroscopy. Macromolecules 31:1466–1472

Mellott MB, Searcy K, Pishko MV (2001) Release of protein from highly cross-linked hydrogels of poly(ethylene glycol) diacrylate fabricated by UV polymerization. Biomaterials 22:929–941

Meng F, Hennink WE, Zhong Z (2009) Reduction-sensitive polymers and bioconjugates for biomedical applications. Biomaterials 30:2180–2198

Menyo MS, Hawker CJ, Waite JH (2013) Versatile tuning of supramolecular hydrogels through metal complexation of oxidation-resistant catechol-inspired ligands. Soft Matter 9:10314–10323

Moon JR, Kim JH (2006) Preparation of biodegradable thermo-responsive polyaspartamides with *N*-isopropylamine pendent groups (I). Bull Korean Chem Soc 27:1981–1984

Moon JR, Kim JH (2008) Biodegradable thermo- and pH-responsive hydrogels based on amphiphilic polyaspartamide derivatives containing N,N-diisopropylamine pendants. Macromol Res 16:489–491

Moon JR, Kim JH (2010) Biodegradable stimuli-responsive hydrogels based on amphiphilic polyaspartamides with tertiary amine pendent groups. Polym Int 59:630–636

Moon JR, Kim BS, Kim J (2006) Preparation and properties of novel biodegradable hydrogel based on cationic polyaspartamide derivative. Bull Korean Chem Soc 27:981

Moon JR, Park YH, Kim JH (2009) Synthesis and characterization of novel thermo- and pH-responsive copolymers based on amphiphilic polyaspartamides. J Appl Polym Sci 111:998–1004

Moon JR, Kim MW, Kim D, Jeong JH, Kim JH (2010) Synthesis and self-assembly behavior of novel polyaspartamide derivatives for anti-tumor drug delivery. Colloid Polym Sci 289:63–71

Moon JR, Jeon YS, Chung DJ, Kim D, Kim JH (2011) In situ gelling and drug release behavior from novel temperature-sensitive polyaspartamides. Macromol Res 19(5):515–518

Morita Y, Kaetsu I (1992) Synthesis of stimuli-sensitive hydrogels. Int J Radiat Appl Instrum C-Radiat Phys Chem 39:473–476

Nakato T, Yoshitake M, Matsubara K, Tomida M, Kakuchi T (1998) Relationships between structure and properties of poly(aspartic acid)s. Macromolecules 31:2107–2113

Nakato T, Oda K, Yoshitake M, Tomida M, Kakuchi T (1999) Synthesis and characterization of poly(aspartic acid) and its derivatives as biodegradable materials. J Macromol Sci Pure Appl Chem 36:949–961

Némethy A, Solti K, Kiss L, Gyarmati B, Deli MA, Csányi E, Szilágyi A (2013) pH- and temperature-responsive poly(aspartic acid)-*l*-poly(*N*-isopropylacrylamide) conetwork hydrogel. Eur Polym J 49:2392–2403

Neri P, Antoni G, Benbenuti F, Colola F, Gazzei G (1973) Synthesis of α,β-poly [(2-hydroxyethyl)-DL-aspartamide], a new plasma expander. J Med Chem 16:893–897

Park JH, Moon JR, Hong KH, Kim JH (2011) Photo-crosslinked polyaspartamide hybrid gel containing thermo-responsive Pluronic triblock copolymer. J Polym Res 18:273–278

Peng F, Li G, Liu X, Wu S, Tong Z (2008) Redox-responsive gel-sol/sol-gel transition in poly (acrylic acid) aqueous solution containing Fe(III) ions switched by light. J Am Chem Soc 130:16166–161667

Petty GW (2006) A first course in atmospheric radiation, 2nd edn. Sundog Publishing

Phan L, Jessop PG (2009) Switching the hydrophilicity of a solute. Green Chem 11:307–308

Pitarresi G, Licciardi M, Craparo EF, Calderaro E, Spadaro G, Giammona G (2002) Microgels of polyaspartamide and poly(ethylene glycol) derivatives obtained by γ-irradiation. Radiat Phys Chem 65:159–167
Pitarresi G, Palumbo FS, Giammona G, Casadei MA, Moracci FM (2003) Biodegradable hydrogels obtained by photocrosslinking of dextran and polyaspartamide derivatives. Biomaterials 24:4301–4313
Pitarresi G, Casadei MA, Mandracchia D (2007) Photocrosslinking of dextran and polyaspartamide derivatives: a combination suitable for colon-specific drug delivery. J Control Release 119:328–338
Pitarresi G, Palumbo FS, Fiorica C, Calascibetta F, Giammona G (2010) Electrospinning of α, β-poly(N-2-hydroxyethyl)-dl-aspartamide-graft-polylactic acid to produce a fibrillar scaffold. Eur Polym J 46(2):181–184
Pitarresi G, Fiorica C, Palumbo FS, Rigogliuso S, Ghersi G, Giammona G (2014) Heparin functionalized polyaspartamide/polyester scaffold for potential blood vessel regeneration. J Biomed Mat Res A 102A:1334–1341
Pivcova H, Saudek V, Drobnik J, Vlasak J (1981) Nmr study of poly(aspartic acid). I. α- and β-peptide bonds in poly(aspartic acid) prepared by thermal polycondensation. Biopolymers 20:1605–1614
Pivcova H, Saudek V, Drobnik H (1982) ^{13}C nmr study of the structure of poly (aspartic acid). Polymer 23:1237–1241
Place ES, George JH, Williams CK, Stevens MM (2009) Synthetic polymer scaffolds for tissue engineering. Chem Soc Rev 38:1139–1151
Ramírez-Fuentes YS, Bucio E, Burillo G (2008) Thermo and pH-sensitive copolymer based on acrylic acid and *N*-isopropylacrylamide grafted onto polypropylene. Polym Bull 60:79–87
Rosiak JM, Ulanski P, Pajewski LA, Yoshii F, Makuuchi K (1995) Radiation formation of hydrogels for biomedical purpose. Some remarks and comments. Radiat Phys Chem 46:161–168
Ross RJ, Mazo GY, Mazo J (2001) New methods in the synthesis of thermal poly (aspartates). ACS symposium series, vol 786, Chap 11, pp 172–181
Safranj A, Kano S, Yoshida M, Omichi H, Katakai R, Suzuki M (1995) Functional polymeric microspheres synthesized by radiation polymerization. Radiat Phys Chem 46:203–206
Schmaljohann D (2006) Thermo- and pH-responsive polymers in drug delivery. Adv Drug Deliv Rev 58:1655–1670
Scranton AB, Bowman CN, Peiffer RW (1997) Photopolymerization: fundamentals and applications. ACS symposium series, vol 673
Son CM, Jeon YS, Kim JH (2011) Preparation and properties of biodegradable superabsorbent gels based on poly(aspartic acid)s with amino acid pendants. Polym-Korea 35:558–564
Soppimath KS, Kulkarni AR, Aminabhavi TM (2001) Chemically modified polyacrylamide-*g*-guar gum-based crosslinked anionic microgels as pH-sensitive drug delivery systems: preparation and characterization. J Control Release 75:331–345
Sun S, Cao H, Su H, Tan H (2009) Preparation and characterization of a novel injectable in situ cross-linked hydrogel. Polym Bull 62:699–711
Tachibana Y, Kurisawa M, Uyama H, Kakuchi T, Kobayashi S (2003a) Biodegradable thermoresponsive poly(amino acid)s. Chem Commun, 106–107
Tachibana Y, Kurisawa M, Uyama H, Kakuchi T, Kobayashi S (2003b) Thermoresponsive hydrogels based on biodegradable poly(amino acid)s. Chem Lett 32:374–375
Takeuchi Y, Uyama H, Tomoshige N, Watanabe E, Tachibana Y, Kobayashi S (2006) Injectable thermoreversible hydrogels based on amphiphilic poly(amino acid)s. J Polym Sci A-Polym Chem 44:671–675
Tang Y, Wheeler AP (2001) Environmental factors that influence biodegradation of thermal poly (aspartate). ACS symposium series, vol 786, pp 157–171
Thakur VK, Thakur MK (2015) Recent advances in green hydrogels from lignin: a review. Int J Biol Macromol 72:834–847

Thakur VK, Thakur MK (2014) Recent trends in hydrogels based on psyllium polysaccharide: a review. J Clean Prod 82:1–15

Thombre SM, Sarwade BD (2005) Synthesis and biodegradability of polyaspartic acid: a critical review. J Macromol Sci A 42:1299–1315

Tomida M, Nakato T, Kuramochi M, Shibata M, Matsunami S, Kakuchi T (1996) Novel method of synthesizing poly(succinimide) and its copolymeric derivatives by acid-catalysed polycondensation of L-aspartic acid. Polymer 37:4435–4437

Torma V, Gyenes T, Szakács Z, Zrínyi M (2010) A novel potentiometric method for the determination of real crosslinking ratio of poly(aspartic acid) gels. Acta Biomater 6:1186–1190

Tran BN, Bui QT, Jeon YS, Kim JH (2015a) Preparation and characterization of CO_2-responsive poly(amino acid) derivatives with guanidine group. Polym Bull 72:2605–2620

Tran BN, Kim JY, Kim YC, Kim YJ, Kim JH (2015b) CO_2-responsive swelling behavior and metal-ion adsorption properties in novel histamine-conjugated polyaspartamide hydrogel. J App Polym Sci 133:43305

Tripodo G, Pitarresi G, Cavallaro G (2009) Controlled release of IgG by novel UV induced polysaccharide/poly(amino acid) hydrogels. Macromol Biosci 9:393–401

Umeda S, Nakade H, Kakuchi T (2011) Preparation of superabsorbent hydrogels from poly (aspartic acid) by chemical crosslinking. Polym Bull 67:1285–1292

Vashist A, Vashist A, Gupta YK, Ahmad S (2014) Recent advances in hydrogel based drug delivery systems for the human body. J Mater Chem B 2:147–166

Wang B, Jeon YS, Park HS, Kim YJ, Kim JH (2015) Mussel-mimetic self-healing polyaspartamide derivative gel via boron-catechol interactions. Express Polym Lett 9:799–808

Ward JH, Peppas NA (2001) Preparation of controlled release systems by free-radical UV polymerization in the presence of a drug. J Control Release 71:183–192

Wolk SK, Swift G, Paik YH, Yocom KM, Smith RL, Simon ES (1994) One- and two-dimensional nuclear magnetic resonance characterization of poly(aspartic acid) prepared by thermal polymerization of L-aspartic acid. Macromolecules 27:7613–7620

Yan Q, Zhao Y (2014) Block copolymer self-assembly controlled by the "green" gas stimulus of carbon dioxide. Chem Commun 50:11631–11641

Yang HM, Park CW, Woo MA, Kim MI, Jo YM, Park HG, Kim JD (2010) HER2/neu antibody conjugated poly(amino acid)-coated iron oxide nanoparticles for breast cancer MR imaging. Biomacromol 11:2866–2872

Yoshimura T, Ochi Y, Fujioka R (2005) Synthesis and properties of hydrogels based on polyaspartamides with various pendants. Polym Bull 55:377–383

Zhao L, Zhu L, Liu F, Shan D, Wang Q (2011) pH triggered injectable amphiphilic hydrogel containing doxorubicin and paclitaxel. Int J Pharm 410:83–91

Zhao C, He P, Xiao C, Gao X, Zhuang X, Chen X (2012) Photo-cross-linked biodegradable thermo- and pH-responsive hydrogels for controlled drug release. J Appl Polym Sci 123:2923–2932

Zhou H, Zhang WZ, Wang YM, Qu JP, Lu XB (2009) N-heterocyclic carbene functionalized polymer for reversible fixation-release of CO_2. Macromolecules 42:5419–5421

Zrinyi M, Gyenes T, Juriga D, Kim JH (2013) Volume change of double cross-linked poly(aspartic acid) hydrogels induced by cleavage of one of the crosslinks. Acta Biomater 9:5122–5131

Chapter 16
Hydrogels from Catechol-Conjugated Polymeric Materials

Saad Moulay

Abstract Digging into in vivo phenomena is not always a vain task. Its merits will lighten sooner or later. Indeed, in the last two decades, Nature has unveiled to scientists the adhesiveness of proteins secreted within the mussel feet, to a large spectrum of substrata, and, unexpectedly, in aqueous environment. The secret behind this bio-adhesiveness lies on the synergetic adhesive action of 3, 4-dihydroxyphenyl-L-alanine (L-DOPA) and lysine, two amino acid residues in protein skeleton. Mimicking the mussel feet protein (*mfp*), a plethora of synthetic and natural polymers functionalized with catechol-containing molecules such as L-DOPA were considered as platforms for hydrogel making. Hydrogels tackled in this chapter include those based on poly(alkene oxide)s including poly(ethylene glycol) (PEG) and Pluronics (PEO/PPO/PEO), polyacrylics, alginate, chitosan, gelatin, hyaluronic acid, polypeptides, polyamides, polyesters, polyurethane, poly(vinyl alcohol), and polyallylamine. The applications thereof, in tune with the properties of polymer–catechol conjugates, are propitiously highlighted.

Keywords Adhesive · Catechol · Hydrogels · Mussels · Polymers

1 Introductory Aspects

Of the innumerable applications and uses of macromolecular materials is hydrogel making. The in vitro hydrogels are viewed as swollen aqueous systems of lightly crosslinked hydrophilic polymers, giving rise to networks that are able to hold a substantial amount of water (Laftah et al. 2011; Ahmed 2015; Thakur and Thakur 2014a, 2015). Moreover, polymers that are uncrosslinked and capable of hydrogen-bond creating may also lead to hydrogel formation by freezing–thawing

S. Moulay (✉)
Laboratoire de Chimie-Physique Moléculaire et Macromoléculaire,
Département de Chimie Industrielle, Faculté de Technologie,
Université Saâd Dahlab de Blida, B. P. 270 Route de Soumâa, 09000 Blida, Algeria
e-mail: polymchemlab@hotmail.com; saadmoulay@univ-blida.dz

V. K. Thakur and M. K. Thakur (eds.), *Hydrogels*, Gels Horizons: From Science to Smart Materials, https://doi.org/10.1007/978-981-10-6077-9_16

process as clearly demonstrated for poly(vinyl alcohol) (Willcox et al. 1999; Hassan and Peppas 2000; Thakur and Kessler 2015). Other means to provoke hydrogel formation were via chemical reactions such as thiol-norbornene reactions (Fairbanks et al. 2009), Michael-type additions (Phelps et al. 2012), Diels–Alder cycloadditions (Alge et al. 2013), and strain-promoted azide–alkyne cycloadditions (Zheng et al. 2012). Polymers of ionic monomers like sodium salt of 2-acrylamido-2-methylpropane sulfonic acid give ionic or polyelectrolyte hydrogels, characterized with high water uptake. Polyelectrolyte hydrogels can be also formed from carboxylic acid-containing polymers with pH- and ionic strength dependencies. For polymers with electron-donating groups, crosslinking inducing polymeric networks for hydrogels setting can be promoted not only covalently but also coordinatively, the electron-donating groups acting as chelating sites. Other facets of hydrogel making and the related morphologies were elsewhere described (Peak et al. 2013). Tailored properties of hydrogels are so widespread that the field of their applications is so vast (Peak et al. 2013; Peppas et al. 2006). Indeed, the special merits of polymer-based hydrogels lie on the vast field of their applications in Today's advanced technology, as actuators, sensors, surgical sealants, drug delivery devices, and tissue engineering scaffolds (Ghobril and Grinstaff 2015). Being genuine materials, hydrogels ought to have basic requisite properties, namely appreciable elasticity modulus (G') and swelling extent (S) (Okay 2009). Both of these two properties depend tightly on the charge and crosslink densities of the polymer network and on the crosslinked polymer concentration. Polymeric materials devised for hydrogel making can be in form of homopolymers, copolymers, star-shaped polymers, dendrimeric polymers, or chemically modified polymers. Polymers, chemically functionalized with appropriate conjugates, would bring up specific properties for hydrogels. Hydrogels from several chemically modified poly (vinyl alcohol)s were judiciously employed in biological and biomedical applications, adsorption and separation technology (Moulay 2015). Those from end-group modified poly(ethylene oxide)s (PEGs) proved, through their good biocompatibility, to be particularly useful in biomedical uses (Zhu 2010).

Of the peculiar conjugates grafted onto polymer matrixes are catechol or pyrocatechol (1,2-dihydroxybenzene) and its amino derivatives such as 3,4-dihydroxyphenyl-L-alanine (DOPA) and 4-(2-aminoethyl)-1,2-benzenediol (dopamine, DA) (Faure et al. 2013). Such graftings (Fig. 1) were inspired from mussel feet proteins, *mfp*, which were distinct by their ability to adhere to nearly all surfaces under wet conditions (Moulay 2014). Recently, the adhesive property in wet environment of the mussel-inspired poly[*N*-(3,4-dihydroxyphenethyl) methacrylamide-*co*-2-methoxyetheyl acrylate] was cleverly exploited in dentistry, doing away with complications related to saliva (Lee et al. 2015a). Surprisingly, the electrical conductivity of polypyrrole, a well-known conducting polymer, was substantially enhanced when functionalized with catechol and its derivatives (DA, DOPA) (Zhang et al. 2015). Several schools from all over the world have been striving to decipher this natural phenomenon, and their reaping has been immensely thriving: School in University of California headed by Professor Herbert Waite, School in Northwestern University (USA) headed by Professor Philip Messersmith, School in Purdue

University (USA) headed by Professor Jonathan Wilker, School in Pohang University of Science and Technology (South Korea) headed by Professor Hyung Joon Cha, School in Advanced Institute of Science and Technology (South Korea) headed by Professor Haeshin Lee, and School in Michigan Technological University headed by Bruce P. Lee (USA). One of the best upshots of this particular research is that the bio-adhesiveness of the mussel proteins results through the oxidation/crosslinking phenomenon at acidic pHs of the DOPA units of *mfp* (Moulay 2014; Fullenkamp et al. 2014), a finding that was advantageously employed for hydrogel making; the crosslinking happens through the formation of bonds C–C (C_6–C_6) and C–O–C between two catechol moieties as illustrated in Fig. 2. In the case of a catecholamine, a crosslinking takes place also through a C–N bond formation between the amino group and the benzene ring of catechol, a phenomenon termed "quinone tanning" (Hong et al. 2014). Crosslinking of DOPA-containing *mfp* occurs aerobically and in the presence of mushroom tyrosinase enzyme in vivo, but it can be, in vitro, inferred either coordinatively with transition metal ions (M_T^{n+}) (Fig. 3), resulting in coordinative crosslinking, or oxidatively with oxidizing agents such as sodium periodate ($NaIO_4$), giving rise to covalent crosslinking (Fig. 2). The former crosslinking type displays stiffness and strength near a covalent bond, depending strongly on the coordination state and type of metals (Xu 2013); the metal-mediated crosslinking builds up through *mono*, *bis*, and *tris* (catecholato) metal complexes with color shifting, by varying the pH from acidic to alkaline values. Moreover, the propensity of metal ion to form complex with catechol ligand through chelation is contingent with the stability constant of the complex. As with some macromolecules and supramolecules (Yang and Urban 2013), metal-induced mussel-inspired hydrogels exhibit unusual self-healing property that can be highly valorized (Krogsgaard et al. 2015). Besides, this self-healing characteristic emerges in wet conditions after a whatsoever damage, whereas the remaining polymers do not exhibit such property. Their applications have been focused mainly on biomedical and environmental fields (Li et al. 2015a, b).

The rheological properties, i.e., the storage or elastic modulus G' and the loss modulus G'', of the below-discussed hydrogels were experimented by strain sweep at frequency, unless otherwise mentioned, of 1 Hz.

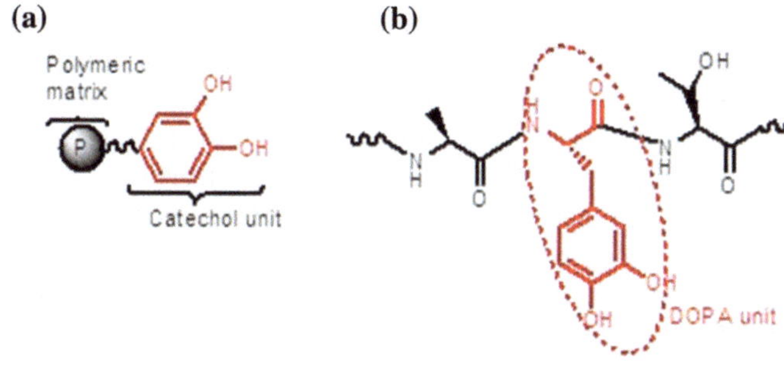

Fig. 1 **a** Catecholic polymer; **b** mussel feet protein (mfp)

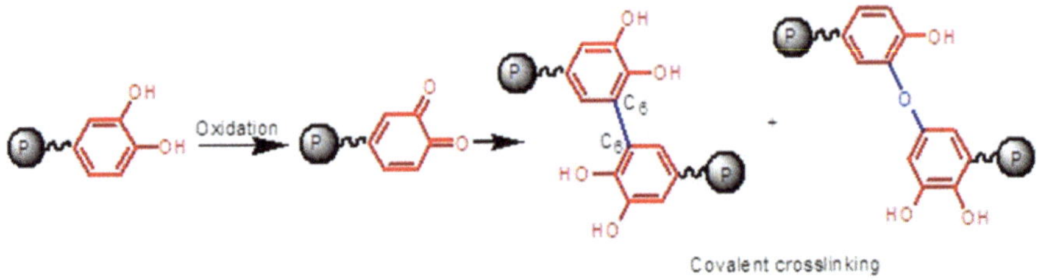

Fig. 2 Covalent crosslinks (blue bond) ensuing adhesion on substratum under wet conditions

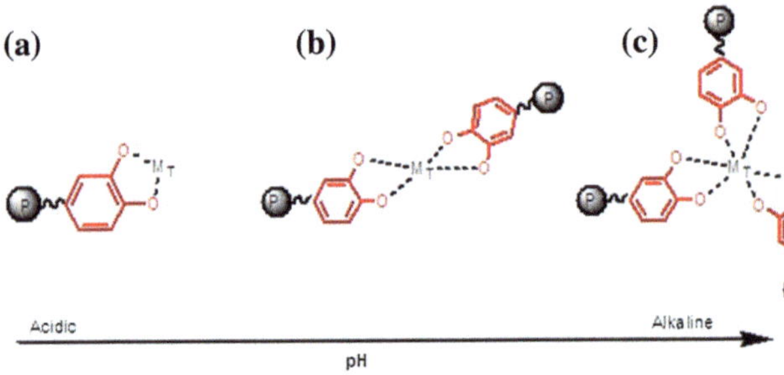

Fig. 3 Coordinative crosslinking. M_T = Transition metal: **a** *mono*(catecholato)-M_T; **b** *bis*(catecholato)-M_T; **c** *tris* (catecholato)-M_T

2 Nature-Related Polymers

2.1 Chitosan

Chitosan, the partially *N*-deacetylated chitin, and its modifications have served as unshakable platforms for hydrogel making, particularly in drug release mainly because of their good biocompatibility and biodegradability (Bhattarai et al. 2010; Riva et al. 2011; Thakur and Thakur 2014b). Grafting catechol entities onto chitosan backbone, in the aim of devising hydrogels for widespread applications, has recently withdrawn the interest of several workers (Ryu et al. 2014a, 2015). By applying the reduction amination approach (Bosch reduction of Schiff base), Yavvari and Srivastava (2015) anchored catechol units onto chitosan matrix by reaction with 4,5-dihydroxybenzoic acid (4,5-DHB) in a modification extent varying from 18 to 80 mol% (Fig. 4a). The gel formation was found to be dependent on metal salt concentration and extent of catechol grafted on chitosan matrix. Strongly colored hydrogels from chitosan functionalized with 66 mol% catechol were set via coordinative way using Fe(III) or Cu(II) at pH = 5. Oxidative gelation of systems with high catechol contents occurred with low sodium periodate ($NaIO_4$) concentration. Mechanical strength, that is storage modulus (G') and the loss modulus (G''), of the Fe(III)-made hydrogel increased with increasing Fe(III) content; G' rose more rapidly than G''. However, G' of the hydrogel provoked by oxidation at the same pH was found to be greater than that induced by Fe(III), 2373 against 278 Pa. Moreover, the hydrogels with covalent crosslinking showed higher brittleness and lower energy absorbing capacity; the absorbed energy or toughness (shear stress × shear strain) was computed to be 5.14×10^3 and 0.078×10^3 J m^{-3} for the hydrogels by Fe(III) and $NaIO_4$, respectively. Hydrogels from both approaches showed a good self-healing to a certain extent, as demonstrated by their ability to recover their initial mechanical strengths (G' and G'') after being subjected to degradation either by amplitude sweep or by applying a high sudden strain; yet, the hydrogels from coordinative pathway self-healed better than those from oxidative one. Similar work was undertaken by Ren's group (Guo et al. 2015) and about the same research outcome was observed.

Hydrocaffeic acid (HCA) (3,4-dihydroxyhydrocinnamic acid) units were introduced to chitosan chains (Fig. 4b) at 14.5 mol% by virtue of the 1-ethyl-3-(3-dimethylaminopropyl)-carbodiimide hydrochloride (EDC) method, and the hydrogel was made therefrom at physiological pH (7.4) in the presence of hematin-grafted chitosan, as biocatalyst surrogate of horseradish peroxidase (HRP), and hydrogen peroxide (Ryu et al. 2014b); iron(IV) in the hematin moiety promoted the catechol oxidation, and the reduced iron Fe(III) can be converted to Fe(IV) by H_2O_2. It was revealed that the simultaneous presence of both hematin-grafted chitosan and H_2O_2 was required for the formation of hydrogel from chitosan–catechol conjugate b under physiological conditions. The gelation time declined by increasing the hydrogen peroxide concentration; it was 130 s and 18 min for H_2O_2 concentration of 100 and 10 mM, respectively. The elastic moduli G' of the

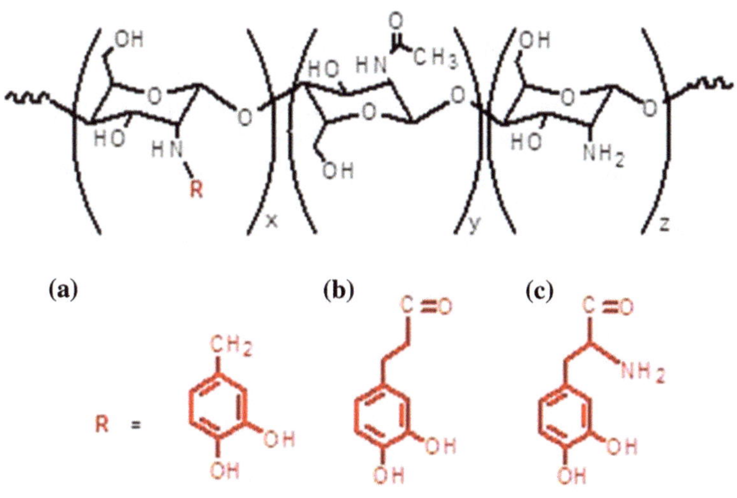

Fig. 4 Chitosan–catechol conjugates

as-prepared hydrogels were in the range of 10 kPa. These hydrogels showed low cytotoxicity ($\sim$100% cell viability), and their adhesive ability to human tissue was clearly demonstrated (a detachment stress of about 33 kPa). The water solubility of chitosan was substantially enhanced when conjugated with 12.7% catechol (Fig. 4b) for pHs up to 7, for making hydrogels (Kim et al. 2013); the virgin chitosan was insoluble in the PBS solution. The formation of hydrogel from 3% of a solution of chitosan-12.7% catechol conjugate was possible regardless the base (NaOH) concentration. The elastic modulus of this hydrogel was in the range of 2.5 kPa, suggesting a soft material. Hydrogels from chitosan-catechol conjugate with 12% catechol (Fig. 4b) were also triggered by vanadyl species (VO^{2+}) (Park et al. 2015); a quantity of vanadium as small as 10% was required for this purpose, compared to the needed quantity of Fe(III) because of the enhanced organic radical formation ability of vanadium as revealed by EPR analysis (electron paramagnetic resonance). The latter organic radical that was responsible for covalent crosslinking within the catechol units was generated from the intermediate vanadium *tris*-complex (Fig. 3, M_T = V). It was observed that the storage modulus of hydrogel increased with reducing vanadium amount or decreased with increasing iron one. For concentrations of vanadium and iron solutions of 110 mM, G's of their respective hydrogels were 6.5 Pa (sol state) and 13,530 Pa (gel state) and were 1454 Pa (gel state) and 536 Pa (sol state) for their concentrations of 5.5 mM. The cytotoxicity test showed that the cell viability for the hydrogel formed with 5.5 mM of vanadium was more than 90% and for that formed with 110 mM of iron was only 50%.

Birkedal and his co-workers (Krogsgaard et al. 2014) elaborated hydrogels from chitosan–DOPA conjugate (Fig. 4c) using Fe(III) as mediator. Chitosan–DOPA conjugate was realized with a grafting density of $\sim$3.3% through EDC/NHS chemistry. As the pH rose from acidic to basic values, the color of chitosan–DOPA conjugate/Fe(III) system changed color from green, to blue, and finally to red, assigned to the *mono*, *bis*, and *tris*(catecholato) iron (III) complexes, corresponding to the observed absorption, $\sim$720, $\sim$547, and $\sim$460 nm. The state of the system varied with varying pH; that is, the gel was observed at pH near 8 ($G' > G''$) after which the system became liquid-like ($G' < G''$); the storage modulus G' of the hydrogel at pH 8 was as high as 30 kPa for a frequency of 25 Hz. Increasing pH from acidic to alkaline values would intensify the coordinative crosslinking and a reduction of polymer charge which lowered its hydrophilicity, and finally the collapse of the hydrogel at a pH of higher than the optimal one.

In the aim at enhancing the mucoadhesive property of chitosan, hydrogels were conceived either by physically mixing chitosan with catechol derivatives (DOPA, HCA, DA) (Xu et al. 2012a) or by chemically conjugating it with them (Xu et al. 2015). Hydrogels from chitosan/DOPA and chitosan/DA mixtures swelled in water better than that from chitosan/HCA; after two hours and at physiological conditions (pH = 7.4), the swelling extent was nearly 35 and 30% for the first two hydrogels, respectively, and only $\sim$10% for the second one. The difference in swelling profile was attributed to electrostatic interactions between the chitosan groups and the catechol derivative. The study of catechol release from hydrogel revealed that at a pH of 5.5, DOPA and DA were freed from hydrogels more rapidly than HCA; an almost complete release was observed for the former two catechol derivatives after only half-hour, but about 50% of the latter one was liberated after two hours; this fact was explained in terms of ionization states of the catechol derivatives at this pH: DOPA and DA are cationic, and HCA is anionic. The mucoadhesion strength of the applied hydrogels on intestinal mucosa was quantified in terms of maximum detachment force (MDF). After a contact time of 3 min, the MDF of the hydrogels from chitosan/catechol derivatives fluctuated between $\sim$0.06 and 0.076 N/cm^2 and that of hydrogel from chitosan alone was 0.035 N/cm^2, that is, twofold enhancement of mucoadhesion with the former hydrogels. The hydrogels oxidized by $NaIO_4$ exhibited an improvement in MDF for chitosan alone (0.056 N/cm^2) and chitosan/HCA (0.101 N/cm^2) and no enhancement in the case of chitosan/DOPA and chitosan/DA. The significant increase of MDF for chitosan/HCA hydrogel was explained by the possible formation of covalent bonds between the oxidized catechol units and cysteine of the glycoprotein in the mucus layer.

However, a mucoadhesive hydrogel from covalently attached catechol to chitosan, that is, chitosan–catechol conjugates (Fig. 4b), was produced using genipin (methyl (1*R*, 2*R*, 6*S*)-2-hydroxy-9-(hydroxymethyl)-3-oxabicyclo[4.3.0]nona-4,8-diene-5-carboxylate), a natural crosslinker, for buccal drug delivery (Xu et al. 2012a); a local anesthesia drug was embedded in the as-made hydrogel capped with a protective layer of ethylcellulose. The erosion test showed that this hydrogel (chitosan/catechol/genipin) underwent a slower weight loss than the hydrogel from chitosan crosslinked with genipin (chitosan/genipin), when they were immersed in PBS buffer solutions (pH = 6.8) for 3 days. Also, lower drug release was observed with hydrogel of catechol-bound chitosan/genipin when compared with a hydrogel of chitosan/genipin, the pores of the former hydrogel being smaller than those of the latter one. The elastic modulus G' of the former hydrogel (with a 19 mol% grafted catechol) was lower than that of chitosan/genipin-made hydrogel, 439 against 1403 Pa. The hydrogel based on catechol-bound chitosan/genipin adhered to the buccal mucosa surface much longer (6 h) than the hydrogel from chitosan/genipin (1.5 h), revealing the superior mucoadhesion of the former hydrogel. In addition, the drug was substantially released from the former hydrogel within a few hours, whereas that from the latter hydrogel was almost nil. This catechol-related enhancement of mucoadhesive property of chitosan was endorsed further by the work reported by Kim et al. (2015).

2.2 *Alginate*

Alginate, a polysaccharide and a major component of algae, proves to be a useful biomaterial in food and biomedical uses, mainly because of its good biocompatibility and its facile gelation (Lee and Mooney 2012). Chemical derivatization of alginate has been a pertinent pathway to realize the coveted properties while keeping some of those of the parent alginate (Lee and Mooney 2012). Hydrogels formulated with catechol–tethered alginate (Fig. 5) were of a great value for certain applications (Pawar and Edgar 2012; Kastrup et al. 2012; Lee and Lee 2014). Alginate–catechol conjugate (Fig. 5), synthesized through EDC/NHS protocol with dopamine, was mixed with PBS solution/sodium periodate solution to yield a light brown-colored hydrogel (Kastrup et al. 2012). The adhesion strength of this hydrogel to vascular cells and tissue was measured to be 12 Pa and the ensued coating of gel on the blood vessels lasted for about a month. Besides, this hydrogel successfully healed arterial incisions. A hydrogel of dopamine-functionalized alginate (Fig. 5), prepared in phosphate buffer saline (PBS), was effective in hosting pyrocatechol violet as a pH-indicating dye (Kastrup et al. 2012). The thus-set hydrogel was a pH colorimetric sensor; that is, it was sensitive to strongly acidic and alkaline solutions as they became red and greenish blue, respectively. The storage and loss moduli G' and G'' of the hydrogel increased with pH, from about 1 to 100 kPa for a pH increasing from 3 to 10. To circumvent the problems encountered with hydrogels from ionically crosslinked alginate (with Ca^{2+}), namely the fast gel dissolution in encapsulation systems, Lee's school (Hong et al. 2015) conceived a strategy which they called "Stable Alginate gel Prepared via Linkage Exchange from ionic to covalent bonds" (STAPLE) as illustrated in Fig. 6; the covalent bond was secured by the slow oxidation of catechol to quinone, and the physicochemical integrity of hydrogel was maintained. The hydrogel from the alginate–catechol (Fig. 5) and a solution of Ca(II) incubated for 30 h PBS (pH = 7.4) was stable, and no gel–sol transition occurred; the color of the hydrogel shifted from white to brown, an indication of oxidation of catechol units. The swelling extent of the alginate–catechol hydrogel reached nearly 660% after 40 h of contact, whereas that of the hydrogel from the bare alginate did not exceed 350%. The physical stability of the alginate–catechol hydrogel was corroborated with the storage modulus variation with time: while that of the alginate hydrogel decreased from 535 to 2 kPa over 30 h, suggesting a gel–sol transition that of the alginate–catechol hydrogel gel declined to about 11 kPa.

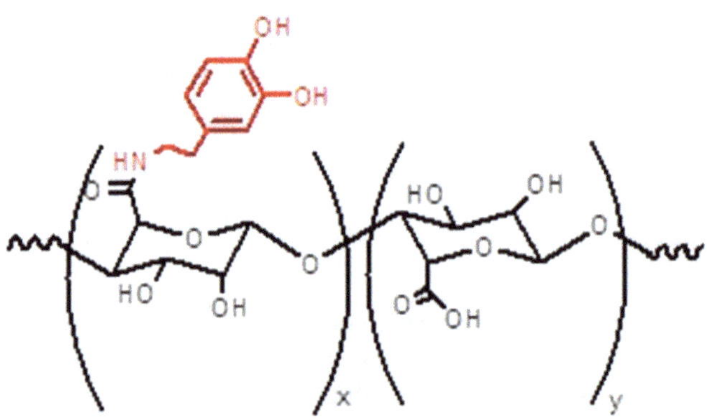

Fig. 5 Alginate–catechol conjugate

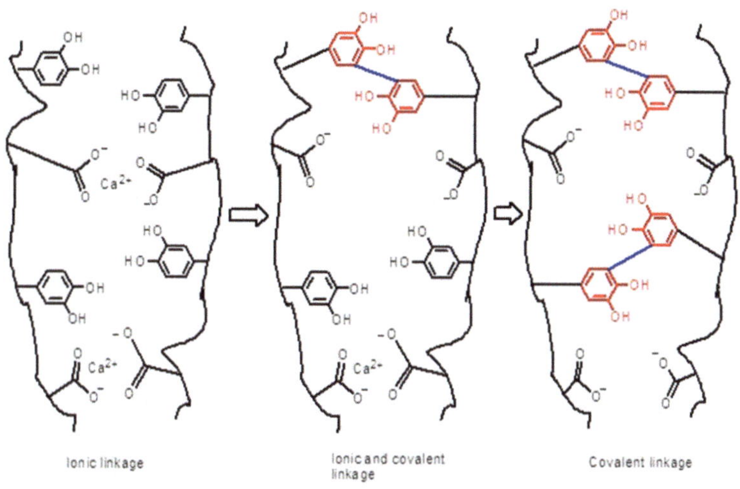

Fig. 6 Linkage exchange from ionic to covalent bonds

2.3 Hyaluronic Acid

Hyaluronic acid, a naturally occurring polysaccharide in living organisms, consists structurally of D-glucuronic acid and D-*N*-acetyl-glucosamine linked by alternating β-1,4 and β-1,3 glycosidic bonds. The different properties and biomedical applications of this biomaterial and chemical derivatizations have been reviewed (Necas et al. 2008; Schante et al. 2011). Hyaluronic acid–catechol conjugates (Fig. 7) have been made generally by coupling dopamine via EDC/NHS approach. Layer-by-layer methodology (LBL) has been applied for making multilayer films based on chitosan and catechol-bound hyaluronic acid to form polymeric coatings for biomedical uses, taking advantage of the adhesive property of catechol units (Neto et al. 2014). Hydrogels from these hyaluronic acid/catechol conjugates via oxidative crosslinking using $NaIO_4$ swelled in PBS solution at 37% to extents of 181% (for a degree of substitution DS of 4.5%) and 124% (for a DS of 8.8%) (Shin et al. 2015). Mechanical test demonstrated that the storage moduli G' of the hydrogels were greater than their loss moduli G'', hinting at its stability and its viscoelastic behavior. The storage moduli G' were 492 and 1321 Pa for hyaluronic acid/catechol with a DS of 4.5% and hyaluronic acid/catechol with a DS of 8.8%, respectively, an indication of softer hydrogel. The adhesion strengths of the hydrogels to liver tissue were measured to be 0.804 and 1.356 kPa, respectively, suggesting that the higher catechol content, the better the adhesion. Such adhesion lasted about a month.

A hydrogel through a coacervation between dopamine-conjugated hyaluronic acid and lactose-modified chitosan in 1:1 ratio was reported (Oh et al. 2012). The coacervation occurred as a result of electrostatic interactions between the two polymeric systems and of Michael addition reaction of catechol units at alkaline pH.

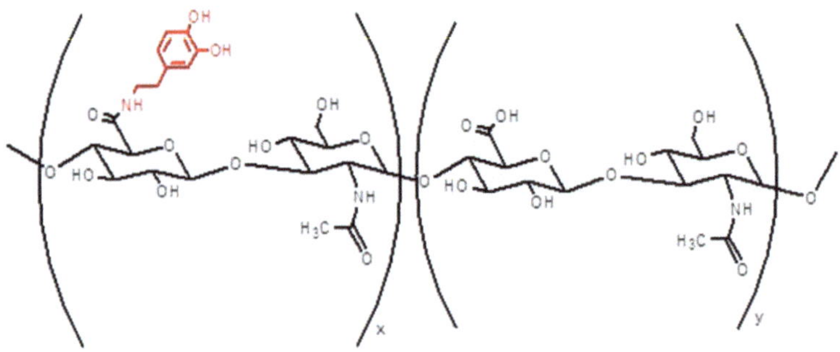

Fig. 7 Hyaluronic acid–catechol conjugate

Indeed, at neutral pH, the coacervate complex was formed via ionic interactions. However, at pH of 8.5, catechol moieties were oxidized to the corresponding quinones, and subsequent Michael addition reaction provided crosslinking between the two polymeric chains; hence, the physical stability of the hydrogels was enhanced. The zeta potential (ζ-potential) of the thus-made hydrogels was found close to zero. The spectral analyses suggested a coacervate formation via catechol–amine interaction, and the swelling test confirmed the impact of catechol-inducing crosslinking on the coacervation complex formation. As the rheological measurement revealed, the coacervate complex behaved as an aqua material-like, that is, a material exhibiting characteristic features of both hydrogel and viscous fluid.

2.4 Proteins/Polypeptides

Hydrogels from peptides and proteins have been well-documented (Jonker et al. 2012; Dooling and Tirrell 2013) and found most uses and applications in biomedicals (Altunbas and Pochan 2012). A tripeptide with a terminal catechol unit (Fig. 8a) was designed, and a hydrogel was made therefrom using metalloprotease as a triggering enzyme (Xu et al. 2014); no gel was formed in the absence of this enzyme. The storage and loss moduli of the hydrogels were found nearly 10 and 1 kPa, respectively, the first modulus being almost tenfold the second one. The cell adhesion of the hydrogel was proved by the great number of adhering cells and by the high cell viability (>98%).

A dipeptide Fmoc-DOPA-DOPA (Fig. 8b) was able to form a nanostructured hydrogel through self-assembly property (Fichman et al. 2014). The storage modulus (G') of the hydrogel was larger than the loss modulus (G''), proving the elastic behavior of the hydrogel. The hydrogel formation was rapid at elevated temperatures; G' of the hydrogel formed at 25-37 °C was found to be 40-fold higher than that for a hydrogel formed at 4 °C. In the presence of silver nitrate, a color change occurred, from semi-transparent to dark brown, after several hours to a few

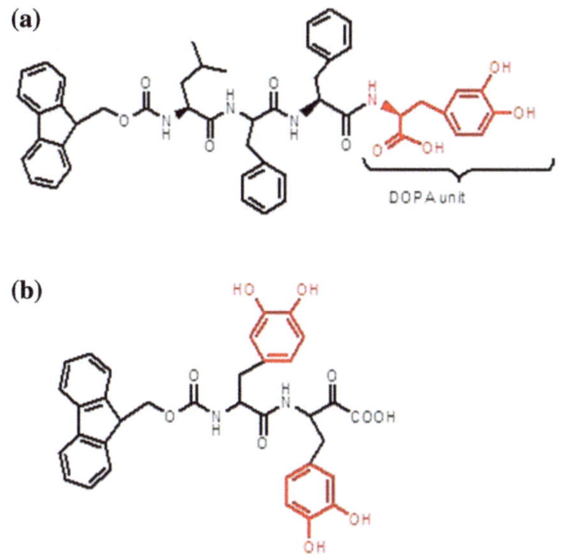

Fig. 8 **a** Tripeptide-containing catechol: Fmoc-Leu-Phe-Phe-DOPA; **b** dipeptide-containing catechol: Fmoc-DOPA-DOPA. (*Fmoc* 9-fluorenylmethoxycarbonyl; *Leu* leucine; *Phe* phenylalanine)

days, a result assigned to the oxidation of DOPA and to the scattering by silver nanoparticles.

Catechol-bearing gelatin was prepared by oxidizing tyrosine residues of gelatin into DOPA units using tyrosinase (Choi et al. 2014). An instantaneous dark brown hydrogel was ensued by treating the brown DOPA-conjugated gelatin solution with Fe(III) solution at 37 °C for 24 h; the formation of *tris*(catecholato) iron(III) complex was evidenced by Raman spectroscopy (a characteristic band at 500–700 nm). The elastic and loss moduli of DOPA-Fe(III) gelatin hydrogel were higher than those of DOPA-gelatin and gelatin, proving the impact of the crosslinking via catechol units and Fe(III) on the mechanical properties; at frequency of 1 Hz, they were measured to be ~3.3 and ~10.20 Pa, ~0.42 and ~0.58 Pa, ~0.04 and ~0.112 Pa, respectively. It was found that cytotoxicity was insignificant with DOPA-Fe(III) gelatin hydrogel. Also, this hydrogel showed a good hemostatic activity; that is, the bleeding of the liver sample was significantly reduced upon injection of the hydrogel; thus, its use as a tissue adhesive in surgical applications is justified.

Recombinant mussel adhesive protein (r*fp*-1) (12 tandemly repeated decapeptides of *fp*-1) obtained by using *Escherichia coli* was treated with mushroom tyrosinase affording r*fp*-1-DOPA conjugate (Kim et al. 2014); for a gelation occurrence, about 37% of the r*fp*-1 tyrosine residues were converted into DOPA molecules. The formation of hydrogels from r*fp*-1-DOPA conjugate was promoted by treatment with sodium periodate in PBS solution (pH = 7.4) or by complexation with Fe(III). With the first treatment, the color of the gel changed to dark brown. However, with the second treatment, the color shifted to green for a pH of 4.5 (*mono*-complex DOPA-Fe(III)), to purple for a pH of 5.5-6.9 (*bis*-complex DOPA-Fe(III)), and finally to pink for a pH higher than 8.2 (*tris*-complex DOPA-Fe (III)). Rheological study revealed the effect of the gelation inducer on the mechanical properties, the storage and loss moduli. The adhesive strength of the Fe (III)-mediated noncovalent *rfp*-1 gel against porcine skin was nearly 130 kPa under alkaline pH ($\sim$8.2) and that of the quinone-mediated covalent one was about 200 kPa. This difference in the adhesive strength was rationalized in terms of elastic moduli gap that of the latter system being higher than that of the former one.

3 Synthetic Polymers

3.1 Poly(oxyalkylene)s

Poly(ethylene glycol) (PEG), a versatile poly(ethylene oxide), indubitably stands as the leading synthetic polymer in medical applications because of its outstanding pertinent properties: good biocompatibility, nontoxicity, and minimal immunogenicity. Recent applications of PEG and its derivatives were reported (Hutanu et al. 2014), and their versatile uses were prevalently in form of hydrogels (Zhu 2010). In several cases and for specific applications, PEG has been chosen as the hydrophilic component of copolymeric materials. Numerous PEGs end-capped with catechol, inspired from mussel feet proteins, were developed and valorized in duly applications; a common observation is that the one-armed PEG–catechol conjugate (Fig. 9) cannot provoke gel formation in several cases, but the multi-armed ones gelled rapidly (Moulay 2014). Besides, PEG–catechol conjugates, designed for hydrogel making, were either functionalized with polymerizable monomers or parts of copolymers. Hydrogels from the functionalized with acrylic monomers have been chiefly directed toward medicals as effective sealants. In fact, their sealing property was excellent for fetal membrane repair with efficient, nondisruptive, and nontoxic bonding (Bilic et al. 2010). For this peculiar application, the hydrophilic PEG segment of PEG–catechol conjugates proved to ease the adhesive interactions between hydrogel-bound catechol residues and tissue/implant surfaces.

Hydrogels were readily formed upon complexing DOPA-bound PEG's (Fig. 10) with Fe(III) (Zeng et al. 2000) or oxidizing them with $NaIO_4$, horseradish peroxidase/H_2O_2, or mushroom tyrosinase/O_2 (Lee et al. 2001, 2002). In the case of oxidative crosslinking, the gelation time was rapid and depended on the type and

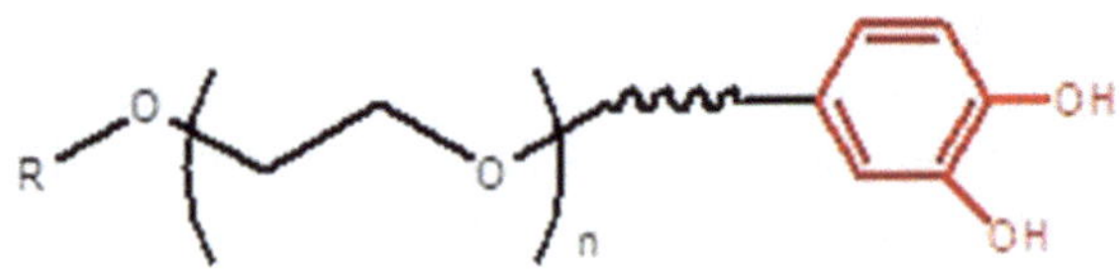

Fig. 9 One-armed PEG–catechol conjugate

concentration of the employed oxidizing agent, as well as on the polymer architecture (linear versus branched). The storage and loss moduli of the hydrogel from Fig. 10a (R = H) using sodium periodate as oxidizing agent were found to be ~13 and ~0.4 kPa, respectively.

A four-armed PEG conjugated with DOPA (Fig. 10a, R = Boc) was employed to produce biomimetic hydrogels using periodate-containing liposomes in form of vesicles; the latter materials were realized by hydration of thin films of phospholipids with sodium aqueous periodate solution in PBS (pH = 6) (Burke et al. 2007). Hydrogels were engendered by heating mixture of PEG-DOPA conjugate with periodate-containing vesicles at 37 °C; at this temperature, $NaIO_4$ was gradually released and triggered the gelation by oxidative crosslinking. The gels formed after nearly 50 min presented shear moduli G' of 6 kPa. The gels showed lap shear strengths of about 34 kPa against a porcine dermal skin surface.

A four-armed PEG-based hydrogel containing catechol groups was produced instantaneously upon addition of aqueous sodium periodate into PBS saline solution of catechol-modified four-armed PEG (n = 55) (Fig. 11a). A color shift was observed from colorless to brown. The hydrogel was valorized as adhesive in extrahepatic tissue islet transplantation (Brubaker et al. 2010) and as membrane sealant for iatrogenic preterm premature rupture of membranes (iPPROM) (Haller et al. 2011).

The four-armed PEG–catechol conjugate of Fig. 11a formed a hydrogel by complexing with 1,3-benzenediboronic acid in PBS solution under basic pH = 9

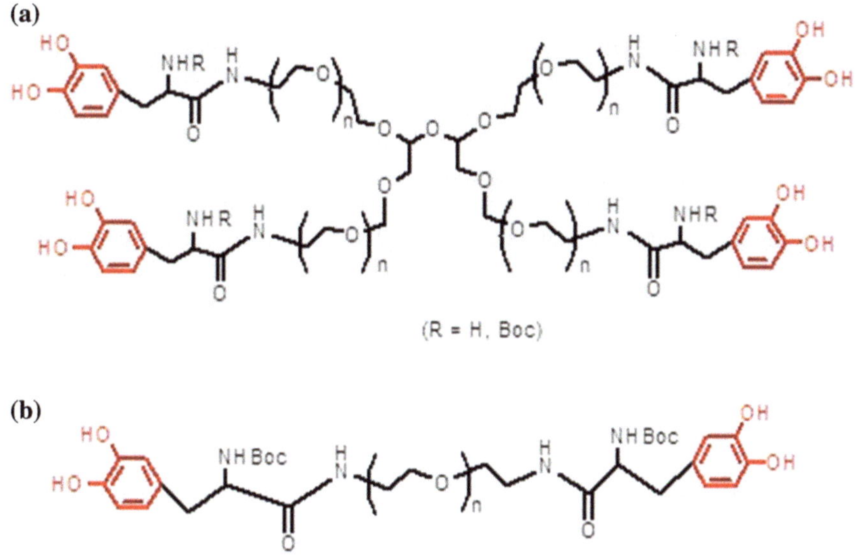

Fig. 10 PEG–catechol conjugates synthesized via carbodiimide chemistry: **a** a four-armed conjugate; **b** a two-armed conjugate. *Boc tert*-butoxycarbonyl (a protecting group)

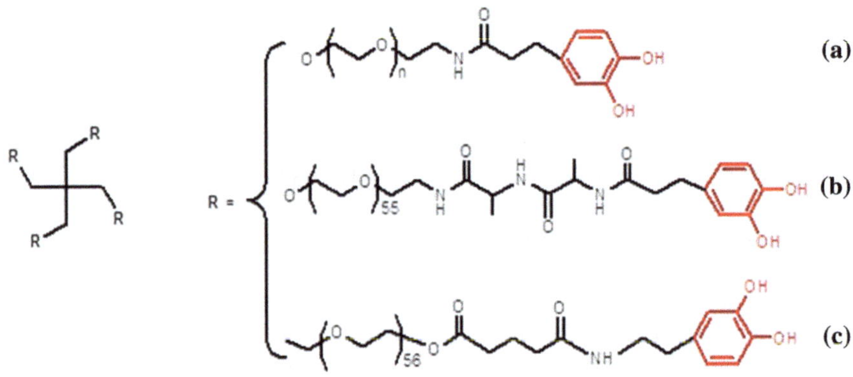

Fig. 11 Four-armed PEG–catechol conjugates

and at 20 °C within 48 h. A tetrahedral borate ester complex led to a three-dimensional polymeric network (Fig. 12) as evidenced by ^{11}B and ^{1}H NMR analyses and UV–visible absorption band at 490 nm (He et al. 2011). The acidic treatment of the hydrogel (pH = 3) resulted in its disruption through the gel-sol transition, with a color shift from red to yellow. A storage modulus of the obtained sticky hydrogel was in the range of 1.5 kPa at 37 °C. Its self-mending property was demonstrated by oscillatory rheology; this property was believed to be attributable to a reversible complexation between boronic acid and catechol.

To develop an insight into the role of the metal coordination crosslinking in providing in vivo self-healing and wet adhesion of soft biological substances, Messersmith's group (Holten-Andersen et al. 2014) investigated the formation of polymer networks from the reaction of four-armed PEG–catechol conjugate (Fig. 11a) with $FeCl_3$, VCl_3, and $AlCl_3$ at the oceanic pH of 8. UV–visible analysis showed a conspicuous band at λ_{max} = 525 nm for the product with $FeCl_3$ (purple color), denoting the existence of *bis*-catechol–Fe coordination; a band at ~395 nm and a broad one at 600–700 nm were seen for the product with VCl_3 (blue color), indicating *tris*-catechol–V complex. However, the absorption of the product with $AlCl_3$ was very weak. While resonance peaks in Raman spectra of the gels with Fe(III) and V(III) were evident at 490–696 cm^{-1} (metal-oxygen interaction) and

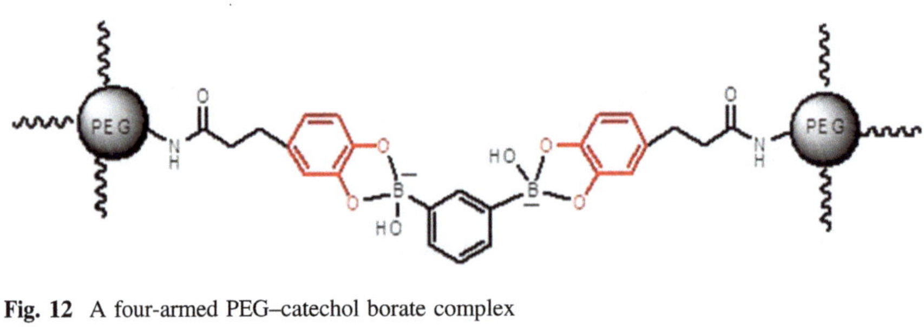

Fig. 12 A four-armed PEG–catechol borate complex

1190–1519 cm^{-1} (catechol ring vibrations), no peaks were observed for the gel with Al(III), suggesting no catechol–Al complexation. Mechanical analysis demonstrated that the hydrogel with V(III) exhibited a stability tenfold higher than that of hydrogel with Fe(III) under the experimental conditions.

Antifouling hydrogels were prepared via oxidative crosslinking of four-armed PEG–catechol conjugate (Fig. 11a) using silver nitrate or sodium periodate as oxidizing agent (Messersmith et al. 2014). The as-made hydrogels presented significant antibiofouling performance for fibrinogen adsorption substrates; about 524 ng/cm^2 of adsorbed fibrinogen mass on bare TiO_2 surface at pH = 7.4 and at a temperature of 37 °C, but only 78 ng/cm^2 of adsorbed fibrinogen mass were estimated when this surface was coated with the hydrogel. The coated hydrogel was resistant to mammalian cell and bacteria cell adhesion; 99% of the *S. Epidermidis* and 99.9% of 3T3 fibroblast cell adhered to the substrate were suppressed after one to two days of exposure at 37 °C.

Another four-armed PEG–catechol conjugate was conceived by implementing a dipeptide alanine-alanine (ala-ala) in-between (Fig. 11b) for making an enzyme-degradable adhesive hydrogel for soft tissue adhesive and medical sealant (Brubaker and Messersmith 2011). The hydrogel formed via oxidative crosslinking (with sodium periodate in water at a physiological pH = 7.4) turned brown upon this treatment. The swelling ratio of the hydrogel was nearly 34%. Its mechanical properties, G' and G'', were appropriate for soft tissue applications and were, respectively, 10 kPa and <1 kPa. The adhesive performance of the hydrogel was demonstrated by its lap shear strength of 34 kPa. In vivo degradation of the hydrogel was assessed with neutrophil elastase and was positive over several months, justifying its application for soft tissue injury, nonstatic tissues (gut and vasculature), and for adhesion in a chronically inflamed site.

van Hest's team (Jonker et al. 2015) adopted the strategy of strain-promoted oxidation-controlled cyclo-octyne-1,2-quinone cycloaddition (SPOCQ) to engineer hydrogels. To this end, the mixture of four-armed PEG–catechol (Fig. 13a) and four-armed PEG functionalized with bicyclo[6.1.0]non-4-yne (Fig. 13b) was oxidized with sodium periodate as shown in Scheme 1. The hydrogel was formed instantaneously as per the fast kinetics of SPOCQ. Gels with low polymer concentrations (10–30 mg/L) were soft, and those with higher ones were stronger; the storage modulus increased with increasing polymer concentration: 65 Pa (for 10 mg/mL), 454 Pa (for 20 mg/mL), and 1375 Pa (for 30 mg/mL).

To assess the effect of pH on the curing rate and adhesive property of mussel-inspired adhesives, a four-armed PEG end-capped with dopamine was adopted as candidate (Cencer et al. 2014). The hydrogel curing rate was found to decrease with increasing pH for $NaIO_4$/dopamine molar ratio up to 0.5 and to increase beyond. In addition, the average molecular weight between crosslinks in the hydrogel, M_c, followed the same profile, supporting the pH dependence. The elastic modulus G' rose with increasing pH: 96.5 kPa (pH = 5.7), 158 kPa (pH = 6.7), 212 kPa (pH = 7.4), 242 kPa (pH = 8). The lap shear strength for gluing two pieces of bovine pericardium with hydrogel was maximum (about 8 kPa) at pH = 7.4.

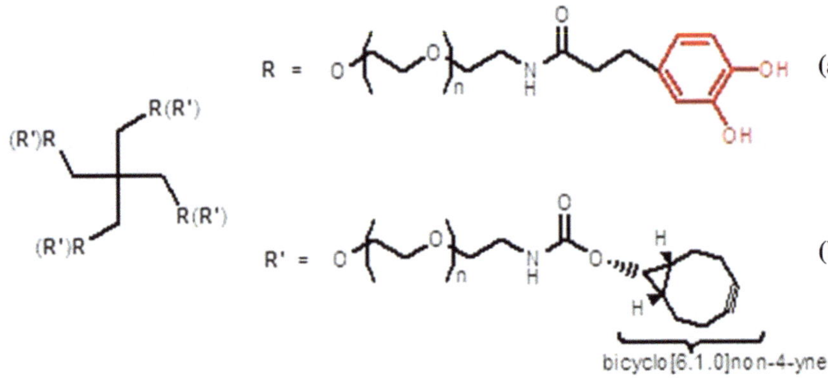

Fig. 13 **a** Four-armed PEG–catechol; **b** four-armed PEG functionalized with bicyclo[6.1.0] non-4-yne

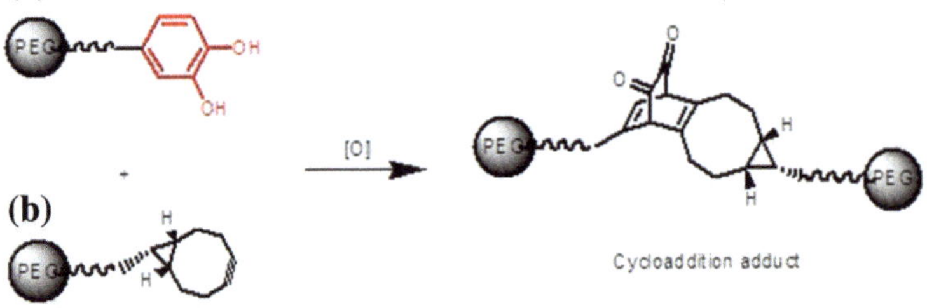

Scheme 1 SPOCQ reaction

An injectable nanocomposite hydrogel for tissue was realized by oxidative curing of a four-armed PEG–dopamine conjugate (Fig. 11c) in PBS solution to which Laponite, a synthetic nanosilicate with a chemical formula of $Na_{0.7}Si_8Mg_{5.5}Li_{0.3}O_{20}(OH)_4$, was incorporated (Liu et al. 2014); Laponite in the formulation of hydrogels has been widely used for the last decade, to ensure physical crosslinking and enhance the mechanical properties. Crosslinking was claimed to be promoted also by an interfacial binding between dopamine and Laponite. Increasing Laponite concentration led to a gelation rate decrease for a $NaIO_4$/dopamine molar ratio of 0.5; in the absence of Laponite, the curing time was 1.81 min and declined to 0.33 min when Laponite was added in a concentration of 2%. Rheological parameters G' and G'' of the nanocomposite hydrogel measured at 1 Hz were greater than those of Laponite-free hydrogels: ~1.5-fold and ~three-fold higher, respectively. The LSS and W_{adh} (adhesion work) for adhering two pieces of bovine pericardium with hydrogel containing 2% Laponite were about 8 kPa and 17 J/m^2, respectively, values higher than those for Laponite-free hydrogels (3.5 kPa and 6.7 J/m^2). The incorporation of Laponite in the

PEG-dopamine hydrogel heightens its bioactivity and adhesive properties. It did not affect the hydrogel degradation and its rate upon incubation in PBS solution for seven days. The cell viability for hydrogel with 2% Laponite neared 100%.

Injectable mussel-inspired tissue bioadhesive hydrogels containing citrate group were made by inducing crosslinking of the prepolymer (Fig. 14) via oxidation with sodium periodate in PBS solution; the latter prepolymer was made by a one-step polycondensation of citric acid, PEG, and dopamine under vacuum and at a temperature of 140–160 °C (Mehdizadeh et al. 2012). The gelation time was found to be function of molecular weight of PEG, the DOPA content, and the concentration of sodium periodate; longer gelation times with high molecular weights of PEG, and shorter ones with high concentrations in DOPA and sodium periodate. Tensile strength, elongation at break, and tensile modulus of the dry hydrogel were larger than those of the swollen one. Indeed, for systems made of PEG with a molecular weight of 200 Da and crosslinked with a 8% sodium periodate solution, they were 8 515 kPa, 397%, and 36 MPa, respectively, for a dry state, against 242 kPa, 210%, and 0.2 MPa for a swollen one. The wet adhesion of these hydrogels to a biological membrane was relatively appreciable; the measured lap shear strength (LSS) oscillated between 40 and 123 kPa, values 2.5–8.0 stronger than that with fibrin glue (15.4 kPa). As to the biological tests, their in vivo cytotoxicity and their degradation products showed good cell viability (66–78%). Incision healing using the hydrogel took no longer than 2 min: The bleeding was ceased and the wound was closed within this lapse of time. After wound healing, the polymers were totally biodegraded within a month.

Six-armed PEG–catecholamine (Fig. 15) was prepared by a one-step reductive amination between a six-armed PEG-amine and 3,4-dihydroxybenzaldehyde (DHBA) in *N*-methylpyrrolidone (NMP) in the presence of sodium cyanoborohydride ($NaBH_3CN$) at room temperature (Hong et al. 2014). The treatment of the PBS solution (pH = 8) of the six-armed PEG–catecholamine with $NaIO_4$ yielded a hydrogel within few seconds ($\sim$2 s); the gelation time for six-armed PEG–catechol was longer ($\sim$12 s), indicating the effect of amine group on the hydrogel

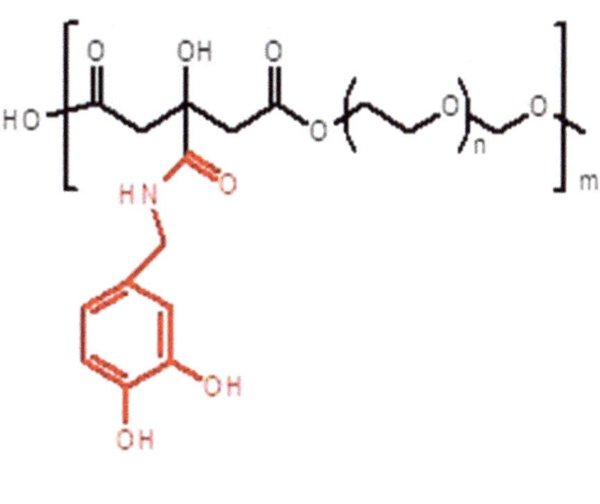

Fig. 14 Injectable citrate-based mussel-inspired tissue bioadhesives

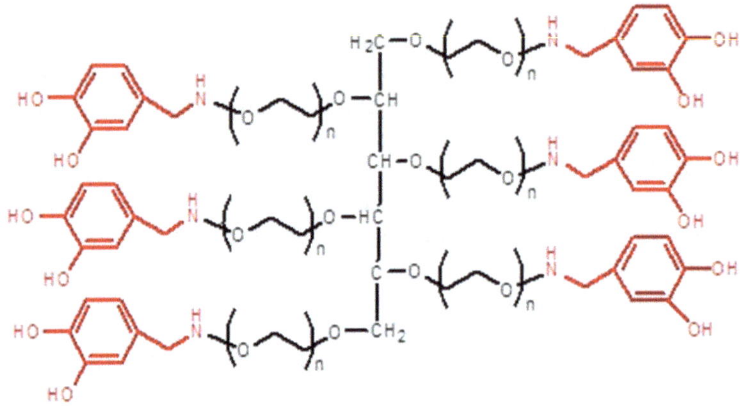

Fig. 15 Six-armed PEG–catecholamine conjugates

mechanism process. Indeed, it was demonstrated chromatographically that both amine and catechol groups participate in the crosslinking process, creating more crosslinking junctions than with catechol alone, as it occurs in vivo. The effect of the amine group manifested also on the mechanical property of the hydrogel: The elastic modulus of the hydrogel from six-armed PEG–catecholamine was twice that of the hydrogel from six-armed PEG–catechol, about 1 kPa against 0.500 kPa.

A sturdy porous hydrogel was conceived by combining dopamine-conjugated eight-armed PEG with a poly(amido amine) dendrimer G4 in a ratio of 2 and induced by oxidation with $NaIO_4$ in PBS solution (Wang et al. 2015a). Dopamine was affixed to the PEG via a disulfide link. To improve the biocompatibility, the dissolution process, and the conjugation feasibility of bioactive peptides, the dendrimer was initially functionalized with maleimide-PEG-succinimidyl carbonate ester and methoxy-PEG-succinimidyl carbonate ester. No hydrogel was formed in the absence of dopamine-conjugated eight-armed PEG, hinting at the role of oxidized dopamine units to afford porous 3D networks. The gelation time was up to 30 min and depended on the amounts of dopamine-conjugated eight-armed PEG and $NaIO_4$ concentrations. The size of the hydrogel pores ranged from 5 to 30 μm. The swelling degree of the hydrogel was as low as 245%, compared to that of hydrogel from dopamine-conjugated eight-armed PEG (1511%), indicating the significant impact of the dendrimer on the properties of the hydrogels. The storage and loss moduli of the hydrogels were ~77–84 and ~12–15 kPa, respectively; those of the hydrogel from dopamine-conjugated eight-armed PEG were lower, ~36 and ~10 kPa. These mechanical properties give an insight into the robustness of the hybrid hydrogels, which was the result of synergetic effects of the high crosslinking density and the dendrimer architecture. The hydrogels were found to biodegrade instantaneously in the presence of 10 mM glutathione and after 50 days of incubation with 4 μM glutathione. The cytotoxicity test proved that the hydrogels were not toxic after 72 h against living cells, and the cell viability was in the range of 160%.

The study of the incidence of the chloro- and nitro substituents on the properties of hydrogels from PEG–catechol derivative conjugates was undertaken (Cui et al. 2014). A four-armed PEG was end-capped with dopamine, 6-chloro-dopamine, 6-nitrodopamine, norepinephrine (β-hydroxy dopamine derivative), and 6-nitro-norepinephrine, and hydrogels therefrom were made by oxidative ($NaIO_4$) and coordinative crosslinkings (Fe^{3+}). Films were obtained by coating hydrogels on quartz or silicon substrates. The deposition kinetics and the film thickness were found to be substituent-related. Oxidation-inducing hydrogels of Cl-dopamine-functionalized star-PEG were evaluated for their antifouling capacity toward gram-negative bacterium *Escherichia coli* (García-Fernández et al. 2013). The shear moduli of these hydrogels fluctuated between 6 and 10 kPa. The hydrogel-coated glass surfaces presented indeed antifouling properties toward this bacterium: About 80% decrease in bacteria density at the surface was observed with increasing concentration of Cl-dopamine fraction in the gel, whereas the dopamine-modified PEG gel did not prevent bacteria attachment.

The copolymerization of photocurable *N*-methacrylated DOPAs with poly (ethylene glycol) diacrylate using photoinitiators such as 4-(dimethylamino)-benzoic acid (DMAB), 2,2-dimethoxy-2-phenyl-acetophenone (DMPA), camphorquinone (CQ), and fluorescein sodium salt (FNa_2) (Scheme 2), afforded DOPA-containing hydrogels (Lee et al. 2004). The gel was red colored, indicating the presence of catechol units. The elastic modulus E of the thus-prepared hydrogel was measured to be 47–51 kPa, a modulus lower than that of PEG-DA (72.3 kPa); such elastic modulus was within the values required for medical applications. The photopolymerization of the macromonomer bearing DOPA moiety (Fig. 16) triggered the adhesive hydrogel formation with a DOPA content of 10 mol% in a few seconds, suggesting an insight into a tissue sealing application (Lee et al. 2006). Adhesion work W_{adh} for the interaction of hydrogel with titanium (Ti) surface was measured as nearly as 410 mJ/m^2. The oxidation of the DOPA units of the hydrogel with $NaIO_4$ significantly reduced W_{adh}, indicating that DOPA in its reduced form (catecholic form) was responsible for the strong water-resistant adhesion to Ti surface. The elastic modulus of bare hydrogel was about 43 kPa and increased to about 55 kPa after treatment with periodate.

Poly(ethylene glycol) served as soft segment in mussel-inspired polyurethane hydrogel (Sun et al. 2014). To achieve such a goal, PEG with different molecular weights (1, 2, and 3 kDa) was allowed to react with 1,6-hexamethylene diisocyanate ended up with a polyurethane prepolymer (PEG/PU) that underwent polycondensation with lysine-dopamine (LDA), affording PEG/PU-LDA (Scheme 3). Coordinative crosslinking of the PEG/PU-LDAs with Fe(III) at a gelation pH (alkaline) ensued red-colored hydrogels instantaneously (1–2 s), the color indicating the formation of *tris*-complex (Fig. 17); hydrogels were not formed under acidic conditions.

Its storage modulus was 3.2 kPa and could be recovered after a gel failure under 1000% strain. Also, a self-healing test for PEG/PU-LDA hydrogel promoted by Fe (III) was positive. However, the hydrogels resulted by oxidative pathway with $NaIO_4$ were orange (Fig. 18), and the gelation time was found to vary with pH,

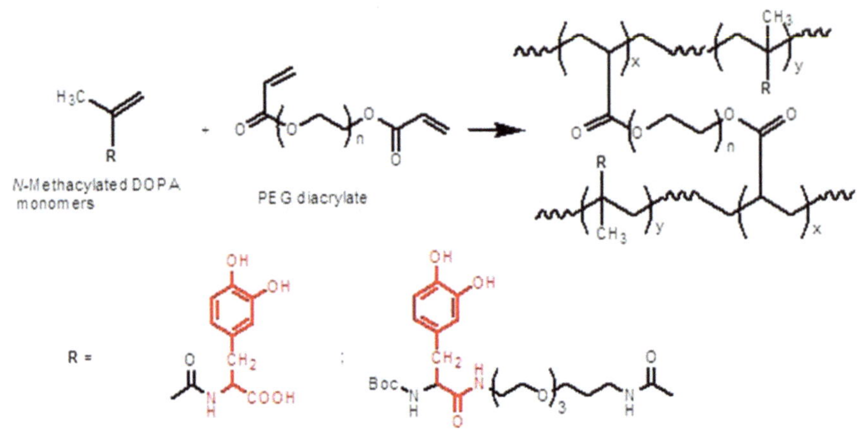

Scheme 2 Hydrogels by photopolymerization

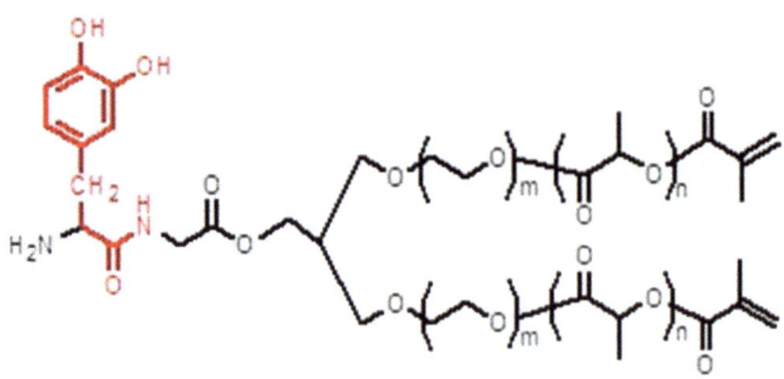

Fig. 16 PEG diacrylate monomer with DOPA unit

[$NaIO_4$]/[DA] molar ratio, and molecular weight of PEG segment in a nonsystematic fashion. A gelation time of 10 s was observed for a pH of 7.4, a [$NaIO_4$]/[DA] molar ratio of 0.5, and molecular weight of PEG of 2 kDa. When the latter parameter was 3 kDa and keeping the remaining parameters at these values, the gelation time was 9 min 10 s.

Pluronics or poloxamers, poly(ethylene oxide-*b*-propylene oxide-*b*-ethylene oxide)s abbreviated as PEO-PPO-PEO, underwent functionalization with catechol derivatives after chemical activation with *N,N*-disuccinimidyl carbonate in the presence of 4-(dimethylamino)pyridine (DMAP) (Huang et al. 2001, 2002, 2012). The DOPA-modified PEO-PPO-PEO polymers (Fig. 19a) aggregated into micelles at their critical micellization temperatures (*cmt*). The *cmt* and the gelation temperature (T_{gel}) varied with the composition of the DOPA-functionalized polymer, its molecular weight, and its aqueous solution concentration; that is, a 3 °C increase in *cmt* for a 2.5 wt% increase in polymer concentration, and for the same composition and molecular weight. T_{gel} was in the range of 20 and 46 °C. The 22 wt% solutions

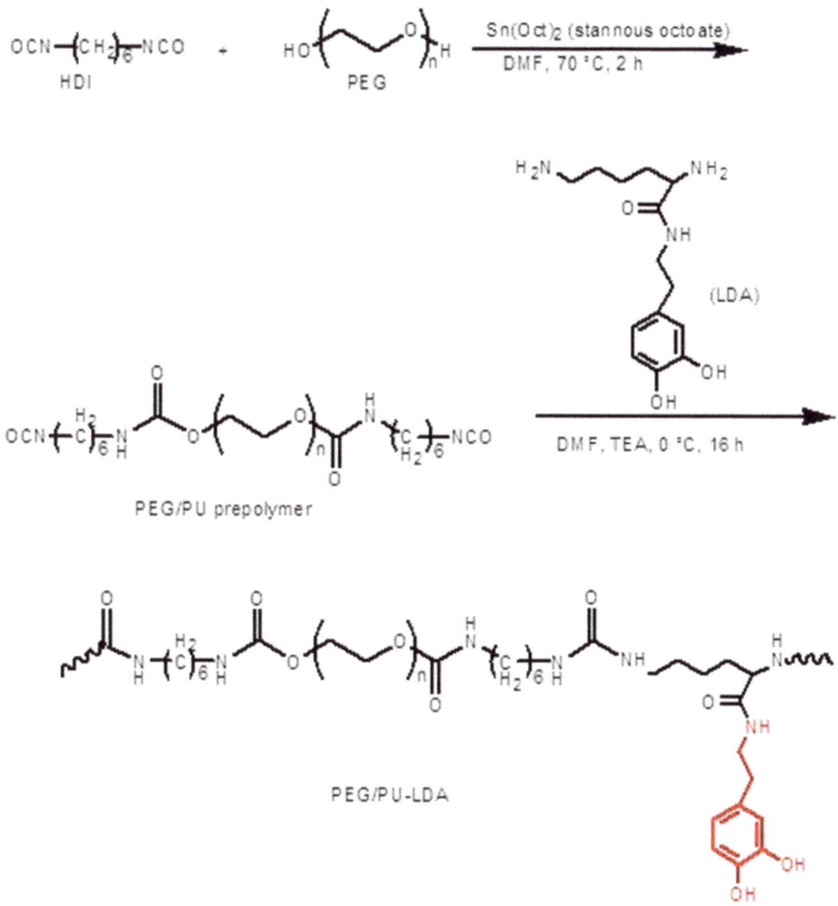

Scheme 3 Synthesis of PEG-PU-LDA polymers

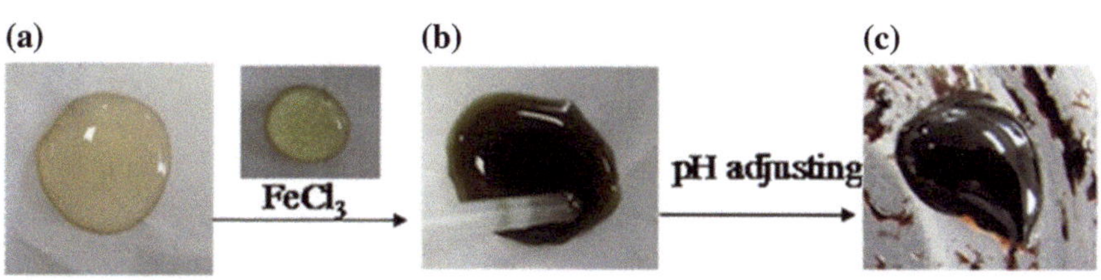

Fig. 17 State and color of the hydrogel formed by treating PEG/PU-LDA solution: **a** in water only (colorless); **b** mixed with $FeCl_3$ (green); **c** pH adjusting with NaOH (dark red) (Sun et al. 2014). Reprinted (adapted) with permission from Sun et al. (2014). Copyright (2016) American Chemical Society

Fig. 18 State and color of the hydrogel formed by treating PEG/PU-LDA solution: **a** in water only; **b** mixed with $NaIO_4$ (orange); **c** gelation in progress; and **d** the vial in (**c**) inverted (Sun et al. 2014). Reprinted (adapted) with permission from Sun et al. (2014). Copyright (2016) American Chemical Society."

of these DOPA–functionalized Pluronics formed transparent gels at 23 °C and lowering the polymer concentration to 18% resulted in a gelation temperature of approximately 31 °C. The hydrogels from Pluronics–DOPA conjugates showed some solidity to flow over long periods of time. Their storage modulus G' increased rapidly at the gel temperature (20.6 °C). The mucoadhesive property of these DOPA-modified Pluronics toward mucin, a natural carbohydrate substance, was estimated via the bioadhesion viscosity of the mucin/DOPA-copolymer system. Its higher viscosity, compared to that of the mucin/unmodified copolymer solution, was imputed to the catechol units of DOPA residues, being able to create intermolecular hydrogen bonding between hydroxyl and different carbonyl groups of mucin.

Injectable hydrogels made of chitosan–catechol conjugate (10% catechol) and thiol-terminated Pluronic (78% thiolation) were crosslinked, at 37 °C and alkaline pHs, by reaction of oxidized catechol with the primary amino groups of the chitosan and with thiol groups of the thiolated Pluronic via a Michael addition reaction or a Schiff base formation reaction (Ryu et al. 2011). While the brown color of the hydrogel was attributed to the formation of *o*-quinones, it was advanced that the

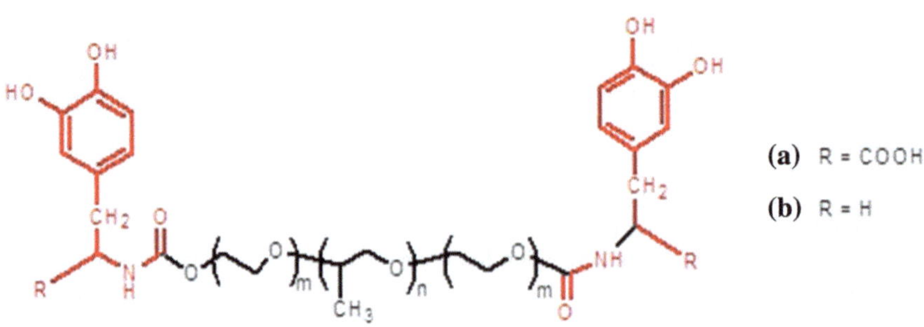

Fig. 19 Pluronic-DOPA/Dopamine conjugates

rapid gelation of the system was mainly due to the physical crosslinking between the Pluronic chains and not to the oxidation of catechol because of its slowness. Rheological experiments indicated an elastic modulus (G') of 20.8 kPa for chitosan–catechol conjugate/thiol-terminated Pluronic hydrogels, modulus larger than those of chitosan–catechol conjugate hydrogels (0.943 kPa) and those of hyaluronic acid–catechol conjugate/thiol-terminated Pluronic ones (4.48 kPa).

3.2 Polyacrylics

The world of polymers puts polyacrylics, among others, forward as leading synthetic materials. Their uses and applications in our daily life are astonishingly widespread, including biomedicals, adhesives, paints, and cosmetics. Polyacrylics are specifically presented in form of hydrogels (Laftah et al. 2011). Water-borne polymers, including several polyacrylics, are of a paramount importance as far as green environment is nowadays of major concern.

Catechol derivatives have been introduced onto polyacrylics or their copolymers via chemical modification or via polymerization of catechol-bearing acrylic monomers, the products of which have been served as bases for hydrogels. By virtue of the carbodiimide chemistry, DOPA entity was anchored onto a triblock copolymer consisting of poly(methyl methacrylate)-poly(methacrylic acid)-poly (methyl methacrylate) (PMMA-PMAA-PMMA) (Fig. 20a) (Guvendiren et al. 2008). A self-assembly hydrogel of the thus-modified triblock copolymer was envisaged through solvent exchange, that is, exhibition of a solution of the copolymer in an organic solvent (DMSO, DMF, or NMP) to saturated water vapor. By doing so, the hydrophobic PMMA blocks aggregated into spherical domains, and the water-soluble PMAA midblocks formed bridges between these zones. The effect of DOPA incorporation manifested in the decrease of swelling in water of the hydrogel at pHs of 6 and 7.4, when compared to swelling of the hydrogel from nonmodified copolymer. At these pHs, the DOPA units became protonated and hence hydrophobic. However, at pHs higher than pKa_1 of DOPA (9.40), the hydrogel swelled significantly. The swollen gel was light brown, an indication of partial oxidation of DOPA since it is deprotonated at these pHs, hence prone to oxidation. The elastic modulus of the hydrogel was found to increase with DOPA content up to 5 kPa. Hydrogels with the fully protonated DOPA units (at pHs of 6 and 7.4) strongly adhered to TiO_2-coated surface after a contact time of 20 min; the cohesive fracture energy of the hydrogel was quantified through the interfacial fracture energy ~ 2 J/m^2. However, at pH of 10, the adhesion of the hydrogel was less effective, and the proneness of DOPA units to oxidation would have decreased its adhesive capacity.

The copolymer of *N*-(3,4-dihydroxyphenethyl) methacrylamide and styrene (Fig. 20b), prepared by free radical copolymerization of the *N*-2-(3′,4′-ditriethylsilyloxyphenyl)ethyl methacrylamide and styrene followed by deblocking of the triethylsilyloxy groups with tetra-*n*-butylammonium fluoride (TBAF), afforded

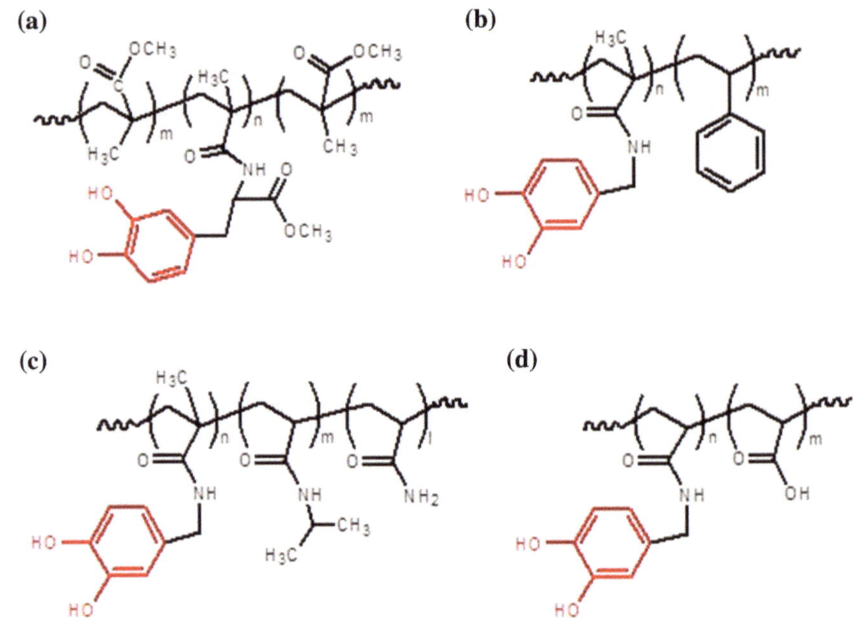

Fig. 20 **a** PMMA-PMAA-PMMA triblock–catechol conjugate; **b** poly[*N*-(3,4-dihydroxyphenethyl) methacrylamide-*co*-styrene]; **c** poly[*N*-(3,4-dihydroxyphenethyl) methacrylamide-*co*-*iso*propylacrylamide-*co*-acrylamide]; **d** poly(acrylic acid)–dopamine conjugate

hydrogels by a dual crosslinking, that is, coordinative and oxidative crosslinking (Xu et al. 2012b, 2013). The latter ones were triggered by iron(III) trifluoromethanesulfonate and sodium methoxide, respectively. Upon the first treatment, the color changed from blue, to blue-violet, and finally to red wine, indicating the formation of *mono*-, *bis*, or *tris*(catecholato) iron(III) complexes, respectively. And, addition of sodium methoxide solution to the product of the first treatment afforded instantaneously a dark purple precipitate, hinting at the occurrence of the two already mentioned mechanisms of crosslinking. By treating a mixture composed of copolymer (Fig. 20c), $FeCl_3$, and diphenyliodonium chloride with sodium bicarbonate solution (pH ~5.6), a purple hydrogel was obtained which turned out to a viscous solution upon UV irradiation ($\lambda = 254$ nm), proving a gel–sol transition upon photolysis (White et al. 2013). This transition was rationalized in terms of a dismantlement of the coordinative crosslinking. The shear modulus G' of the hydrogel before UV exposure was greater than its loss modulus G''. After UV irradiation, however, both moduli decreased and G'' became larger than G', hinting at the collapse of the gel state. Also, the lap shear adhesion test showed that the shear strength of the gel dropped from 1.68 (before irradiation) to 0.35 kPa (after UV irradiation).

Elsewhere, aqueous solution of poly[*N*-(3,4-dihydroxyphenethyl) methacrylamide-*co*-*iso*propylacrylamide] gelled upon adding an adequate amount of boric acid (H_3BO_3) and adjusting the pH at alkaline values (Vatankhah-Varnoosfaderani et al. 2014). At pH = 9, the hydrogel with *bis* (catecholato)borate complex was formed in a few minutes as confirmed by the appearance of a peak at $\delta = -4.9$ ppm in its ^{11}B NMR spectrum and of a band at $\lambda_{max} = 490$ nm in its UV–visible spectrum. The gel was found to self-heal rapidly as demonstrated by a full recovery of its mechanical properties after being subjected to a high shear. Waite's research group (Ahn et al. 2014) reported a fascinating demonstration of self-healing of poly (catechol acrylate) and poly(catechol methacrylate) in the absence of metal ion such as Fe(III), proving the catechol units were unshakably responsible for the repair through intermolecular hydrogen bonding (Wilker 2014).

Hydrogels as ionoprinting actuators were made by irradiation of DMSO/water solution of a copolymer of *N*-(3,4-dihydroxyphenethyl) methacrylamide, *N,N'*-methylene-*bis*acrylamide, *N*-hydroxyethyl acrylamide, and 2,2-dimethoxy-2-phenylacetophenone (Lee et al. 2014, 2015b; Lee and Konst 2014). The last two species acted as bifunctional crosslinker and photoinitiator, respectively. The actuation effect of the hydrogel was assured by the catechol–Fe(III) chelates. In another work (Meng et al. 2015), and following the same procedure, the hydrogel from copolymer of *N*-(3,4-dihydroxyphenethyl) methacrylamide, acrylamide, *N*, *N'*-methylene-*bis*acrylamide, and in the presence of 2,2-dimethoxy-2-phenylacetophenone was studied for the generation of hydrogen peroxide, an oxidation product of the catechol units, and for its impact on cell viability and proliferation.

Poly(acrylic acid) (PAA) was conjugated with ~30 mol% catechol entities by reaction with dopamine under *N*, *N'*-dicyclohexylcarbodiimide/*N*-hydroxysuccinimide chemistry (Fig. 20d) (Wang et al. 2015b). A coacervate was immediately formed upon addition of Zn(II) solution to a solution of PAA-dopamine (pH = 4). Yet, at this pH, the complexation of catechol with Zn(II) is weak because of lower charge density (compared to Fe(III) which higher charge density). The obtained coacervate was fluid as demonstrated further by rheological study: a shear modulus lower than loss modulus. Interestingly, this PAA–dopamine–Zn(II) coacervate showed an adhesive ability to aluminum substrate greater than cyanoacrylate glue (super glue) and white glue (Elmer's glue). However, raising the pH to about 9, the formed coacervate turned out to a gel as commonly observed for Fe(III)-mediated catechol-containing systems. At alkaline pHs, the created phenate forms of catechol units would easily chelate Zn(II), and consequently provoking a coordinative crosslinking; the mixture of PAA and Zn(II) solution at alkaline pH did not gel. The elastic modulus of the gel is larger than the loss modulus, proving its gel network. Excellent self-healing property of the PAA-dopamine-Zn(II)-based hydrogel was mechanically proved; the recovery of its initial elastic modulus (13.1 kPa) after shearing test asserted the reversible coordinative crosslinking of the (catecholato) zinc(II) complex.

Interfacial adhesion enhancement of the nanocomposite hydrogels based on polyacrylamide conjugated with nitrodopamine (0.5 mol% catechol) (Fig. 21) was reported (Ding et al. 2015). The nanocomposite hydrogels were made in the presence of Laponite in aqueous buffer solutions for two days. The swelling capacity or the water content at equilibrium of the hydrogel was near 95% a swelling degree lower than that of polyacrylamide, due to the strong interfacial binding between polymer-bound catechol and Laponite. The storage and loss moduli, *G′* and *G″*, of the nanocomposite hydrogels with polyacrylamide–nitrodopamine and polyacrylamide–dopamine were one to twofold greater than those of nanocomposites from polyacrylamide and Laponite, an indication for their weak interaction. *G′* and *G ″* of the hydrogel with polyacrylamide–nitrodopamine were measured to be 4.2 and 1 kPa, respectively, at pH = 5, and decreased at higher pHs. However, those of the hydrogel with dopamine were 1.7 and 0.5 kPa, respectively, at a pH range of 3–7.4, and dropped to 1 and 0.3 kPa at pH of 9. The moduli of the former hydrogel were significantly higher than those of the latter one, basically due to an increased solidity toward oxidation of nitrodopamine. The same school reported earlier a similar work (Skelton et al. 2013); Laponite was used in the nanocomposite hydrogel formulation consisting of poly[*N*-(3,4-dihydroxyphenethyl) methacrylamide] and poly(acrylamide) crosslinked with *N*, *N′*-methylene-*bis*acrylamide. Increasing both Laponite and *N*-(3,4-dihydroxyphenethyl) methacrylamide concentrations lowered the swelling degree of the nanocomposite hydrogel and improved its mechanical properties. However, the adjunction of Laponite in the polyacrylamide-based hydrogel did not affect its swelling degree (~93% in PBS, pH = 7.4). Such differences were imputed to the strong catechol–Laponite binding. Indeed, for nanocomposite gels with 2 wt% Laponite, the maximum stress, the maximum toughness, the elastic modulus *G′*, and the viscous modulus *G″* increased from 208 kPa, 49 kJ m^{-3}, 1.1 kPa, 0.226 kPa to 3240 kPa, 772 kJ m^{-3}, 13.4 kPa, and 4.22 kPa, respectively, when *N*-(3,4-dihydroxyphenethyl) methacrylamide concentration increased from 0 to 3 wt%.

Terpolymer-based hydrogel was fashioned by UV irradiation of a methanol solution of a mixture of *N*-(3,4-dihydroxyphenethyl) acrylamide, *N*-*iso*propylacrylamide, poly (ethylene glycol)-triacrylate (as a crosslinker) in the presence of Irgacure 2959 as photoinitiator (Ai et al. 2014). By increasing the *N*-(3,4-dihydroxyphenethyl)

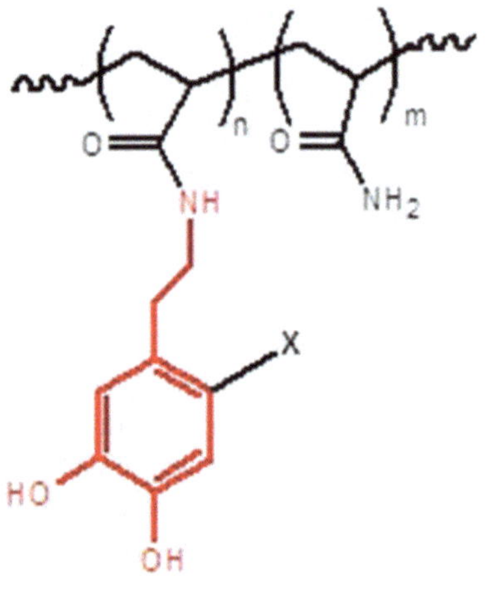

Fig. 21 Poly(acrylamide)–catechol conjugates. **X** = H, poly(acrylamide)–dopamine conjugate; **X** = NO_2, poly (acrylamide)–nitrodopamine conjugate

acrylamide in the system, the brown color of the hydrogel deepened and its swelling degree decreased as a result of increasing *o*-quinone content, oxidized form of catechol. The swelling degree for hydrogel dropped from 110 to ~15% when [*N*-(3,4-dihydroxyphenethyl) acrylamide]/[*N*-*iso*propylacrylamide] molar ratio varied from 28 to 8:2 at 25 °C. At 37 °C, the swelling ratios of the hydrogels were lower than those at 25 °C, which was rationalized in terms of increased hydrophobicity of the hydrogel by dehydration, forming a compact gel structure. At room temperature, the strength of adhesion of the hydrogel (molar ratio of 8:2) to a gelatin coating was nearly 2 MPa against about 1 MPa for hydrogel without *N*-(3,4-dihydroxyphenethyl) acrylamide. The gels were found to be biocompatible and less cytotoxic toward the growth of mouse fibroblast cells.

The hydrogel from a triblock copolymer ABA consisting of poly[(*N*-*iso*propylacrylamide)-*co*-(*N*-3,4-dihydroxyphenethyl acrylamide)]-*b*-poly(ethylene oxide)-*b*-poly[(*N*-*iso*propylacrylamide)-*co*-(*N*-3,4-dihydroxyphenethyl acrylamide)] was conceived as an injectable thermosensitive self-healing material with antibiofouling property, via a self-assembly of the copolymer in metal-free aqueous environment (Li et al. 2015b); poly(*N*-*iso*propylacrylamide) and poly(ethylene oxide) assured the hydrophobic and hydrophilic characters, respectively, for the hydrogel. Also, its excellent antibiofouling was imputed, besides to catechol moieties, to the poly (ethylene oxide) segments. The hydrogel was demonstrated to possess a rapid thermo-responsive sol–gel transition and self-repaired from several damages, poly (*N*-*iso*propylacrylamide) being the thermosensitive block. The sol–gel transition temperature of the hydrogel, the crossover point ($G' = G''$) of the two curves $G' = f(T)$ and $G'' = f(T)$, was about 16 °C. Also, the sol–gel transition was reversible at high temperature, proving that the hydrogel was built through hydrophobic interactions from poly(*N*-*iso*propylacrylamide) rather than *o*-quinone crosslinking. These mechanical properties were recovered after severe strain deformation. The excellent self-healing of the hydrogel was attributed to hydrogen bonding and aromatic π-π stacking interactions between catechol groups.

3.3 *Miscellaneous Polymers*

Poly(vinyl alcohol), a water-borne and versatile polymer, has been widely functionalized for a large spectrum of applications, and particularly for hydrogel making (Moulay 2009, 2015). Poly(vinyl alcohol) functionalized with DOPA (8–80 mol%) (Fig. 22a) through an esterification reaction gave hydrogels either by dissolution in a buffer solution at pH 9 at room temperature or by addition to Fe(III) aqueous solution and adjusting the pH at 9 (Shi et al. 2015). The formation of the metal-free hydrogel (white aspect) was claimed to be the result of the hydrogen-bond interactions between the hydroxyl, catechol, and amine groups. The second type of hydrogel, that with Fe(III), was dark, suggesting a coordinative crosslinking. The elastic and loss moduli (G', G'') of the metal-free hydrogel with 8 mol% DOPA were in the range of 10 and 2.8 kPa, respectively. For higher DOPA content, G' and

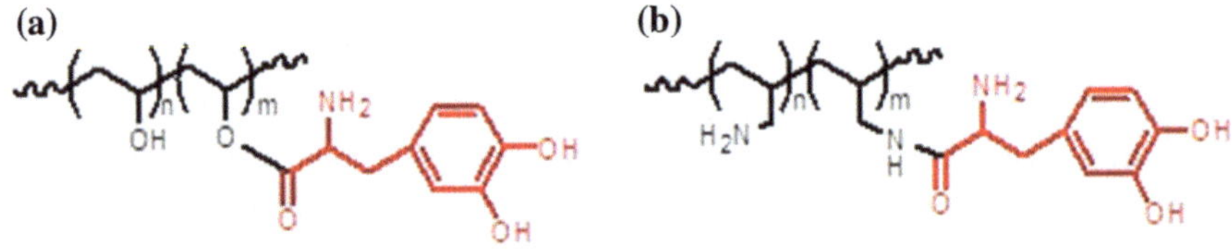

Fig. 22 **a** Poly(vinyl alcohol)–DOPA conjugate; **b** polyallylamine–DOPA conjugate

G'' decreased to values as low as 100 Pa, stemming from the promoted intramolecular interactions between the DOPA units in disfavor of intermolecular interactions, hence lowering the crosslinking extent. The storage modulus of the hydrogel increased by increasing pH; at basic pH, the catechol groups of DOPA units would be oxidized and more crosslinking would have been created by Michael reaction of the formed quinone with amine. This hydrogel presented self-healing capacity; the storage modulus decreased to 6.5 kPa upon fracture of the hydrogel and reached its initial value (10 kPa) after 270 s of healing process. However, G' of the Fe(III)-mediated hydrogel with 8 mol% DOPA was 21 kPa, a value almost twofold higher than that for the metal-free hydrogel. The elastic modulus of this metal-mediated hydrogel decreased to 15 kPa after breaking, and its initial value was recovered after 600 s of repair operation.

Birkedal's school worked on self-healing hydrogels from DOPA-functionalized polyallylamine (Fig. 22b), induced by Fe(III) (Krogsgaard et al. 2013) and by Al (III), Ga(III), and In(III) (Krogsgaard et al. 2014). The EDC/NHS protocol was adopted for the functionalization of polyallylamine. Mechanical properties of the hydrogels with Fe(III) were pH-responsive; that is, they increased from acidic pH to a pH approaching pK_a of the polyallylamine site (9.3–9.7) and decreased beyond. The maximum storage modulus G' of the hydrogel was measured as 7 kPa at frequency of 25 Hz for a pH near 9.3, that of the unmodified polyallylamine/Fe(III) was nil at all pHs, presenting an almost viscous solution. Increasing pH led to an increase in coordinative crosslinking and a reduction of polymer charge which lowered its hydrophilicity and finally the collapse of the hydrogel at pH higher to isoelectric point (pI) of polyallylamine. However, it was claimed that the crosslinking was purely coordinative, and not catechol–catechol covalent as per oxidation. Also, by increasing pH, the color of hydrogel shifted from green, to purple, and then to red, confirming the formation of *mono*, *bis*, and *tris*(catecholato) iron(III) complexes, respectively. Moreover, the hydrogel with Fe(III) self-healed within 10 s, restoring its initial G' value. However, hydrogels with Al(III), Ga(III), and In(III) were lightly orange colored, and the color was found to deepen at higher pHs due to the quinone tanning phenomenon. These hydrogels showed self-healing properties irrespective to pH; that is, the integrity of the hydrogels was recovered after 2–7 min after cleavage test. The maximum shear moduli were observed at a pH range of 8.1–8.8, pH's lower than the pK_a of polyallylamine; they were 18.5 kPa for hydrogel-Al(III) (pH = 8.8), 13 kPa for hydrogel-Ga(III) (pH = 8.1), and 12 kPa for hydrogel-In(III) (pH = 8.5) for a frequency of 25 Hz. This storage modulus decreasing is in parallel with the decreasing metal hardness (Pearson's HSAB principle), the catechol ligand being a hard base. These findings would hint at the role of the coordinating metal in conjunction with that of the polymer. The hydrogels collapsed when the pH rose 12 and the *G's* dropped down to almost nil.

Agarose composite hydrogel beads with a mean size of 2–4 mm were engineered by adding agarose to the poly(L-DOPA)-poly(allylamine) polyelectrolyte (Fig. 23) suspension, followed by a dropwise addition of the mixture to nitrogen liquid to induce gelation, and evaporation of the nitrogen liquid afterward (Yu et al. 2015). The as-made hydrogel beads were exploited in the adsorption of heavy metal and ionic organic dyes. Kinetic isotherms of adsorption of Pb(II), Cd(II), and Cu(II) fitted best Langmuir model, indicating a monolayer adsorption. The maximum adsorption capacity of the hydrogel toward the metals was 1625, 657, and 388 mg/g, respectively. The high value for Pb(II) would point out to a strong binding of polyelectrolyte hydrogel for this metal ion. The adsorption of the metal ions was believed to occur at the catechol sites and between carboxylate ions and amine

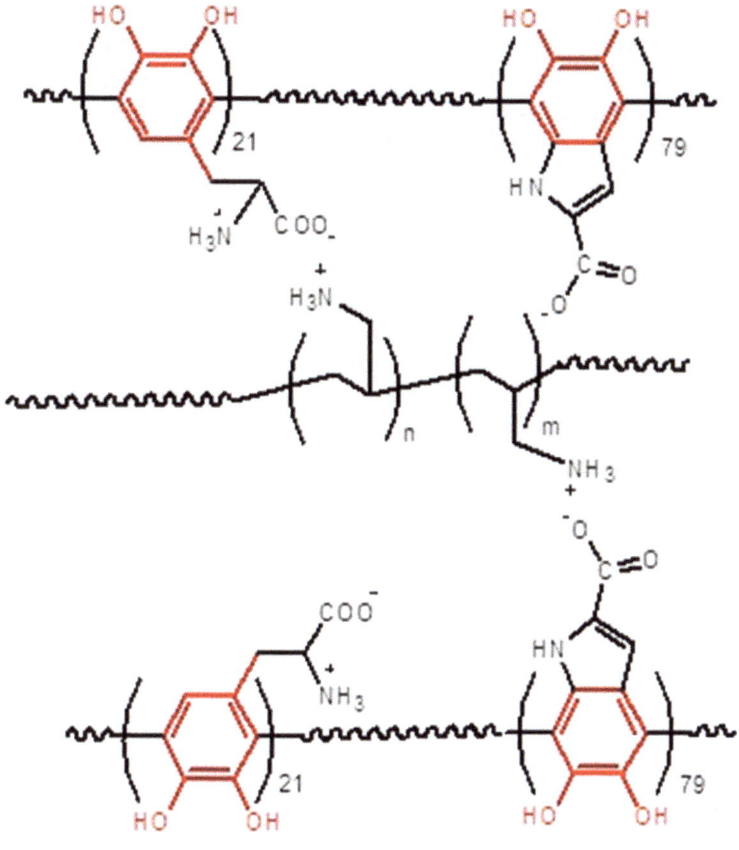

Fig. 23 Poly(DOPA)–polyallyamine polyelectrolyte complex

groups. However, the maximum adsorption capacity of the hydrogel toward methylene blue (positively charged) and methyl orange (negatively charged) was 467 and 870 mg/g, respectively, and their adsorption isotherms obeyed also the Langmuir model; this adsorption difference would hint at the effect of the molecular nature of ionic organic dye. The maximum adsorption uptakes of the agarose hydrogel beads toward Pb(II), Cd(II), Cu(II), methylene blue, and methyl orange was 5.6, 4.3, 4.1, 4.5, and 5.2 mg/g, respectively, values lower than those of agarose/polyelectrolyte hydrogel beads by 2 or 3 order of magnitude.

Wang et al. (2015c) developed a hydrogel via a boron-based coordinative crosslinking using catechol-bearing polyaspartamide derivative (Fig. 24a); the latter polymeric material was achieved by reaction of dopamine and ethanolamine with a synthesized polysuccinimide (M_w ~ 130 kDa). The dopamine content varied from 22 to 27%. pH-responsive hydrogels were ensued by dissolution of the polymer in water in the presence of boric acid: (1) At pH of 8, the system was yellow liquid, suggesting *mono* (catecholato) borate complex, (2) at pH of 9, the system became a dark red gel, indicating a *bis*-complex (λ_{max} = 490 nm), and 3) at pH of 12, a dark brown gel was obtained, in which quinone-mediated covalent crosslinking prevailed. At acidic pH, boric acid does not complex with catechol. It was observed that G' and G'' of the hydrogel were 20-fold higher for coordinative crosslinking (pH 9) than for covalent crosslinking (pH 12). G' and G'' varied also with [dopamine]/[boric acid] molar ratio and, overall, fluctuated between 1–10 kPa, and 0.15–10 kPa, respectively. Full recovery of the hydrogel after fracture occurred, indicating a self-healing property of the hydrogel by restoring (catecholato) borate complex. Elsewhere, Kim's school (Seo et al. 2015) reported the synthesis of dopamine-functionalized polysuccinimide (M_w ~ 190 kDa) (Fig. 24b) and valorized its hydrogel form in metal ions removal. The hydrogel was made by crosslinking dopamine-functionalized polysuccinimide with diaminobutane in DMSO (organogel), followed by hydrolysis at pH of 9 for 72 h. The maximum water uptake of the hydrogel neared 450 g/g and could be monitored by the crosslinker content. The maximum adsorption of Fe(III) by the hydrogel was about 270 mg/g of gel. The swelling degree declined in the presence of metal ion with high valency.

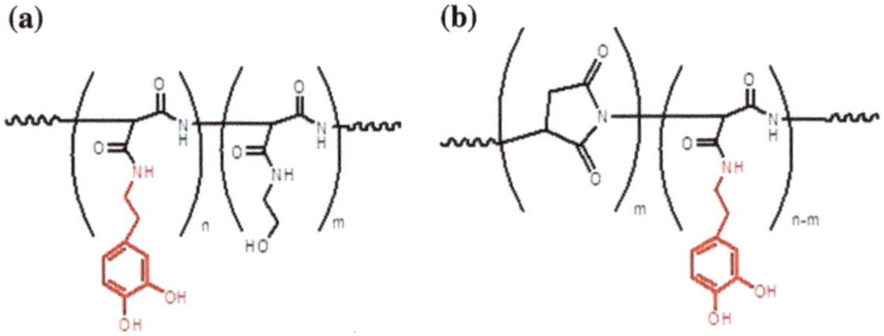

Fig. 24 **a** Polyaspartamide derivative–catechol conjugate; **b** polysuccinimide–catechol conjugate

4 Conclusion and Future Perspectives

According to the above-delineated survey, hydrogels composed of catechol-containing polymers arise as materials with brilliant perspectives, particularly in biomedical uses. The noticeable self-healing and adhesion properties of such mussel-inspired hydrogels would secure safer surgical sutures. The variety of polymeric scaffolds fosters further the proprieties of these hydrogels and enlarges therefore the field of applications. The extant catechol functionality in these materials exhibits a unique crosslinking-related chemistry. The effects of pH, temperature, crosslinking-mediated oxidizing agent, and crosslinking-mediated metal ion on the formation of hydrogels from the catechol-bearing macromolecules are importantly manifest.

References

Ahmed EM (2015) Hydrogel: preparation, characterization, and applications: a review. J Adv Res 6:105–121

Ahn BK, Lee DW, Israelachvili JN, Waite JH (2014) Surface-initiated self-healing of polymers in aqueous media. Nat Mater 13:867–872

Ai Y, Nie J, Wu G, Yang D (2014) The DOPA-functionalized bioadhesive with properties of photocrosslinked and thermoresponsive. J Appl Polym Sci 131:10p. https://doi.org/10.1002/APP.41102

Alge DL, Azagarsamy MA, Donohue DF, Anseth KS (2013) Synthetically tractable click hydrogels for three-dimensional cell culture formed using tetrazine-norbornene chemistry. Biomacromol 14:949–953

Altunbas A, Pochan DJ (2012) Peptide-based and polypeptide-based hydrogels for drug delivery and tissue engineering. Top Curr Chem 310:135–167

Bhattarai N, Gunn J, Zhang M (2010) Chitosan-based hydrogels for controlled, localized drug delivery. In: Amidi M, Hennink WE (eds) Chitosan-based formulations of drugs, imaging agents and biotherapeutics. Adv Drug Del Rev 62:83–99

Bilic G, Brubaker C, Messersmith PB, Mallik AS, Quinn TM, Haller C, Done E, Gucciardo L, Zeisberger SM, Zimmermann R, Deprest J, Zisch AH (2010) Injectable candidate sealants for fetal membrane repair: bonding and toxicity in vitro. Am J Obstet Gynecol 202:85.e1–85.e9

Brubaker CE, Messersmith PB (2011) Enzymatically degradable mussel-inspired adhesive hydrogel. Biomacromol 12:4326–4334

Brubaker CE, Kissler H, Wang LJ, Kaufman DB, Messersmith PB (2010) Biological performance of mussel-inspired adhesive in extrahepatic islet transplantation. Biomaterials 31:420–427

Burke SA, Ritter-Jones M, Lee BP, Messersmith PB (2007) Thermal gelation and tissue adhesion of biomimetic hydrogels. Biomed Mater 2:203–210. https://doi.org/10.1088/1748-6041/2/4/001

Cencer M, Liu Y, Winter A, Murley M, Meng H, Lee BP (2014) Effect of pH on the rate of curing and bioadhesive properties of dopamine functionalized poly(ethylene glycol) hydrogels. Biomacromol 15:2861–2869

Choi YC, Choi JS, Jung YJ, Cho YW (2014) Human gelatin tissue-adhesive hydrogels prepared by enzyme-mediated biosynthesis of DOPA and Fe^{3+} ion crosslinking. J Mater Chem B 2:201–209

Cui J, Iturri J, Paez J, Shafiq Z, Serrano C, d'Ischia M, del Campo A (2014) Dopamine-based coatings and hydrogels: toward substitution-related structure-property relationships. Macromol Chem Phys 215:2403–2413

Ding X, Vegesna GK, Meng H, Winter A, Lee BP (2015) Nitro-group functionalization of dopamine and its contribution to the viscoelastic properties of catechol-containing nanocomposite hydrogels. Macromol Chem Phys, 11 p. https://doi.org/10.1002/macp.201500010

Dooling LJ, Tirrell DA (2013) Peptide and protein hydrogels. In: Loh XJ, Scherman OA (eds) Polymeric and self-assembled hydrogels: from fundamental understanding to applications (Chap. 5). Royal Society of Chemistry, pp 93–124

Fairbanks BD, Schwartz MP, Halevi AE, Nuttelman CR, Bowman CN, Anseth KS (2009) A versatile synthetic extracellular matrix mimic via thiol-norbornene photopolymerization. Adv Mater 21:5005–5010

Faure E, Falentin-Daudré C, Jérôme C, Lyskawa J, Fournier D, Woisel P, Detrembleur C (2013) Catechols as versatile platforms in polymer chemistry. Prog Polym Sci 38:236–270

Fichman G, Adler-Abramovich L, Manohar S, Mironi-Harpaz I, Guterman T, Seliktar D, Messersmith PB, Gazi E (2014) Seamless metallic coating and surface adhesion of self-assembled bioinspired nanostructures based on di-(3,4-dihydroxy-L-phenylalanine) peptide motif. ACS Nano 8:7220–7228

Fullenkamp DE, Barrett DG, Miller DR, Kurutz JW, Messersmith PB (2014) pH-dependent cross-linking of catechols through oxidation *via* Fe^{3+} and potential implications for mussel adhesion. RSC Adv 4:25127–25134

García-Fernández L, Cui J, Serrano C, Shafi Z, Gropeanu RA, Miguel VS, Ramos JI, Wang M, Auernhammer GK, Rit S, Golriz AA, Berger R, Wagner M, del Campo A (2013) Antibacterial strategies from the sea: Polymer-bound Cl-catechols for prevention of biofilm formation. Adv Mater 25:529–533

Ghobril C, Grinstaff MW (2015) The chemistry and engineering of polymeric hydrogel adhesives for wound closure: a tutorial. Chem Soc Rev 44:1820–1835

Guo Z, Ni K, Wei D, Ren Y (2015) Fe^{3+}-induced oxidation and coordination cross-linking in catechol-chitosan hydrogels under acidic pH conditions. RSC Adv 5:37377–37384

Guvendiren M, Messersmith PB, Shull KR (2008) Self-assembly and adhesion of DOPA modified methacrylic triblock hydrogels. Biomacromol 9:122–128

Haller CM, Buerzle W, Brubaker CE, Messersmith PB, Mazza E, Ochsenbein-Koelble N, Zimmermann R, Ehrbar M (2011) Mussel-mimetic tissue adhesive for fetal membrane repair: a standardized *ex vivo* evaluation using elastomeric membranes. Prenat Diagn 31:654–660

Hassan CM, Peppas NA (2000) Structure and applications of poly(vinyl alcohol) hydrogels produced by conventional crosslinking or by freezing/thawing methods. Adv Polym Sci 153:37–65

He L, Fullenkamp DE, Rivera JG, Messersmith PB (2011) pH responsive self-healing hydrogels formed by boronate-catechol complexation. Chem Commun 47:7497–7499

Holten-Andersen N, Jaishankar A, Harrington MJ, Fullenkamp DE, DiMarco G, He L, McKinley GH, Messersmith PB, Lee KYC (2014) Metal-coordination: using one of nature's tricks to control soft material mechanics. J Mater Chem B 2:2467–2472

Hong S, Lee H, Lee H (2014) Controlling mechanical properties of bio-inspired hydrogels by modulating nano-scale, inter-polymeric junctions. Beilstein J Nanotechnol 5:887–894

Hong SH, Shin M, Lee J, Ryu JH, Lee S, Yang JW, Kim WD, Lee H (2015) STAPLE: stable alginate gel prepared by linkage exchange from ionic to covalent bonds. Adv Healthcare Mater, 5 p. https://doi.org/10.1002/adhm.201400833

Huang K, Lee B, Messersmith PB (2001) Synthesis and characterization of self-assembling block copolymers containing adhesive moieties. Polym Prepr 42:147–148

Huang K, Lee B, Ingram DR, Messersmith PB (2002) Synthesis and characterization of self-assembling block copolymers containing bioadhesive end groups. Biomacromol 3:397–406

Huang K, Chen JS, Liu Y (2012) Bioinspired synthesis of cross-linking of dopamine-containing block copolymers to form thermosensitive covalent hydrogels. Key Eng Mater 531–532:238–241

Hutanu D, Frishberg MD, Guo L, Darie CC (2014) Recent applications of polyethylene glycols (PEGs) and PEG derivatives. Mod Chem Appl 2:6p. https://doi.org/10.4172/2329-6798.1000132

Jonker AM, Löwik DWPM, van Hest JCM (2012) Peptide- and protein-based hydrogels. Chem Mater 24:759–773

Jonker AM, Borrmann A, van Eck ERH, van Delft FL, Löwik DWPM, van Hest JCM (2015) A fast and activatable cross-linking strategy for hydrogel formation. Adv Mater 27:1235–1240

Kastrup CJ, Nahrendorf M, Figueiredo JL, Lee H, Kambhampati S, Lee T, Cho SW, Gorbatov R, Iwamoto Y, Dang TT, Dutta P, Yeon JH, Cheng H, Pritchard CD, Vegas AJ, Siegel CD, MacDougall S, Okonkwo M, Thai A, Stone JR, Coury AJ, Weissleder R, Langer R, Anderson DG (2012) Painting blood vessels and atherosclerotic plaques with an adhesive drug depot. Proc Natl Acad Sci USA 109:21444–21449

Kim K, Ryu JH, Lee DY, Lee H (2013) Bio-inspired catechol conjugation converts water insoluble chitosan into a highly water soluble, adhesive chitosan derivative for hydrogels and LbL assembly. Biomater Sci 1:783–790

Kim BJ, Oh DX, Kim S, Seo JH, Hwang DS, Masic A, Han DK, Cha HJ (2014) Mussel-mimetic protein-based adhesive hydrogel. Biomacromol 15:1579–1585

Kim K, Kim K, Ryu JH, Lee H (2015) Chitosan-catechol: a polymer with long-lasting mucoadhesive properties. Biomaterials 52:161–170

Krogsgaard M, Behrens MA, Pedersen JS, Birkedal H (2013) Self-healing mussel-inspired multi-pH-responsive hydrogels. Biomacromol 14:297–301

Krogsgaard M, Hansen MR, Birkedal H (2014) Metals & polymers in the mix: fine-tuning the mechanical properties & color of self-healing mussel-inspired hydrogels. J Mater Chem B 2:8292–8297

Krogsgaard M, Nue V, Birkedal H (2015) Mussel-inspired materials: self-healing through coordination chemistry, Minireview. Chem Eur J 21:1–15

Laftah WA, Hashim S, Ibrahim AN (2011) Polymer hydrogels: a review. Polym Plast Technol Eng 50:1475–1486

Lee BP, Konst S (2014) Novel hydrogel actuator inspired by reversible mussel adhesive protein chemistry. Adv Mater 26:3415–3419

Lee YK, Lee SY (2014) A colorimetric alginate-catechol hydrogel suitable as a spreadable pH indicator. Dyes Pigments 108:1–6

Lee KY, Mooney DJ (2012) Alginate: properties and biomedical applications. Prog Polym Sci 37:106–126

Lee BP, Dalsin JL, Messersmith PB (2001) Enzymatic and non-enzymatic pathways to formation of DOPA-modified PEG hydrogels. Polym Prepr 42:151–152

Lee BP, Dalsin JL, Messersmith PB (2002) Synthesis and gelation of DOPA-modified poly (ethylene glycol) hydrogels. Biomacromol 3:1038–1047

Lee BP, Huang K, Nunalee FN, Shull K, Messersmith PB (2004) Synthesis of 3,4-dihydroxyphenylalanine (DOPA) containing monomers and their co-polymerization with PEG diacrylate to form hydrogels. J Biomater Sci Polym Edn 15:449–464

Lee BP, Chao CY, Nunalee FN, Motan E, Shull KR, Messersmith PB (2006) Rapid gel formation and adhesion in photocurable and biodegradable block copolymers with high DOPA content. Macromolecules 39:1740–1748

Lee BP, Liu Y, Konst S (2014) Novel hydrogel actuator based on biomimetic chemistry. In: MRS Proceedings, vol. 1710. Materials Research Society. http://dx.doi.org/10.1557/opl.2014.511

Lee BP, Lin MH, Narkar A, Konst S, Wilharm R (2015a) Modulating the movement of hydrogel actuator based on catechol-iron ion coordination chemistry. Sens Actuator B Chem 206:456–462

Lee SB, González-Cabezas C, Kim KM, Kim KN, Kuroda K (2015b) Catechol-functionalized synthetic polymer as a dental adhesive to contaminated dentin surface for a composite restoration. Biomacromol 16:2265–2275

Li L, Smitthipong W, Zeng H (2015a) Mussel-inspired hydrogels for biomedical and environmental applications. Polym Chem 6:353–358

Li L, Yan B, Yang J, Chen L, Zeng H (2015b) Novel mussel-inspired injectable self-healing hydrogel with anti-biofouling property. Adv Mater, 5 p. https://doi.org/10.1002/adma.201405166

Liu Y, Meng H, Konst S, Sarmiento R, Rajachar R, Lee BP (2014) Injectable dopamine-modified poly(ethylene glycol) nanocomposite hydrogel with enhanced adhesive property and bioactivity. ACS Appl Mater Interfaces 6:16982–16992

Mehdizadeh M, Weng H, Gyawali D, Tang L, Yang J (2012) Injectable citrate-based mussel inspired tissue bioadhesives with high wet strength for sutureless wound closure. Biomaterials 33:7972–7983

Meng H, Li Y, Faust M, Konst S, Lee BP (2015) Hydrogen peroxide generation and biocompatibility of hydrogel-bound mussel adhesive moiety. Acta Biomater 17:160–169

Messersmith PB, Rivera J, Gong YK (2014) Antifouling hydrogels, coatings, and methods of synthesis and use thereof. US Patent 8,796,394 B2, 5 Aug 2014

Moulay S (2009) Poly(vinyl alcohol): what a material! what a chemistry! In: Pandalai SG (ed) Recent research and developments in applied polymer science, Part II, vol 4. Research Signpost, Kerala, India, pp 391–496

Moulay S (2014) Dopa/catechol-tethered polymers: bioadhesives and biomimetic adhesive materials. Polym Rev 54:436–513

Moulay S (2015) Review: poly(vinyl alcohol) functionalizations and applications. Polym Plast Technol Eng 54:1289–1319

Necas J, Bartosikova L, Brauner P, Kolar J (2008) Hyaluronic acid (hyaluronan): a review. Vet Med Czech 53:397–411

Neto AI, Cibrão AC, Correia CR, Carvalho RR, Luz GM, Ferrer GG, Botelho G, Picart C, Alves NM, Mano JF (2014) Nanostructured polymeric coatings based on chitosan and dopamine-modified hyaluronic acid for biomedical applications. Small, 11 p. https://doi.org/10.1002/smll.201303568

Oh YJ, Cho H, Lee H, Park KJ, Lee H, Park SY (2012) Bio-inspired catechol chemistry: a new way to develop a re-moldable and injectable coacervate hydrogel. Chem Commun 48:11895–11897

Okay O (2009) General properties of hydrogels. In: Gerlach G, Arndt KF (eds) Hydrogel sensors and actuators. Springer Series on chemical sensors and biosensors, vol. 6. Springer, Berlin Heidelberg, pp 1–14

Park JP, Song IT, Lee J, Ryu JH, Lee Y, Lee H (2015) Vanadyl-catecholamine hydrogels inspired by ascidians and mussels. Chem Mater 27:105–111

Pawar SN, Edgar KJ (2012) Alginate derivatization: a review of chemistry, properties and applications. Biomaterials 33:3279–3305

Peak CW, Wilker JJ, Schmidt G (2013) A review on tough and sticky hydrogels. Colloid Polym Sci 291:2031–2047

Peppas NA, Hilt JZ, Khademhosseini A, Langer R (2006) Hydrogels in biology and medicine: from molecular principles to bionanotechnology. Adv Mater 18:1345–1360

Phelps EA, Enemchukwu NO, Fiore VF, Sy JC, Murthy N, Sulchek TA, Barker TH, García AJ (2012) Maleimide cross-linked bioactive PEG hydrogel exhibits improved reaction kinetics and cross-linking for cell encapsulation and in situ delivery. Adv Mater 24:64–70

Riva R, Ragelle H, des Rieux A, Duhem N, Jérôme C, Préat V (2011) Chitosan and chitosan derivatives in drug delivery and tissue engineering. Adv Polym Sci 244:19–44

Ryu JH, Lee Y, Kong WH, Kim TG, Park TG, Lee H (2011) Catechol-functionalized chitosan/pluronic hydrogels for tissue adhesives and hemostatic materials. Biomacromol 12:2653–2659

Ryu JH, Jo S, Koh MY, Lee H (2014a) Bio-Inspired, water-soluble to insoluble self-conversion for flexible, biocompatible, transparent, catecholamine polysaccharide thin films. Adv Funct Mater, 8 p. https://doi.org/10.1002/adfm.201402250

Ryu JH, Lee Y, Do MJ, Jo SD, Kim JS, Kim BS, Im GI, Park TG, Lee H (2014b) Chitosan-g-hematin: enzyme-mimicking polymeric catalyst for adhesive hydrogels. Acta Biomater 10:224–233

Ryu JH, Hong S, Lee H (2015) Bio-inspired adhesive catechol-conjugated chitosan for biomedical applications: a mini review. Acta Biomater 27:101–115

Schante CE, Zuber G, Herlin C, Vandamme TF (2011) Chemical modifications of hyaluronic acid for the synthesis of derivatives for a broad range of biomedical applications. Carbohydr Polym 85:469–489

Seo JH, Park SJ, Um SH, Nam SW, Kim YJ, Kim JH (2015) Swelling and metal-ion adsorption properties of dopamine-conjugated polyaspartate hydrogel. J Polym Environ 23:90–96

Shi D, Liu R, Dong W, Li X, Zhang H, Chen M, Akashi M (2015) pH-Dependent and self-healing properties of mussel modified poly(vinyl alcohol) hydrogels in a metal-free environment. RSC Adv 5:82252–82258

Shin J, Lee JS, Lee C, Park HJ, Yang K, Jin Y, Ryu JH, Hong KS, Moon SH, Chung HM, Yang HS, Um SH, Oh JW, Kim DI, Lee H, Cho SW (2015) Tissue adhesive catechol-modified hyaluronic acid hydrogel for effective, minimally invasive cell therapy. Adv Funct Mater 25:3814–3824

Skelton S, Bostwick M, O'Connor K, Konst S, Caseya S, Lee BP (2013) Biomimetic adhesive containing nanocomposite hydrogel with enhanced materials properties. Soft Matter 9:3825–3833

Sun P, Wang J, Yao X, Peng Y, Tu X, Du P, Zheng Z, Wang X (2014) Facile preparation of mussel-inspired polyurethane hydrogel and its rapid curing behavior. ACS Appl Mater Interfaces 6:12495–12504

Thakur VK, Kessler MR (2015) Self-healing polymer nanocomposite materials: a review. Polymer 69:369–383

Thakur VK, Thakur MK (2014a) Recent trends in hydrogels based on psyllium polysaccharide: a review. J Clean Prod 82:1–15

Thakur VK, Thakur MK (2014b) Recent advances in graft copolymerization and applications of chitosan: a review. ACS Sustain Chem Eng 2(12):2637–2652

Thakur VK, Thakur MK (2015) Recent advances in green hydrogels from lignin: a review. Int J Biol Macromol 72:834–847

Vatankhah-Varnoosfaderani M, Hashmi S, GhavamiNejad A, Stadler FJ (2014) Rapid self-healing and triple stimuli responsiveness of a supramolecular polymer gel based on boron-catechol interactions in a novel water-soluble mussel-inspired copolymer. Polym Chem 5:512–523

Wang Y, Zhao Q, Luo Y, Xu Z, Zhang H, Yang S, Wei Y, Jia X (2015a) A high stiffness bio-inspired hydrogel from the combination of a poly(amido amine) dendrimer with DOPA. Chem Commun 51:16786–16789

Wang W, Xu Y, Li A, Li T, Liu M, von Klitzing R, Ober CK, Kayitmazer AB, Li L, Guo X (2015b) Zinc induced polyelectrolyte coacervate bioadhesive and its transition to a self-healing hydrogel. RSC Adv 5:66871–66878

Wang B, Jeon YS, Park HS, Kim YJ, Kim JH (2015c) Mussel-mimetic self-healing polyaspartamide derivative gel via boron-catechol interactions. eXPRESS Polym Lett 9:799–808

White EM, Seppala JE, Rushworth PM, Ritchie BW, Sharma S, Locklin J (2013) Switching the adhesive state of catecholic hydrogels using phototitration. Macromolecules 46:8882–8887

Wilker JJ (2014) Self-healing polymers: sticky when wet. Nat Mater 13:849–850

Willcox PJ, Howie DWJr, Schmidt-Rohr K, Hoagland DA, Gido SP, Pudjijanto S, Kleiner LW, Venkatraman S (1999) Microstructure of poly(vinyl alcohol) hydrogels produced by freeze/thaw cycling. J Polym Sci B Polym Phys 37:3438–3454

Xu Z (2013) Mechanics of metal-catecholate complexes: The roles of coordination state and metal types. Sci Rep 3: 2914, 7 p https://doi.org/10.1038/srep02914

Xu J, Soliman GM, Barralet J, Cerruti M (2012a) Mollusk glue inspired mucoadhesives for biomedical applications. Langmuir 28:14010–14017

Xu H, Nishida J, Ma W, Wu H, Kobayashi M, Otsuka H, Takahara A (2012b) Competition between oxidation and coordination in cross-linking of polystyrene copolymer containing catechol groups. ACS Macro Lett 1:457–460

Xu H, Nishida J, Wu H, Higaki Y, Otsuka H, Ohta N, Takahara A (2013) Structural effects of catechol-containing polystyrene gels based on dual cross-linking approach. Soft Matter 9:1967–1974

Xu JX, Zhou Z, Wu B, He BF (2014) Enzymatic formation of a novel cell-adhesive hydrogel based on small peptides with a laterally grafted L-3,4-dihydroxyphenylalanine group. Nanoscale 6:1277–1280

Xu J, Strandman S, Zhu JXX, Barralet J, Cerruti M (2015) Genipin-crosslinked catechol-chitosan mucoadhesive hydrogels for buccal drug delivery. Biomaterials 37:395–404

Yang Y, Urban MW (2013) Self-healing polymeric materials. Chem Soc Rev 42:7446–7467

Yavvari PS, Srivastava A (2015) Robust, self-healing hydrogels synthesised from catechol rich polymers. J Mater Chem B 3:899–910

Yu L, Liu X, Yuan W, Brown LJ, Wang D (2015) Confined flocculation of ionic pollutants by poly(L-dopa)-based polyelectrolyte complexes in hydrogel beads for three-dimensional, quantitative, efficient water decontamination. Langmuir 31:6351–6366

Zeng X, Westhaus E, Eberle N, Lee B, Messersmith PB (2000) Synthesis and characterization of DOPA-PEG conjugates. Polym Prepr 41:989–990

Zhang W, Pan Z, Yang FK, Zhao B (2015) A facile in situ approach to polypyrrole functionalization through bioinspired catechols. Adv Funct Mater, 10 p. https://doi.org/10.1002/adfm.201403115

Zheng J, Callahan LAS, Hao J, Guo K, Wesdemiotis C, Weiss RA, Becker ML (2012) Strain-promoted cross-linking of PEG-based hydrogels via copper-free cycloaddition. ACS Macro Lett 1:1071–1073

Zhu J (2010) Bioactive modification of poly(ethylene glycol) hydrogels for tissue engineering. Biomaterials 31:4639–4656

MIX
Papier aus verantwortungsvollen Quellen
Paper from responsible sources
FSC® C105338

If you have any concerns about our products,
you can contact us on
ProductSafety@springernature.com

In case Publisher is established outside the EU,
the EU authorized representative is:
Springer Nature Customer Service Center GmbH
Europaplatz 3, 69115 Heidelberg, Germany

Printed by Libri Plureos GmbH
in Hamburg, Germany